OCEANOGRAPHY

CONTEMPORARY READINGS IN OCEAN SCIENCES

OCEANOGRAPHY
CONTEMPORARY READINGS IN OCEAN SCIENCES

Edited by R. GORDON PIRIE

UNIVERSITY OF WISCONSIN-MILWAUKEE

NEW YORK · OXFORD UNIVERSITY PRESS · LONDON 1973 TORONTO

To Everyone
Interested in Exploring
and Understanding Our World's Oceans

Preface

In collecting these readings, I have emphasized the wide range of topics and the interrelationships of the many disciplines commonly gathered under the term oceanography. These include biology, chemistry, geology, physics, meteorology, navigation, economics, law, conservation, engineering, and aeronautics. Both overview-type articles, which survey some of the generalities of an ocean-science discipline, and in-depth articles, which present a more detailed study within a discipline, are represented. These readings are directed toward two audiences: readers who have a general interest in science and the environment and specialists who wish to broaden their understanding of oceanography outside their own area.

An explanation of why this collection was prepared is in order. In teaching oceanography to science and non-science majors, I discovered that my students frequently wanted additional reading to supplement the typically brief explanations provided in the lectures and by the textbooks; I also discovered that no up-to-date collection of readings was available. There were several solutions, the most obvious being to prepare a book myself.

I have selected nearly 50 introductory and intermediate level articles, most of them published within the past five years; any major changes in the articles were made by the author, or with the author's approval, as indicated at the beginning of each article. In keeping with the international character of oceanography, a number of articles by ocean scientists from other nations have been included to remind the reader to guard against attitudes that engender the parochialism sometimes found in oceanographic research.

Necessarily, some duplications between the articles and oceanography textbooks were unavoidable, but some were intentional—so that the collected readings would also be useful in the absence of an introductory text. Level and depth of coverage, length of article, availability, and the length of the collection itself have all influenced the selection process. Unfortunately, no article on oceanographic engineering could be found to represent this diversified field; those available were either too specific or too general. Also, a note to the non-oceanographer is in order here: archeological findings are considered resources, and thus the article on such findings appears under ocean resources.

In selecting and editing a collection of readings, there are numerous articles from which to choose. For this reason, it is unlikely that any two editors' selections would be similar. I accept full responsibility for any omissions or failings in this collection.

Sincere thanks are extended to the authors of the articles selected, edited, and republished in this collection. Their full cooperation, willing advice, and friendly encouragement are most appreciated. Obviously, no collection would have existed without the authors' unselfish assistance. The original publishers also cooperated fully in communicating with the author and editor and in obtaining copyright permissions and cost estimates. In many cases, these were tedious tasks.

I should also like to thank my friends and colleagues at universities and oceanographic institutions who aided me in my selection and editing and provided stimulating discussion, advice, and encouragement in many ways and under many circumstances. It is not possible to name all those who helped, but I want specifically to thank M. Grant Gross, Marine Sciences Research Center, State University of New York, Stony Brook, New York; Walter H. Munk, Institute of Geophysics and Planetary Physics, Scripps Institute of Oceanography, University of California, San Diego, California; and David A. Ross, Woods Hole Oceanographic Institution, Woods Hole, Massachusetts, for their assistance. Dr. Ross critically reviewed the manuscript at various stages; Judy Mortonson and Ann Reed critically reviewed the final manuscript and made many helpful suggestions.

I should like to extend special thanks to Ellis H. Rosenberg and Carol Miller of Oxford University Press for their assistance on some special organizational problems. Maureen Ballsieper, Diane Kotsubka, and Mary Polzin typed and proofread the manuscript and ably handled volumes of correspondence.

University of Wisconsin—Milwaukee   R. G. Pirie
Milwaukee, Wisconsin

Foreword

Oceanography is generally thought to have begun with the *Challenger* expedition (1872-76); thus, it is about 100 years old. In the early years of oceanography, the limited technology only allowed descriptions of the physical and chemical characteristics of the ocean and some of the different forms of life it contained. By the late 1920's, echo-sounders had been developed and were being used to determine the shape of the ocean bottom. Later, in the 1950's and the 1960's, more sophisticated geophysical devices made the study of the structure below the ocean floor possible. As oceanography evolved, four main disciplines emerged: Biological Oceanography; Chemical Oceanography; Marine Geology and Geophysics; and Physical Oceanography. These divisions are for convenience only, since the ocean is highly complex. For example, sediments in the deep ocean basins are generally composed of the shells of organisms that previously floated in the surface and near-surface waters. These organisms flourished in response to a favorable physical and chemical environment. Thus, a study of their shells can yield information concerning past biological, chemical, and physical conditions in the oceans.

However, within the last decade or so, Man's interest in and dependence on the ocean environment has increased many-fold, due in part to a desire to exploit the mineral and food resources of the sea, to solve the problems of chemical and physical marine pollution, and to protect the ocean from dangerous exploitation. Rarely does a week go by without concern being expressed in the press or on television about overfishing, the threatened extinction of some marine species, or a new source of marine pollution. There are also constant reminders that the oceans hold great mineral resources (gas and oil, especially) and food resources, which are needed for our ultimate survival.

What should be done to ensure safe and wise use of the oceans? Laws and regulations are presently being formulated that will affect their future use. Unfortunately, many critical decisions are being made by people who have vested interests in uncontrolled exploitation, little or no understanding of oceanography, or both. The following articles have been selected to provide basic information to help the reader develop a better understanding of the oceans and their effective management.

Other articles discuss food chains, the physics and chemistry of sea water, and the effect of the ocean on the weather. Background information is given in articles on exploring the sea and the sea floor and navigation and positioning; scientific theories of the origins of the oceans in articles on marine geology.

This book of readings, like all others, cannot be comprehensive without extending into many volumes. However, Dr. Pirie has gathered together many solid articles that cover most aspects of oceanography. The significance and the usefulness of oceans—subjects vital to everyone—are well represented.

Woods Hole Oceanographic Institution   David A. Ross
Woods Hole, Massachusetts
October 1972

Contents

1

EXPLORING
THE SEA AND SEA FLOOR

The Early History of Diving

ROBERT E. MARX 1971

Sitting before my typewriter only ten miles from where the Apollo 14 rocket was successfully launched this morning, I find it all too easy to forget that there are regions of the earth that have not been fully explored. Man has learned a great deal about his natural environment, but much remains to be learned, particularly about the waters that cover more than two-thirds of the earth's surface. In spite of important advances in science and technology, there are many regions of the sea that are still a mystery. For centuries men have struggled to penetrate these regions, struggled against difficulties as formidable as those the astronauts face today.

No one knows who were the first men to penetrate the depths of the sea, nor can anyone be sure why they dived, though it is fairly safe to assume that the pursuit of fish, mollusks, crustaceans and other food was an important reason. Archeologists have established that men were diving as early as 4500 B.C.; Mesopotamian excavations have yielded shells that could only have been recovered from the sea floor by divers. Then there is a gap in diving history of more than a thousand years, until the era of the Sixth Dynasty Thebes in Egypt, around 3200 B.C. The vast number of carved mother-of-pearl ornaments discovered on many archeological sites indicates that diving was widespread.

The Cretans, who flourished around 2500-1400 B.C., worshipped a diving god named Glaucus; he remained the patron god of Greek divers, fishermen and sailors until modern times. A mortal before being exalted to divinity, Glaucus was a fisherman in Anthedon, a village famed for its inhabitants' love of diving. The legendary cause of his immortalization was his discovery of a plant that possessed magical properties. Though there is no evidence that anyone else had his good luck, the early Greek free divers found their work extremely profitable, for they provided the ancient world with most of its sponges and their descendants have done the same to this day. At that time sponges were used for a variety of purposes: soaked in water, they served as canteens for soldiers and travelers; saturated with honey, they were given to infants to

Robert F. Marx is an explorer-archeologist specializing in marine archeology. He is currently with Seafinders, Inc., California, locating and salvaging shipwrecks in deep ocean water. He has written over one-hundred scientific reports and popular articles.

From *Oceans Magazine*, Vol. 4, Nos. 4 and 5, pp. 66-74 and pp. 24-34, 1971. Reprinted with light editing and by permission of the author and *Oceans Magazine*.

stop their crying; soldiers used them as pad-
ding under their heavy armor; as coverings
for wounds, they became makeshift ban-
dages; and, of course, women found many
uses for sponges in the home, just as they do
today.

Besides sponges, the early free divers of
the Mediterranean brought up valuable shells
from the sea bottom. Among them were
mother-of-pearl shells, oyster shells containing pearls, and murex shells, the source of
Tyrian purple, a popular dye. More highly
prized than any of the shells was the red
coral used for jewelry and ornaments. Red
coral was largely responsible for the establishment of trade between the Mediterranean
cultures and China: it was the main item
desired by the Chinese, who were well aware
of treasures from the sea—as early as 2200
B.C. the Emperor Yü was receiving pearls
acquired by divers as tribute from the coastal regions of his realm.

The divers of the ancient world must have
been remarkably skillful, for red coral is
seldom found in water shallower than 100
feet and diving to such a depth without any
breathing apparatus is no mean feat. Gulping
air into their lungs, they jumped into the sea
with no aids other than a stone weight
grasped in their arms to carry them down
and guide ropes from the boat tied to their
waists to bring them up again. Once on the
bottom, they jettisoned the weights; when
they ran out of air, they either tugged at the
ropes and were pulled up to the surface, or
ascended hand over hand by themselves.
How deep these divers were able to go and
how long they were able to remain submerged is not known, but a dive of 100 feet
for red coral must have taken at least two or
three minutes, and it is very likely that the
ancient divers were able to go deeper and
remain under longer.

Today the women pearl divers of Japan
are able to make continuous dives to depths
of 150 feet and remain below as long as five
minutes. Several years ago on the island of
Tobago, in the Caribbean, I met a 58-year-
old Negro diver named Big Anthony, who

could free dive as well, or better, than the
Japanese divers. Although he was only diving
for sport and to break records, his deepest
free dive was 190 feet, and the longest I
clocked him staying underwater on his own
breath was six minutes and ten seconds. To
accomplish these fantastic feats he wore
nothing more than a face mask—he did not
even use any weights. In recent years world
records have been established in deep free
diving and breath holding in swimming
pools, but when these feats are compared
with the capabilities of the ancient free
divers, the Japanese women pearl divers, or
Big Anthony, they do not really amount to
much, considering the preparations made
and the methods used.

The accomplishments of the ancient free
divers are especially remarkable because they
usually worked without any protection for
their eyes, and anyone who has ever attempted to open his eyes in salt water knows
how difficult it is to see underwater. There is
evidence that these divers used some curious
aids, curious because modern science has
shown they were no use at all: for example,
oil poured into the ears was supposed to
protect the eardrums from water pressure,
and a sponge held in the mouth supposedly
enabled the diver to remain submerged
longer.

The divers' work did not stop with acquiring valuables from the bottom of the
sea. They dived for other reasons as well:
construction work in rivers and harbors, to
recover sunken treasure, and for military
purposes. Divers were occasionally assigned
bizarre tasks, as an amusing story of the
Greek historian Plutarch reveals. Cleopatra
persuaded her lover, the Roman general
Mark Antony, to participate in a fishing contest. Knowing himself to be a poor fisherman and not wanting to lose face before a
lady he was most eager to impress, he hired a
diver to keep his hook well supplied with
fish. Unfortunately for him, Cleopatra's
spies discovered the trick, and Antony's face
must have been very red indeed when his
first catch on the second day of the contest

turned out to be a large dead fish—dried, salted and ready for the frying pan.

So many divers were engaged in the underwater construction business and so keen was the competition among them that they began to form corporations to obtain contracts for jobs. During the second century B.C., the Emperor of Rome granted a concession for conducting all diving operations along the Tiber River to one of these corporations.

The earliest account of divers in quest of sunken treasure is mentioned by Herodotus, a Greek historian who wrote around the middle of the fifth century B.C. According to one legend, about thirty years earlier Xerxes, the King of Persia, had employed a Greek diver named Scyllias and his daughter Cyane to recover an immense treasure from several Persian galleys sunk during a battle with a Greek fleet. After they had brought up the treasure, Xerxes refused to give them the reward he had promised and instead detained them aboard his galley, doubtless for other diving jobs he had in mind. Seething at this treachery, Scyllias and Cyane jumped overboard in the midst of a storm and gained revenge by cutting the anchor cables of the Persian ships, which caused many of them to collide. Angered in their turn, the Persians pursued the culprits as soon as order had been restored, but Scyllias and Cyane managed to escape by swimming to Artemisium, a distance of nine miles, completely underwater. Herodotus expressed some doubt about the last part of the story, since swimming such a distance underwater was a feat unheard of at the time. However, later historians were not so skeptical, believing that the swimmers might have held hollow reeds in their mouths to breathe air from the surface—an early version of the snorkel that millions of divers use today. In fact, it was Pliny the Elder, the Roman scholar and naturalist of the first century A.D., who first mentioned any such breathing device, but it could well have been used earlier.

By the third century B.C., diving for sunken treasure was so common among the Greeks that they passed special laws regarding the division of the finds. Treasure recovered in two cubits of water or less (a cubit was equal to 1½ feet) entitled the diver to a tenth of its value; treasure recovered from between two and eight cubits gave the diver a third of its value; and treasure recovered in depths in excess of eight cubits provided a reward equal to half its value. The part of the treasure not given to the diver was the property of the original owner or, in the event of his death it became the property of the ruler from whose waters it was recovered.

The earliest record of divers engaged in military operations was Homer's mention in the *Iliad* of their use during the Trojan War (1194-1184 B.C. is the traditional date). What their tasks were Homer did not say, but they probably included sabotage of enemy vessels—boring holes in their bottoms or cutting their anchor cables. As in the case of Scyllias and his daughter, these divers may have used snorkels: concealment underwater would have helped them surprise their quarry.

The Greek historian Thucydides gave a contemporary account of divers in warfare in the late fifth century B.C. A band of Spartans on the island of Sphacteria off the southwestern shores of Greece, found themselves besieged by their enemies, the Athenians. Cut off from the outside on which their survival depended, the Spartans used divers who, under cover of darkness, swam underwater past the besiegers' vessels; they returned the same way, carrying animal skins full of victuals that their allies had sent to them. The Athenians, expecting starvation to force the Spartans into quick surrender, were bewildered at first, but soon caught on and stationed sentinels to capture the divers.

About a decade later, when the Athenians tried to attack the harbor of Syracuse in Sicily, they discovered that the Syracusans had constructed underwater barriers to prevent the besiegers' ships from approaching the shore. The Athenians sent divers armed with saws and axes to cut down the obstructions and tie the fragments to towlines so they could be pulled out of the way. Yet as fast as the Athenians removed the obstructions, the Syracusan divers replaced them. In the end the Athenians, who had come as attackers, were attacked themselves when allies of the Syracusans arrived; they lost many ships and more than 40,000 men.

There are many other historical accounts of free divers engaged in military operations. Many concern sabotage of enemy vessels and others recount the reconnaissance work of divers who swam into enemy ports to ascertain naval strength or to eavesdrop on the conversations of enemy leaders aboard their vessels. Divers undoubtedly turned the tide of many a sea battle, and by the fifth century B.C., they were considered so important in warfare that the Romans, ever a war-minded people, took precautions against them: all anchor cables of Roman vessels were made of iron chain to prevent their being cut, and special guards of the fleet, who were divers armed with trident spears, were on duty round the clock to stop enemy divers from infiltrating the defenses underwater.

In 332 B.C., shortly after Alexander the Great had begun his conquest of the world, the island stronghold of Tyre offered such a long and fierce resistance that both attackers and defenders had frequent recourse to divers. On one occasion the defenders used them to destroy a dike of timber that the Macedonians had erected as a blockade. The Macedonians, after reducing the fighting spirit of the defenders, gave their divers the task of destroying the boom defenses of the port.

The most interesting feature of this battle to students of diving history is a legend that Alexander himself descended into the sea in some sort of a container to watch the destruction of the boom defenses. Unfortunately, no further details of this notable event have been preserved. A thirteenth-century French manuscript contains an illustration showing Alexander inside a glass barrel brilliantly lit by two candles. Numerous

species of marine life surround him and a large whale dominates the scene, but the painting is an imaginary reconstruction, as are all later illustrations, and no one knows what the container really looked like.

Scanty as the details are, Alexander's descent is historically important as a milestone in man's struggle to master the sea. However, it was not the first such technical advancement. Three decades earlier, in 360 B.C., Aristotle's *Problemata* mentioned a device to aid divers:

... in order that these fishers of sponges may be supplied with a facility of respiration, a kettle is let down to them, not filled with water, but with air, which constantly assists the submerged man; it is forcibly kept upright in its descent, in order that it may be sent down at an equal level all around, to prevent the air from escaping and the water from entering. ...

It is not known how long such devices had been in existence before Aristotle described them or how widespread was their use. Such diving devices disappear from recorded history until the year 1250 A.D., when Roger Bacon's *Novum Organum* mentioned Alexander's container, but no others. It is the consensus of opinion that few diving aids were in general use before the sixteenth century, which began the age of invention in diving.

Only one other aid to divers discovered during the age of free diving is worthy of note—goggles. Who first invented them and when they were first used is unknown. A ceramic Peruvian vase of the second century A.D., now on exhibit in the American Museum of Natural History in New York City, depicts a diver wearing goggles and grasping two fish in his hands (Figure 1-1). The first historical reference to diving goggles occurs in the reports of fourteenth-century travelers returning to Europe from the Persian Gulf, where pearl divers used goggles with lenses of ground tortoiseshell. The Polynesians, whose underwater exploits are legendary, also used them for several centuries before European ships first visited their waters. Yet

Figure 1-1. Peruvian vase from the second or third century A.D. depicts a diver wearing goggles and holding two fish. (Courtesy of American Museum of National History.)

the free divers of the past never made extensive use of goggles, probably for the same reason divers do not often use them today. Although goggles provide much better underwater vision, the deeper a diver descends, the harder the water pressure forces them against his eyes: his eyesight could suffer severe permanent damage.

Nearly as famous as the brilliant diving exploits of the early free divers of the Old World were those of the Indian and Negro pearl divers of the Caribbean, who remained free divers long after various types of diving equipment were invented and brought into widespread use. Though advanced diving technology has put these divers more or less out of business in recent years, as late as the 1940's they were vigorously plying the trade their ancestors had been engaged in for centuries.

Even before Columbus' discovery of the New World in 1492, many of its aborigines practiced diving, primarily because, like the Mediterranean divers, they found the sea floor a valuable source of food. North American Indians used diving as a basic hunting technique: swimming underwater and breathing through reeds, they were able to approach unwarry fowl and game and capture or kill them with nets, spears or their bare hands. The Mayans of Mexico, like the ancient Greeks, venerated a diving god, and a fresco of this deity may be seen today in a temple known as the Temple of the Diving God at the archeological site of Tulum on the eastern coast of Yucatán.

The Lucayan, Carib and Arawak tribes of the Caribbean dove for pearls on a small scale but it did not become a major occupation until the coming of the white man. One

Figure 1-2. Sponge divers, such as these depicted in the Mediterranean about 1200-1400 A.D., wore no breathing devices. Rigorous training developed their lungs to enable them to work underwater for long periods. (Courtesy of the Library of Congress.)

day in 1498, during Columbus' third voyage of exploration, his fleet anchored at the island of Cubagua, near the coast of Venezuela, to obtain a supply of fresh water and fruit. While some of his men were ashore, they noticed a Carib Indian woman wearing a pearl necklace. They made inquiries and informed Columbus that the natives of the island possessed great quantities of valuable and exquisite pearls, which they had found in the adjacent waters. Columbus sent Indian divers in search of oysters, and the result confirmed the story of his men. Immediately after his return to Spain, Columbus reported his find to the king, who ordered that a pearl fishery be established on Cubagua at once. Other large oyster beds were found during the next few years in areas near Cubagua, including Margarita Island, which eventually became the center of the pearl industry, a position it maintains today. Thus Columbus, already assured of his place in history, has another discovery to his credit; over the centuries the Caribbean pearl fisheries furnished Spain with a source of wealth surpassed only

by the gold and silver it took from the New World.

Soon after the opening of the pearl fisheries around Cubagua and Margarita, the local supply of divers was exhausted. Many died from diseases that the Spaniards had brought from Europe: others died from overwork at the hands of their greedy employers, who forced them to dive as many as 16 hours a day. The next source of divers tapped by the Spaniards was the Lucayan Indians of the Bahamas, who were then considered the best divers in the New World. The Spanish historian Oviedo, writing in 1535, gave an account of a visit to the pearl fisheries of Margarita where he watched the Lucayan divers in action. He marveled at their fantastic abilities, stating that they were able to descend to depths of 100 feet, remain submerged as long as 15 minutes, and unlike the Carib Indians, who had less stamina, could dive from sunrise to sunset seven days a week without appearing to tire. As the divers of the Old World (Figure 1-2) had done for thousands of years, they descended by grasping stone weights in their arms and dove completely naked except for a net bag around their necks, in which they would deposit the oysters they found on the bottom. So great was the demand for the Lucayan divers in the pearl fisheries, that in only a few decades all of their people were enslaved and the Bahamas were bereft of their former inhabitants, the first natives seen by Columbus on his epic voyage of discovery.

Like the Carib Indians before them, the supply of Lucayans was soon exhausted, and by the middle of the sixteen century the Spaniards were hard pressed for divers. They solved the problem by importing Negro slaves from Africa, most of whom had never dived and in many cases had never seen the sea until their enforced voyage. Amazingly, they adapted at once and became as good divers as their predecessors had been. Women slaves were preferred to men, probably because the extra layers of body fat in their tissues prevented chilling and enabled them to work longer hours. Yet, regardless of sex, the average length of time these divers could work with efficiency was seldom more than a few years. Like the Indians, they suffered much from overwork, fell prey to disease and, as though they did not have enough troubles already, cannibalistic Carib Indians made frequent raids on the pearl fisheries and carried off great numbers of the divers for meals.

The most interesting fact about these early Caribbean divers was their ability to remain submerged for long periods of time. Oviedo's mention of 15 minutes sounds greatly exaggerated, but there are at least six later accounts of travelers in the Caribbean who witnessed the Margarita pearl divers in action during the sixteenth, seventeenth and eighteenth centuries, and all of them made the same statement. Did these travelers merely echo Oviedo, or did the Caribbean divers possess some long-lost secret that enabled them to stay underwater for so long?

Several contemporary accounts mention that the divers themselves claimed they owed their remarkable endurance to (of all things) tobacco. Both the Indian and Negro pearl divers were very heavy smokers, and a letter written by the governor of Margarita in 1617 informed the King of Spain that when the island recently ran out of tobacco the divers went on strike. The governor first resorted to punishment to induce them to go back to work, but finally gave up and sent a ship to Cuba for a new supply.

Though it seems difficult to believe in the tobacco legend, the divers of the Caribbean were said to possess a secret that is less of a strain on our credulity. In 1712 another governor of Margarita wrote that during his fifty years of residence on the island sharks and other monsters had devoured many of the divers, but that only a few years earlier the men had discovered a certain mineral that, when rubbed over their bodies before diving, would repel sharks. Unfortunately for modern man, who is still trying to discover a satisfactory shark repellent, the governor did not identify the mineral.

The Spaniards, who knew when they were on to a good thing, soon found a use for their divers' talents as important as pearl diving—salvage work. Every year from 1503 on, ships from Spain carried supplies across the Atlantic to sustain its new settlements in the New World. On the return voyages these ships carried the treasures and products of the colonies back to the mother country. Because of frequent storms or careless navigation, a great many of them were lost at sea. In major colonial ports like Havana, Veracruz, Cartagena and Panama, teams of native divers were kept aboard salvage vessels, which were ready to depart on short notice to attempt the recovery of sunken treasure. From the sixteenth century to the end of the eighteenth, these divers recovered from Spanish wrecks more than 100,000,000 ducats ($1,250,000,000 in modern currency); they saved the monarchs of Spain from bankruptcy on more than one occasion.

Ironically, when other European nations began colonizing the West Indies, these same divers were instrumental in depleting the Spanish exchequer: their new employers made use of them in salvaging Spanish wrecks, but this time the profits went into English, French and Dutch treasuries.

By the middle of the sixteenth century, when the Spaniards were forced to send their ships across the Atlantic in organized convoys called *flotas* because of increasing attacks by pirates and enemy naval fleets, each vessel carried Indian or Negro divers, who proved invaluable throughout the voyage. Before the *flota* left port, the divers inspected the ships and made necessary repairs underwater. So highly regarded was their opinion that an adverse report from them on a ship's condition was enough to prevent her from sailing, and there is no doubt that they were responsible for saving many a ship and her cargo. Once the *flota* was under way, the divers were in constant demand: since the ships habitually sailed dangerously overloaded, their seams would open in any kind of heavy weather, causing them to leak badly. Once the location of the leak was ascertained, the divers were lowered over the side, where they would seal the leak with wedges of lead or nail large planks over it. Neither method was easy, for the ships had to keep moving so they would not fall behind the protection of the convoy. In 1578 an admiral of a *flota* wrote to the king recommending that a noble title be conferred on one of his divers who, through diligent efforts to keep a number of vessels from sinking, deserved much of the credit for the successful arrival of 12,000,000 ducats in Spain that year.

Another task assigned to the divers, and a profitable one, was a kind of underwater customs inspection. When the *flotas* reached Spain laden with treasures from the New World, a great deal of smuggling went on because of high import taxes; these smugglers developed devious methods to outwit the customs officers of the king. One of the most common was to have dishonest divers attach contraband to the underwater parts of the ship's hull. Another was to throw the smuggled valuables overboard before the customs officers arrived and have divers recover them later. On one occasion a diver found that an enterprising captain had replaced the lower part of the rudder with one he had had made of solid silver while he was in one of the New World ports: the fraud was detected because the paint concealing the silver had worn off during the voyage. Presumably the diver was richly rewarded. Divers engaged in this kind of work generally were well paid, sometimes enough to buy their way out of slavery.

Due to the fact that virtually all of the ships lost in the New World were in relatively shallow water, diving bells were not widely used over the centuries. Their earliest appearance in the New World was in 1612, when an Englishman named Richard Norwood used a primitive type of "diving bell" to hunt for sunken treasure: his device was nothing more than an inverted wine barrel with weights attached to carry him to the bottom. Having decided there was not enough money in his regular line of work—piracy—to suit him, he went in search of some treasure-laden wrecks that reputedly lay in the vicinity of Bermuda. Failing to locate them, he left for the West Indies to look for wrecks there, but history does not say if he met with any success. He probably didn't: a man who could not make a go of it as a pirate in an age when piracy flourished, would hardly have been likely to succeed in a newer, more challenging profession.

Actually it was an American from Boston named William Phips, who made the greatest single treasure recovery. In 1641 a Spanish treasure *flota* totaling over thirty ships was struck by a hurricane in the Florida Straits and nine of them were dashed to pieces on the coast of Florida. The *almiranta,* or vice-admiral's galleon, had lost all of her masts and was full of water. For a week the galleon was carried along at the mercy of the wind and currents. She was finally wrecked on a reef, now known as Silver Shoals, about ninety miles north of the coast of Hispaniola. Only a few of the 600 persons aboard ever reached land—the others perished from thirst and hunger on a small sandpit nearby or in makeshift rafts at sea. The Spaniards searched unsuccessfully for this wreck for years before they finally gave up hope of locating it.

Phips had been bitten by the treasure bug when he listened to sailors' tales as a child. In 1681 he took the savings he had earned as a shipwright and later in his own shipping business and made his first treasure hunt in the Caribbean. Although he did not recover the vast treasure of his dreams, he did locate several wrecks in the Bahamas which more than covered his expenses—a promising beginning for any treasure hunter. Sure that he was destined to hit a real jackpot soon, he decided to go after a Spanish galleon, reputedly laden with gold, that had been lost near Nassau. Phips wanted the best ship, men and equipment possible for the venture. Failing to raise the money in Boston, he went to London in the spring of 1682, hoping to enlist the aid of King Charles II. It was 18 months before he was granted a royal audience, but Phips, a determined man, waited it out, talked to Charles and persuaded him to back the expedition in return for a large share of the profits.

Phips wasted a month on the wreck near Nassau, which turned out to be barren of treasure. He was determined to find another wreck. His crew, however, had other ideas: irate at not having received the expected shares of treasure, they rebelled and attempted to take over the ship and get rich as pirates. With only a handful of loyal men, Phips managed to quell the mutiny and bring the ship into Port Royal, where he was able to sign on a new crew. It was there he first learned of the lost *almiranta* on Silver Shoals.

In 1685, after an unsuccessful search for the *almiranta* in the waters north of Hispaniola, Phips reluctantly returned to England— empty-handed—only to find that the king had died during his absence. The new monarch was not as interested in tales of sunken Spanish treasure and apparently Phips even

spent some time in jail. Other eager backers, however, soon persuaded the king to grant permission for the expedition. In 1686 Phips sailed from England with *two* ships.

After arriving in the Caribbean, he detoured to Jamaica and picked up about two dozen Negro pearl divers, refugees from the pearl fisheries of Margarita. Phips sent out one of his ships to search Silver Shoals; he drove himself and his men almost to the breaking point, particularly the divers, who scoured the miles of reef from dawn to dusk. Perseverance was at last rewarded when the wreck was found. One of the divers, sighting a large sea fan, dove down to retrieve it for a souvenir; upon surfacing, he reported sighting traces of a sunken galleon. Phips burst into tears of joy when he heard the news. The next weeks were a constant struggle to bring up the treasure while fighting off pirates who had received news of the windfall. More than 32 tons of silver, vast numbers of coins, gold, chests of pearls, and leather bags containing precious gems were recovered before bad weather and exhausted provisions put an end to the salvage operations. The total value of Phips' recovery in today's currency was over $3,000,000. He received a share large enough to make him one of the richest men in America. After he returned to England, the king knighted Phips and later made him governor of the Massachusetts colony.

Pearl diving, making repairs, detective work, treasure hunting—there seems to be no limit to the jobs entrusted to the Caribbean divers of the past. The memoirs of a French missionary, Père Labat, contains an account of a diver who performed an unusual task that ranks, for sheer excitement, with any adventure tale:

While visiting the Island of St. Kitts, I learned from some people, whom I trust to tell the truth, that in 1676 a large hammerhead shark bit off the leg of a young boy who was swimming in the harbor and this resulted in the death of the boy. A Carib Indian, who was a local diver very skilled in spearing fish underwater, volunteered to kill the shark. To understand the danger of this undertaking, one must first realize that the hammerhead shark is one of the most voracious, powerful, and dangerous fish in the sea. The father of the child who had been killed by the shark was glad of the opportunity of having the monster killed and thus offered the diver a good reward to obtain this poor consolation.

The diver armed himself with two good well-sharpened bayonets and, after raising his courage by drinking two glasses of rum, he dived into the sea. The shark, which had acquired a taste for human flesh, attacked the diver as soon as he saw him. The diver allowed it to approach without doing anything until the moment it was on the verge of making its rush. But at the instant it charged he dived underneath it and stabbed it in the belly with both bayonets. The result of this was at once made apparent by the blood which tinged the sea all red around the shark. Each time the shark rushed the diver, he repeated this same tactic and repeatedly stabbed the shark. This scene was enacted seven or eight times, and then at the end of half an hour the shark turned belly up and died.

After the diver had come ashore, some people went out in a canoe and tied a rope to the shark's tail, and then the shark was towed to the beach. It proved to be twenty feet long and its girth was as large as a horse. The child's leg was found whole in its stomach.

When the age of invention in diving began early in the sixteenth century, its herald was Leonardo da Vinci, the Italian artist famous for his inventions even in his own day. When the Venetians asked him for a device that would aid the divers they were using in their war against the Turks, he designed a snorkel breathing tube more advanced than anything the world had ever seen. It was attached to a leather helmet fitting over the diver's head: the helmet even had glass windows so the diver could see underwater. Leonardo also designed swim fins for the diver's hands and feet to enable him to swim faster and cover longer distances. This gear was the forerunner of the basic skin diving equipment millions of divers use today.

The Venetian Senate, however, rejected his snorkel on the grounds that the Turks would be able to see the end of the tube above the water—the divers would lose the element of surprise. They asked him to design a breathing device which would allow the divers to approach the enemy totally concealed. Leonardo responded to the challenge and designed the first scuba. It resembled his first design, but differed in one important respect: the end of the tube, instead of protruding above the water, was attached to a leather bag containing air. To enable the diver to walk on the sea floor, Leonardo discarded fins in favor of a complete diving suit. The diver, encased in leather from head to toe, with a bag of air on his chest, was supposed to descend carrying a heavy weight, which he would abandon when he wished to come up again.

The idea never got past the drawing board. Though Leonardo was sure it would work—he claimed the diver could remain submerged for four hours—he never tested his invention and refused to divulge it to the Venetians. In his memoirs, written years later, he explained his change of heart:

How and why I do not describe my method of remaining under water for as long a time as I can stay without eating: and I do not publish or divulge these by reasons of the evil nature of man, who would use these as a means of murder at the bottom of the sea, by breaking the bottom of ships and sinking them altogether with the men in them.

Leonardo's humanitarian scruples merit respect, but they were unnecessary, for experts today know that his invention could never have worked. Far from the four hours that Leonardo envisioned, the diver could not have lasted more than a few minutes, and only then in shallow water.

A few years later, in 1511, several illustrations which appeared in a popular book on warfare revealed two breathing devices by an anonymous inventor. One picture that depicted a diver wearing a scuba very much like Leonardo's device, attracted widespread attention. The helmet had no windows, however, and was therefore useless for anything but an underwater game of blindman's buff; in addition, the air bag was much smaller and would have given the diver even less breathing time. Testing probably revealed the worthlessness of the invention. In any case, no more was heard of it for a long time.

The second invention was just as unworkable as the first. The diver wore the same outfit except for the air bag. Instead, a long tube protruded from the top of the helmet, its open end kept afloat by two small air bags on the surface. Water pressure would only have permitted this device to work in less than two feet of water. Amazingly, the busy inventors of the sixteenth and seventeenth centuries never seem to have realized the importance of water pressure. Designs similar to those followed thick and fast, all showing fantastically long tubes or snorkels, most accompanied by claims stating the great depths that divers could reach with them.

The devices featuring snorkels and breathing tubes belonged to the realm of science fiction, but inventors did achieve a measure of practical success with another device—a diving bell which functioned on a principle described by Aristotle over 1,800 years before. He described a kettle or barrel that was inverted to imprison air and then held level as it descended to keep the air from escaping. It enclosed either a diver's head or his entire body—and it often worked.

The first appearance of a bell in more modern times was in 1531, when divers used one in Lake Nemi, near Rome, to locate two of Emperor Caligula's pleasure galleys, which reputedly had carried treasure when they sank. The barrel-shaped bell, the invention of an Italian physicist, Guglielmo de Lorena, covered the diver's head and torso and was raised and lowered by ropes. The diver within was able to walk upon the lake bed and stay below for nearly an hour at a time before surfacing for fresh air. Within a few weeks he found both wrecks, for which free divers had searched in vain for many years. Unfortunately, locating them was

only half the battle: for the next three centuries many unsuccessful attempts were made to raise them. The Italian government finally succeeded, in the late 1920's, only by draining the lake dry.

In 1538 two Greeks designed and built a diving bell, then demonstrated its use in Toledo, Spain, before the king, Emperor Charles V, and more than ten thousand spectators. Unlike de Lorena's bell, it was large enough to hold both inventors, who sat inside with a burning candle. Much to the astonishment of the audience, the flame was still burning when they returned to the surface.

News of the Toledo bell spread like wildfire all over Europe, and many similar bells were built. Their practical use sometimes differed from the Spanish demonstration in one important respect: divers, discovering that remaining inside the bell prevented them from working efficiently, relied on it as a kind of air bank. They swam outside to get the job done and returned to the bell at intervals to gulp air into their lungs. Many bells seem to have served their purpose, no matter how they were employed. Some, on the other hand, were useless, like two diving devices invented by an Italian, Niccolò Tartaglia, in 1551. One consisted of a wooden frame shaped like an hourglass on which the diver stood. The glass bowl enclosing his head assured him of good vision but not much breathing time, since it could not have provided him with more than a few minutes of air. The second invention featured the same hourglass frame, but this time the bowl, which seemed to have no opening, enclosed the diver's whole body. There is no indication that either device was ever built.

In 1616, a German, Franz Kessler, produced a bell that was not attached by ropes to anything on the surface: the diver descended with weights and released them when he wished to ascend. The bell consisted of a wooden barrel-shaped chamber, long enough to reach the diver's ankles. It was covered with leather to keep it watertight and possessed several glass windows at eye level (Figure 1-3). Though tests proved

Figure 1-3. A German, Franz Kessler, invented this diving bell in 1616; unlike many earlier designs, it had no connection with the surface.

that Kessler's invention worked, it was clearly ahead of its time and was not widely used.

In 1677 an enormous wooden bell, thirteen feet high and nine feet wide across the rim, was built in Spain and used in the port of Cadaqués to salvage two shipwrecks containing a vast sum of money. The two Moorish divers stayed down for more than an hour at a time. They claimed they could have stayed even longer, but the terrific heat created in the bell by their own breathing forced them to surface to permit the heat to dissipate. They did their work inside the bell and managed to recover several million Spanish pieces of eight. The divers were paid in a novel way: each time they surfaced with treasure, they were permitted to keep as much of it as they could hold in their mouths and hands.

A French physicist, Denis Papin, made a significant improvement in the diving bell in 1689. He devised a way to use a pump or large bellows to supply diving bells with fresh air from the surface. His invention had four great virtues: it permitted divers to remain below for any period of time they chose; it eliminated the danger of carbon dioxide in the air supply which could kill the divers; it made working conditions more comfortable inside the bell, since the fresh air forced out the heat; and, most important of all, it enabled the divers to reach depths of up to seventy feet, since pumping air into the bell under presssure kept the water out. It was not until almost one hundred years later that better pumps were invented and divers could go deeper in the bells.

At about the same time that Dr. Papin came up with his device and even before he tested it, Edmund Halley, the famous English astronomer for whom Halley's comet was named, built a diving bell that used a different method to provide the diver with a continuous air supply. He had the air lowered in leaden casks with valves and tubes attached to them; the divers inside the bell could pull in the end of the tube, open the valve and get as much air as they needed (Figure 1-4). This method was more primitive than Papin's device but it worked better, since the pumps of the period, besides being unable to exert much force, were constantly breaking down.

Halley's bell had several important advantages. More than sixty cubic feet in volume, it was larger than the bell used for salvage in Spain in 1677. It was made of wood with a lead covering that kept it watertight and prevented it from overturning because of unevenly distributed weights. There were glass viewing ports on the sides and an exhaust system on the top to release the hot air produced by the diver's breathing as it rose upward. Halley also provided a way to extend the radius of maneuverability underwater. The diver wore a full diving suit with a helmet to which a flexible tube was attached; another diver inside the bell held the other end of the tube and kept his partner supplied with air. As long as there was enough air lowered in the casks to keep water from filling the bell, Halley's invention could reach depths beyond the capabilities of any other bells in existence. Casks became the principal means of providing fresh air for diving bells until 1788, when an English engineer, John Smeaton, constructed a truly reliable pump that made Dr. Papin's inven-

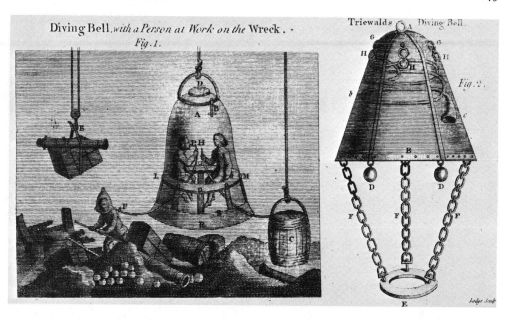

Figure 1-4. The noted English astronomer Edmund Halley invented this diving bell in 1720. Divers received a continuous air supply through tubes attached to a cask lowered from the surface. (Courtesy of National Maritime Museum, Great Britain.)

tion a reality. By the end of the eighteenth century diving bells were so commonplace all over Europe that few large ports were without them. Not only were they used for many different types of underwater tasks, but also to take tourists on sightseeing trips. During a state visit to England in 1818, Archduke Maximilian went down in a bell in Plymouth Harbor and picked up a stone from the bottom as a souvenir.

The earliest design for a diving chamber appeared in 1578 in a book written by an Englishman, William Bourne. It consisted of a wooden chamber, supposedly watertight, made from the hull of a small vessel; in reality, it appears to be the forerunner of the submarine. It was never constructed and very little interest was shown in it.

In 1715 another Englishman, John Lethbridge, obtained a patent for what he called a "diving machine." It resembled a cross between a diving chamber and an armored suit, consisting of an irregular metal cylinder six feet long and 2½ feet in diameter at the head and 1½ feet at the foot. Armholes in the sides allowed the diver to work with his hands and a glass port enabled him to see. The device had no air supply other than what was trapped inside the chamber before it was closed. It was raised and lowered by cables, but could not work deeper than fifty feet, because the increased water pressure caused leaks in the sealings around the armholes, viewing port and entrance hatch. Because Lethbridge's invention failed to extend the limit of the depth divers could reach at that time, it did not attract much attention.

In 1772 yet another Englishman, John Day, built a diving chamber that had no connection with the surface; he asserted that on his first test in a pond he had reached a depth of thirty feet and remained submerged for 24 hours. His story was believed, despite the fact that the chamber could not have provided him with sufficient air for more than five or ten minutes, and he found backers to support another test. This time he planned to go to 130 feet and remain below for twelve hours. With several thousand spectators on hand, the chamber descended in Plymouth Harbor on June 29, 1774, and neither it nor its inventor was ever seen again.

The next attempt to reach great depths occurred in 1831, when a Spaniard named Cervo constructed a small wooden chamber and claimed he could reach a depth of 600 feet with it. On the first trial, to 200 feet, he failed to come up again. Pieces of the chamber floated to the surface, indicating that water pressure had crushed it. This was the last recorded attempt to reach great depths in wooden diving chambers.

In 1849 two Americans, Richards and Wolcott, designed a sphere-shaped metal diving chamber that was meant to be lowered and raised by chains connected to the surface. However, they failed to provide a means of supplying it with air. Their invention was never constructed because of lack of funds, but news of the idea circulated around the world and inventors everywhere turned their efforts to the construction of metal diving chambers.

The first to come up with a workable device was a French engineer named Ernest Bazin. In 1865 he built a chamber to search for treasure in Vigo Bay, Spain. Like the Richards-Wolcott chamber, it did not have a supply of fresh air other than what was trapped before it submerged. This lack did not prevent it from reaching the amazing depth of 245 feet, almost three times the depth any other device had reached, and remaining submerged for a hour and a half. A Venetian named Toselli solved the air supply problem in 1875 by improving Bazin's chamber. He provided it with a huge cylinder of compressed air capable of sustaining the diver inside the chamber for fifty hours underwater.

The way had been cleared: inventors continued to labor, constructing chambers of thicker and stronger metals, making them safe and more comfortable for the men within. In 1934 two Americans, Beebe and Barton, descended in a diving chamber called

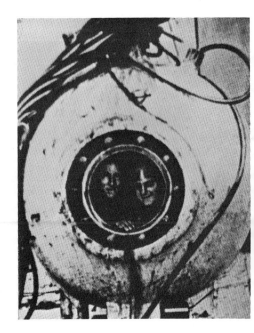

Figure 1-5. Dr. William Beebe and Otis Barton pioneered deep diving in their bathysphere, when they descended 3,028 feet in 1934.

a bathysphere, which functioned on the same principle as Bazin's chamber but was more highly evolved (Figure 1-5). They reached the fantastic depth of 3,028 feet, a record broken over and over again in the following decades.

The possibility of a man being able to walk freely on the sea floor dressed in a diving suit and breathing air from the surface through tubes, for centuries as fantastic an idea as walking on the moon, became a reality in 1715, when an English inventor named Becker gave a demonstration of his new invention in London. It consisted of a full leather diving suit and a large globular metal helmet with a window. Three tubes led from the helmet to the surface—one for the diver's expended air and the others for the fresh air that several large bellows pumped down. During the demonstration the diver was able to stay submerged for an hour, but the depth he reached was not recorded. That the invention was put to practical use is known from an account written by a traveler visiting England in 1745. He witnessed divers clad in the Becker gear trying to salvage a recently lost warship; a pump supplied air to them rather than the more primitive and unreliable bellows.

During the eighteenth century the system of pumping air down to divers from the surface seems to have been in use only in England. In fact, while the English had taken this great step forward in diving technology, inventors on the Continent were still turning out drawings and designs that showed divers trying to suck air down through long tubes. Since the pumps used by the English until the beginning of the nineteenth century could only provide air to divers working no deeper than twenty feet, it is not too surprising perhaps that the rest of the world failed to take much notice. It was only with the invention of powerful pumps called air compressors, which could force air down to greater depths, that the kind of diving known as helmet diving became popular.

Augustus Siebe, a German who lived in England, is sometimes called the father of helmet diving. In 1819 he invented the diving suit and helmet that evolved into the standard helmet diving rig now in use all over the world. His invention consisted of a brass helmet, into which air was pumped from the surface by hand-driven air compressors, and a suit made of leather with a canvas overlay to prevent it being cut underwater. The excess air, as well as the diver's expelled breath, was forced out of the suit through vents near the waist. These vents were a great disadvantage because they compelled the diver to remain upright constantly, just like a diving bell, or else all the air pumped into his helmet would rush out through the waist vents.

In 1834, an American named Norcross improved the Siebe diving suit by eliminating the waist vents and placing an exhaust vent for the excess and stale air on top of the helmet. The diver could now bend over or even lie down on the bottom without endangering his air supply. Augustus Siebe,

Figure 1-6. Today's "hard hat" diving helmet evolved from an 1819 invention of a German, Augustus Siebe, which consisted of a brass helmet fitted on to a canvas suit. Excess air in the rig escaped through waist vents.

probably not wanting to be outdone by an American, modified his own diving suit in 1837, making it completely watertight and adding Norcross' exhaust vent (Figure 1-6). He went one step further: the diver could now control the vent and expel air only when he thought it necessary. This improvement had two great advantages. First, when the diver wished to reach great depths, he could build up air pressure inside the suit to resist the increasing water pressure outside; otherwise, it would press the suit tightly against his body, thus stopping his circulation and causing temporary paralysis. Second, in case the diver could not communicate with the surface and get pulled up, he could inflate his suit with air until it filled up like a balloon, then surface by himself. This safety factor has saved the lives of many divers over the years.

The first opportunity for the Siebe diving suit to prove its worth came very soon after its inventor had tested it and put it on the market. Half a century before, the 108-gun *Royal George,* the largest and most important ship in the British Navy, had sunk at Portsmouth with the loss of almost a thousand lives. Although the wreck lay in only 65 feet of water, all previous attempts to raise it or destroy it had failed. It was a great hazard to navigation in this busy port, so in 1834 the Admiralty enlisted the services of a civilian diver named Deane to see what he could do about it. Using the Siebe open diving suit, he spent three years surveying the wreck, raised many items, including thirty cannon, and finally came to the conclusion that it could not be raised, since marine worms had eaten through much of the wood. Army engineers later used the Siebe closed diving suit with even greater success to plant explosives that finally demolished the wreck.

When the Siebe rig was first put into general use, nothing was known of the bends or other dangers a diver faced in deeper water. Miraculously, there were no fatalities among the divers working on the wreck of the *Royal George*. There were, however, quite a few accidents, the most serious occurring when a diver's hose broke and all the air rushed out of his helmet. Luckily for him, his plight was noticed on the surface and he was pulled up immediately, his face and neck swollen and bleeding from the ears, eyes and mouth. He spent more than a month in the hospital and was never able to dive again.

The United States Navy did not begin to use helmet diving rigs until late in the 1870's. It inaugurated the first helmet diving school at Newport, Rhode Island, in 1882. For many years official interest in this new branch of naval operations was lukewarm. Divers were given an inadequate two weeks

of training, hampered by regulations that limited the depth they could penetrate to sixty feet; they spent most of their time recovering spent torpedoes. Meanwhile, British divers were working in depths of 130 feet, and it was due to their pioneering efforts that so much was learned about the perils besetting divers in deep water. In 1898, when the USS *Maine* was blown up and sunk in Havana Harbor, American divers had their first chance to show what they could do and they came through with flying colors: by recovering the ship's cipher code and the keys to the munitions magazine, they prevented them from falling into enemy hands. In 1914 a naval enlisted man, Chief Gunner George Stillson, established a depth record by diving to 274 feet.

An Englishman named W. H. Taylor invented the first armored diving suit with articulated arms and legs in 1838, the year following the appearance of Siebe's closed diving suit (Figure 1-7). At that time English helmet divers had been working at depths less than 100 feet; Taylor's suit was built to reach 150 feet. Within a decade, others were being made to reach depths of 200 to 300 feet, and by the turn of the century 600 feet. The bends and nitrogen narcosis had finally been recognized as dangers to helmet divers, and interest in these suits grew. However, they never came into widespread use because they have several drawbacks. The most important is that their weight and bulk made underwater mobility impossible: they must depend on surface attendants to move them around on the sea floor. Another is the expense involved in their construction and maintenance. Still another is that if they should develop leaks, the diver within, whose air is at atmospheric pressure, would be drowned instantly as water rushed in.

The desire for total concealment underwater, which had prompted the Venetians to reject Leonardo da Vinci's snorkel, motivated many later inventors to devise versions of scuba that would allow divers to approach enemy ships without any telltale tubes or

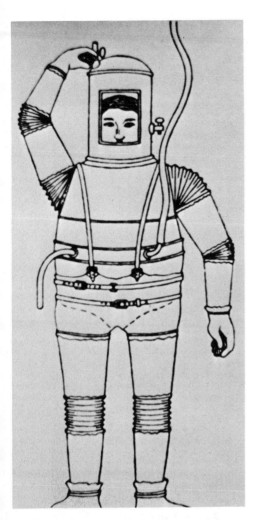

Figure 1-7. In 1838 W. H. Taylor invented the first armored diving suit. It had articulated arms and legs and allowed divers to descend to 150 feet—more than 50 feet deeper than Siebe's suit.

reeds. These inventors were no more successful than Leonardo had been, but they continued to labor, undoubtedly because they knew that a workable device would make its inventor rich.

In Spain, the most war-minded nation in Europe in the sixteenth and early seventeenth centuries, inventors were particularly busy; Spanish archives are full of early scuba designs, some with possibilities, some ludicrous. In 1631, when Dutch supremacy on the high seas was driving Spain to bankruptcy, a desperate King Philip IV offered 10,000 ducats to anyone who could invent a usable scuba. Spies had advised him that the leading Dutch ports of Amsterdam and Rotterdam sometimes harbored as many as 500 ships at a time; he planned to send divers into both ports during a storm to cut anchor cables, thus causing the ships to collide. Since the Dutch had sentinels stationed on all their ships, the scheme could only work if the divers were totally concealed. A special committee was appointed to consider the designs that came in from every corner of the Spanish Empire. Not one of them was found feasible, and the plan to deal a death-blow to the Dutch had to be abandoned.

Although none of the inventions were suitable, several were very interesting, among them three submitted by a Fleming, Florencio Valangren. Two were not scuba at all. The first was a long breathing tube, similar to those that so many inventors were turning out regularly—they didn't work unless they were used close to the surface. This design had an original touch, however: an exhaust valve on the bottom of the tube for the exhalation of foul air. The second design also featured a tube, with the end above the surface connected to a bellows. This was the first known instance of air pumped down to a diver, predating Dr. Papin by many years. The device could have worked in shallow water and, according to the letter Valangren sent along with his designs, similar devices were then in use in Flanders and Holland, mainly for repair work on ships' bottoms. The third design, a very primitive scuba, consisted of a large animal hide pumped full of air by a bellows and dropped by weights to the sea floor. Connected to the hide was a tube with a mouthpiece; the diver, swimming freely in the water, could open it when he needed to take a breath. Valangren's letter stated that this device was also in current use, and there is no reason to doubt him: an air bag made from a large cowhide and lowered in about twenty feet of water would have supplied the diver with at least 15 minutes of air.

Another invention sent to Philip, this one anonymous, is even more interesting, for it took a further step toward the development of practical scuba. It consisted of a tubular air reservoir worn like a belt and probably made of an animal's intestines. At one end was a mouthpiece, and at the other a bellows which would allow the diver to obtain air from the surface without assistance.

In 1679 an Italian, Giovanni Borelli, made the next important attempt at developing a workable scuba. His invention was based on the principle of air regeneration: the diver using it could supposedly breathe the same air over and over again after his spent air was passed through a filter and made fresh. The diver in Borelli's drawing wore a goatskin suit, swim fins and a metal helmet ·with a window. The air reservoir was the helmet itself, and a copper tube (the filter system) led from the front to the back. Borelli claimed that condensation would collect in the tube and, when the diver exhaled, this condensation would filter out the carbon dioxide from the spent air. His invention probably never got past the drawing board. Almost a hundred years later, in 1776, a Frenchman named Sieur actually built an air-regeneration system that was virtually a duplicate of Borelli's device, and discovered that the filter did not purify the air.

In 1808 a German, Friedrich von Drieberg, invented an apparatus he called the "Triton," which somewhat resembles the scuba of today. It consisted of a metal cylinder filled with air and worn on the diver's back, with a breathing tube leading to his mouthpiece. Most of the time the Triton was not self-contained, because tubes led from it to a pump on the surface. However, if the pump broke down or some other emergency arose, the diver would have sufficient air to stay below for quite some time (Figure 1-8).

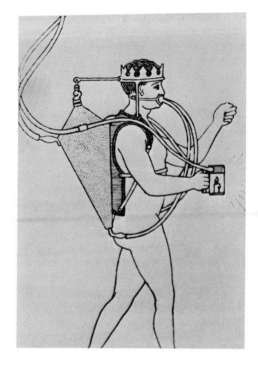

Figure 1-8. An even earlier scuba rig, the Triton, designed by Friedrich von Drieberg in 1808.

An Englishman, William James, devised the first workable and practical scuba in 1825. The air reservoir was an iron cylindrical belt extending from the diver's waist to his armpits. It contained compressed air at a pressure of thirty atmospheres, or about 450 PSI, and the diver breathed through a hose connected to his helmet. He wore a full suit, carried weights, and wore boots instead of fins, so the rig was obviously designed for walking on the sea bottom rather than for swimming (Figure 1-9).

Around the same time, a Brooklyn factory worker, Charles Condert, also developed a workable scuba, which differed considerably from James' in appearance. It consisted of a full diving suit and attached hood, both made of heavy cloth overlaid with gum rubber. At the top of the hood was a hole the size of a pinhead to release spent air. The air reservoir—a four-foot length of copper

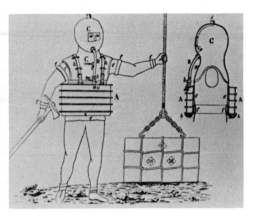

Figure 1-9. The first practical self-contained underwater breathing apparatus (scuba), invented by William James, an Englishman, in 1825.

tubing, six inches in diameter, bent like a horseshoe around the diver's body—had a safety feature that the James gear lacked: a valve to emit air directly into the diver's suit at his discretion, thus preventing water pressure from forcing the suit against his body. Condert successfully used his apparatus to depths of twenty feet in various jobs around the Brooklyn docks until his death in August 1832, when the valve cock between his air reservoir and his suit broke down as he was trying to surface and he became entangled in some underwater obstruction.

The next major advance came in 1865, when two Frenchmen, Benoit Rouquayrol, a mining engineer, and Lieutenant Auguste Denayrouze, a naval officer, designed an apparatus that they called a "self-contained diving suit." In actuality it was not completely self-contained, since it depended on an air supply pumped down from the surface to the metal cylinder on the diver's back; as in the case of Drieberg's Triton, however, it could be disconnected from the surface and function for short periods of time on the air in the cylinder which was constantly maintained at a pressure of forty atmospheres.

The most important feature of the Rouquayrol-Denayrouze invention was an attachment to the air reservoir that regulated the amount of air passing through the tube leading to the diver's mouthpiece. Inside the regulator was a membrane subjected to the outside water pressure; the membrane controlled a valve permitting air at equal or greater pressure to pass into the body of the diver's suit as water pressure increased. Because the regulator sent air into the diver's mouthpiece only when he took a breath rather than constantly, he wasted none of the precious air. This rig could be used in greater depths than the helmet diving rigs of the period, because it required less air and had the safety factor of the reserve air supply in the cylinder. The regulator, a very precise mechanism, played a more important part than any other device in the development of the modern scuba.

Originally the Rouquayrol-Denayrouze apparatus was designed to supply air to a naked diver wearing goggles, but the inventors soon recognized the danger to the diver if his goggles became flattened against his eyes. They replaced them with a face mask, which was attached to the hose leading to the cylinder. The invention was found so satisfactory that many helmet divers began to wear air reservoirs on their backs to free themselves from any connection to the surface. The apparatus was manufactured commercially in 1867 and most of the world's navies soon adopted it officially. Just five years after its invention, Jules Verne immortalized it in his prophetic *Twenty Thousand Leagues Under the Sea.*

In 1876 an English merchant marine officer, Henry Fleuss, invented a simple, compact and lightweight self-contained diving rig, which used pure compressed oxygen rather than compressed air and functioned on an air-regeneration system that worked. Because of the danger of oxygen poisoning below 33 feet, divers using it could not descend to even half the depth they could reach with the Rouquayrol-Denayrouze apparatus; it did make divers totally independent of any surface air supply, however, just as the inventions of James and Condert had. Its main advantage lay in the three-hour breathing supply it provided. The Fleuss apparatus (generally referred to as a closed-circuit oxygen rebreather) closely resembled the Rouquayrol-Denayrouze rig. The components were an air reservoir, regulator, air hose and the filter system that was the heart of the invention: the caustic soda in the filter system absorbed all the carbon dioxide and the diver was able to breathe the same oxygen over and over again. The diver could use it whether he wore a suit or swam naked, whether he chose goggles and mouthpiece, a mask, or even a helmet—versatility was one of its virtues (Figure 1-10).

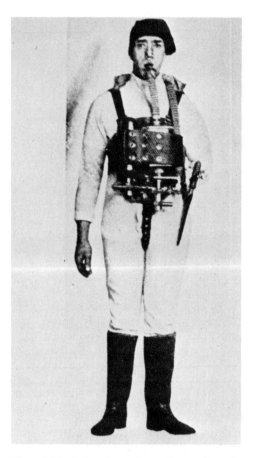

Figure 1-10. A three-hour air supply was the main advantage of this versatile closed-circuit oxygen rebreather invented by an Englishman, Henry Fleuss, in 1876.

Four years after its invention, the Fleuss rebreather proved its value in a crisis caused by the flooding of a tunnel under construction beneath the Severn River near Bristol, England. Before the tunnel could be pumped dry, a large iron door had to be closed. Divers wearing the standard helmet gear were unable to do the job because of obstacles in the tunnel which would have snagged or cut their air hoses. The Rouquayrol-Denayrouze apparatus was unworkable too: disconnected from the surface, it would not have supplied enough air for the duration of the job. Fleuss hired a well-known diver, Alexander Lambert, and taught him how to use the rebreather. Lambert succeeded in closing the door, accomplishing what the builders of the tunnel had almost given up as a lost cause. Three years later, when the same tunnel flooded a second time, Lambert was again successful, but this time nearly lost his life due to the oxygen poisoning that resulted from his extreme physical exertion.

In 1900, Louis Boutan, a French professor of biology and underwater photography pioneer, finding no type of diving equipment to his liking, devised his own scuba. Along with the standard helmet diving suit, he wore a huge cylinder of compressed air on his back, capable of sustaining him for three hours at a depth of seventy feet. As far as anyone knows, no one else ever used his invention.

In 1925, a French naval officer, Yves Le Prieur, invented a scuba unit which was widely used until the development of the Cousteau-Gagnan unit. His unit consisted of a steel cylinder of compressed air worn on the diver's back. An air hose from a regulator at the top of the cylinder connected to a mouthpiece which the diver held in his mouth. He controlled a valve which regulated the amount of air he required; goggles and nose clips completed the unit. The cylinder was quite small and only provided enough air for the diver to remain below for twenty minutes at 25 feet.

After several years of testing the unit, Le Prieur found that the goggles hurt the diver's eyes too much and he replaced them and the nose clips with a full face mask. This wasted more air, but he overcame the problem by using large air clyinders. In 1933, when the new sport of spear fishing or skin diving was becoming popular in the Mediterranean, he obtained a patent and began manufacturing the unit commercially for the growing band of enthusiasts. That same year he invented the first underwater spear gun, propelled by compressed air, and in 1935 the first wet suit.

One of the most puzzling facts in the history of scuba design is the complete neglect of the Rouquayrol-Denayrouze demand regulator for so many years. Both Boutan and Le Prieur made use of a simple regulator that provided the diver with a pressure equal or greater than the ambient water pressure, but both failed to use a demand regulator which would only deliver air when the diver breathed. A great volume of air was being wasted and the diver's bottom time was decreased considerably. Not until 1942 was it brought back into use by the two Frenchmen—Jacques-Yves Cousteau, a close friend of Le Prieur, and Emile Gagnan, an engineer—who created the celebrated aqualung, which millions of people are using around the world today.

The Voyage of H.M.S. Challenger

MARGARET DEACON   1971

Early in the summer of 1871, in a lecture given at the Royal Institution, W. B. Carpenter* called on Her Majesty's Government not to allow Britain's present lead in marine science to go by default.[1] News had come from the United States of a projected cruise in the Atlantic and Pacific Oceans to be led jointly by Jean Louis Agassiz (1807-1873) and Count Pourtalès.[2] The Germans were planning an expedition in the Atlantic and Sweden had sent two ships to the Arctic. *Nature* commented that if an opportunity were not found for following up the discoveries already made it would be a blow to national prestige and unfair to the scientists whose efforts had given the country its commanding position.[3]

Carpenter had his own very definite ideas about what the next step should be. As early as 1869 the idea had taken shape in his mind of persuading the government to send out a voyage of circumnavigation which would take the techniques and concepts developed on board the *Lightning* and *Porcupine* and put them to work in the oceans of the world.[4] Now, in June 1871, he wrote to G. G. Stokes, Secretary of the Royal Society, suggesting that the Society should consult with other leading scientific bodies and draw up a joint plan for marine research which could be submitted to parliament.[5] In a further correspondence, with George Goschen (1831-1907), First Lord of the Admiralty, he received an assurance that the government would give a favorable consideration to an application with this backing.[6]

The British Association discussed the need for further action at its summer meeting and adopted the resolution that

* W. B. Carpenter and C. Wyville Thomson had taken part in voyages, initiated by them, in H. M.S. *Lightning* and *Porcupine*, in 1868, 1869, and 1870 for the purpose of exploring the deep sea.

1. *Nature* (8 June 1871). Vol. 4, p. 107.
2. *Nature* (1 June 1871). Vol. 4, p. 87.
3. *Nature* (8 June 1871). Vol. 4, p. 107.

4. Harold L. Burstyn, "Science and government in the nineteenth century: the Challenger expedition and its report," *Bull. Inst. océanogr. Monaco* numéro spécial 2 (1968). Vol. 2, pp. 603-611.
5. *Report on the Scientific Results of the Voyage of H.M.S. Challenger during the years 1873-1876* (ed. C. Wyville Thomson and John Murray). H.M.S.O., London, *Narrative* (1885). Vol. 1, part 1, pp. l-li.
6. *Ibid.* p. li.

Margaret Deacon was most recently a research historian with the Science Studies Unit of the University of Edinburgh, Great Britain.

From *Scientists and the Sea 1650-1900*, pp. 333-365, 1971. Reprinted with light editing and by permission of the author and Academic Press Inc. Limited, London.

the President and Council of the British Association be authorised to co-operate with the President and Council of the Royal Society, in whatever manner may seem to them to be best, for the promotion of the circumnavigation expedition specially fitted out for carrying the physical and biological Exploration of the Deep-sea into all the great oceanic centres.[7]

Carpenter, on board H.M.S. *Shearwater*, wrote again to the Royal Society from Malta urging action and revealing the existence of his correspondence with Goschen and its results.[5] At the same time, in a letter to *Nature*, he referred to the

Scientific Circumnavigation Expedition, which (I have every reason to expect) will be fitted out by Her Majesty's Government next year.[8]

If the expedition was to be ready to go in 1872 there was no time to be lost. The Royal Society set up a committee consisting of officers and council members and the people most closely involved in the planning of the previous voyages; it included Carpenter, Wyville Thomson, Gwyn Jeffreys, Captain Richards, T. H. Huxley, Sir William Thomson and J. D. Hooker.[5] On its recommendation the Council asked the government to send out an expedition which would make a scientific study of the oceans.[9] The application was successful, consent being given in April 1872, and preparations for departure began immediately.

The ease and rapidity with which the matter was settled was deceptively simple and gives the impression of open-handedness where science was concerned that was in fact far from true in general. Two factors worked in its favor. One was that the expedition was a single event, one that it reflected favorably upon the country as a whole. In assisting the expeditions sent out to observe the solar eclipses of December 1870 and 1871 the government had already shown itself ready to support specific enterprises of obvious scientific value where there was no question of incurring repeated expenditure.[10] When quite modest financial aid was sought for a long term project the outcome was very different.

In 1867 the British Association had set up a committee, under the leadership of Sir William Thomson, to improve the harmonic analysis of tides.[11] The committee, which consisted of a number of leading scientists and people such as T. G. Bunt who had made special contributions to the study of tides, was financed at a rate of £100 a year. Harmonic analysis represented the greatest advance in this branch of science since Lubbock and Whewell had given it renewed impetus in the 1830s. Briefly, it enabled the complex forces creating tides to be split into simple components whose effects could be individually determined. The new insight which this analysis gave into the complicated structure of the actual tides experienced in any one place enabled the process to be reversed and led a few years later to the construction of Sir William Thomson's tide-calculating machine, which when set with variables involved, would integrate them to predict the height of tide to be expected at the same point over future periods.[12]

The useful applications of his research were many but in spite of this an application from the British Association to the Treasury in 1871 for a grant to relieve the charge on its annual budget, which was gradually falling off from a peak attained in the late 1860s, was turned down flat.[13] In the letter

---

7. *Nature* (10 August 1871). Vol. 4, p. 290. *Nature* (17 August 1871). Vol. 4, p. 313.
8. *Nature* (12 October 1871). Vol. 4, p. 468.
9. *Challenger Report. op. cit.* pp. li-liii.

10. "The government and the eclipse expedition," *Nature* (22 September 1870). Vol. 2, pp. 409-410. Also (3 November 1870). Vol. 3, p. 13 and (17 November 1870). Vol. 3, p. 52; (24 August 1871). Vol. 4, p. 324.
11. *Report of the 37th meeting of the British Association for the Advancement of Science (1867)*, (1868), pp. lxi-lxii.
12. Silvanus P. Thompson, *The Life of William Thomson, Baron Kelvin of Largs*, 2 vols. London (1910). Vol. 2, pp. 729-731.
13. *Report of the 41st meeting of the British Association for the Advancement of Science*

explaining the decision the Treasury pointed out that if they were to give the £150 a year asked for one kind of scientific work, however deserving, representatives of all the other sciences would feel themselves entitled to similar treatment and an impossible situation would arise.[14] *Nature's* editorial refrained from commenting on the "narrow stupidity" betrayed by this point of view to attain the higher object of illustrating, by this pointed example, the present condition of State science in England.[15]

A voyage to consolidate and extend the work done on the *Lightning* and *Porcupine* did not seem likely to open the door to extra commitments and much use was made of the argument that, seeing that a naval vessel would be used, the cost of the expedition would not be very much more than the government would be spending anyway to keep the ship in commission and pay the officers and men.[16] But it is clear that much of the groundwork done by W. B. Carpenter was of crucial importance. Recent research has shown how he was able to use his social contacts with Gladstone and his ministers to get the idea of the voyage accepted with the minimum of delay.[17]

During the summer of 1872 the expedition took shape. The ship chosen was the steam corvette H.M.S. *Challenger* and the Admiralty appointed Captain Nares to take command. Fitting out took place at Sheerness and here the *Challenger's* guns were removed and laboratories and extra cabins built in the space this set free. Gradually the officers and crew were assembled. Second in command was J. L. P. Maclear (1838-1907) and the lieutenants were Pelham Aldrich (1844-1930), A. C. B. Bromley, G. R.

Bethell (1849-1919) and Thomas Henry Tizard (1839-1924).[18]

The task of appointing the civilian scientists fell to the Royal Society. As Carpenter had already declared his intention of not going to sea again, Wyville Thomson was the obvious choice to head the team. He temporarily resigned his professorship and was awarded as compensation a salary of £1,000 a year for the duration of the voyage.[19] Three naturalists were chosen to assist him. Henry Nottidge Moseley (1844-1891) had studied science at Oxford and went on to take medicine at University College, London, but in 1871 he had abandoned the course to go to Ceylon with the eclipse expedition. The second, William Stirling, resigned before they set out and was replaced by a young German biologist who had met Wyville Thomson when he visited Edinburgh with the German North Sea Expedition in 1872. This was Rudolph von Willemöes-Suhm (1847-1875) who died tragically while the *Challenger* was in the Pacific.[20] The last, and ultimately the most important as far as the development of oceanography was concerned, was John Murray (1841-1914). He had been born in Canada but came to Scotland to be educated and eventually entered Edinburgh University, ostensibly to study medicine. Instead, he joined a whaler as surgeon for a seven-month voyage to the Arctic. Like Edward Forbes and Wyville Thomson before him, Murray took no examinations and on his return studied as he pleased. In 1872 he was working under Peter Guthrie Tait (1831-1901), Professor of Natural Philosophy, and it was Tait who recommended him to Wyville Thomson.[21]

(1871), (1872), p. lxxiii. The Association also called for observations to be made on the coast of India, *ibid.* p. lxxiv. This suggestion was taken up.
14. "The Tides and the Treasury," *Nature* (27 June 1872). Vol. 6, pp. 157-158.
15. *Ibid.* p. 158.
16. Daniel Merriman, "A posse ad esse," *J. Mar. Res.* (1948). Vol. 7, pp. 139-146.
17. Harold L. Burstyn, *op. cit.* p. 608.

18. *Challenger Report, op. cit. Narrative.* Vol. 1, part 1, pp. 2-19. Charles Wyville Thomson, *The Voyage of the Challenger. The Atlantic. A preliminary account of the general results of the exploring voyage of H.M.S. Challenger during the year 1873 and the early part of the year 1876,* 2 vols, London (1877). Vol. 1, pp. 9-59.
19. *Challenger Report, op. cit. Narrative.* Vol. 1, part 1, p. 22.
20. *Ibid.* p. 33.
21. Biographical material about Sir John Murray is to be found in obituary notices in *Proc. R. Soc.,*

Equally individualistic but in a totally different way was John Young Buchanan (1844-1925) who was appointed chemist to the expedition. He came from a well-to-do Scottish family and had studied widely on the continent. In addition he was technically very able, which was a strong reason for his being chosen for a voyage where much would depend on the ability of the scientists to improvise or make do. Unfortunately he also seems to have been a rather hard person to get on with, extremely reserved and impatient with those whose ideas did not keep pace with his own.[22] The team was completed by John James Wild, artist and Wyville Thomson's secretary.

By the beginning of December 1872, the scientists and their equipment had been assembled. The instruments were largely those which had been tried and tested on the earlier voyages. For measuring temperature they had the Miller-Casella thermometers, each with its pressure correction determined individually by Staff-Commander Davis. Unfortunately, bereft of the guidance of W. A. Miller who had died in 1870, Davis had admitted an error into his calibrations by failing to allow for the rise in temperature inside his apparatus and the corrections were all considerably too large.[23] This led to some confusion as the matter was not finally cleared up until some years after the voyage was over (see p. 41). They also took with them Siemens's electrical resistance ther-

mometer and his photometric apparatus [24] and some Johnson's metallic thermometers.[25]

As before they used hemp lines for putting out the nets, trawls and dredges, and for sounding but in June 1872 Sir William Thomson had made a sounding of 2,700 fathoms on board his yacht *Lalla Rookh* using steel piano wire and one of his machines was taken on the *Challenger*.[26] The device was still in its infancy at this stage and when it was tried out the drum collapsed.[27] For sampling the sea floor they used for preference the modification of the *Hydra* sounding machine named after its inventor, Lieutenant C. W. Ballie. The German North Sea Expedition had suggested using a slip water bottle to collect sea water from the bottom when they visited Leith.[28] Buchanan designed and later improved a stopcock water bottle which would bring up samples from intermediate depths and retain the gases dissolved in them.[29]

After the Lords of the Admiralty and the Royal Society committee had inspected the ship, and a farewell dinner party had been held on board,[30] the *Challenger* left for Portsmouth and it was from there that she finally set sail, on 21 December 1872, on a voyage that was to last nearly three and a half years.

It was confidently expected that she would bring back the answers to the questions which the work in the North Atlantic

series B (1915-1916). Vol. 89, pp. vi-xv; *Proc. R. Soc. of Edinburgh* (1914-1915). Vol. 35, pp. 305-317, with bibliography; *Nature* (26 March 1914). Vol. 93, pp. 88-89. *See also:* William A. Herdman, *Founders of Oceanography* (1923), and William N. Boog Watson, "Sir John Murray—a Chronic Student," *University of Edinburgh Journal* (1967-1968). Vol. 23, pp. 123-138. Tait had been a colleague of Wyville Thomson's at Belfast before moving to Edinburgh in 1860.
22. For obituaries of J. Y. Buchanan, *see: Proc. R. Soc.* series A (1926). Vol. 110, pp. xii-xiii; *Proc. R. Soc. Edinb.* (1924-1925). Vol. 45, pp. 364-367; *Nature* (14 November 1925). Vol. 116, pp. 719-720. The last two are by H. R. Mill.
23. J. E. Davis, "On deep-sea thermometers," *Proc. Meteorol. Soc.* (1869-1871). Vol. 5, pp. 305-342. Also in: *Nature* (14 December 1871). Vol. 5, pp. 124-128.

24. *Challenger Report, op. cit.* pp. 95-97.
25. Preserved in the *Challenger* collection in the British Museum (Natural History).
26. Sir William Thomson, "On deep-sea sounding by piano-forte wire," *Proceedings of the Philosophical Society of Glasgow* (1873-1875). Vol. 9, pp. 111-117. Silvanus P. Thompson, *Life of Lord Kelvin, op. cit.* Vol. 2, p. 719.
27. *Challenger Report, op. cit.* pp. 70-72. John Murray and Johan Hjort, *The Depths of the Ocean. A general account of the modern science of oceanography based largely on the scientific researches of the Norwegian Steamer Michael Sars in the North Atlantic*, London (1912), p. 27.
28. C. Wyville Thomson, *op. cit.* Vol. 1, pp. 34-37.
29. *Ibid.* pp. 37-40.
30. C. Wyville Thomson, *op. cit.* Vol. 1, p. 59. *Nature* (12 December 1872). Vol. 7, pp. 109-110.

had brought to mind. Would living creatures be found in the greatest depths of the sea? Was the floor of the ocean entirely covered by modern chalk and by the exciting proto-plasmic substance *Bathybius*? Would it be possible to measure by direct and indirect means the vast slow-moving currents which, according to the theory of oceanic circula-tion, carried the icy water of the polar seas along the sea bed to the equator? These were the possibilities which had been suggested by the discoveries made in the *Lightning* and *Porcupine*.

During the first weeks of the voyage, as the *Challenger* left the tempestuous seas of the northern winter behind and made for the Canary Islands, her crew of sailors and scien-tists began to settle down to the routine procedure of scientific stations and tried out their instruments and apparatus. Her officers and men did the actual work of sounding and dredging, current measuring and water sampling. Tizard, who in H.M.S. *Shearwater* eighteen months before had measured the currents in the Strait of Gibraltar for Car-penter, was now in charge of the deep sea temperature measurements made with the Miller-Casella thermometers, in addition to his work as navigating lieutenant. Bethel operated the Siemens thermometer.

As well as bearing the brunt of the ocean-ographic work, the *Challenger*'s crew were expected to carry out the normal duties of a survey ship, to make meteorological, mag-netic and other routine observations, to pro-vide information for improving Admiralty charts and to make soundings on routes for possible submarine telegraph cables.[31]

The naturalists' work was to deal with the contents of the trawls, nets and dredges as they came in, a repetitive task which Moseley came to find frankly boring, much preferring his botanical and anthropological work on shore at their numerous and often

exotic ports of call.[32] Buchanan's work was also largely routine. He had to determine the specific gravity of the sea water samples, which he did with a hydrometer, and to extract and analyze the gases which they contained and the carbonic acid content. Samples for full chemical analysis were col-lected to be sent home. In addition Buchanan had several research projects of his own.[33] Tizard tells us that the after-dinner smoking circle provided the forum for offi-cers and scientists to compare results and discuss their latest ideas.[34]

The voyage proper began when they left Teneriffe, heading for the West Indies. Every two or three days, oftener near land, they hove to, for observations. At every station they sounded and measured the temperature of the surface and bottom water and fairly frequently of the intermediate layers as well. Usually they made one or more hauls with the dredge or trawl and tow-nets and mea-sured the surface current, and on some occa-sions the subsurface currents also.[35]

They had not been long at sea when a curious change took place in the character of the samples which the sounding machine was bringing up from the sea bed. As the water got deeper the familiar pale grey globigerina ooze which they had thought might cover

31. See Nares's instructions from the Hydro-grapher, *Challenger Report, op. cit.* pp. 34-39. Al-so, the scientific instructions prepared by the Royal Society, pp. 23-33.

32. Henry Nottidge Moseley, *Notes by a Natural-ist. An Account of Observations made during the Voyage of H.M.S. Challenger,* new ed., London (1892), pp. 1, 501-502.
33. John Young Buchanan, "Report on the specif-ic gravity of samples of ocean water, observed on board H.M.S. *Challenger* during the years 1873-76," *Report on the Scientific Results of the Voyage of H.M.S. Challenger. Physics and Chemis-try* (1884). Vol. 1, 46 pp.
John Young Buchanan, "On the absorption of car-bonic acid by saline solutions," *Proc. R. Soc.* (1873-1874). Vol. 22, pp. 192-196.
See also Scientific Papers, Cambridge (1913). Vol. 1 and *Accounts Rendered of Work Done and Things Seen,* Cambridge (1919).
34. Edwin Ray Lankester *et al.,* "The scientific results of the *Challenger* Expedition," *Natural Sci-ence* (July 1895). Vol. 7, no. 41, pp. 7-75.
35. *Challenger Report, Narrative.* Vol. 1, and *A Summary of the Scientific Results* (1895). Vols 1 and 2.
Also, "The *Challenger* Collections," *Nature* (18 January 1877). Vol. 15, pp. 254-256.

the entire floor of the ocean, gradually darkened until it became unrecognizable. Examination under a microscope showed that the foraminifera remains which constituted the greater part of the ordinary ooze became fewer and fewer.[36] Finally on 26 February 1873 they found themselves in over 3,000 fathoms of water and at the end of what they soon realized had been a transition. Wyville Thomson wrote:

The depth was 3,150 fathoms; the bottom a perfectly smooth red clay, containing scarcely a trace of organic matter.[37]

Thomson's first reaction was not that here was an area too deep for the foramini-America.[38] He expected it to be a local phenomenon and as the *Challenger* entered the rather shallower area in mid-Atlantic, which Nares called the Dolphin Rise after the U.S. Coast Survey ship which had discovered it, the globigerina ooze returned only to vanish again as they came into deeper water where the red clay reappeared.[39] On 7 March the dredge brought up not only clay but also "a number of very peculiar black oval bodies about an inch long." Thomson thought that perhaps they were fossils but when Buchanan examined them he found that they were composed of almost pure manganese peroxide.[40] The provenance of these mysterious nodules could only be guessed at but a few days later the dredge brought up clay inhabited by marine worms which Willemöes-Suhm identified as tube-building annelids.[41] This was more like what they were looking for. The find showed that living creatures could exist in the greatest depths of the sea—the discovery had been

made in nearly 3,000 fathoms depth of water—and since they had sailed over the central parts of the ocean without finding a much greater depth it seemed safe to assume that the greatest depth could not be much more.[42] "It affords, in fact," Thomson wrote

conclusive proof that the conditions of the bottom of the sea to all depths are not only such as to admit of the existence of animal life, but are such as to allow of the unlimited extension of the distribution of animals high in the zoological series, and closely in relation with the characteristic faunae of shallower zones.[42]

After leaving the West Indies, however, he had to revise some of his conclusions. Only 90 miles north of St. Thomas they found themselves sounding in water 3,875 fathoms deep and the self-registering thermometers on the end of the line were fractured by the enormous pressure to which they had been subjected.[43] This great depth (in the Puerto Rico Trench) turned out to be exceptional but they had deep water all the way to Bermuda, their next port of call, and the sea bed turned out to be composed almost entirely of red clay. It was no longer possible to think of this as a localized phenomenon,[44] and they began to realize that the clay was characteristic of the deeper areas. Though they continued to speak of it as red clay this coloring was in fact local as Wyville Thomson had originally suggested, due to the sediment poured into the sea by the South American rivers, and in other areas darker browns and grays were the general rule.

From Bermuda the *Challenger* sailed for Halifax, Nova Scotia, and on the way made a wide detour to examine the Gulf Stream. The current was already much better known than any of the other ocean currents because of the work of the United States Coast Sur-

36. C. Wyville Thomson, *op. cit.* Vol. 1, p. 183.
37. *Ibid.* p. 182.
38. *Ibid.* pp. 226-227.
C. Wyville Thomson, "Notes from the *Challenger*," *Nature* (8 May 1873). Vol. 8, pp. 28-30.
39. C. Wyville Thomson, *The Voyage of the Challenger.* Vol. 1, pp. 193, 223-225.
40. *Ibid.* pp. 195-196.
C. Wyville Thomson, "Notes from the *Challenger*," *Nature* (15 May 1873). Vol. 8, pp. 51-53.
41. C. Wyville Thomson, *The Voyage of the Challenger.* Vol. 1, pp. 201-203.

42. C. Wyville Thomson, "Notes from the *Challenger*," *op. cit.* Vol. 8, p. 53.
43. *Ibid. Nature* (5 June 1873). Vol. 8, pp. 109-110.
44. *Nature* (31 July 1873). Vol. 8, pp. 266-267.

vey. The *Challenger* serial temperature measurements showed features which were already familiar to physical geographers—the relatively shallow body of warm water which formed the current and cold water below it which rose to the surface on its western flank and which contrasted sharply with the thick layer of warm water on the eastern side.[45] They also made direct current measurements both at the surface and at different depths below it.

On the Atlantic crossing several attempts had been made to see if subsurface movements of water in a contrary direction to the surface current could be detected. To track the undercurrents they used a drogue of the same construction that Nares and Carpenter had had in H.M.S. *Shearwater,* consisting of four vertical canvas fins stretched on iron bars, the whole thing moored by a line to a buoy or small boat.[46] On the first few occasions it was used the surface current was superficial and the movement of the lower layers so sluggish as to render the conclusion indeterminate.[47]

Nares had been instructed to use the drogue in the neighborhood of the equator. His orders ran

There is reason to believe that the depth of the Atlantic equatorial region does not exceed 2000 fathoms, which is easily within reach both of the sounding lead and of the dredge, and it is hoped that by means of anchoring a boat or beacon you will be able to ascertain to what depth the surface current extends, and what are the conditions of the circulation in the lower strata of the ocean.[48]

At station 106, near the equator, they found a westward flowing current of 2 m.p.h. on the surface.[49] At 15 fathoms the westward movement had diminished to three-quarters

and at 50 fathoms to half a mile and at 75 fathoms there was no current at all, showing they said, "how very superficial the Equatorial Current is." It was a pity that they did not consider it worthwhile to spend more time on these observations and continue them at greater depths. Had they done so the discovery of the equatorial undercurrent might perhaps have been brought forward by a dozen years.

As the *Challenger* pursued her crossing from the Cape Verde Islands to Brazil a significant alteration appeared in the pattern of submarine temperature. In the deepest water encountered during the first part of this stage of the cruise the minimum temperature had been 36°F but after passing St. Paul's Rocks they began to find 34°F at a comparable depth. In his report to the Hydrographer, Nares speculated that perhaps the Atlantic was divided down the center by a bank or series of banks extending from the Dolphin Rise in the north to Ascension Island or even St. Helena in the south and that this served to separate what he called the "cold stream" on the west from the warmer water to the east.[50] For the moment, however, this speculation about the cause of the change stayed in the realm of hypothesis. When they recrossed the Atlantic further south, from Rio to the Cape of Good Hope, the differentiation had vanished.[51] The exploration of what this implied was left until the *Challenger* returned to the Atlantic in 1876.

During the first few months of the voyage Bethell had made a number of observations on the temperature of sea water at different depths using Siemens's' electrical resistance thermometer. He reported that it worked well, giving results which on the whole correlated with those obtained with the self-registering mercury thermometers. However,

45. C. Wyville Thomson, *The Voyage of the Challenger.* Vol. 1, pp. 371-375.
Challenger Report, Narrative. Vol. 1, part 1, pp. 154-158.
46. *Ibid.* pp. 79-82.
47. *Ibid.* pp. 123-124, 154, 157.
48. *Ibid.* p. 36.
49. *Ibid.* p. 193.

50. George S. Nares, *H.M.S. Challenger. Reports of Captain G. S. Nares, R.N. with abstract of soundings and diagrams of ocean temperature in North and South Atlantic Oceans,* Admiralty (1873), pp. 10-12.
51. *Ibid.* p. 13.

in the opinion of the heads of the expedition, its disadvantages continued to outweigh its usefulness. It was still ill adapted structurally for use at sea and difficult to read accurately.[52] They felt that though it was potentially a great advance on existing methods for the moment they had done as much with it as was necessary for present purposes. Accordingly when they reached the Cape, late in 1873, Nares had it sent home,[53] a decision which must soon have caused him some annoyance.

On leaving the Cape of Good Hope the *Challenger* sailed along the perimeter of the Indian Ocean calling at the remote and desolate Marion and Prince Edward Islands and the Crozet group before turning southwards to Kerguelen. From there they went on past Heard Island to the edge of the ice pack, reaching a southern latitude of more than 65° before turning north again. In these high latitudes they discovered to their surprise that beneath the cold water on the surface there lay a thick layer of slightly warmer water. At station 153, their furthest south, on 14 February 1874, in 1,675 fathoms of water, the temperature of the surface layer was 29·5°F but at 300 fathoms it was 32°F and at 500 fathoms 32·8F°.[54] This was nothing more than the temperature inversion measured by Wales and Bayly a century before but the *Challenger* scientists seem to have been totally unprepared for it. Lieutenant Spry was amused to see their consternation at the find, "putting their whole theory out of gear" as he thought.[55] In fact it was not difficult to understand the probable cause of the inversion. As they saw it two things contributed to keep down the temperature of the surface water, cooling by the air in winter and by the presence of water from the melting ice in summer. Because ice is fresh it would keep down the salinity as well as the temperature of the surface water as it melted so that its density would still be less than that of the warmer layer below. This they identified with the lower layer of the poleward flow from the equator, less affected at this depth by the factors in operation at the surface and therefore still retaining some of its warmth.[56]

The difficulty rose over the remaining 1,300 fathoms of water. Because the Miller-Casella thermometers could only register extremes of temperature they were only satisfactory in conditions where temperature decreased with depth. Their observations showed that the temperature of the bottom layers lay somewhere between the extremes of the cold and warmer layers near the surface and that was all that could be said about it.

Meanwhile an important stage had been reached in the revision of ideas about the deep sea floor. John Murray was in charge of the biological collections made at the surface and at intermediate depths and of the sea floor samples. He used a tow net to catch samples of plankton at depths ranging from 0 to 100 fathoms. His findings threw new light on the diurnal migration of plankton [57] and on the controversy about the habitat of the foraminifera whose remains made up globigerina ooze.

During the 1860s new evidence had been produced by Major Owen to support the view that the relevant species of foraminifera live in the surface layers of the ocean. [58] Gwyn Jeffreys was convinced but Wyville Thomson and Carpenter felt that the in-

52. *Ibid.* p. 2.
The Voyage of the Challenger. Vol. 1, pp. 242-244.
53. G.S. Nares, *op. cit.* p. 15. He also sent back the photometric device but does not say if it had been tried out.
54. *Challenger Report, Narrative.* Vol. 1, part 1, p. 399; *Summary.* Vol. 1, p. 496.
55. William J. J. Spry, *The Cruise of H.M.S. Challenger,* 10th ed., London (1884), p. 106.

56. George S. Nares, *H.M.S. Challenger. No. 2. Reports on ocean soundings and temperature, Antarctic Sea, Australia, New Zealand* (1874), pp. 10, 15-18.
57. C. Wyville Thomson, "Notes from the *Challenger,*" *Nature* (8 May 1873). Vol. 8, pp. 28-30. Diurnal migration was discovered by Bellingshausen. *See* Ch. 11.
58. Samuel Owen, "On the surface-fauna of mid-ocean," *J. Linn. Soc.* (1865). Vol. 8 (Zoology), pp. 202-205. Vol. 9, Zoology, (1868). pp. 147-157.

stances quoted might have been exceptional and held to their original conviction which was that the creatures lived on the sea bed.[59] Very soon after the *Challenger* set sail Murray had collected enough evidence to show that Owen was in fact right. His catches with the tow net revealed living in the upper layers of the sea just those foraminifera whose remains predominated in the ooze in the sea bed below.[60]

Successive observations, made over the full extent of the ocean, corroborated the theory. Murray found that globigerina live in almost all parts of the ocean but that the distribution and sizes of the different species varied with latitude. These changes and the relative abundance of globigerina and other kinds of foraminifera were faithfully recorded in the varying composition of the globigerina ooze below. For example, south of Kerguelen they found only one species of globigerina, *Globigerina bulloides,* in the tow net and its remains made up the entire foraminifera content of the ooze in this region. As they approached the ice this too disappeared and instead they found diatoms living in the surface water and their remains in the bottom deposit, which they logically named diatom ooze.[61]

Carpenter tried to save his theory by arguing that perhaps the globigerina were pelagic for part of their lives and sank to the bottom as they developed,[62] but Wyville Thomson was completely convinced by Murray's conclusions although they made necessary yet a further revision of his views concerning the clay deposit found in the deepest

areas.[59, 63] He had supposed when red clay turned out to be widespread that it was sediment collecting in places too deep for globigerina to survive. Murray's results showed that their remains must fall on all parts of the ocean alike so why was there no sign of their presence in these regions? Examination of samples from different depths showed that in the deeper samples the calcareous skeletons of foraminifera, molluscs and other creatures progressively disappeared. Clearly some chemical reaction took place in these depths which removed the carbonate of lime that made up 98 per cent of the ooze. Thomson supposed that the red clay was the residue of this process. Buchanan pointed to an increase in carbonic acid in sea water at these depths as the cause and attempted to simulate the process in the laboratory. He removed the carbonate of lime from some globigerina ooze and analyzed the remainder, finding silica, alumina and red oxide of iron. The substance certainly seemed to look like red clay.

In March 1874 the *Challenger* reached Australia and during a two-month stay at Sydney the members of the expedition enjoyed in their various ways a change from the routine of life on board ship. Wyville Thomson, Murray and Aldrich went to Queensland to study the fresh-water fauna of its rivers. Moseley went on hunting expeditions and returned with the opossums and fruit-eating bats which he shot. For the rest there was a round of balls and excursions culminating in a dredging picnic and farewell party given by the expedition for their hosts.

As the explorers relaxed, on the other side of the world the controversy over the cause of currents had renewed as a result of their unwitting efforts and raged with unabated vigor. As will be remembered the argument had been broken off, though not

59. C. Wyville Thomson, "Preliminary notes on the nature of the sea-bottom procured by the soundings of H.M.S. *Challenger* during her cruise in the 'Southern Sea' in the early part of the year 1874," *Proc. R. Soc.* (1874-1875). Vol. 23, pp. 32-49; p. 34.
60. *Ibid.* pp. 33-39.
61. *Ibid.* pp. 47-48.
62. William B. Carpenter, "Remarks on Professor Wyville Thomson's preliminary notes on the nature of the sea-bottom procured by the soundings of H.M.S. *Challenger,*" *Proc. R. Soc.* (1874-1875). Vol. 23, pp. 234-245.

63. C. Wyville Thomson "The *Challenger* Expedition," *Nature* (1874-1875). Vol. 11 (3 December 1874) pp. 95-97; (10 December 1874) pp. 116-119.

concluded, in 1871 with W. B. Carpenter unable to see that the motive force lay not exclusively in the superior weight of one column of water over another but in the presence at the same level in the ocean, of water masses of different densities which set up internal pressure gradients, and consequently unable satisfactorily to dispose of James Croll's criticism that the system depended on the difference in level between one part of the ocean and another and that a sufficient difference did not exist.

In February 1874 Croll broke his long silence with an article in the *Philosophical Magazine* in which he again attacked the theory of vertical oceanic circulation due to difference in temperature which Carpenter had put forward.[64] This time however Croll did not restrict himself to demonstrating the fallaciousness, as he saw it, of the theory in general. No one could now seriously deny that undercurrents existed in the Strait of Gibraltar and the Bosporus, where Wharton's work in H.M.S. *Shearwater* in 1872 had completely overturned Spratt's earlier conclusion.[65] Nor was it possible to deny that in some form water of polar origin was penetrating the depths of the oceans. Some alternative force had to be located to account for these movements if the effect of differences in density were disallowed.

Croll now tried to show that they followed equally well if one took wind stress as the motive power. For example, as he saw it, the system of currents in the Strait of Gibraltar was caused because water drifted eastward by westerly winds and pouring into the Mediterranean created a head of water there and so set up an undercurrent in the

opposite direction.[66] In the ocean the concrete physical barrier of the land which would operate in gulfs and partially enclosed seas would be replaced by the opposition of contrary wind and water movements. Here a current would take the line of least resistance and this might mean dipping below the surface. In this way the polar current which flowed southward along the coast of Labrador ultimately became an undercurrent below the Gulf Stream and the Gulf Stream itself disappeared below the surface on reaching the Arctic.

Carpenter replied by reiterating his theory that if constant sources of heat and cold are applied to the surface of the sea at different points, disturbance of equilibrium and continual interchange of water must ensue.[67] The *Challenger* temperature observations unequivocally showed the presence of polar water of low salinity not far below the surface of the sea at the equator and this, for him, satisfactorily demonstrated the existence of upwelling polar water at a point where the theory led him to expect it. If Croll were to be believed, the cold deep water in the North Atlantic was attributable to a reflux from the Gulf Stream which would be nowhere near sufficient to put such a large body of water in motion. The *Challenger* observations in the South Atlantic showed that cold water there was much nearer the surface than in the North Atlantic, which accorded well with his theory because, since there was freedom of communication between the Southern Ocean and the South Atlantic, the density circulation was correspondingly more vigorous. Croll would not be able to account satisfactorily for this activity because there was only a comparatively weak wind-driven surface current flowing southwards from the equator and no land barrier to act as the source of a return current. Croll's arguments

64. James Croll, "On ocean currents—Part III. On the physical cause of ocean-currents," *Phil. Mag.* (January-June 1874). Series 4, Vol. 47, pp. 94-122.
65. William J. L. Wharton, "Observations on the currents and undercurrents of the Dardanelles and Bosphorus, made by Commander J. L. Wharton, of H.M. Surveying-Ship *Shearwater*, between the months of June and October, 1872," *Proc. R. Soc.* (1872-1872). Vol. 21, pp. 387-393. See Ch. 14.

66. James Croll, *op. cit.* pp. 168-190.
67. William B. Carpenter, "Ocean currents," *Nature* (2 April 1874). Vol. 9, pp. 423-424.

attempting to show that the warm water in the surface layers of the Atlantic was derived from the Gulf Stream were equally unsound.[68]

Croll replied that Carpenter had misrepresented his point of view. He had not said that the cold water in the North Atlantic was the reflux of the Gulf Stream, nor was it necessary that the undercurrent of limited extent which he had in mind should keep the whole body of water in motion. A cold current only a fraction of the size of the Gulf Stream would keep the deep water of the Atlantic at Arctic temperatures since the amount of heat gained through the earth's crust had been shown to be negligible.[69] He would not allow that the *Challenger* results favored Carpenter's views at the expense of his own. Winds and currents prevailing in the Atlantic tended to transport water from south to north so it was logical that cold water came nearer to the surface in the South Atlantic and furthermore since the layer of warm water was so thin at the equator the slope between the equator and the poles must be even smaller than they had supposed and could not exceed 2½ feet. [70] To these points Carpenter found himself unable to make any convincing reply.[71]

Meanwhile the *Challenger* had left Australia to spend the last part of 1874 and the early months of 1875 threading the network of seas between the islands of the East Indies and the mainland of Asia. Many of these seas were cut off from the main ocean below a

certain depth by underwater ridges connecting the islands. There they found the phenomenon noticed a year or two before by Chimmo who had shown that in each of these basins the temperatures stopped falling at the greatest depth of the surrounding sill and then stayed the same all the way to the bottom. For example the Celebes Sea below 700 fathoms maintained a constant temperature of $39°F$ and the Sulu Sea of $50·5°F$ below 400 fathoms although both were more than 2,500 fathoms deep.[72] This illustration of what happened in bodies of water cut off from the influence of polar currents afforded a classic example for the exponents of a general oceanic circulation.

In 1875 the controversy reached its climax. This year saw the publication of two major works, both definitive of their different points of view. Joseph Prestwich's survey of deep sea temperature observations and theories of ocean circulation appeared in the *Philosophical Transactions*[73] and Croll brought out a book called *Climate and Time* in which he brought together the arguments contained in his earlier papers.[74] Carpenter welcomed the evidence which Prestwich produced to show that previous observations had already induced Lenz and other scientists to adopt a viewpoint similar to his own and to draw the same conclusions over such points as the presence of cold water near the surface at the equator.[75] He was also able to

68. William B. Carpenter, "On the physical cause of ocean-currents," *Phil. Mag.* (January-June 1874). Series 4, Vol. 47, pp. 359-362. See also his "Further inquiries on oceanic circulation," *Proc. R. geog. Soc.* (1873-1874). Vol. 18, pp. 301-407.
69. James Croll, "On the physical cause of ocean currents," *Phil. Mag.* (January-June 1874). Series 4. Vol. 47, pp. 434-437.
70. James Croll, "Ocean currents," *Nature* (21 May 1874). Vol. 10, pp. 52-53.
71. William B. Carpenter, "Ocean Circulation," *Nature* (28 May 1874). Vol. 10, p. 62.
William B. Carpenter, "Lenz's doctrine of ocean circulation," *Nature* (2 July 1874). pp. 170-171.
See also: "Ocean circulation—Dr. Carpenter and Mr. Croll," an anonymous letter defending Croll in *Nature* (4 June 1874). Vol. 10, pp. 83-84.

72. Thomas H. Tizard, *H.M.S. Challenger, No. 3. Reports on ocean soundings and temperature, New Zealand to Torres Strait, Torres Strait to Manila and Hong Kong* (1874), p. 7.
"Remarks on the temperatures of the China, Sulu, Celebes and Banda Seas," *H.M.S. Challenger, No. 4. Report on ocean soundings and temperatures, Pacific Ocean, China and adjacent seas* (1875), pp. 4-9.
Challenger Report, Summary. Vol. 1, pp. 771, 782.
73. Joseph Prestwich, "Tables of temperatures of the sea at different depths beneath the surface, reduced and collated from the various observations made between the years 1749 and 1868, discussed," *Phil. Trans.* (1875). Vol. 165, pp. 587-674.
74. James Croll, *Climate and Time in their Geological Relations: a Theory of Secular Changes of the Earth's Climate,* London (1875).
75. William B. Carpenter, "Lenz's doctrine of

utilize the *Challenger*'s observations in the East Indies and the work of the U.S. Survey Ship *Tuscarora* in the North Pacific in 1874[76] for further corroborative details. Meanwhile, however, Croll was planning to deliver what he confidently expected would be the *coup de grâce* to Carpenter's theory.

The *Challenger* temperature sections in the North Atlantic had shown the great extent of the unusually deep body of warm water there, about one thousand fathoms in all, which had first been revealed by observations made early in the century. Croll regarded this warm water as the end product of the Gulf Stream but Carpenter cited it as proof of ocean circulation. An independent point of view came from J. Y. Buchanan who explained it as the result of seasonal changes in the ocean. The heat of summer would evaporate the surface water, leaving a layer of warm saline water. As this cooled down in autumn it would sink through the warm but less saline water underneath and distribute its heat to the layers below.[77]

Croll produced his argument to end all arguments at the British Association meeting at Bristol in the summer of 1875.[78] Using the *Challenger* temperatures and tables giving the thermal expansion of sea water he calculated that the vertical extent of the warm water in the North Atlantic must raise its surface above that of the equator where the layer of warm water was shallower. As he understood it the gravitational theory of currents would only work if there were a

downhill slope between the equator and the poles so that this discovery showed, as far as he was concerned, that density currents in the ocean were a practical impossibility. Furthermore, he had come to think that the winds performed the very function that Carpenter had ascribed to density differences. By generating the currents which carried water away from the equator to higher latitudes, they indirectly served to build up a head of water which would cause an outflow from the base of the heavier column along the sea bed in the same way that Carpenter had described.[79]

Croll was in fact right in his deductions about the relative height of the North Atlantic and also in that the pattern of winds over the ocean at large could produce differences in water level, but the rest of his arguments were based on a misconceived idea of the theory he was attacking. Yet it was largely due to Carpenter's own misapprehension that the situation had arisen and Carpenter even now could do no more than try and cast doubt on Croll's figures and show that there was in reality a slope favorable to his hypothesis.[80] Not surprisingly Croll was nettled at Carpenter's refusal to concede the victory outright and felt "rather astonished at the nature of the objections which he advances." He wrote in reply to Carpenter's objection that water was so near to being a perfect fluid that the size of the slope did not matter

What possible connection can "viscosity" have with the crucial test argument? Suppose water to be a perfect fluid and absolutely frictionless: this would not in any way enable it to *flow up-hill*.

The crucial test argument brings the question at issue, in so far as the North Atlantic

ocean circulation," *Nature* (2 July 1874) Vol. 10, pp. 170-171.
76. William B. Carpenter, "Summary of recent observations on ocean temperature made in H.M.S. *Challenger*, and U.S.S. *Tuscarora;* with their bearing on the doctrine of a general oceanic circulation sustained by difference of temperature," *Proc. R. geog. Soc.* (1874-1875). Vol. 19, pp. 493-514.
77. John Young Buchanan, "Note on the vertical distribution of temperature in the ocean," *Proc. R. Soc.* (1874-1875). Vol. 23, pp. 123-127.
78. James Croll, "On the *Challenger*'s crucial test of the wind and gravitation theories of oceanic circulation," *Report of the 45th meeting of the British Association (1875)* (1876). Part 2, pp. 191-193.
ibid., Phil. Mag. (July-December 1875). Series 4, Vol. 50, pp. 242-250.

79. James Croll, "The wind theory of oceanic circulation.—Objections examined," *Phil. Mag.* (July-December 1875). Series 4, Vol. 50, pp. 286-290.
80. William B. Carpenter, "Remarks on Mr. Croll's 'crucial-test' argument," *ibid.* pp. 402-404.
William B. Carpenter, "Further remarks on the 'crucial-test' argument," *ibid.* pp. 489-491.
William B. Carpenter, "Ocean circulation," *Nature* (23 September 1875). Vol. 12, pp. 454-455.

is concerned, within very narrow limits. The point at issue is now simply this: *Does it follow, or does it not, from the temperature-soundings given in Dr. Carpenter's own section, that the North Atlantic at lat. 38° is above the level of the equator?* If he or anyone else will prove that it does not, I shall at once abandon the crucial test argument and acknowledge my mistake; but if they fail to do this, I submit that they ought at least in all fairness to admit that in so far as the North Atlantic is concerned, the gravitation theory is untenable.

The Atlantic column is lengthened by heat no less than eight feet above what it would otherwise be were the water of the uniform temperature of 32°F., whereas the equatorial column is lengthened only four feet six inches. The expansion of the Atlantic column below the level of the bottom of the equatorial not being, of course, taken into account. How then is it possible that the equatorial column can be above the level of the Atlantic column? And if not, let it be explained how a surface-flow from the equator pole-wards, resulting from gravity, is to be obtained.[81]

But seeing the mounting evidence which still seemed only to be satisfactorily explained by the theory of density currents Carpenter could not let it go because of a mechanical difficulty and tried to find a way round it. He argued that the level of the Atlantic would only be above that of the equator in actuality if the ocean were in a state of equilibrium, but it was not. In any case the low salinity of the water at the equator would tend to neutralize the effect due to temperature further north. And because water was so nearly a perfect fluid very small differences in level would suffice to maintain vertical circulation. He wrote

my position is, that the void created by the slow descent of water chilled by the surface-cold of the Polar area will be so speedily replaced by the inflow of water from the circumpolar area, and this again by inflow from the temperate region, as to produce a

continual upper-flow of equatorial water towards the pole, without the gradient which Mr. Croll persistently asserts to be necessary.[82]

Croll agreed that the ocean was not in a state of static equilibrium but this, he argued, had the effect of keeping the equatorial column even lower. He admitted, and this was a considerable departure from his previous statements, that if no other agencies were at work difference in temperature would result in circulation, but as things were, the warm water at the equator was borne away by the winds before it could accumulate sufficiently to set the system going.[83]

It is impossible to escape the conclusion that Croll had managed to appropriate to himself a good deal of the moral superiority which Carpenter's side of the argument had enjoyed in the early days and, while the crew of the *Challenger* and other observers were amassing information which established the existence of density circulation beyond all reasonable doubt, in the context of the private feud Croll was now getting the best of it. The reviewer of *Climate and Time* for the *Philosophical Magazine* wrote

The manner in which Mr. Croll meets and combats Dr. Carpenter's theory of oceanic circulation is very characteristic of the unwearying patience, acuteness, and courage of his investigations. It is not our purpose here to express any opinion on this unfinished controversy, further than to say that Mr. Croll follows Dr. Carpenter into every one of his positions with a resolution and tenacity of purpose which, were it not so really calm and passionless, might almost be looked on as cruel and unmerciful. When, according to his own view, he has completely overthrown any of Dr. Carpenter's arguments by one special line of attack, he opens fire upon it from another point with the same hearty goodwill he might be expected to exhibit in leading a gallant onslaught against any

81. James Croll, "Oceanic circulation," *Nature* (7 October 1875). Vol. 12, p. 494.

82. William B. Carpenter, "Ocean circulation," *Nature* (21 October 1875). Vol. 12, p. 533.
83. James Croll, "Oceanic circulation," *Nature* (25 November 1875). Vol. 13, pp. 66-67.

hitherto unassailed position. This he does not do with the intention of thrice slaying the slain, but simply because he has in an eminent degree the faculty of examining a problem from many sides; and he is concerned, not for dialectic triumphs, but for the presentation of the truth in its entirety. No sooner has the active and versatile Doctor marshalled new arguments from new facts in favour of his vast conception, than Mr. Croll is ready to take up his challenge, and show that all the facts can be read off in a light which harmonizes with his adopted wind theory. Mr. Croll has not the literary dexterity and skill which eminently characterize Dr. Carpenter's pen; but he, as much as any man, knows the manifold physical bearings of the vexed question; and while on the one side the exposition of an argument is adorned with literary grace, on the other the real scientific value of facts and observations is more promptly apprehended.[84]

A little while before the currents controversy had become active again there had been a shorter but more acrimonious wrangle in the columns of *Nature* over the naming of the Miller-Casella thermometer. When the *Lightning*'s cruise was under discussion in 1868 leading instrument-makers had been circularized with a view to testing the reaction of their thermometers to the effects of pressure at sea. One firm, Negretti and Zambra, had not replied. This was the firm which had made protected thermometers for Admiral Fitzroy in 1857. Now they complained that the distinction which should rightly have been theirs had been wrongfully pre-empted by Casella in the naming of his thermometer in a way that would never had been sanctioned by W. A. Miller.[85] He had been quick to acknowledge that theirs had priority when he learned of

its existence, some time after the appearance of his own.[86]

Having established their case, Negretti and Zambra again turned their attention to the problems of marine instrumentation, this time to the need for a device which could record the absolute temperature at any depth required and so overcome the limitations of the maximum and minimum thermometer. They invented a device which could be used to turn the thermometer upside down as soon as it began to ascend in the water. The thread of mercury then broke and the amount involved in the measurement was trapped in the second arm of the thermometer where it could be read off at the surface.[87] Of course the device immediately became known as the Negretti and Zambra reversing thermometer but no one supposed that they were trying to steal the glory which perhaps belonged to Aimé who had invented a similar device thirty years before. Perhaps Casella had after all had a point when he said that the use that was made of an instrument counted for more in its naming than strict considerations of originality.

The new thermometers reached the *Challenger* at Hong Kong where she had arrived in November 1874. They had not been very well packed and many were broken but the survivors were used on a number of occasions during the remainder of the voyage, the first time being on 28 January 1875 in the Sulu Sea.[88] After some initial difficulties had been ironed out they worked well but the results were mostly rather higher than those obtained with the Miller-Casella thermometers and Tizard suspected that

84. Review of *Climate and Time*, *Phil. Mag.* (July-December 1875) Series 4. Vol. 50, pp. 322-324. According to his obituary Croll had "maintained with great force and with general approbation the position for which he contended," *Nature* (25 December 1890). Vol. 43, pp. 180-181. He was made an F.R.S. in 1876.
85. Henry Negretti and Joseph Warren Zambra, "Deep-sea soundings and deep-sea thermometers," *Nature* (23 October 1873). Vol. 8, p. 529.

86. See correspondence in *Nature* (1873-1874). Vol. 9 (6 November 1873) p. 5; (20 November 1873) pp. 41-42; (27 November 1873), pp. 62-63; (11 December 1873) p. 102.
87. Henry Negretti and Joseph Warren Zambra, "On a new deep-sea thermometer," *Proc. R. Soc.* (1873-1874). Vol. 22, pp. 238-241. "A new thermometer," *Nature* (19 March 1874). Vol. 9, pp. 387-388.
88. *Challenger Report, Narrative*, Vol. 1, part 2, p. 655.

they were insufficiently protected against the effects of pressure.[89]

The stay at Hong Kong saw a change in the expedition's leadership. Nares had been chosen to head the expedition which was then being planned to the Arctic and he went home, accompanied by Pelham Aldrich. The *Challenger*'s new commander was Captain Frank Tourle Thomson and Aldrich's place was taken by Lieutenant Alfred Carpenter.[90]

The year 1875 saw the *Challenger* in the Pacific. By now she was not alone in the field. In 1874 the U.S.S. *Tuscarora* had made soundings in the North Pacific along a proposed telegraph cable route between America and Japan. This voyage had been more than a mere survey. The *Tuscarora*'s captain, George Belknap tried new kinds of sounding machines and obtained good results with an improved version of Sir William Thomson's machine for sounding with wire. He also made surface and subsurface current measurements.[91] The Germans had sent out a round the world expedition of their own in the S.S. *Gazelle* and in 1875 she too was in the Pacific.[92]

The combined efforts of these ships showed the Pacific to differ considerably from the other oceans. Not only did it cover a far greater area but it was uniformly deeper, much of the sea bed being at a depth greater than 2,000 fathoms and covered with the characteristic deep sea clay. Associated with the clay were larger mineral particles— quartz, mica and pumice in considerable quantities, pieces ranging in size from a pea to a football. This led John Murray to dispute Wyville Thomson's view that the clay

was a residue of decayed organic matter and to point instead to volcanic material as its source. The number of volcanoes in the Pacific area, the discovery of pumice associated with them and its ability to float long distances, as they had seen it do, all confirmed him in this belief.[93, 94]

Almost always present with the clay were manganese nodules in far greater quantities than they had found before. The characteristic form which they took was of a concretion round some object which served as a nucleus. This was commonly a shark's tooth or the ear bone of a whale. Such remains were plentiful in the sediment, some only lightly coated with manganese, and showed, said Murray, by their abundance here and scarcity in other types of deposit, how very slowly the clay was being formed in comparison with the oozes in shallower water. Some of the shark's teeth were very large one being 4 inches long, and this was later identified as having belonged to a gigantic fossil species, *Carcharodon megalodon*.[95]

It was difficult to account for the presence of the manganese. Murray thought that it too was of volcanic origin and that manganiferous rocks were decomposed giving carbonate of manganese which was then oxidized by sea water. Buchanan on the other hand suggested that organic remains on the sea bed reacted with the sulfates in sea water to produce sulfides which in turn combined with the iron and manganese which were then changed into oxides by the sea water.[94,96] Later on the French geologist

---

89. *Challenger Report, Narrative.* Vol. 1, part 1, pp. 88-95.
90. *Challenger Report, Narrative.* Vol. 1, part 2, p. 638.
91. "Deep-sea soundings in the Pacific Ocean," *Nature* (3 September 1874). Vol. 10, pp. 356-357. "Soundings and currents in the North Pacific Ocean," *Nature* (15 October 1874), pp. 484-485.
92. "Temperatures and ocean currents in the South Pacific," *Nature* (11 January 1877). Vol. 15, p. 237.

93. John Murray, "Preliminary reports to Professor Wyville Thomson, F.R.S., Director of the civilian staff, on work done on board the *Challenger*," *Proc. R. Soc.* (1875-1876). Vol. 24, pp. 471-544.
94. John Murray, "On the distribution of volcanic débris over the floor of the ocean—its character, source, and some of the products of its disintegration and decomposition," *Proc. R. Soc. Edinb.* (1875). Vol. 9, pp. 247-261.
95. John Murray and A. Renard, *Report on Deep-Sea Deposits, Challenger Report* (1891), pp. 268-270.
96. John Young Buchanan, "On the occurrence of sulphur in marine muds and nodules, and its bearing on their mode of formation," *Proc. R. Soc. Edinb.* (1890-1891). Vol. 18, pp. 17-39.

Dieulafait put forward the idea that manganese was precipitated during a reaction which took place at the surface of the sea between the sea water and the air.[97] But none of these explanations, said their critics, could account for the formation of the nodules.

From time to time the exploring ships came across places in the Pacific which were yet still deeper than the considerable average depth of the ocean as a whole. At one place not far from the coast of Japan the *Tuscarora* had found no bottom in 4,643 fathoms of water.[98] On 23 March 1875, in latitude 11° 24. N, longitude 143° 16' E, the *Challenger* sounded in 4,475 fathoms and again the thermometers broke under a pressure which it had never been expected that they would have to bear.[99]

At depths of this magnitude yet another change came over the character of the bottom deposits. Wyville Thomson announced in a letter written to T. H. Huxley that here the deep sea clay was replaced by a deposit containing a large proportion of Radiolarians, hence Radiolarian ooze. Radiolarians lived in all the oceans and at all depths but they were most plentiful in the Pacific. They had siliceous skeletons which did not decay at great depths as did the limey skeletons of foraminifera and so it was easy to see that where there was the greatest depth of water their remains would be most in evidence.[93, 100]

Also in this historic letter, Wyville Thomson announced that the mystery surrounding *Bathybius* had at last been solved. *Bathybius haeckelii* had been discovered by Huxley in 1868 when he was re-examining the sea bed

samples which had been collected by H.M.S. *Cyclops* in 1857 and preserved in spirit. It appeared as a gelatinous substance which had organic properties and Huxley believed that he had lighted upon a form of protoplasm.[101] The interest of the discovery was enlarged by its possible significance for arguments current at the time about abiogenesis or spontaneous generation. One of the foremost exponents of this hypothesis was Ernst Haeckel (1834-1919), hence Huxley's choice of name.[102]

During the first two and a half years of the *Challenger*'s voyage the watery substance from the sea floor samples was examined under the microscope for "hours at a time" without finding any sign of life nor did tests reveal anything resembling the appearances described by Huxley and Haeckel. Then one day someone noticed that a jelly-like substance was to be seen in the samples preserved in spirit but not in the jars containing specimens kept in water.[103] Buchanan examined the mysterious substance and discovered that it did show the chemical appearances they had described. In subsequent experiments he showed beyond doubt that what had caused so much excitement in the scientific world was nothing more than the result of precipitation of calcium sulfate from the sea water in the deposit by the alcohol in which it had been preserved.[103] Years later he wrote

I had established the mineral nature of *Bathybius* before the *Challenger* arrived at

97. L. Dieulafait, "Le manganése dans les eaux des mers actuelles et dans certains de leurs dépots; conséquence relative à la craie blanche de la période secondire," *C.r. hebd. Séanc. Acad. Sci. Paris* (1883). Vol. 96, pp. 718-721.
98. "Deep-sea soundings in the Pacific Ocean," *Nature* (3 September 1874). Vol. 10, pp. 356-357.
99. *Challenger Report, Narrative.* Vol. 1, part 2, pp. 734-735.
100. T. H. Huxley, "Notes from the *Challenger*," *Nature* (19 August 1875). Vol. 12, pp. 315-316.

101. T. H. Huxley, "On some organisms living at great depths in the North Atlantic Ocean," *Q. J. microsc. Sci.* (1868). Vol. 8, pp. 203-212.
102. E(dwin) R(ay) L(ankester), "Ernst Haeckel on the mechanical theory of life and spontaneous generation," *Nature* (2 March 1871). Vol. 3, pp. 354-356.
103. *Report on the Scientific Results of the Voyage of H.M.S. Challenger, Narrative.* Vol. 1, part 2, pp. 939-940.
John Young Buchanan, "Preliminary report to Professor Wyville Thomson, F.R.S., director of the civilian scientific staff, on work (chemical and geological) done on board H.M.S. *Challenger*," *Proc. Roy. Soc.* (1875-1876). Vol. 24, pp. 593-623; pp. 605-606.

Japan, where she made a stay of nearly two months. Feeling that it would be satisfactory if a chemist at home had a statement from the chemist on board of the method adopted in this demonstration I wrote to my friend and former chief, Professor Crum Brown, in Edinburgh. . . . Professor Crum Brown has related to me how, after receiving my letter, he interested his friends in Edinburgh by showing them how to *make Bathybius.* [104]

On learning of these developments from Wyville Thomson, Huxley freely admitted that he had made a howler. [105] As Buchanan observed, no acknowledgement could have been franker. But Hugh Robert Mill (1861-1950) in his obituary of Buchanan, written fifty years later, tells us that his biological colleagues were very reluctant to believe that Huxley could have been wrong. [106] In an extreme case, fully three years after the return of the expedition, George Allman, speaking through a heavy cold, told the British Association in his presidential address

It is not easy to believe, however, that the very elaborate investigations of Huxley and Haeckel can be thus disposed of. These, moreover, have received strong confirmation from the still more recent observations of the Arctic voyager, Bessels, who was one of the explorers of the ill-fated *Polaris,* and who states that he dredged from the Greenland seas masses of living undifferentiated protoplasm. Bessels assigns to these the name of Protobathybius, but they are apparently indistinguishable from the Bathybius of the *Porcupine.* Further arguments against the reality of Bathybius will therefore be needed before a doctrine founded on observations so carefully conducted shall be relegated to the region of confuted hypotheses. [107]

This unsolicited expression of support put Huxley in an awkward position since he was seconding the vote of thanks but he managed to extricate both himself and the President with a tactful disclaimer which saved both sides from loss of face. [108]

Yet this was far ahead as the expedition sailed on through the Pacific, saddened by the illness and death of Rudolf von Willemöes-Suhm. [109] At last, early in 1876, the *Challenger* passed through the Magellan Straits and regained the Atlantic.

When his routine work permitted, Buchanan had been working on the problems involved in using pressures measured at the sea bed to give an independent source of information on ocean depths. Before the *Challenger* set sail he had made initial observations with an instrument resembling a barometer but recording the pressure of water instead of air, a piezometer. His first task on board ship was to establish the relative compressibility of the fluids to be employed. The hydraulic press which had been sent with them proved unreliable and Buchanan finally adopted the expedient of attaching his instruments to the end of the sounding line along with a deep-sea thermometer and used the measure of depth given by the line to calculate the pressure. In a series of observations made between Tahiti and Valparaiso he collected data on the compressibility of distilled water and then went on to determine the relative compressibility of sea water and mercury. Back in the Atlantic he was at last able to begin getting independent results. He used two instruments, a water piezometer and a mercury piezometer. Mercury alters its volume rapidly in

104. John Young Buchanan, *Scientific Papers,* 1913. Vol. 1, p. ix.
105. T. H. Huxley, "Notes from the *Challenger,*" *Nature* (19 August 1875). Vol. 12, pp. 315-316.
106. Hugh Robert Mill, "Obituary of J. Y. Buchanan," *Nature* (14 November 1925). Vol. 116, pp. 719-720.
107. George J. Allman, "Presidential address to the British Association," *Nature* (21 August 1879). Vol. 20, pp. 384-393.

Report of the 49th meeting of the British Association for the Advancement of Science (1879), (1879). Part 1, pp. 1-30.
108. "The British Association at Sheffield," *Nature* (leader) (28 August 1879). Vol. 20, pp. 405-407. The writer commented: "Surprise has been expressed in some quarters that Prof. Allman should have ventured to found such momentous speculations, even partly, on so unstable a basis as 'Bathybius.' "
109. *Challenger Report, Narrative.* Vol. 1, part 2, pp. 769-771.

response to changes in temperature but very little as a result of increases in pressure. Buchanan used the depth as given by the sounding line to correct the mercury instrument for pressure, giving the temperature at the bottom. He used this figure to correct the reading of the water piezometer, giving the depth. By reapplying this more accurate figure again to the reading of the mercury he was able to get a yet closer approximation to the true temperature. He realized that the ability which the technique conferred of discovering temperature without the limitations imposed by the self-registering thermometers, would enable them to measure temperatures in a situation like the one encountered in the Southern Ocean, where warm and cold layers of water near the surface had made it impossible for maximum and minimum thermometers to give information about the deeper layers. [110]

Buchanan's independent measurements of temperature made it clear that the pressure corrections which the expedition were applying to the results obtained with the Miller-Casella thermometers were unnecessarily large. When the *Challenger* got home Wyville Thomson turned them over to Professor Tait for a reappraisal of the effects of pressure. [111] In 1877 he wrote to Tizard that the temperature tables would all have to be redone and that it would probably be best to omit corrections altogether since, whatever they turned out to be, they would certainly be very small. [112] Tait had to have a press

specially made for the new observations and was not able to start work on them until 1879. He found that Davis had neglected to allow for the increase of heat due to pressure inside his apparatus and that the thermometers were in fact much better able to withstand pressure than they had been given credit for. The true correction varied from instrument to instrument and was not more than $0.14°F$ per mile of depth, very much less than the previous figure of half a degree. Because of the smallness of the changes of temperature in the depths of the sea and their significance in understanding the internal mechanism of ocean circulation, science, said Tait, could not afford to be mistaken about such small but possibly crucial differences. [113]

Back in the Atlantic, the *Challenger* explored more thoroughly the anomalies which had been noticed earlier in the voyage. In 1873 the temperature sections made across the southern parts of the ocean had shown a difference between the eastern and western sides, the bottom temperatures to the east being slightly warmer than the others. Nares had postulated the existence of a ridge in the center of the ocean keeping the two bodies of water distinct (see p. 30). Now, in 1876, as the *Challenger* sailed northwards from Tristan da Cunha to Ascension Island, the soundings showed that she was over an area considerably shallower than was normal for the ocean as a whole. [114] Gradually the daring idea emerged of a continuous mountain range, running parallel to the continental outlines right down the Atlantic from north to south. [115] From the alterations in the deep temperatures on either side of the ridge Tizard deduced the existence of subsidiary

110. John Young Buchanan, "Preliminary note on the use of the piezometer in deep-sea sounding," *Proc. R. Soc.* (1876-1877). Vol. 25, pp. 161-164. John Young Buchanan, "Laboratory experiences on board the *Challenger*," *J. Chem. Soc.* (1878). Vol. 33, pp. 445-469. The Challenger's scientists believed that the maximum and minimum thermometers were incapable of recording temperature inversions but when the final corrections were applied it turned out that in fact they had. This depended on their being reeled out or in quickly enough to avoid being affected by the intervening layers of water.
111. *Challenger Report. Narrative.* Vol. 1, part 1, pp. 98-102.
112. Daniel and Mary Merriman, "Sir C. Wyville Thomson's letters to Staff-Commander Thomas H.

Tizard, 1877-1881," *J. Mar. Res.* (1958). Vol. 17, pp. 347-374; pp. 354-355.
113. Peter Guthrie Tait, "The pressure errors of the *Challenger* thermometers," *Report on the Scientific Results of the Voyage of H.M.S. Challenger, Narrative* (1882). Vol. 2, appendix A, 42 pp.
114. C. Wyville Thomson, *The Voyage of the Challenger: The Atlantic* (1877). Vol. 2, p. 256.
115. *Ibid.* pp. 290.

ridges branching out from the mid-ocean range and linking it to the continental margins. One of these ran from a point somewhere near Tristan da Cunha to the African mainland (the Walvis Ridge) and prevented the cold Antarctic bottom water from entering the eastern basin of the South Atlantic; on the other side but much further north, another one linked the ridge with the coast of South America, keeping the coldest water out of the depths of the North Atlantic. [116]

Far from accepting that this provided proof of the system of general oceanic circulation as proposed by Carpenter, Wyville Thomson now put forward a quite different explanation of his own for the temperature structure which the *Challenger* observations had shown that all the major oceans had in common. He did not deny that cold Antarctic water was moving northward in the depths of the Atlantic, Pacific and Indian Oceans but he reasoned that it was as compensation for the transport of water vapor by the atmosphere in the opposite direction. Excess of evaporation in the northern hemisphere and of precipitation in the southern created an imbalance in the ocean and it was sensible to think of its movements as part of a combined circulation of ocean and atmosphere produced by water being taken out of the sea in one area and put back in another. There was not, he said,

the slightest ground for supposing that such a thing exists as a general vertical circulation of the water of the ocean depending upon differences of specific gravity. [117]

Carpenter was none too pleased by this new departure and expressed the view that Wyville Thomson should have waited until

he got home and had a chance to catch up on recent developments before committing himself to such views. [118] He referred to Froude's paper on friction given at the British Association in 1875 and quoted Sir William Thomson as having said that circulation was a matter "not of opinion but of irrefragable necessity." Wyville Thomson was nevertheless unrepentant. In an address to the British Association in 1878 he reiterated his views. The ocean, he said, might be regarded as consisting of two layers. From the surface to about 500 fathoms its movements were governed by the winds. At this depth

we arrive at a layer of water at a temperature of $40°$F., and this may be regarded as a kind of neutral band separating the two layers. [119]

Below this curious survival from the past was the Antarctic underflow.

The incident demonstrated that the *Challenger* had failed in what, in Carpenter's eyes at least, had been its major objective, in that it had not produced incontrovertible proof of the system of ocean circulation which he advocated. It was not so much that people like Croll and Wyville Thomson remained unconvinced since there are always two sides to an argument and their opinions did not alter the convictions of physicists and meteorologists here and overseas. The real difficulty lay in the fact that while basically simple in theory, in practice ocean circulation is a highly complicated process depending upon a large number of variables and resulting in a vast network of interrelated movements which bear only a distant resemblance to the simple pattern of the kind Carpenter used as his model. The picture lay hidden in the vast mass of data on tempera-

116. *Ibid.* pp. 291.
Thomas H. Tizard, "Report on temperatures" and "General Summary of Atlantic Ocean Temperature," *H.M.S. Challenger, No. 7. Report on ocean soundings and temperatures, Atlantic Ocean* (1876), pp. 5-19.
117. C. Wyville Thomson, "Preliminary report to the Hydrographer of the Admiralty on some of the results of the cruise of H.M.S. *Challenger* between Hawaii and Valparaiso," *Proc. R. Soc.* (1875-1876). Vol. 24, pp. 463-470; p. 470.

118. William B. Carpenter, "Oceanic circulation," *The Athenaeum* (13 May 1876). No. 2533, pp. 666-667. (Wyville Thomson's views had been reported on 29 April, No. 2531, p. 600.)
119. C. Wyville Thomson, Presidential address to Section E (Geography) of the British Association, 1878, *Nature* (22 August 1878). Vol. 18, pp. 448-452.

ture and specific gravity which the *Challenger* had collected but no one immediately connected with the voyage had the skills necessary for sorting out the pieces of the puzzle and putting them together. Of all the criticisms which have been made of the voyage the most telling is that the expedition ought to have included a physicist who could have paid special attention to this aspect of the work, which was originally intended to rank equally with the biological program.

As it was, the expedition had been overweighted on the biological side and most of the physical observations had been left to the ship's officers who already had plenty to cope with. The result was that apparatus like the current drogue and the electrical and reversing thermometers were not used to their full potential. J. Y. Buchanan obtained some interesting results in the course of his determinations of specific gravity and of the air dissolved in sea water. He noticed vertical changes in the distribution of salinity which he found to decrease with depth to about 1,000 fathoms and then increase slightly toward the bottom.[120] Similarly, the oxygen content of sea water decreased from the surface to a depth of 300 fathoms and then increased again. In neither instance does he seem to have considered that the vertical changes might be due to horizontal movements. He saw the oxygen deficiency entirely as the result of consumption by animal life, still plentiful at those depths whereas the oceanic plant life is restricted by its dependence on sunlight to the first hundred fathoms.[121] Tizard who was respon-

sible for the temperature observations seems to have had some interesting ideas about the possibility of a southerly outflow from the North Atlantic but there is no evidence that he developed them.[122] Apart from anything else the uncertainty about the size of the pressure correction made it difficult to generalize. The only person who attempted to give an immediate synthesis of the physical results was Wild who published his book *Thalassa* in 1877.[123]

In the other main aspects of the voyage's work there was no such ambiguity and the general results could be better appreciated. The pioneer work done by Buchanan and Murray on deep-sea deposits during the voyage had opened up an almost entirely new field which Murray was to make peculiarly his own in years to come. The biological aims of the cruise too had been amply fulfilled. The existence of life in the abysses of the ocean had been established and, though much remained to be done both as regarded the individual creatures caught and, for the future, in developing more accurate methods of sampling animal populations and relating them, if free swimming or pelagic, to the actual depths at which they lived, some generalizations could already be made. The deep-sea fauna had turned out to be remarkably uniform throughout the world. As for its connection to more ancient forms of life, Wyville Thomson wrote

the relations of the abyssal fauna to the faunae of the older tertiary and the newer

120. John Young Buchanan, "On the distribution of salt in the ocean as indicated by the specific gravity of its waters," *Proc. R. geogr. Soc.* (1876–1877). Vol. 21, pp. 255-257.
121. *Nature* (26 July 1877). Vol. 16, p. 255. J. Y. Buchanan, "Oxygen in seawater," *Nature* (27 December 1877). Vol. 17, p. 162. The first of these references is a note on Buchanan's paper to the Royal Society of Edinburgh (*see* ref. 122); the second a reply by him to a criticism based on the former made by Wyville Thomson in *The Voyage of the Challenger: The Atlantic*, Vol. 2, p. 267, in which he said that Buchanan's findings were contrary to experience.

122. John Young Buchanan, "Note on the specific gravity of ocean water," *Proc. R. Soc. Edinb.* (1875-1878). Vol. 9, pp. 283-287.
Thomas H. Tizard, abstract of paper to the British Association, "On the temperature obtained in the Atlantic Ocean, during the cruise of H.M.S. *Challenger*," *Nature* (28 September 1876). Vol. 14, pp. 490-491.
123. John James Wild, *Thalassa: an essay on the depth, temperature, and currents of the ocean*, London (1877). Wild also used his drawings, apart from those which appeared in the Report, to write an illustrated account of the voyage: *At Anchor. A narrative of experiences afloat and ashore during the voyage of H.M.S. Challenger from 1872 to 1876*, London (1878).

Mesozoic periods are much closer than are those of the faunae of shallow water; I must admit, however, that these relations are not so close as I expected them to be—that hitherto we have found living only a very few representatives of groups which had been supposed to be extinct.[124]

Before the results of the voyage could be fully assessed the enormous collections of birds, plants, marine animals, rocks, oozes and sea water samples which had been sent home at intervals during the voyage would have to be sorted and sent to the individual scientists best qualified to work on them. Then their findings would have to be published as part of the expedition's report. It was his plans for this undertaking that were uppermost in Wyville Thomson's mind as the *Challenger* headed for home after an absence

of three and a half years. No amount of foresight could have enabled him to see how much trouble this was to involve and that the whole procedure was to take five times as long as the voyage itself.

For the moment all was serene. On 24 May 1876 the *Challenger* arrived at Spithead and finally docked at Sheerness where leading British scientists and foreign visitors to the conferences being held in connection with the Loan Exhition of Scientific Apparatus, then going on in London, came to see her.[125] Queen Victoria conferred a knighthood on Wyville Thomson and by early July he, Buchanan and Murray were back in Edinburgh where their return was celebrated by a civic banquet.[126] The great expedition was over.

124. C. Wyville Thomson, Presidential address to Section E (Geography) of the British Association, 1878, *Nature, op. cit.* p. 452.

125. *Nature* (1 June 1876). Vol. 14, pp. 119-120.
126. "Dinner to the *Challenger* Staff," *Nature* (13 July 1876). Vol. 14, pp. 238-239.

The Future of Research Habitats

JOHN B. TENNEY, JR.   1971

A year ago *Tektite 2* was resting on the sandy bottom of Lameshur Bay and John H. Perry Jr., developer of *Hydro-Lab,* reviewed the state-of-the-art front pages of *Oceanology International.*

Today, *Tektite 2* rests upon keel blocks in Philadelphia while *Hydro-Lab* is being used for a series of experiments in the Bahamas. Perhaps some new thoughts on habitats are in order.

In 1969, Gerhard Haux, chief engineer for Dragerwerke, published the best account to date of known underwater laboratory projects. He noted that ". . . it is by no means easy to define the concept of an underwater laboratory as clearly as might be desired" and concluded ". . . some generosity in defining will do no harm."

We shall consider an underwater habitat to be any system which permits men to live and work in the sea at ambient pressure while employing saturation diving techniques.

Habitats are more remarkable for their differences than for their similarities. The magnitude of these differences should cau- tion us from making facile comparisons between systems for determining which features are "best."

Saturation diving techniques are widely used by commercial and military divers. Equipment for this work is advanced and improving rapidly.

In commercial work, emphasis is placed on completing a job and returning to the surface. There is no premium placed on non-working time on the bottom. The sooner the diver completes the job and leaves the water, the better.

Outstanding examples of saturation diving systems have been developed by many firms. The COMEX (French) system has been successfully used for open ocean dives to 840 ft. Packaged systems such as the ADS-IV are designed to permit work at 1,000 ft. One of the most ingenious underwater saturation systems is Taylor Diving's underwater welding habitat, designed to straddle a submerged pipe and permit the diver to perform repairs under dry conditions.

All of these systems show a bold and

John B. Tenney, Jr., was project engineer of the Tektite 1 habitat and program engineer on Tektite 2 while working under the General Electric Company's Ocean Systems Programs.

From *Oceanology International,* Vol. 6, No. 8, pp. 23-25, 1971. Reprinted with light editing and by permission of the author and *Oceanology International.*

imaginative approach to saturation diving. In each case the diver is returned to the surface once the job is completed.

In some marine scientific work, divers must consider saturation diving. However, geological investigations in the Great Lakes by the University of Michigan used helmet diving, and bounce diving is appropriate for many studies. *Sublimnos,* located in Lake Huron, demonstrated that an underwater station is beneficial when saturation is not required. Other classes of study place greater importance on time spent on the bottom.

For some classes of work, the PTC-DDC systems used by commercial firms work best. But, for most types of marine science requiring saturation diving, the scientist benefits by living on the bottom.

Tektite 2 used more than 50 scientific experiments. Archeologists also amass considerable bottom time, but in the past no habitat system was available for them. With the development of small, inexpensive habitats this situation should change.

The scientific advantages of saturation diving are manifold. It eliminates time-consuming decompression, permits the scientist's time spent in the water to be a function of his job and stamina, and usually reduces transit time to and from his work area.

Also, the saturation laboratory facilitates both serial observation and sampling. Psychologically, the diver is a captive, and is totally involved in his environment. Chemists and biologists who collect large volumes of water samples can analyze them in the stable, weather-proof habitats instead of transporting them to the surface. Future habitats probably will be equipped with computer terminals to permit data collected to be immediately stored and analyzed.

Trial Depths of Known Habitat Projects*

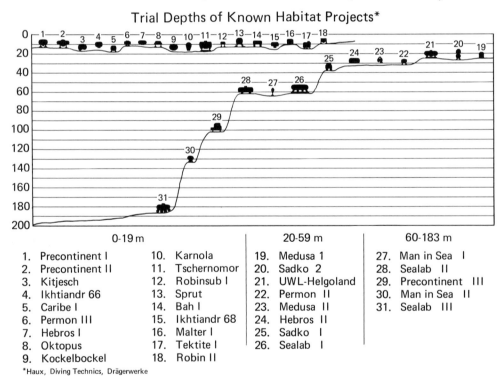

0-19 m		20-59 m	60-183 m
1. Precontinent I	10. Karnola	19. Medusa 1	27. Man in Sea I
2. Precontinent II	11. Tschernomor	20. Sadko 2	28. Sealab II
3. Kitjesch	12. Robinsub I	21. UWL-Helgoland	29. Precontinent III
4. Ikhtiandr 66	13. Sprut	22. Permon II	30. Man in Sea II
5. Caribe I	14. Bah I	23. Medusa II	31. Sealab III
6. Permon III	15. Ikhtiandr 68	24. Hebros II	
7. Hebros I	16. Malter I	25. Sadko I	
8. Oktopus	17. Tektite I	26. Sealab I	
9. Kockelbockel	18. Robin II		

*Haux, Diving Technics, Drägerwerke

Figure 1-11. Trial depths of known habitat projects

A frequent request from *Tektite 2* scientists is to repeat the experiment in the certainty that the second would be better planned. Most scientists in a saturation habitat project are participating for the first time, and have no precedent for planning or conducting their experiment. This will change in time, but for the present it is important to expose large numbers of young marine scientists to this environment.

A good overview of collective experience to date is shown in the accompanying table, adapted from the Haux survey. It shows that 80 per cent of all projects have occurred at depths less than 100 ft—a consequence of the narcotic effect of N_2-O_2 at greater pressures.

Helium, as an inert diluent, extends man's diving capability. (The most recently reported saturated chamber dives were to 1,700 ft.) Hydrogen may lower this boundary, but the working depths achieved by habitats always will lag behind chamber tests.

In addition to increased problems from depth, low water temperature requires increased power for heating and special insulating techniques to reduce heat loss.

Habitat requirements which planners saddle designers with include the alleged need for autonomy, mobility, deep operating capability, multi-depth capability, and the ability to double as a decompression chamber. These goals—coupled with other factors not related to the habitat itself—frequently result in a system with large support crews, complex surface systems, and high costs.

AUTONOMY

Autonomy infers that the habitat and its support system are capable of independent operation for protracted periods. A nuclear sub equipped with an ambient pressure laboratory and lock-out capability satisfies this requirement, but, for scientific purposes, is too expensive. Work on the French *Argyronete,* which would have possessed a high degree of autonomy, has been discontinued because of cost.

Self-contained power supplies in the form of batteries are too bulky, and fuel cells and nuclear systems are too costly. Power requirements for practical laboratories are such that, with all respect to the highly sophisticated technology of fuel cells, their use on *Hydro-Lab* and *SDM-Malta* must be considered as tours de force.

The battery-powered mesoscaph *Ben Franklin* has on-board power and life support equipment with no umbilical ties to the surface. Yet, it is not autonomous because it does not operate without surface support.

For this reason, there is no need to fit non-essential functions in the subsurface element of the system if they can better be left on the surface. Put more directly, diesel-electric power is hard to beat. The problem for most habitats is not to replace surface generators, but to find simple, inexpensive, and reliable ways to lay, connect and recover the umbilical cable, and to manage it at the naviface in bad weather.

In the immediate future, habitats will use surface facilities for power, gas monitoring, safety, and decompression operations. The goal should not be autonomous habitats, but autonomous systems.

MOBILITY

Mobility can be considered only in the light of intended use and emplacement technique. *Aegir* possesses a high degree of mobility, but is not self-propelled and requires a surface vessel for towing and surface support. Weighing 160 tons (fully ballasted), *Tektite 2* was designed to maximize involvement by Navy Seabee divers and to be transported by LSDs. Yet, this non-mobile habitat has made two round-trips between Philadelphia and the Virgin Islands.

Minitat's problems stemmed from instability caused by a "free surface" condition in the pontoons (added for mobility).

For a habitat intended to be used recurrently, it is advantageous to be able to lift it clear of the water for maintenance. Its ballast may be fixed or removable (the choice will impact on mobility).

DEEP OPERATING CAPABILITY

Beyond 100 ft, helium must be added to nitrogen and oxygen as an inert diluent. The addition of helium causes considerable problems in the area of heat loss, speech distortion, and permeation of electronic components. Also, it becomes necessary for the scientists to possess a degree of diving skill beyond that of most working scientists at this time.

MULTIPLE-DEPTH CAPABILITY

This requirement creates numerous unsuspected problems. Control equipment for many internal systems (e.g. refrigeration equipment) depends upon differential pressure switches. Fixed pressure designs are much easier and cheaper to design, test, and operate than those which can go anywhere. There are few advantages in a hull designed to go to 200 ft unless the environmental control system is compatible with heliox mixtures.

DUTY AS DECOMPRESSION CHAMBER

Some programs have been hampered by the requirement that the habitat also serve as a decompression chamber. This dual role was performed by *Conshelf III* and *Helgoland.* It was a requirement on *Minitat,* and in the early stages considered as a requirement for *Tektite 1.*

It appears to enhance flexibility, but in some cases this is illusory. If no surface chamber is required, support divers must rely on the occupied habitat, which must be capable of treating diving casualties. Such a habitat may be required to withstand both internal and external pressures.

MINIMIZING NUMBER OF SUPPORT PERSONNEL

In contrast to the high number of surface personnel used in some diving projects, the recent design for the Japanese Man-in-the-Sea program calls for a four-man habitat which will be submerged for periods up to 30 days at 300 ft, and living quarters for an 11-man support crew on the surface.

Support functions for shallow-water habitats can be collected on inexpensive barges, or on simple hulls such as the surface ship for Russia's *Chernomor.*

More elegant solutions, such as the surface buoy Fustchen (named for designer Dr. Hans Fust) employed at *Helgoland,* are approaching minimal surface support prior to cutting the cord. The surface buoy had an outstanding performance record.

Programs such as *Sublimnos* and student projects such as *Edelhab I* and SDM-Malta (an inflatable underwater habitat built and emplaced off the island of Malta in 1969 by a group of British students) prove that underwater stations can be inexpensive.

Increased use of simple systems in shallow depths will have two consequences. It will create a base of trained marine scientists who can take advantage of technological advances at greater depths with more involved equipment; and it will increase understanding and identify applications in areas we now but dimly perceive.

There are many areas common to most habitat systems where immediate improvements can be made. Material transfer methods are high on this list.

In *Conself II,* Cousteau used simple pressure cookers to transport mail and small items from the surface. In *Sealab II,* the U.S. Navy devised a more elaborate dumbwaiter system in which transfer pots were raised and lowered from the surface. On *Sealab III,* this system would have been replaced by the UUDD (underwater upside down davit).

On shallow-water programs such as *Tektite 1* and *2,* continued use was made of simple pressure containers, including commercially available paint spray pots. While these permitted the program to continue successfully, they were for the most part heavy, time-consuming, and potentially dangerous.

A second area for improvement is the general field of surface support equipment.

Emphasis should be placed on a simple, reliable, and inexpensive support device that can operate unattended for long periods of time.

A third improvement needed is a simple and inexpensive diver recovery system. In its most simple form, this system consists of a personnel transfer capsule, a winch for lowering the capsule, and a surface decompression chamber to permit saturated aquanauts to enter the PTC underwater. Future advances in diving physiology may shorten decompression time for saturated divers, but for the present, designers should consider making decompression safer, more comfortable, and less expensive.

There is room for much innovation in the design of the habitat as a pressure chamber. *Helgoland* had a lock-out capability which proved valuable when the habitat was being refurbished between missions.

Improvements in closed-cycle breathing systems and in diver heating devices hold promise that scientists can work in the sea as they work in their laboratories.

Plans are under way to use *Aegir* in three saturation experiments progressing to 600 ft under Navy sponsorship. These dives are scheduled to begin in September off Hawaii.

In Texas, planners at the Marine Biomedical Institute will place *Tektite* at a depth of 80 to 90 ft on the Flower Garden Reef in the Gulf of Mexico. Surface support at this open-ocean location will come from a platform similar to a production platform.

Japan is developing its program to send four oceanauts to 300 ft for 30 days.

Germany plans new programs using *Helgoland.*

In North Carolina, a group of universities is considering new roles for *Minitat.* In Florida, the Naval Ship Research and Development Laboratory has placed *Sealab I* in 65 ft of water one mile offshore. The University of Michigan has suggested using a portable habitat in Grand Traverse Bay.

Further in the future, we can anticipate increasingly complex habitat systems using sophisticated materials and diving techniques. Architectural schools are rediscovering the ocean, and many of the student designs are bold and innovative.

To conservative scientists, these predictions may sound strange and wild. But they have only to read of the amazement with which scientists viewed the contents of the first plankton nets and bottom dredges a hundred years ago to conclude that data gathered by diving scientists will represent a quantum jump in marine science.

Habitats will provide the means to this end.

Remote Sensing of the Oceans

KURT R. STEHLING   1969

Surface and subsurface features of the oceans can often be seen more clearly from a height than from the surface. This fact, combined with the remoteness of much of the ocean and the high cost of shipboard operation and data collecting—typically $4000 per day—makes spacecraft observation economically practical. A strong case can be made for an ocean-observation satellite whose prime purpose would be global ocean scanning, using within its payload and data-handling limitations, every remote-sensing tool technology can provide. For the immediate future, however, practical problems, chiefly in developing sensors, which in many cases are now no more than breadboard circuits, and funding will probably boil down the sensor array to microwave and video detectors, with possible ancillary infrared sensing.

Within easy sight are the waves and swells, biological growth and currents, floating objects and pollutants which occur in the top few centimeters of the water. In very clear water on a sunny day, features can be discerned below the surface to depths of 50-100 ft, often more clearly from some height. Aircraft observers have sighted many shipwrecks, shoals, reefs, and the like, invisible to mariners.

Remote-sensing of many of these features has been proved. Currents have been located from photographs much as the unaided eye does, by noting how their color differs from that of the surrounding water. Currents can also be traced by sensors that detect temperature and temperature difference. The Nimbus high-resolution infrared radiometer outlined the Gulf Stream (Figure 1-12) and other thermally differing water masses. The Gulf Stream has also been seen from space by noting cloud bank boundaries. Aircraft-borne infrared thermometers and radiometers mapped thermal distributions near Iceland in 1968, helping that country determine where to find fish. From Earth-synchronous altitude, ATS-1 delineated the Peru Current, and the Equatorial South Pa-

Dr. Kurt R. Stehling has been in the forefront of astronautics for the past twenty years. He is now an advisor for the Manned Undersea Science and Technology with the National Oceanic and Atmospheric Administration, Maryland. He has written over one-hundred books and technical articles.

From *Astronautics and Aeronautics*, May pp. 62-67, 1969. Reprinted with author's revisions and light editing and by permission of the author and *Astronautics and Aeronautics*, a publication of the American Institute of Aeronautics and Astronautics.

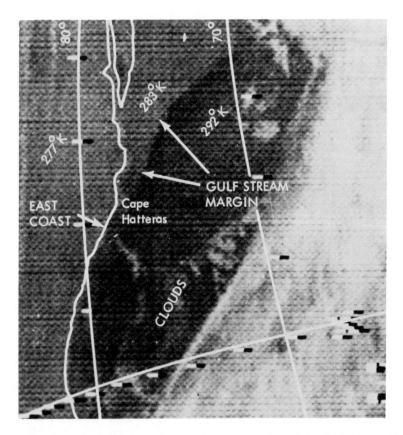

Figure 1-12. Nimbus-11 high-resolution infrared picks out the edge of the Gulf Stream. (Courtesy of National Aeronautics and Space Administration.)

cific Current. In its video pictures could also be seen cloud patterns following the global air currents.

Examining video pictures for other kinds of evidence, analysts have been able to tell from the Sunglitter in Tiros-II pictures, and from aircraft photos, whether the sea was rough or smooth. A similarly crude measure of sea-state is given by aircraft-borne radar which can detect the characteristic signature of rough seas.

Various different degrees of sea roughness have been detected by microwave radiometer. Each produced a different shape curve when traces of brightness temperature were plotted against viewing angle. Some of these studies will follow up recent NASA

and other flight tests showing that a rough sea with foam may seriously prejudice microwave thermal mapping by causing confusing emissivity changes. Much testing will be necessary, perhaps at longer wavelengths (10 cm and beyond) to resolve the anomaly. Further theoretical studies on ocean microwave radiation anomalies (polarization, "sky" temperature, emissivity changes related to sea-state, etc.) are planned or underway at the National Oceanographic and Atmospheric Administration (NOAA), NASA, and within industry.

Life also places its mark upon the waters. Fish schools leave characteristic oily patches, form dark patterns, and tend to congregate near upwellings which can be spotted by their color or thermal difference. These clues have been exploited by the Bureau of Commercial Fisheries of the Department of the Interior, in conjunction with NASA and the Navy Oceanographic Office, who have done important work on fish-school detection from aircraft. NASA aircraft-borne radar has detected kelp beds near Point Loma, California. With image enhancement of color infrared photographs taken by low-flying aircraft, Itek Corp. has spotted algae blooms in the Potomac River.

A polar-orbiting satellite carrying sensors to detect these features could cover the globe twice a day. A synchronous spacecraft could cover an entire ocean continuously. They could not make subsurface physical-chemical measurements and other *in situ* quantitative measurements done routinely by oceanographic research vessels. Furthermore, many sensors see only during the day and when skies are clear, whereas oceanographers generally want all-weather and 24-hr observations. This limitation can be obviated by combining a variety of instruments with different cloud-penetration or low-light abilities, or perhaps, devising a single sensor with all-weather, 24-hr vision. The latter is a utopian concept if high resolution imagery is desired, although feasible for certain spectrophotometric or radiometric narrow-band scanning.

The exact shape of a remote ocean-scanning system will be formed by pressures from ocean and marine users—scientists, engineers, fishermen, and meteorologists.

Amalgamating and sorting what these users would like to see, and what probably can be seen, we arrive at the following list of "observables":
—Sea state.
—Long-period swells.
—Seismic waves (tsunamis).
—Ocean currents (direction, and hopefully, velocity).
—Thermal distribution (also related to currents).
—Plankton and other similar organisms, and sea "weed."

—Fish schools, and large herds of mammals.
—Bioluminescence.
—Pollutants, such as oil, phenols, etc.
—Mud and silt.
—Saline—fresh-water distribution in deltas and estuaries.
—Icebergs and sheets.
—Air/sea interactions (violent storms, fog, clouds, precipitation).
—Shoals, reefs, sunken vessels, etc. (in clear, shallow water).

It may also be possible to infer from surface observations after much experience certain subsurface phenomena, including deep currents, upwellings, and the like.

The most useful set of sensors within the payload, power, and data-handling capacities of an unmanned, polar-orbiting spacecraft of the early '70s may be a microwave radiometer/scatterometer (for sea-state) and video with infrared radiometry for cloudless seeing.

A microwave radiometer for the 5-10 cm wavelength region detects emissions from the surface of water. It can measure surface temperature within $\pm 1°C$ and can do better in the future. Sea-state may be recognized from radiometer data when emissivity anomalies are understood. Chemical-water differences, between salt and saline, clean and polluted, for example, may possibly be distinguishable. Mapping of sea-ice distribution has already been achieved from aircraft with remarkable results as shown by the comparison of radiometry with a photograph (Figure 1-13). Measurements of air/sea interaction such as "weather" fog and low clouds are possible, especially if a dual-wavelength radiometer were used.

Spatial resolution depends on satellite altitude and antenna aperture. Resolution of 20 nautical miles (n. mi.) should be achieveable from 300 n. mi. orbit with a 10,000 cm^2 antenna at 5-cm wavelength. A 600 n. mi. footprint or swath should be possible for the line-trace image formed.

Microwave radiometry and scatterometry is a rapidly developing technology. Aircraft tests have proved feasibility with existing

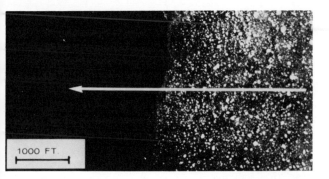

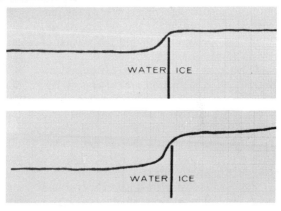

Figure 1-13. Microwave radiometer traces indicated the ice/sea water boundary in the Goose Bay, Labrador, area as the aircraft flew the track drawn on the photo. Courtesy of National Aeronautics and Space Administration.)

equipment. By 1970 or so, a microwave system could be operational if a spacecraft is available. Atmospheric absorption above 3-cm wavelength is low. Rain drops and some types of clouds could result in some scattering or absorption of ocean radiation. In general, a microwave radiometer has all-weather, day-and-night ability to gather data.

Since "conventional" satellite/ground/user links can be used, new data management systems would not be required. Continuous real-time readout is possible, but usually unnecessary for marine purposes. Onboard raw-data storage and later dumping is readily feasible.

If a microwave system were combined with a video sensor of the sort yielding reso-lutions of about 400-500 n. mi. on Tiros, or ITOS, we would have a versatile and useful all-weather sensor system. The microwave sensor would furnish data on the ocean surface in almost all weather, and if multiband, on the atmosphere as well, while the video would show concurrent cloud distribution and probably some ocean features. If the video resolution could be further improved, its footprint reduced, and low-light scanning reached, the satellite would be even more useful. Another big step forward in sophistication would be taken by changing to color or multi-waveband video.

A polar orbit is preferred. Synchronous, or 24-hr, and an equatorial spacecraft could complement the polar one. The synchronous satellite could provide continuous (albeit low-resolution for the near future) surveillance of about 1/3 of the globe, such as most of the Pacific Ocean, with video. Air/sea or weather features rather than intrinsic oceanographic features would be seen. The equatorial vehicle could furnish more continuous detail of the interesting tropical ocean/weather pheonomena such as hurricanes.

In general, the lower the satellite altitude, the greater the spatial resolution. That is, more detail can be seen, since a smaller area or field of view is covered. However, the intrinsic angular resolving power of the sensor, whether optical or microwave, derives from the aperture of the entrance lens, or antenna.

Other sensors than the above microwave/video combination are, of course, feasible and available, and in some cases, have special virtues that make them useful for certain ocean-sensing applications. However, because of their general over-all limitations, such as clear-weather operation, they are only briefly listed here.

Film cameras give the highest spatial and color resolution, at least for the foreseeable future. Gemini photos taken with a "simple" camera are examples. On the other hand, they require either a manned spacecraft, a retrieval system, or cumbersome onboard film development and scanning and are limited to strictly daylight and clear skies.

Infrared radiometry or imagery would be excellent for meteorological purposes, as already demonstrated on various spacecraft and aircraft, and could be a useful companion reference radiometer on board microwave ocean-scanning spacecraft. Atmospheric absorption, cloud cover, rain, etc., limit it.

Radar—synthetic aperture (for greater resolution), radar altimeter, or scatterometer—is an active rather than passive technique. It detects sea-state characteristics, waves, and swells, the sea-land interface, and in the sub-3-cm region, air/sea interactions. Its drawbacks are bulk, heavy power demands, data management, and interpretation. Laser

radar, with its potential for even greater resolution, may lie over the horizon.

Spectroscopy by multi-wavelength radiometry or photography in the visible or near-infrared spectral regions can detect many substances including biota and oil slicks. It could pick up surface phosphorescence and fluorescence. It cannot penetrate cloud cover, so is strictly clear-weather-limited.

Polarimetry has some utility for analysis of sea-surface state, in visible light, only in clear weather and sunlight.

These represent the most important actual or possible sensors. The rapid development of solid-state technology, image analysis, and communications technology no doubt will result in a variety of high-resolution, relatively low-cost and versatile detectors that will become available in the future.

No orbital ocean-sensing ought to be considered before organizing a complementary, broad, ground-based measuring system. The most promising would be a widespread buoy array, which would yield surface and subsurface data for reference. Buoys and other platforms such as ships could also serve as data-relay links. The retrieval and management and analysis of the data will be a serious matter, considering the limited availability of ground receivers and networks, and competing demands. An orbital video or photographic scan of the oceans will yield thousands of pictures during any one mission. Various schemes have been proposed and tried for photometrically dissecting black and white pictures into digital elements, permitting rapid analysis and compact storage. It is also possible to extract otherwise undetectable features from black and white pictures by such techniques as "multi-spectral analysis." This scheme has been applied by Philco-Ford and I^2S Company scientists in Palo Alto, California, to various black and white high-altitude ocean photos. By enhancing the contrast already present in the emulsion of the photographic print, they have detected fish schools, sub-

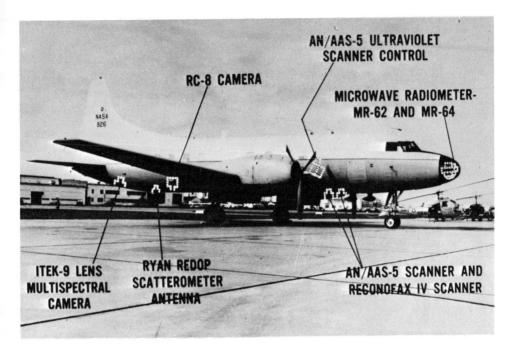

Figure 1-14. NASA earth-resources survey aircraft carries a large array of sensors. (Courtesy of National Aeronautics and Space Administration.)

surface features, and biota not otherwise apparent to the observer.

As can be seen, the technology is in hand, although by no means fully worked out in every case. If it is to be put to use, potential users in the marine community must more actively argue their need for spacecraft sensor data and decide which data will satisfy the needs of the largest number of users. When they do so, and work more closely with federal agencies that have the money in shaping a satellite to their needs, they may be able to pry loose the funds now lacking.

Among groups helping to get users interested and involved are the interagency Spacecraft Oceanographic Committee (SPOC) and the NAS/NASA Woods Hole Summer Study Panels. They have focused various interests and have examined needs and purposes of the federal/industrial/institutional developer and the user community of potential users of a satellite.

The task of the federal agencies remains to continue developing sensor and spacecraft technology which will demonstrate to users that an ocean-scanning satellite could be useful, as was done with the first communications satellite. NOAA has already demonstrated the Tiros, and NASA, the Nimbus and ATS, with ESSA participation. NASA works in this direction by providing almost the only funding for continued sensor technology development and for a clearinghouse of user needs. Figure 1-14 shows the wide variety of sensors carried on its research aircraft. The U.S. Coast Guard has led the way in detecting ice by aircraft sensor. The Navy's Oceanographic Office (NAV-OCEANO) has pioneered initial activities and provided staff and facilities for NASA contract administration. The Department of the Interior, Bureau of Commercial Fisheries (BCF), and the Geological Survey have tackled Gemini and Apollo photo analysis, among other endeavors. The National Council on Marine Resources and Engineering Development has followed closely on the heels of these activities and helped to stimulate further national and international interest to trigger a new round of activity.

Turning to future federal activity, the Barbados Meteorological Experiment (BOMEX) was significant for including spacecraft coverage for the first time in such a project and requiring wide cooperation. It studied air/sea interactions in the Caribbean via buoys, aircraft, ships and some spacecraft (ATS, Tiros) coverage. This large-area (several hundred thousand square miles) experiment is both a U.S. (NOAA, Coast Guard, U.S. Navy, etc.) and an international venture. The technical and managerial base is developing, and is well on the road to becoming a reality.

SUGGESTED READINGS

Stehling, K. R., "Earth Scanning from an Orbital Vehicle," in the Proceedings of the International Astronautical Federation, 1953.

"Oceanography from Space," in Proceedings of Conference on the Feasibility of Conducting Oceanographic Explorations from Aircraft, Manned Orbital and Lunar Laboratories, held at Woods Hole, Mass., August 24-28, 1964, Ref. No. 65-10, edited by Ewing, G. C., Chairman.

"Man's Geophysical Environment—Its Study from Space," a report to the Administrator of ESSA, prepared by an ESSA Task Force, chaired by J. P. Kuettner, U.S. Printing Office, Washington, D.C. 20402.

"United States Activities in Spacecraft Oceanography," a publication prepared for the National Council on Marine Resource and Engineering Development, U.S. Printing Office, Washington, D.C. 20402.

"Final Report on the Space/Oceanographic Study." prepared for the National Council on Marine Resources and Engineering Development by the General Electric Co., Missile and Space Div., Valley Forge Space Center, P. O. Box 8555, Philadelphia, Pa.

Annual Report, Oct. 1965-Sept. 1966, of Spacecraft Oceanographic Project (SPOC), U.S. Naval Oceanographic Office, Washington, D.C.

Stevenson, R. E. and Nelson, M., "An Index of Ocean Features, Photographed from Gemini Spacecraft," Contribution No. 253, Bureau of Commercial Fisheries, Biological Laboratory, Galveston, Texas, prepared for NASA MSC.

Pierson, W. J., et al. "Some Applications of Radar Return Data to the Study of Terrestrial and Ocean Phenomena," Proceedings of the Third Goddard Memorial Sumposium on Scientific Experiments for Manned Orbital Flight, 1965, pp. 87-137.

Alexiou, A. G., "Spacecraft Oceanography— Fact not Fiction," 1966 Honolulu Conference Proceedings, in Proceedings of the IEEE.

Stogryn, A., "The Apparent Temperature of the Sea at Microwave Frequencies," IEEE Transactions on Antennas and Propagation, Vol. AP-15, No. 2, March 1967.

Deep Sea Drilling for Scientific Purposes: A Decade of Dreams

TJEERD H. VAN ANDEL   1968

INTRODUCTION

The development of marine geology has, in a sense, been the inverse of that of the geologic study of the continents. Classical land geology, concerned in the beginning mainly with the earth's history, has included recent geological events in its scope at a relatively late date, notwithstanding early recognition of their relevance.[1] Marine geologists have had to content themselves with describing and interpreting the features of the present sea floor and the processes which form them. Nevertheless, investigation of the geological history of the deep ocean offers an obvious path to the study of our planet as a whole. Here, in a protected environment covering almost three-quarters of the surface of the earth, the record of geologic events on and in the crust of the earth is most likely to be preserved with minimum disturbance.

Here also, according to recent concepts,[2] major events take place in the crust and mantle which control the distribution, nature, and shape of continents and oceans.

On the continents, the combined forces of mountain building and erosion by wind and water have exposed much of the geologic record dating back at least 3.5 billion years. Even though this record is far from complete, a century and a half of observation in the field, aided in the last 50 years by commercial drilling and by geophysical studies, has reconstructed a comprehensive and detailed historical record. The results suggest convincingly that oceans and seas have been an integral part of the world from the beginning of recorded time.

The ocean floor is subjected far less intensely to erosion and deformation and is covered with an almost continuous blanket

1. J. Playfair, *Illustrations of the Huttonian Theory of the Earth* (facsimile ed., 1802; Univ. of Illinois Press, Urbana, 1956).

2. H. H. Hess, in *Petrologic Studies; A Volume to Honor A. F. Buddington* (Geological Society of America, New York, 1962), p. 509; F. J. Vine, Science 154, 1405 (1966).

Dr. Tjeerd H. van Andel is professor of oceanography at Oregon State University and research associate at Scripps Institution of Oceanography, California; he recently served as science advisor and member of the Atlantic Site Panel with the Deep Sea Drilling Project. He has written over fifty articles.

From *Science*, Vol. 160, No. 3835, pp. 1419-1424, 1968. Reprinted with light editing and by permission of the author and the American Association for the Advancement of Science. Copyright 1968 by the American Association for the Advancement of Science.

of modern sediments, so that outcrops of older rocks are rare and restricted in their occurrence. Since the marine geologist is separated from the objects of his research by thousands of meters of water, he is almost entirely dependent on remotely controlled instrumentation. Consequently, much energy has been devoted to the development of tools to penetrate the sea floor. This development has progressed from devices propelled by gravity or explosives to piston corers; but the best of these can penetrate at most 20 to 30 meters, which represents only a small slice of geologic time.

Study of the data so obtained and the piecing together of this fragmentary record by geophysical methods have permitted preliminary conclusions regarding the genesis and history of the ocean basins and their relation to the evolution of our planet. One of the more remarkable results is that evidence for a great age of the oceans, comparable to that of the continents, appears to be completely lacking. This peculiar paradox is the basis of current hypotheses regarding the evolution of the crust of the earth, and it forms a large part of the scientific justification for drilling in the deep ocean.

It is natural that the dreams of marine geologists interested in the history of the oceans have been directed toward the use of deep-sea drilling as a tool. Developments in offshore drilling technology by the petroleum industry have now reached a stage in which the hardware and know-how are available in principle to undertake drilling projects in deep water with some hope of success. Experience in water depths exceeding 1000 meters is, however, still limited.

EVOLUTION OF THE IDEA

The use of drilling to study the ocean was first considered seriously 10 years ago in Project Mohole, which was supported by the National Science Foundation. This project generated many of the ideas concerned with drilling projects in the sea, and it stimulated the development of some of the necessary technology. Project Mohole also produced, early in 1961, the first successful drilling and coring operation in deep water. The barge *Cuss I*, equipped with a large drilling rig, drilled five experimental holes, in depths around 1000 meters, off San Diego, and five more, in depths near 3600 meters, off Guadalupe Island on the west coast of Mexico, with bottom penetrations to 150 meters. Since anchoring in such depths is not feasible, the position of the vessel was controlled dynamically by taut-wire buoys. The feasibility of scientific deep-sea drilling was thus demonstrated.[3]

It was then decided that Project Mohole would proceed directly to its primary objective to sample the earth's mantle. A long period of time would be required to develop the tools, systems, and drilling platform necessary to penetrate several kilometers of hard rock to a total depth of 10 to 12 kilometers below the surface of the ocean. During those several years no ocean drilling was to be undertaken by Project Mohole. In the meanwhile, however, many geologists advocated a broad investigation of the sedimentary layer of the oceans. This concept was distinct from that of Project Mohole, and its technical requirements were quite different. For a sedimentary drilling program, a vessel capable of a speedy transit between widely spaced drilling sites would be required, and the relatively shallow core holes needed to penetrate only the sedimentary layer could be accomplished with existing drilling equipment and techniques. Such a program, advocated by sediment-oriented geologists, would yield valuable technological experience and at the same time would produce data on the historical record of the deep-sea sediments.

This project, notwithstanding its relative simplicity, would still be too complex and costly for one individual or even one institu-

3. American Miscellaneous Society, Experimental Drilling in Deep Water at La Jolla and Guadalupe Sites (National Academy of Sciences - National Research Council, Publication 914, Washington, D. C., 1961).

tion to initiate and manage independently. Consequently, the next few years were devoted to finding a suitable organizational form to prepare plans, and to gathering the scientific input, creating confidence, and acquiring the funds to carry them out. In all this time, there never was any lack of excellent proposals for drilling goals and sites; long unresolved was the creation of an organization that would possess the qualities needed to assume responsibility for the execution of these ideas, a problem underlying many of the struggles in the preparation and execution of "big science" projects.

Early in 1962, after a proposal of Cesare Emiliani, from the Dorothy H. and Lewis Rosentiel School of Marine and Atmospheric Sciences, University of Miami, to charter a drilling vessel for work in the Caribbean and western Atlantic, a committee (LOCO) was formed consisting of two scientists each from Miami, Lamont-Doherty Geological Observatory of Columbia University, Woods Hole Oceanographic Institution, Scripps Institution of Oceanography of the University of California, and Princeton University. The LOCO committee, realizing that a formal organization was needed, considered a nonprofit corporation of individuals or institutions, but failed to agree on its charter. Later that year, Maurice Ewing of Lamont, J. B. Hersey of Woods Hole, and R. R. Revelle of Scripps formed such a corporation (CORE), which, in February 1962, submitted a proposal to carry out a drilling program as visualized in the intermediate phase of Project Mohole. This proposal was not endorsed by LOCO and was not funded, and both groups faded away.

Subsequently, on the initiative of C. Emiliani and the late F. F. Koczy and supported by a grant from the National Science Foundation, the University of Miami carried out a moderately successful shallow-water drilling and coring program from R. V. *Submarex* on the Nicargua Rise off Jamaica,[4] in

spite of adverse weather. Finally, in the first half of 1964, scientists from Miami, Lamont-Doherty, Woods Hole, and Scripps decided to attempt once more to form an organization to initiate and carry out large drilling projects in the ocean. In May of that year, the directors of these four institutions signed a formal agreement called JOIDES (Joint Oceanographic Institutions Deep Earth Sampling) to cooperate in deep-sea drilling. It was the intent that JOIDES should prepare and propose drilling programs based on the ideas of broad segments of the oceanographic community. It would designate one of its members to act as the operating institution for such a project and to be responsible to the funding agency for its management. This form of cooperation was chosen because of its flexibility, and because it placed the operating responsibility in the hands of an institution experienced in management. The operating institution would act in trust for the entire scientific community; all scientists actively concerned with the project would be invited to participate; samples and data would be distributed to all interested scientists without regard to affiliation. The agreement provided for an executive committee to establish policy and a planning committee to prepare plans for operations and to collect information regarding possible drilling objectives and sites.[5] During the remainder of 1964, the planning committee compiled a list of suitable drilling sites and collected background data.

Early in 1965, there was an opportunity to use R. V. *Caldrill,* which was under charter to Panamerican Petroleum Company, and was then on its way from California to Newfoundland. The Panamerican Petroleum Company agreed to allow a brief charter for scientific purposes anywhere along the voyage. The Blake Plateau area off southeastern United States was selected, and funds for a 1-month drilling program in

4. H. M. Bolli, C. Emiliani, W. W. Hay, R. J. Hurley, and J. L. Jones, *Bull. Geol. Soc. Amer.* 79, 459 (1968).

5. Joint Oceanographic Institutes Deep Earth Sampling (JOIDES), *Amer. Ass. Petrol. Geol. Bull.* 51, 1787 (1967); *Trans. Amer. Geophys. Union* 48, 817 (1967).

April and May were made available by the National Science Foundation. Under the management of Lamont and with the aid of an *ad hoc* committee from JOIDES and from other institutions, six holes were drilled and cored successfully in water depths from 25 to 1030 meters, with bottom penetrations of 120 to 320 meters.[6]

The success of this program and of the organization behind it stimulated plans for drilling in deeper water. A study made by the planning committee indicated that the technology for drilling in water down to 6000 meters in depth was available for reasonable cost, and that a suitable drilling ship could be found. This study also showed that much more could be learned from an immediate advance into deep water than from continued effort at shallow depths, a domain into which the petroleum industry was moving rapidly, in part for purposes of research.

The National Science Foundation in its budget request to the Congress for fiscal year 1966 asked for, and was granted, funds to initiate a National Research Program of Ocean Sediment Coring. The director of NSF had, in 1964, announced his intention of supporting such a program, conceived as distinct from and complementary to Project Mohole. Subsequently, JOIDES and Scripps together prepared a proposal to the NSF for funds for an ocean-drilling project which Scripps would manage. The nature and scope of the project were decided, and a contract between Scripps and NSF was signed in January 1967. Under the contract, a drilling program in the Atlantic and Pacific oceans for at least 18 months beginning in mid-1968[5] has been planned.

DRILLING IN DEEP WATER

Commercial drilling vessels come in a wide variety of configurations, sizes, and shapes. Vessels for deep-sea drilling for scientific purposes, however, have great basic similar-

6. JOIDES, *Science* 150, 709 (1965).

ity because of their special requirements. Since drilling sites are widely spaced, the vessels must be self-propelled and able to operate without support at great distance from their home base and for long periods of time. Adequate space for scientific facilities and scientists is also required. The operation in very deep water also has special consequences, which tend to simplify the operation compared to commercial deep drilling near shore. The thickness of the sediment in the oceans is generally small: between 100 and 1000 meters of sediment commonly overlie the hard rock basement. Only in a few areas adjacent to the continental margins do sediment thicknesses of several thousand meters occur. The shallow penetration required in the deep sea simplifies engineering requirements.

Marine drilling systems consist of three main components, which have been tested in practice, but are being used here in an unusual arrangement and for a relatively untried operation: (i) the vessel itself, supporting the drilling rig and all facilities; (ii) the positioning system; (iii) the drilling rig and machinery.

With the exception of D. S. *Cuss I,* all vessels used or planned are self-propelled ships varying in size from a few thousand to over 10,000 tons and with drilling capacities ranging from 600 to 6000 meters. In earlier vessels, drilling capacity and endurance were quite limited, as were space and facilities for scientists. On R. V. *Caldrill,* for example, scientists worked under remarkably primitive conditions. The R. V. *Glomar Challenger* (Figure 1-15), planned for future operations, is by comparison spacious, well equipped, and has long range. The price of the increase in capacity is high; day costs range from $5,000 to $25,000, depending on vessel and service.

While drilling is being done the vessel must be kept near a point directly above the hole in the sea floor. In shallow water, this is accomplished by anchoring or by raising the platform on legs. In deep water, these solutions are not practical, and a dynamic posi-

Figure 1-15. R. V. *Glomar Challenger,* the drilling vessel used for the Deep-Sea Drilling Project in the Atlantic and Pacific oceans. (Courtesy of Scripps Institution of Oceanography.)

tioning system is used. This system consists of a means to determine the position of the vessel with respect to the hole, and a means to correct the drift and maintain the desired position. Deviations have usually been determined with respect to anchored buoys equipped with radar and sonar reflectors or to a taut wire extending from the ship to the sea floor. In drilling operations now being planned the reference system will consist of acoustic beacons dropped on the sea floor, which are much easier to position than buoys and are not subject to stresses from wind and current. The signals of these beacons are received by three hydrophones placed in a triangle on the hull and are converted by computer into instructions to propulsion machinery. Four outboard motors placed forward and aft on both sides of the hull have been used; a combination of the ship's main propulsion with transverse thrusters placed in tunnels is planned for the future. Variations in drift and heading are

anticipated by the control system on the basis of past experience, and overcorrecting must be avoided.

The drill string has a great deal of flexibility; drilling in 6000 meters of water with a conventional drill string may be compared to drilling in 100 meters with a piece of baling wire. As a result, the radius within which the vessel must be kept is a function of the depth of the water; it should be of the order of 1 to 3 per cent of the depth. Obviously, maintaining position in water 5000 meters deep is far easier than staying within 5 to 10 meters in 200 meters of water.

A conventional rotary drill is used. As compared to drilling on land or in shallow water for commercial purposes, in the deep sea penetration is shallow, and the rocks are soft. As a result, since drill bits need not be changed frequently, the hole consequently is not reentered, a difficult operation under water. In shallow holes in ocean sediments, a single bit, tipped with tungsten carbide or diamonds, should suffice and should even permit some drilling in the underlying basement.

In commercial deep drilling, high-density drilling fluid is circulated from the pumps, through the pipe, into the hole, and back to the ship through a riser pipe in order to wash out drill cuttings and maintain a stable condition in the hole. In shallow penetrations, drilling can be accomplished with sea water, without complete circulation; the water is flushed out on the sea floor, and the costly and time-consuming installation of the riser pipe can be eliminated. Hence, a deep-ocean drilling operation for scientific work as described here is adequately served with a simplified system at low cost; this gain, however, is partially offset by the cost of operation on the high seas. An additional advantage of the single system is that, in case of approach of bad weather, the hole can be quickly abandoned.

The coring operation, the most important step in the entire procedure, is carried out with core barrels fitted inside the drill pipe. These barrels are pumped down with the drilling fluid and latched in place automatically. After the core has been taken by rotating or punching, a wire line with a catch is used to retrieve the barrel. Cores varying in length from 6 to 12 meters with diameters from 2 to 6 centimeters can be obtained without reentering the hole.

The motions of the vessel can affect the operation of the drilling equipment. Severe roll and pitch make handling of the pipe impossible, even though automated systems are used, and they place excessive stress on the drill string just below the vessel. These motions can be reduced significantly by devices for stabilizing the motion of the vessel (anti-roll devices) and by proper positioning of the vessel with respect to wind and sea, but in all open-sea operations loss of time because of weather is inevitable. Heaving of the vessel tends to pump the drilling fluid in the string, and it bounces the bit on the bottom, which damages both hole

and tools. This is counteracted by the use of bumper-subs: spring-loaded, telescoping sections of drill pipe just above the bit, which will accommodate heave up to several meters. However, an inevitable consequence is that the vertical distance from the ship to the bottom of the hole cannot be determined from the length of drill pipe to better than 3 to 6 meters. Since additional uncertainty is introduced by the fact that the vessel is not usually precisely above the hole, the precise depth of the core remains unknown within relatively large limits.

In a shallow hole in pelagic sediments, where 1 meter may represent as much as 1 million years of time, these inaccuracies are serious. They can be reduced in part, but not entirely, by measuring certain physical parameters of the rock continuously in the open hole by well-logging, and subsequent comparison of this log with measurements of the same parameters on the core. Although this has never been done in deep-sea sediments, such parameters as the variations of natural radioactivity may permit correlation between core and well log.

Several hours elapse between the withdrawal of one core barrel and the arrival at the bottom of the hole of the next one. During this interval, a thick layer of caved sediment may accumulate at the bottom, or the bottom may be scoured, so that ill-defined gaps exist between successive cores. Moreover, the coring process tends to be more effective in certain types of sediment than it is in others and more efficient at depth than near the surface where the deposits are soft. Thus, expected recoveries from the core vary from 40 to 80 per cent of the total interval that is cored.

Consequently, even when an attempt is made to core a hole continuously, the cores ultimately obtained will have gaps and uncertainties. Advanced designs incorporating increased experience will improve this situation, but a complete record such as marine geologists expect from their usual sampling tools is not likely to be obtained.

SOME OCEAN-DRILLING TARGETS

In 1946, Kuenen[7] estimated by various independent methods that the average thickness of sediment in the ocean should be approximately 3 kilometers, if one assumes that the age of the earth is 2 billion years. The average thickness was later found to be about 10 per cent of this figure, and it is now known that the earth is more than 4 billion years old. Thus, the record of ocean sediment must be incomplete, or it does not reach farther back than a few hundred million years. Independent estimates based on sedimentation rates, sediment thicknesses, and consolidation lead to the latter conclusion.[8] This is in agreement with the fact that nearly all of the older rocks recovered from the ocean floor date from the Tertiary or the Upper Cretaceous, a very few from the Lower Cretaceous.[9] Even if the underlying sediment is included by extrapolation, this still terminates the record somewhere in the early Mesozoic or late Paleozoic, or 200 to 300 million years ago. Direct instead of extrapolated data regarding the age of the oldest sediment in the oceans are of great value in confirming this strange observation. Drilling in the oldest parts of the oceans, for example, the western Atlantic and Pacific, would supply such information while at the same time shedding light on the conditions prevailing at that time.

In a different vein altogether, Hess[2] postulated in 1962 that convective flow in the mantle, an old concept, might be a major factor in shaping the floors of the oceans. This hypothesis, recently expanded by Vine,[2] states that a hot mantle current rises to the surface under the mid-ocean ridges and spreads laterally from there to the continental margins where it submerges and completes the convection cell. This hypothesis explains many of the morphological and geophysical features of the ocean floor and offers a mechanism for the process of continental drift.

This concept also explains how old deep-sea deposits are removed continuously from the oceanic realm by transport toward the continents where they are submerged under the continental mass. It thus accounts for the absence of an ancient oceanic record. In consequence, the lower part of the sedimentary sequence should increase in age from the mid-ocean ridges to the continental margin. This fact can be established by drilling a series of holes across the mid-oceanic ridges in the Pacific and Atlantic.

The combined sea-floor spreading and continental drift hypotheses, although far from completely accepted, have stimulated much imaginative research and speculation. A major contribution to testing this line of thought can be made by using deep-sea drilling.

A different group of problems deals with the history of the oceanic circulation and the closely related climatic history of the earth. In both the Atlantic and Pacific, the oceanic circulation consists of a westward-flowing equatorial current with northward and southward return flow around a northern and a southern central water mass. This circulation system, driven by the planetary winds, strongly affects the nature of the pelagic sediments, especially in the Pacific where disturbing influences of sea-floor topography and adjacent continents are minimized. A thick layer of highly fossiliferous carbonates marks the path of the equatorial current, while the sediments under the north and south Pacific central water masses consist of a slowly accumulating, nonfossiliferous clays. Sparse evidence[10] indicates that the boundaries between these sediment provinces have migrated back and forth dur-

7. P. H. Kuenen, *Marine Geology* (Wiley, New York, 1950), p. 386.
8. E. L. Hamilton, *Bull. Geol. Soc. Amer.* 70, 1399 (1959).
9. T. Saito, L. Burckle, M. Ewing, *Science* 154, 1125 (1966); M. Ewing, T. Saito, J. I. Ewing, L. H. Burckle, *ibid.* 152, 751 (1966).

10. W. R. Riedel and B. M. Funnell, *Quart. J. Geol. Soc. London* 120, 305 (1964).

ing the Tertiary in response to changes in intensity and distribution of currents and water masses. Similar relations probably obtain in the regions of eastward return flow at high latitudes. Drilling the complete sediment sequence along a north-south profile in the central Pacific should greatly improve the understanding of the history of ocean circulation since the Cretaceous, and of the related climatic evolution of the earth.

The last 100 million years have been a period of major evolutionary change in marine life and of large changes in the composition of the pelagic fauna and flora. Bramlette[11] has drawn attention to such a major change occurring at the end of the Mesozoic and suggests that it may have been caused by a reduction in the nutrient supply, which, in turn, was the result of the deeply eroded state of the continents at that time. Recently, it has been suggested that abrupt evolutionary changes, including wholesale species extinctions, may have been caused by increased cosmic-ray activity during reversals of the earth's magnetic field.[12] A detailed biostratigraphic record based on reasonably continuous samples of pelagic sediments extending back into the Mesozoic would contribute otherwise unobtainable data for understanding these aspects of the history of life. An important corollary is the improvement that might be achieved in the use of planktonic fossils for biostratigraphic correlation and age dating. The existing biostratigraphic time scale is based almost entirely on sediments deposited in near-continent environments and now exposed on land. These formations are often disturbed and discontinuous, and the correlations from one continent to another are highly controversial. Good stratigraphic sequences in true pelagic environments placed strategically between continents would go far toward resolving these difficulties.

11. M. N. Bramlette, *Science* 148, 1696 (1965).
12. This is a controversial subject; see N. D. Opdyke, B. Glass, J. D. Hays, and J. Foster, *Science* 154, 349 (1966); C. J. Waddington, *ibid.*, 158, 913 (1967).

Many other problem areas such as those related to marine volcanism, the diagenetic change of sediments with time and burial, the history of the major topographic features of the ocean floor, and the history of the continents as reflected in the sediments shed into the oceans could be cited. The brief listing of major issues presented above, however, may suffice to justify the strong motivation for deep-ocean sediment drilling.

THE DEEP-SEA DRILLING PROJECT

The scientific goals described here form the core of the program of the Deep-Sea Drilling Project at Scripps Institution of Oceanography, under contract from the National Science Foundation as part of its National Ocean Sediment Coring Program. The Project has signed a subcontract with Global Marine Inc., of Los Angeles, to provide and operate R. V. *Glomar Challenger,* now under construction, for approximately 18 months beginning in July of this year. First operations are expected to take place in the Gulf of Mexico, the Atlantic, and the Caribbean. With the advice of several panels formed under the auspices of JOIDES, a proposal for drilling objectives and for sites was prepared;[5] the proposed sites and cruise tracks are shown in Figures 1-16 and 1-17. After consideration of further advice and a review of the schedule, in the spring a final drilling program will be established, which will include a large portion of the proposed program; a detailed discussion of this program and its scientific justification can be found in reference 5.

Under the mandate from the National Science Foundation, the project will mobilize its drilling operation, drill the sites, prepare an initial description of the cores for publication, and preserve and store the core material and data before its distribution for research purposes. The initial description of the core is designed to record characteristics of perishable sediment and to provide enough descriptive information to permit scientists to define research programs and to

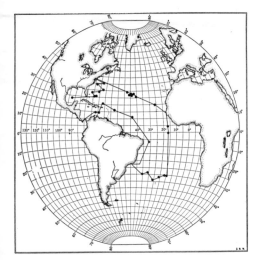

Figure 1-16. Tracks and drill sites for the Deep-Sea Drilling Project in the Atlantic Ocean.

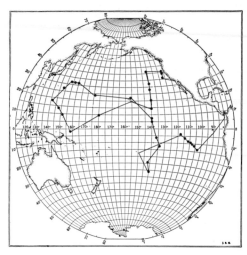

Figure 1-17. Tracks and drill sites for the Deep-Sea Drilling Project in the Pacific Ocean.

select their samples economically and efficiently. Reports should be available approximately 4 to 6 months after the termination of each segment of the drilling cruise. The project welcomes all inquiries and suggestions for additions and improvements in the program.[13]

13. Inquiries may be directed to Project Manager, Deep-Sea Drilling Project Building T-16, Scripps Institution of Oceanography, Univ. of California,

The Deep-Sea Drilling Project is the first large-scale expression of the dreams described in this article. It is close to reality, and a new chapter on deep-sea drilling can soon be written.

La Jolla 92037. In preparing this article I have been helped by my colleagues in various JOIDES committees, in particular by C. L. Drake and W. R. Riedel. Contribution from Scripps Institution of Oceanography.

2

*NAVIGATION
AND POSITIONING*

Oceanography and Navigation

GEORGE E. R. DEACON   1969

Position finding and hydrography—the plotting of coasts, channels and dangers to navigation—are the most obvious applications of science to the oceans, and they are the basis of well-established professions in which the use of accurate methods and the presentation of precise information are of paramount importance. Navigators can also make good use of knowledge of waves, tides and currents, but these are more variable in time and space, and, needing to be understood as well as charted, offer plenty of scope for academic study as well as a professional approach. Together with the study of life in the sea, and of the nature and character of the sea floor, they have always attracted the attention of scholars and sailors, and, especially in the early days navigation and science have gone hand-in-hand. In 1612 Sir Thomas Button, seeking the NW Passage, was instructed to record the tides, eddies and overfalls, as well as the soundings, islands, rocks and trend of the coasts.[1]

Francis Bacon asked for studies of the "History of Ebbs and Flows of the Sea; its Saltness, its various Colours, its Depth; also of Rocks, Mountains and Vallies under the sea."[2] William Oughtred saw the finding of longitude "as the difficultie and Master Piece of Nautical Science" but also insisted that "If the *Master of Ships* and *Pilots* will take the paines in the journalls of their voyages diligently and faithfully to set down in severall columns, not only the Rumbe they goe on, and the measure of the Ships way in degrees, and the observation of latitude and variation of their compasse, but also their conjectures and reasons of their correction they shall make of the aberations they shall find, and the quality or condition of their Ship, and the diversities and seasons of the windes, and the secret motions or agitations of the Seas, when they beginne and how long they continue, how farre they extend, and with what inequality; and what else they shall observe at Sea worthy consideration, and will be pleased freely to communicate,

1. *The Voyages of Captain Luke Fox and Captain Thomas James,* Hakluyt Society, 1894, Vol. 89, pp. 636-7.

2. Sir Francis Bacon, *Works,* ed. Spedding, Ellis, and Heath, 1857-74, Vol. 4, p. 226.

Dr. George E. R. Deacon is the director of the National Institute of Oceanography in Great Britain and is past president of the Royal Institute of Navigation. His work has covered most aspects of physics of the oceans and development of modern instruments and techniques.

From *Journal of the Institute of Navigation,* Vol. 22, No. 1, pp. 77-91, 1969. Reprinted with light editing and by permission of the author and The Royal Institute of Navigation, London.

the same with Artists, such as are indeed skilfull in the Mathematicks, and lovers and inquiries of the truth: I doubt not but there shall in convenient time be brought to light many necessary praecepts, which may tend to the perfecting of navigation, and the help and safetie of such, whose vocations doe enforce them to commit their lives and estates in the vase and wide Ocean to the providence of God."[3]

Such careful observations and records were the basis of our knowledge of the sea, but even in the most practical aspects of navigation there was always some resistance. In 1580 William Bourne wrote, "I doe hope that in these dayes, that the knowledge of the Masters of Shippes is very well mended, for I have knowen within this 20 yeeres that them that were auncient masters of ships hath derided and mocked the that have occupied their Cards and Plats, and also the observatiō of the altitude of the Pole, saying that they care not for their Sheepes skins, for hee could keepe a better account upon a board . . . and when that they did take the latitude, they would cal them starre shooters and Sunne shooters, and would aske if they had striken it."[4] Nearly 100 years later Sir John Narborough demonstrating the value of a Mercator chart wrote, "it is an hard matter to convince any of the old Navigators, from their Method of sailing by the Plain Chart; shew most of them the Globe, yet they will walk in their wonted Road."[5]

Later in the same century Isaac Newton, writing to the Treasurer of Christ's Hospital where the king had established a school of navigation, recommended that a child should be taught more than "the use of Instruments and the bare practise of Seamen in their

beaten road" and went on to say that he should be given the understanding that would assist "in inventing new things & practises, and correcting old ones, or in judging what comes before him" and "let it be further considered whether it be most for the advantage of Sea affaires that the ablest of our Marriners should be but mere Empiricks in Navigation, or that they should be alsoe able to reason well about those figures, forces, and motions they are hourly concerned in." The boys, he wrote, should "be exhorted or obliged by the Governors to communicate to the School (in gratitude to the place of their education) such accurate observations, curious discoveries and select draughts as they shall make abroad in their Voyages." It would be interesting to know how successful this exhortation was. After quoting the appeal made by Oughtred "a Man whose judgement (if any man's) may be safely relyed upon" he remarks, "Thus farr that very good and judicious of Seamen to able Mathematicians at Land, the Land would send able Mathematicians to Sea. It would signify much more to the improvem^t of Navigation and safety of Mens lives and estates on that element."[6,7]

The Land had one opportunity when Halley the astronomer was appointed Master of the *Paramour Pink* which he sailed to the edge of the Antarctic pack ice, making great contributions to our knowledge of the trade winds, the Earth's magnetism and the tides. But administratively the voyage was not a success: innumerable difficulties were wished and pressed on Halley by a first lieutenant whose method of determining longitude he had been obliged to condemn a few years earlier and who now made the most of an unjust sense of grievance of which Halley

3. William Oughtred, *The Circles of Proportion and the Horizontal Instrument newly increased with an Additament for Navigation,* 1633, Part 2, pp. 55-6.
4. *A Regiment for the Sea and other Writings on Navigation,* William Bourne, ed. E.G.R. Taylor, Hakluyt Society, 2nd Series, No. 121, p. 294.
5. Sir John Narborough's journal, 1669-71, An Account of Several Late Voyages and Discoveries to the South and North, 1694.

6. Correspondence of Sir Isaac Newton and Professor Cotes, including letters of other eminent men, now first published from the originals now in the Library of Trinity College, Cambridge, ed. J. Edleston, 1850.
7. Deacon, Margaret (1965). Founders of Marine Science in Britain: the work of the early Fellows of the Royal Society, *Notes and Records of the Royal Society of London,* 20, 28.

had forgotten the cause. But it led to trouble and the navy sent no more mathematicians to sea. Wales and Bayly the astronomers on Cook's second and third voyages were able to measure water temperatures but even the great *Challenger* expedition of 1872-76, devoted entirely to the study of the sea, had no physicist or mathematician: nor did the mathematicians on land take enough interest and she missed the chance of a lifetime. Fifty years later, after the German Antarctic ship *Deutschland* had posted home preliminary results on her outward voyage, it could be seen that the *Challenger* might have outlined the meridional circulation of the subsurface, deep and bottom layers of the Atlantic Ocean much as we know it today if her observations had been used with more confidence and skill.

It was perhaps due to the skill and ingenuity of men like Faraday and the immediate success of their work that scientists took much more interest in experimental, laboratory science, than in large-scale environmental science, especially when it meant going to sea, but during the last world war the quotation from Newton began to appear on the walls of the offices of directors of naval research in the U.K. and U.S.A. The main idea may have been to encourage mathematicians to be as practical as possible, but the mathematicians did go to sea, and they have been largely responsible for the rapid growth of marine science during the past few decades. The careful observations and theoretical studies of earlier sailors and scientists provided the foundation, but the difficulties involved in the theoretical work were sufficient to hinder the development of the effective framework necessary for continuing study. Early workers such as Benjamin Franklin, who made a careful study of the stilling of waves by oil, Froude, the pioneer of experimental ship tanks, and Stokes who provided much constructive theory, realized that the waves they were studying were a combination of simpler waves which traveled at different speeds, and, getting in and out of step, caused high

and low groups of waves analogous to the beating of two adjacent musical notes, but it was not till 1944 that the growing urgency of wartime requirements allowed a small group of scientists in the Admiralty Research Laboratory at Teddington to undertake the first spectral analysis of sea waves. It had been done for light, sound and radio waves, long before, but only the urgency of a defense program could provide the facilities required for the same basic treatment of sea waves, which give the name to all kinds of waves and have their own great practical importance.

The value of the new work was apparent as soon as it was done. It showed how the mixture of waves in a storm behaves as a continuous spectrum of wavelengths, from ripples up to long, low undulations with as much as 3000 ft between successive crests. A 72-knot wind blowing over several hundred miles of deep water will produce wave periods up to 24 seconds and wavelengths up to nearly 3000 ft. The different wavelengths are found to travel almost independently, at speeds proportional to their periods or to the square root of their wavelengths. The longest waves travel at about the same speed as the strongest wind, but the rate at which trains of waves advance across the ocean, from the storm area to distant coasts, is much less, because energy traveling to the front of each wave train goes into building up its wave motion. Visual observation of a fairly uniform group of high waves shows that waves seem to travel through it entering at the back and vanishing in front as they make waves in still water. In deep water half the kinetic energy of the waves reaching the front is converted to potential energy, and the energy as a whole advances across the ocean at a "group velocity" which is only half that of the individual waves. The longest waves generated by 72-knot winds will therefore make their way across the ocean at about 36 knots. All the wavelengths in the spectrum travel independently with their theoretical group velocities and after they leave the storm area they appear as an ever lengthening procession with the long waves getting farther and farther ahead of the shorter. If the nearest coast is 1800 miles away the 24-second waves will take 50 hours to reach it and the slowest waves, likely to have enough energy to arrive and to be recognizable on arrival, may take four times as long. The longest, 24 second waves may be only a few inches high and unrecognizable in deep water, but when waves enter shallow water their speed decreases, and since they must carry as much energy as before their height increases. As they slow down those behind catch up on those in front and long waves that are indistinguishable in deep water become visible in shallow water where they are higher and closer together. This change accounts for the name "ground swell." The appearance of ground swell is usually a reliable indication that higher swell will follow because although the 24 second waves may be only a few inches high the 16 second waves, which contain most of the storm energy and arrive 25 hours after the longest forerunners, may be as much as 10 ft high. In the winter season when most of the swell on our western seaboard comes from the adjacent belt of westerlies there is not much uncertainty but by 1945 it had been shown that long, low swell from the neighborhood of Cape Horn can be measured on the English coast and the attenuation of this, over the whole length of the ocean, is such that it will not build up to more than a low ground swell. In recent years swell arriving on the coast of California has been traced back to generating areas near New Zealand and even to the high latitudes of the Indian Ocean. The rollers of Ascension and St. Helena are no longer a mystery and could be predicted if weather maps of the distant parts of the ocean were available.

Spectral analysis has allowed a clear and numerical separation of locally generated waves from those traveling from distant storm areas, and the separation of bands of swell from different storms, and we understand how they will vary with time. It has been shown that there are only random rela-

tionships between the different wave components and reliable statistical descriptions of the sea surface can therefore be based on relatively few measurements of a wave record and even from wind charts and forecasts. "Freak waves" such as are frequently reported are generally the result of a large number of wave components getting into step at the same time and place and the probability of such extremes can be calculated, though the actual times and places at which they will occur cannot be predicted. Ability to separate the products of different wind systems has led to much improved wave predictions and better estimates of the probabilities of different wave heights, lengths and slopes, and the rapid progress and growing promise of the work stimulated the development of new wave recorders. A useful wave record can be made from a ship provided it is not moving too fast, and a great wealth of research material and statistics has been obtained by these automatic wave recorders in lightvessels and weather ships. By stopping a ship and putting out a special buoy which measures wave slopes in two directions at right angles, we can measure the angular distribution of wave energy as well as the spectrum of wave heights. This is needed for research on ship motion.

As soon as wave analyzing techniques were developed they were applied to simultaneous recordings of the rolling and pitching of ships, and it was found that although ship motion, especially the rolling, tends to be greatest at the natural periods determined by the size, shape and displacement of the vessel, she responds in some degree to all the frequencies present in the wave spectrum, modified when she is under way by her course and speed. The statistical methods used for waves also were applied to ships, and, knowing something of the wave pattern and the response of the ship to particular frequencies, the probability of particular accelerations and undesirable stresses can be calculated. This is leading to an almost new science of ship hydrodynamics, which begins to supplement the well-established work on stability, freeboard and structural strength, and must soon improve our understanding of the sea-keeping qualities of ships in rough water, in which no two ships, not even sister ships, seem to be exactly alike. It may not make much apparent difference to the appearance of ships already designed on long experience, or to the way in which a skillful sailor operates his ship, but it must help to express the problems with which they are concerned as clearly as possible and in a form that will encourage further reasoning, observation and experiment. It will help with new problems that are arising from the use of larger ships, and with new ventures such as the use of hovercraft and hydrofoils. These have already made good use of the new wave studies.

Engineering aspects of navigation that are concerned with shallow water are likely to profit from the new studies. The changes in velocity and momentum which take place in waves as they approach the coast cause appreciable changes in mean water level, and the consequent variations in head of water from place to place give rise to significant currents. Where waves are running up a gently sloping beach the mean water level inside the breaker zone will be higher than that in the less disturbed water offshore, and with a moderately heavy swell the difference can be 2 or 3 ft. When waves break over an off-lying sandbank or are partially reflected from a submarine breakwater the mean water level will be higher inside than outside and the difference is likely to be sufficient to cause a strong current through a narrow opening. There can be dangerous outward currents on beaches where there are large breaking waves. Where there is a gully or depression running up the beach, energy will be focused on to the shallower areas on either side and there may be a strong outward current over the deeper channel. Such currents known as "rip currents" are well known on ocean coasts and are often visible from the top of a cliff or from the air as they carry sand- or foam-laden water out through the breakers. There is no doubt that

more detailed knowledge of such currents would help to save lives. The more general effects of the refraction and energy changes in waves approaching a coast are also of interest and importance. The way in which the longer waves and swell generally seem to approach beaches parallel to the shore in whichever direction the beach is facing is largely due to refraction. Swell entering the English Channel from the Atlantic will be bent round to the English and French coasts because of the shallowing soundings on the two sides, so that an observer on the French side will see it coming from the north, while on the English side it will appear to come from the south. Swell may be refracted round a headland which is affording good shelter from shorter waves. Energy tends to be focused on to a spit or shallow area and dispersed at the head of a deep bay. When tidal streams are concentrated near the coast the changes in current strength will bend waves so that wave conditions on a coast may change with the tidal streams as well as the depth of water. It has been shown that a wave cannot travel against a current that is faster than a quarter of the wave speed, and when waves meet such opposing currents they must lose their energy in overfalls and broken water. When the current is in the same direction as the waves they will travel smoothly. In a wide estuary there are likely to be many different wave conditions depending on all the variations of depth and current, and the pattern will vary with time. A general wave prediction can easily be made for the Strait of Dover, but it would be very difficult to forecast all the changes a hovercraft pilot may meet in the varying depths and currents between Ramsgate and Calais in great detail.

Much of the energy of the troublesome harbor oscillations which make a ship range at her moorings unless special precautions are taken, and of the surgings at offshore moorings comes from waves. One of the most common features observed in ocean swell that has traveled a great distance and, having outdistanced or outlived shorter waves, has only a narrow range of wave-lengths, is its regular alternation of groups of high and low waves. This gives rise to the saying that the seventh wave is always the highest, and although it is not necessarily the seventh or tenth the slightly different wave-lengths tend to get in and out of step at about this interval. In deep water there are small variations in water transport and mean water level associated with the high and low groups. In shallow water these effects in-crease and the changes in momentum that take place on shelving beaches send out long waves of about 1.5- to 2.5-minute period, and with amplitudes about a tenth of those of the incoming swell. Range action is most common near coasts exposed to ocean swell, and particularly in artificial harbors whose dimensions are such as to give them natural periods of oscillation within the 1- to 3-minute range. The outgoing long waves are generally called surf beats.

The disastrous long waves that are usually called tidal waves derive their energy from seismic disturbances of the sea floor or from intense storms traveling at critical speeds across areas of relatively shallow water. The Pacific Ocean is particularly subject to seis-mic waves and they are much studied there. There is a central organization to assemble reports from seismographs and if the re-ported disturbance seems likely to have caused a sea wave the center calls for reports from tide gauges at appropriate stations. These are used to predict the time of arrival of the water wave at all parts of the ocean, and the prediction is usually correct to with-in a few minutes. But the warning is not always as effective as it might be because of the difficulty of predicting the height of the waves. The cause may be a subsidence of the sea floor probably over a long line which gives the source directional properties that are not easily estimated. The traveling water wave, having a wavelength which is long compared with the depth of the ocean, travels at a speed determined only by the depth, and it is refracted by all extensive topographical features in its path, focused in

some areas and dispersed in others, the changes being particularly great in the ap-proaches to land. The energy tends to spread out between great circles starting from the source, and to converge again after traveling a quarter of the way round the globe focus-ing toward an area antipodal from the source; this too, is a significant factor. The effect of local resonance is also important. The oncoming waves have periods something like 10 to 15 minutes and where there are coastal bays, channels and shelves with na-tural periods near this figure, local reso-nances may enhance the effect of the waves. Such complications all somewhat inter-dependent make precise prediction very dif-ficult and predictions, which have to be on the safe side, are sometimes followed by only rather minor disturbances. The public, seeing this, and not wishing to be disturbed, tends to underrate subsequent warnings. Per-haps the most remarkable features of seismic sea waves are their great speed and small amplitude in deep water. Although in the deep ocean they travel at 400 to 500 knots and carry enormous amounts of energy they are only 1 or 2 ft high with 100 miles or so between successive crests so that a ship passing over them could detect nothing. It is only when they reach shallow water where their speed is reduced to a few knots that they become spectacular. To carry as much energy as before they may grow to cata-strophic heights and end in a final rush up the beach and over the coastal fringe carry-ing everything before them. The Chilean earthquake killed 150 people in Japan. The name tsunami, meaning in Japanese "harbor wave," is generally used for seismic waves in the Pacific Ocean.

Storm surges are just as dangerous. In the North Sea they are largely due to the piling up of water in front of the wind in the narrowing funnel of the southern part of the sea, but external surges entering the sea round the north of Scotland and traveling anticlockwise round the coasts play a signifi-cant part. The external surges are caused by high winds blowing at critical speeds over

fairly extensive stretches of relatively shallow water, and it is this type of surge that
causes great disasters in other parts of the
world. Over the deep ocean hurricanes and
typhoons cannot travel as fast as the long
waves they make and the energy put into
long waves is rapidly dissipated. But in 50
fathoms the speed of a long wave ($\sqrt{gh}$
where h is the depth of water, and g the
acceleration due to gravity) falls to little
more than 60 knots, and a storm traveling at
$\sqrt{gh}$ can feed energy continuously to a wave
traveling at about the same speed. The result
is analogous to the supersonic bang produced when an aircraft travels at the speed
of sound. The surge is often of several hours
duration and although the first rush of water
is not as great as that of the seismic waves, it
may rise 10 ft or more above high-tide level
and being accompanied by high wind waves
does much damage. In East Pakistan coastal
plains have sometimes been flooded to a
distance of several hundred miles inland.
There are sometimes minor surges on the
south coast of England, in which the tide
seems to rise and fall several feet in as many
minutes causing some confusion among holiday makers on the beach and among boats
moored in estuaries. The reason is often not
obvious, but careful examination is likely to
show that a weak meteorological front has
crossed the Channel at just the critical speed.
The rapid movement of deep secondary depressions sometimes offers more serious
threats to our west coast ports and very high
water levels have been recorded, without so
far any serious damage.

Navigators and port authorities can find
little wrong with the tide tables, but there
are a few shallow-water ports where greater
detail and accuracy would be useful. As the
draughts of ships and the costs of dredging
and maintaining harbors increase the need
for even minor improvements grows. The
difficulty about shallow-water tides is that
they do not follow the very regular rise and
fall observed in deep water. Because of the
distortion of the tidal wave as it travels over
shallow water and into constricting channels

it can no longer be represented adequately
by a sine curve or a simple combination of
such curves, but requires much more complicated theoretical treatment and numerical
expression. The tide tables are as excellent
for nearly all purposes as they were meant to
be, but to be just as accurate in all the
difficult places where reasonable approxima-
tions have been adequate so far, and to be as
accurate in tabulating hourly heights as they
have been with heights and times of high and
low waters, requires much more science and
money than was previously available. There
is also a growing need to supplement the
astronomical tide prediction which can be
made and printed as far ahead as is necessary
with forecasts of the effects of winds and
variations of atmospheric pressure on the
predicted levels. Everyone knows that the
wind will advance or hold back a tide, but to
express all the factors involved precisely
enough to allow useful forecasts to be prepared in time for advantage to be taken of
longer high waters, and for precautions to be
taken against extremely high and low waters,
is as difficult as any other problem in advanced science. From the point of view of
housing and industry as well as navigation it
is urgent: the high levels cause floods, and
low levels have been known to interrupt
cooling water supplies to power stations.

As more first-rate mathematicians,
theoretical and practical physicists and engineers become interested in marine science,
and as new techniques for computation and
measurement become available fresh progress is made. Tide gauges have been improved and they are calibrated more frequently; new methods of analysis which will
measure the activity at any period have been
devised to supplement those which examined the main astronomical periods and
their harmonics, and they are themselves
supplemented by statistical techniques that
try to squeeze as much information as possible out of the records whether its physical
basis is known or not. The new methods
include analogue as well as digital computers
which allow us to apply what we know of

the effects of wind stress, moving pressure gradients and bottom friction to actual seas, taking account of their irregular shape and bottom topography and of the effect of the earth's rotation which is important even in a large estuary. From them we have learnt how the effects of local and general disturbances build up in space and time and have almost reached the stage where tables of calculated response factors can be fed into meteorological forecasting machines to obtain sea-level forecasts. There is still much to do, especially round the edges of the sea, but by checking numerical computations for past weather patterns as well as we can, and working out typical situations and key problems, we have learnt quite a lot about the response of a complicated sea like the North Sea to the varying natural forces that control it. The work will not be complete till we can extend our observations and analysis across the continental shelf, and into the deep ocean where the tides gain nearly all their energy, and where we begin to see that winds and traveling depressions may produce some tidal modifications large enough to be transmitted to the shelf and inshore tides. It is a question of facing up to the difficulties and refinements that we could leave on one side in the first analysis. It happens in other sciences, even the equations of motions of the moon, the peak of accuracy in astronomy and navigation, are being re-worked to satisfy the requirements of moon-rocket navigation. With the tides we can still occasionally be 2 ft too high or too low at Tower Bridge.

The subject of waves made by ships is perhaps neither oceanography nor navigation, but it is of interest to both. At high speeds a rapidly increasing proportion of the ship's engine power goes into making bow and stern waves, and it is this wave resistance that makes attempts to reach higher speeds uneconomical. The mechanics of the subject is known fairly well, but there is scope for further theoretical and practical study. It may, for example, and in spite of all the objections of sailors and other difficulties,

still prove advantageous to build the propellers at the bow. The long waves made by ships are of particular interest in shoal water and narrow channels. When passing over a shoal the ship will push water ahead of her and squat lower in the water, but if she attains the critical speed of $\sqrt{gh}$ she will climb higher and experience less resistance as she is borne along on a wave of her own making. This was the principle of the fly boats which could be galloped along a canal by a single horse, and Mark Twain describes how Mississippi steamers would call for more steam to lift themselves over a bar. There is fresh scope for theoretical and practical work as deeper draught vessels want to use harbor entrances with depth restrictions, and there are likely to be problems even in the North Sea. When the ship floats in a shallow layer of fresh or brackish water lying above salt water, slow moving waves in the boundary surface between the two layers can offer great resistance to the forward motion of a low-powered ship. It used to be frequently reported in near-Arctic waters and in Norwegian fjords, and the name "dead water" was used to describe it. Nansen wanting to shoot a bear on an ice floe a few miles away took several hours to get there, and the history of the phenomenon is well written by his scientific associate W. Ekman. It goes back to early times when it was supposed that the devil had the boat by the heels. Some yachtsmen sometimes feel the same and the whole subject of the effect of stratification on wave resistance may be worth further study. It may affect the performance of a ship over a measured mile. In all layers of marked density gradient throughout the ocean there are internal waves with periods ranging from a few minutes to the main tidal periods. When the layers are shallow so that the water movements associated with the internal waves can reduce the roughness of the surface waves they give rise to what look like parallel slicks on the surface, and which can be detected on the radar screen if not by eye. When the tidal wave crosses the edge of the continental shelf west of the English

Channel it produces large internal waves in the summer discontinuity between the warm surface water and colder deep water. This is a phenomenon comparable with that of the lee waves produced when a wind blows over a mountain range, and as far as we can see the internal water waves may be responsible for the large sand waves with much the same wavelength and 30 to 40 ft high that are found there. There are very large internal waves in the boundary layer between the Atlantic water flowing into the Mediterranean Sea through the Straits of Gibraltar, and the colder, more saline water flowing out as an undercurrent. They are large enough to be a hazard to submarine navigation, requiring frequent alterations of trim. Internal waves seem to play an important part in mixing the surface and deep waters in the approaches to the continents, increasing the supply of nutrient substances to the well-illuminated layer where photosynthesis can take place, and increasing the plankton crop and general fertility of the continental shelves. At the higher frequencies there is no clear distinction between waves and turbulence, and a better knowledge of both is essential for the better understanding of ocean currents.

There is no very widespread demand from sailors for day-to-day forecasts of ocean currents which are somewhat analogous to the hour-to-hour forecasts of winds needed by air pilots, but it is reasonable to suppose that they could be put to good use if readily available. Some tankers making regular voyages along the east coast of the U.S.A. will tow a device which tells them the temperature at 100 fathoms because this shows them how to find or avoid the strongest part of the Gulf Stream. Most ships rely on the monthly current charts, and although these are most reliable they tell the user only as much as previous observations can show about the probability of currents from different directions. Off our own coasts, and after allowance has been made for the tide, the current depends mainly on the local wind, being almost in the same direction as

the wind at the surface and averaging some 20° to the right of the wind between the surface and 30 ft in deep water and less in shallow water. It is a complex mechanism and in some parts of the oceans, particularly on the western sides, the currents are due more to the distribution of wind stress over the ocean as a whole than to the local winds. The Gulf Stream is part of an ocean-wide circulation for which the trade winds and the westerlies are important, but although it is a permanent current its transport and limits vary considerably and it has such large eddies that a ship may sometimes find that she has been contending for half a day or so with a counter current, while, according to the average information, she should have been getting most help from the current. The monthly atlases give three kinds of chart. The first gives information for each point of the compass, and near our coasts, in the approaches to the Channel for instance, currents to the north-east do not average much stronger or much more prevalent than those in other directions. The second chart tries to make it easier for users by reporting only the direction that has been observed most frequently, and the third plots a rather unimaginative average of all the reported currents. There is nothing wrong with the charts, they are as good as careful observations and straightforward plotting can make them; they take care to warn the user that they give average information, and on the average after very many runs through the area he will find the same. But everyone tends to neglect the explanatory notes on a chart and to take the arrows at more than their face value, and by looking at currents averaged over a long period and over fairly large areas of ocean we have tended to gain the impression that they are steadier than they really are. There is little doubt that we can make serious attempts at day-to-day forecasts, and with the help of a few current measuring and reporting buoys at well-chosen places, the forecasts would soon be as reliable as the wind forecasts. They could be used for predicting the spread of oil or other pollution as well as for navigation, and after some trial and error would probably help fishermen. The work would have an application to meteorology since the causes of some of the transient but important climatic fluctuations are likely to be found in the oceans. This means that we must find out more about the transport of heat in the oceans, its part in making the winds and how these in turn decide the pattern of the currents.

For more than 100 years we have known that the water at the bottom of the tropical Atlantic Ocean was only a few degrees above freezing and have realized that it must flow there as a bottom current from the polar regions, and over the past 50 years or so a remarkable picture has been obtained of ocean-wide exchanges between high and low latitudes and between regions of high and low balance of evaporation over rainfall. Each region tends to produce water with a distinctive range of temperature, salinity and density, and the different types of water tend to spread across the ocean from their source regions at appropriate density levels. As they enter new areas and mix with adjacent water masses they tend to find new levels thus taking part in a complex interwoven circulatory system that involves the whole ocean. Water which sinks round the south-west margin of the Weddell Sea can be traced eastwards round the Antarctic continent and northwards past the equator in the Atlantic, Indian and Pacific Oceans. Water which is diluted as well as cooled all round the Antarctic can be traced beyond the equator at intermediate depths of 500-1000 meters. Water which sinks in the North Atlantic Ocean can be traced southwards as far as the Antarctic continent between its cold surface and bottom waters, and also into the deep and bottom layers of the Indian and Pacific Oceans. Over a very long time, perhaps 1000 years, the oceans are completely mixed. There are cold and warm areas and poorly and highly saline areas, but the salt itself, except for the small amounts of dissolved phosphates, nitrates,

silicates and trace metals that enter into the nutrient cycles of marine plants and animals, is the same all the world over, with the same constituents present in the same proportions. But we do not know much about the speeds and transports of the different branches of the currents, or how they are energized. It is not likely that they are due entirely to horizontal differences of density. Although these must be primarily responsible for the stratification of the ocean, wind stress may be an important factor in moving the water from the source region to the place where it sinks to the new level appropriate to its density.

It is the difficulty of measuring the speeds of the deep currents that has restricted progress. Measurements made by lowering current meters from ships moored by long wires had many uncertainties. Much better progress has been made during the past 10 years by using floats which can be made to have neutral buoyancy at any predetermined depth where they drift with the water and send out sound signals which can be followed by a ship at the surface. In the deep layers as well as at the surface the strongest currents are on the western sides of the oceans and they tend to be about twice as fast at 4000 meters (about 0.2 knots) than at 2000 meters. On one occasion 0.8 knots was averaged over 2 days at 4000 meters north of Bermuda. The observations made so far suggest that there are eddies with a diameter of 50-100 miles or so. If two floats are placed 10 miles apart at the same depth they tend to keep together, if they are placed 50 miles apart they are likely to move in different directions. It has also been found that the general direction at any place is likely to change every week or so. More than a year's observations north of Bermuda gave little or no indication of a mean direction of flow, showing only what might be described as large-scale turbulence. They give some reason to believe that a synoptic current chart of the deep ocean, if we could make such a thing, would look something like a weather chart, though with many more cyclonic and anticyclonic systems each smaller but lasting longer than those in the atmosphere. The ocean-wide movements inferred from the temperature and salinity distributions may be the outcome of these, just as the prevailing winds of the atmosphere are the overall results of the varying day-to-day weather patterns. It has been shown theoretically that a passing depression will cause appreciable movement at the bottom of the ocean as well as at the surface, and although the mathematics, which involves the effect of the earth's rotation as well as wind stress and the effect of moving pressure gradients, is very difficult, it is not surprising that the bottom water should be disturbed because the ocean is, on a large scale, a broad shallow dish, and the horizontal scale of an atmospheric depression is large compared with the depth of water. Neutrally buoyant floats have been followed for periods up to 49 days, but we need longer series of measurements, and current measuring systems, in which vertical series of current meters are suspended below anchored buoys are beginning to give useful results. The most successful use subsurface instead of surface buoys, the buoy and its meters being released from the bottom anchor by an acoustically operated release when the ship wants to recover them. Particularly strong currents have been measured at the bottom of some channels: 4 knots in the outflow from the Straits of Gibraltar and nearly as much where Arctic water flows through the Denmark Strait and the Faroe Bank Channel.

This work is still rather remote from most aspects of navigation but it begins to be related to the surface currents. One of the most interesting studies of recent years is that of the Somali Current which runs only during the SW monsoon. Its speed and transport increases downwind, the increasing volume of water being supplied partly by upwelling of cold water near the coast and partly by an inshore movement at the bottom of the well-mixed surface layer; the SW wind tries to maintain an offshore drift in the upper part. The effect of the earth's rotation on a current makes the density surfaces slope up to the left in the northern hemisphere and this slope increases till there is a balance between this geostrophic force and the gravitational forces which try to make the density surfaces horizontal again. The Somali current runs so fast that such a simple balance becomes no longer possible. In $8°$ N where the speed of the current was found to reach 7 knots conditions seemed to become critical. The inshore temperature was only $12°C$ and the density distribution could apparently do no more to balance the effect of the earth's rotation and the current turned sharply away from the coast. There were very sharp current boundaries and temperature fluctuations between $8°N$ and Socotra, and it is not surprising that the varying currents of this region are notorious during the SW monsoon.

Another feature of recent interest is the equatorial eastward subsurface current. In the Pacific and Atlantic Oceans, and in the Indian Ocean during the NE monsoon there is a very narrow ribbon of subsurface current, extending not more than 1 degree north and south of the equator, and confined to depths between 50 and 100 meters. It flows eastward when all the surrounding water flows west and can be as fast as 2 to 3 knots. Many hypotheses have been put forward to explain it, and its further investigation must do much for the study of water movements.

The growing size of ships wanting to use shallow seas, and the spread of marine technology over the continental shelf, have encouraged fresh study of the details of the sea-bed. The sand and sediments move in response to the waves, tides and currents. Large sand waves are a feature of some areas, and they move about in an apparently complex way. They may be 10 or 15 ft high with wavelengths of 200 to 300 yd, and must therefore limit the accuracy to which such areas can be sounded and charted. They are also likely to be a hazard to structures resting on the sea-bed. Authorities responsible for maintaining navigable channels may well be able to make good use of better understanding of the movements of sandbanks, and like other marine studies this may respond to advanced theoretical and experimental study as well as more detailed and continuous observation. The growing recovery of sand and gravel from the sea-bed may have some bearing on navigation. The beaches and coasts are in some kind of equilibrium with the sea floor, and large-scale dredging may alter the supply and movements of sediments; it may also modify the wave pattern sufficiently to encourage some new effect on the shore. There are many interesting advances in our knowledge and understanding of the floor of the deep ocean, and although these are, for the time being primarily academic, marine technology is beginning to look to the deep ocean as well as the continental margins.

One of the most effective uses of science on shore has been to assist great industries, and there can be little doubt that marine science can help most aspects of seagoing, and it may indeed prove essential in face of growing competition. But science cannot grow to be effective without the stimulus of constant application, and, as in the past, marine science depends largely on seamen.

Hydrographic Surveying

ANTHONY G. STEPHENSON   1970

Hydrography encompasses the detailed examination and recording of every body of water—its depth and movement, bottom and sub-bottom, shoreline configuration, and navigational hazards—for reproduction on charts and in tables.

The original hydrographic surveyors were the navigators who discovered new lands, for, wherever they went, they recorded depths, channels, dangers, and coastline details to help assure that future passages would be safer. Each successive ship's captain entering an area added to the information on his chart and wrote a narrative known as the "Pilot." Since that time charts have been updated continually and the Pilot is amended by a periodic circular known as the "Notice to Mariners," which is in part based on the reports of modern navigators.

Government hydrographic departments date back to the early 19th Century and have developed most of the world charts in use today. In the last fifteen years, however, as industry and commerce moved rapidly into the marine environment, there has been a strong call for contract hydrographic surveys to undertake detailed work at specific developing locations.

The hydrographer's expertise encompasses an almost confounding breadth, requiring him to change roles from surveyor to mathematician, from navigator to electronic technician, and at times, from workhorse to diplomat. The fundamental requirement is a thorough grounding in land survey techniques. In land surveying, the problems vary with the terrain (submerged or exposed), but a hydrographic surveyor, whose work may take him to any part of the world, has to be prepared to combat differing environmental difficulties in a variety that his shorebound counterpart rarely encounters. Hydrographic surveying lays heavy stress on versatility, requiring a man to be as adept at mountain climbing as at navigation and drafting.

The first requirement on any hydrographic survey is to locate the vessel accurately (and continuously); it is equally important to be able to relocate an exact position where there are no buoys or beacons to mark the point. There are a variety of methods of "fixing" (locating) a survey boat—

Anthony G. Stephenson, a hydrographer with Decca Survey Systems, Inc., Houston, Texas, emigrated to the United States from Great Britain in 1963. He has conducted hydrographic surveys for port and harbor development, pipeline location, and water pollution studies throughout the world.

From *Oceanology International,* Vol. 5, No. 6, pp. 35-37, 1970. Reprinted with light editing and by permission of the author and *Oceanology International.*

some recent, some traditional. In all cases the designed precision of instrumentation dictates the accuracy obtainable.

Visual methods have their restrictions, but in many instances prove the most effective for economic reasons and, sometimes, even more accurate than more sophisticated alternatives (dependent on location). Triangulation is one method of visual fixing. It calls for setting up two theodolites on shore with the azimuth and distance between them known, and sighting out to the boat's mast. Both instrument operators, at a signal from the boat, observe the angles. Knowing the baseline distance and measuring the two included angles, they can calculate the boat's position. This method is seldom satisfactory, however; in most cases it covers a limited area, and the fact that the fixing team is separated, requiring long-range coordination, contributes to delays and inaccuracies.

A more satisfactory visual method is to use hydrographic sextants. Sextants for hydrographic work are specially designed, can measure larger angles, are lighter, and have a larger field of view than the navigational type. The method involves the observation from the boat of three well-defined marks on shore whose coordinates are known. Two sextant operators simultaneously observe the two angles between the left and center, center and right marks.

The resultant fix is in the form of a three-point problem, mathematically tedious to solve but graphically simple and quick with the aid of an instrument called a station pointer. It is a precise three-arm protractor that reads to a minute of arc. The two observed angles are applied to the station pointer and the arms moved to intersect the three coordinate points on the sounding chart. The intersection of the three arms is the boat's position. Experienced men can take the fix and plot position in only 10 to 15 seconds. The fact that all people concerned with obtaining the fix are in the boat permits excellent coordination, enabling the survey boat to operate without a shore party.

In areas of considerable habitation, conspicuous marks on shore have usually been surveyed previously. This land survey work frees the hydrographer to change marks frequently and still obtain good fixes without the tedium of onshore control surveying.

There are other variations in the use of a sextant; one important one is steering along an arc of a sextant angle, using the geometric principle that the angle subtended by a chord at the circumference of the circle remains constant. The survey party draws up a plotting sheet with the known marks located and swings arcs from two of the marks across the survey area at the required spacing, and notes the subtended angle. Because the line is an arc, it is preferable to keep the subtended angle as large as possible. (Fitting a 90-deg prism to the hydrographic sextant enables reading angles of up to 210 deg.) The sextant is set to the predetermined angle and the boat conned along the arc; another observer sights a second angle to locate the boat's position on the arc.

Visual methods, though sound, are restricted to near shore, under good visibility in daylight hours. In recent years the demand has been great for accurate fixes far offshore, not limited by visibility or darkness. To accommodate these requirements, electronic positioning systems were introduced into hydrography. These enable not only positions to be located accurately well out of sight of land, but in some cases can be used also to guide a vessel along predetermined lines in any direction and at any required spacing and also to return the vessel to spot or rerun a line in the future. The accuracy of these systems is dependent on range, baseline locations, survey accuracy of the onshore transmitting stations, unobstructed paths between shore stations and the ship, and weather conditions. The trained surveyor knows these factors well and either eliminates them or takes them into careful account by adjusting accordingly.

Not all systems on the market are the same, and what is advantageous for some surveys can be disadvantageous for others. There are a variety of systems, some accurate but limited to line-of-sight work and others less accurate but usable over much greater distances.

The modes in which one can use some of the systems create different lane patterns, known as hyperbolic and range-range. The hyperbolic mode requires three onshore stations and creates a hyperbolic pattern radiating from them. In this mode more than one vessel may use the system, but the area of high accuracy coverage is limited by the expansion of the reference lanes. The same instrumentation may be used in the range-range mode for good accuracy over a larger area because there is no lane expansion, but is limited to the use of one vessel. It takes a trained hydrographic surveyor to evaluate all systems and modes to ensure the most accurate and economic results for the survey in question.

Deriving a fix for a vessel at sea, important as it is to the hydrographic survey, is only a locational tool. To help the hydrographer collect other desired data, instrumentation has been modernized and developed. The echo sounder gives precise depths acoustically and has been developed in several variations, all aiming at different requirements. Some, for example, plumb great depths but forfeit accuracy in shallow water. Thought has to be given to the type of vessel to be used, results required, portability, and so on. Many modern depth recorders use a linear scale giving a true graphic "picture" of the seabed. Earlier models used a rotating arm that swung across the paper in an arc that distorted the record. Nonetheless, some models of this kind are more accurate than the newer, straight-line recorders whose manufacturers apparently have not understood the real requirements of the hydrographer.

Echo sounder record paper that has a scale preprinted on it illustrates this lack of understanding of hydrographic work. The hydrographer has to make a tidal reduction on the record to reduce soundings to datum, hence, the printed scale is much less useful than might be thought. Also some echo sounders show little thought to ease of calibration to allow for variation in salinity and temperature. To eliminate a man standing by the instrument to record each fix, a remote event marker switch is a very good feature. Time interval marks, available on some machines, help to correlate tidal information from one or more tidal gauges located close to the area of sounding.

With the rapid increase in ship sizes and drafts, navigational surveys have assumed great importance. With many ports unable to accommodate the huge new ships, new deep water terminals have to be found. Offshore loading and unloading in some areas are essential but expensive. Such facilities require very careful surveys to bring the ship terminal as close to shore as safety permits (perhaps leaving barely a foot beneath the keel of the largest ship at the low-water spring tides) and ensure an area swept of all obstructions.

Instruments such as the side-scanning sonar help locate obstructions between the actual lines of soundings. This instrument sends a narrow horizontal but broad vertical acoustical beam that scans the seabed at right angles to the vessel's track. Undulations, outcrops, or other hazards appear in the record; the strong signal return off the vertical face of an object shows as a black mark on the record, and the total loss of the returning sound on the far side of the object shows as a white shadow.

With the simultaneous use of the echo sounder, side-scanning sonar, and accurate fixing, it is feasible to sweep for obstructions acoustically, but there are other methods. Two boats linked by a thin wire that passes through two streamlined sinkers and attaches to both boats can sweep to a required depth if handled properly. One end of the fine wire is secured to one boat and the other to a winch in the control boat. By the two boats keeping some 200 to 300 feet apart with a winch operator keeping tension on the wire, they can clear the area between them. To ensure complete coverage of the area, it is usual to run a 50 per cent overlap to each side of the swept path, in effect, sweeping the area twice. Attaching a depth meter to each weight provides a method for monitoring the exact depth of the wire.

The oil industry and engineers involved in offshore construction have an ever-increasing need for knowledge of shallow sub-bottom geologic variations. There are instruments that record shallow sub-seabed stratification with high definition, graphically displaying a continuous profile of the geologic stratification. Typical projects requiring this type of survey are offshore drilling platform footings, underwater tunnels, placing piles for a pier, and pipeline routes.

To obtain absolute knowledge of the variation of the strata, it has been essential to drill core holes. Coring gives discontinuous (spot) indications of changes—comparable to taking lead line soundings—not a continuous record such as a precision depth recorder obtains. With the sub-bottom profiler a few reference core holes suffice to identify the changes in strata evident on the profiler record; beyond these baseline samples, changes in the geologic structure can then be charted on the recorder's continuous profile. For surveying pipeline and tunnel routes, this instrumentation is invaluable, virtually eliminating needless encounters with underlying rock and inadequately consolidated sedimentation.

Another use for the sub-bottom profiler is offshore exploration for minerals. Irregularities in the bedrock can create traps that may hold, in highly mineralized areas, many precious metals. Large enough deposits may be minable at a profit.

Extensive sub-bottom surveying can locate these traps, and core samples can determine their probable values. A mineral survey is conducted in much the same manner as a navigational survey, but here bedrock and overburden as well as the seabed are mapped. The likely extent and volume of mineralization can accurately be determined by this method. In contrast, this type of work, done by arbitrary core sampling, is not only extremely expensive but inconclusive.

In these applications the surveyor must choose the correct acoustic transmission frequency, since penetration varies inversely with frequency but definition varies directly with it and both factors depend on the type of sedimentation to be penetrated. The hydrographer must weigh all the requirements before selecting which type of system to use.

Dredging is another industry that has always required hydrographic surveys, and, as ship size increases, exact determination of channel depth becomes more essential. A survey is undertaken before dredging, to determine the volume of excavation. This estimate must be reasonably accurate because a dredging contractor usually bids on the total volume to be removed, and incorrect estimates lead to contractual difficulties. Simultaneous sub-bottom and echo-sounding profiles can locate underlying rock or could damage the dredge or affect the dredging time. A post-dredging survey ensures that the area has been completely dredged and protects the client from needless costs of overdredging. The side-scanning sonar finds a role in post-dredging operations to reveal undredged peaks and locate any slumping of channel walls into the navigable water lane. It is an advantage to know how often a channel need be dredged and what area can best receive the spoil. Hydrographic studies using silt samples and current measurements can reveal the likely effects. These same methods are useful in the study of coastal erosion and accretion, important factors in the preservation of coastlines.

Movements of currents and tidal streams have been measured for many years, but the increasing problem of water pollution has imparted a new emphasis. Few studies to determine the effects of the effluent on marine life and human recreation have preceded the installations of outfalls. The assumption that dispersion of an effluent in a much larger body of water will be adequate may or may not prove out. A hydrographic survey can spell the difference, but surveys have usually followed installation, after the damage has been done.

The approach to this type of survey is largely dependent on the type of effluent and the area of discharge. Generally two methods are used. Variations of the water movement can be determined with a current meter by measuring flow velocity and direction at different depths. Such studies can reveal the optimum elevation for an outfall pipe to ensure placement in the maximum current to carry the effluent away with the minimum period of slack water. The next step is to find where and how far the water carries the effluent and whether it carries ashore, concentrates, or disperses. This is achieved by tidal stream studies, usually performed with floats having large vanes extending down to the depth of the discharge. The stream carries the floats away until the tide changes, then carries them back. By observing the direction and velocity of float movements through all states of the tide, it is possible to obtain a very good idea of the dispersal rates over the area.

For thermal discharge studies, similar methods apply, supplemented by temperature surveying. Reasonable results have been obtained with dye discharged into the water at the outfall, which remains visible for one tidal cycle. These studies can also determine if hot water outfall is likely to return to a cooling water intake—of help to power station design.

The types of hydrographic surveys described in this article fulfill basic requirements using fundamental methods. The rapid advancement in computer technology and growth of solid state electronics con-

tinue to shrink time requirements and increase accuracies. There are now such innovations as digital echo sounders that can read depths into a shipboard computer for printout onto a chart at the correct location—all while surveying at 30 knots. Or automatic pilots that accept radio fix signals and maintain a vessel along an exact preplotted track. Satellite positioning (in conjunction with other systems) makes possible hydrographic surveys in the deep ocean to an accuracy previously only obtainable no more than 300 miles from land. Although still not as precise as using shore-based stations, the method has opened up the avenues to more accurate knowledge of the oceans.

Pollution studies are also advancing, thanks to digitized computer inputs from continuous reading sensors that monitor dissolved oxygen, salinity, turbidity, temperature, and current velocity and direction. All of these steps help the hydrographer to obtain greater amounts of data in a shorter time and in some cases at a lower cost than do the conventional methods.

Though the instrumental advances in data collection are many, only the correct selection of instrument and procedure can create a meaningful result. Basic data usually require correction, editing, and analysis before presentation in cartographic or tabular format for the client. Throughout these stages the hydrographer ensures that the required accuracy is achieved and that this accuracy is achieved at minimum cost to the client. The client, who is interested in minimum overall cost rather than the minimum daily rate, is coming to realize that, in the oceanographic field as in many others, professional advice pays.

Needs and Future Development

PHILIP M. COHEN   1970

FORECAST OF DEVELOPMENTS

In the years since echo sounders were introduced to general use, charting has undergone radical change. Under the impetus of technological advances in electronics and computers, and spurred by a greater need for ocean bathymetry, programs of bathymetric collection and chart production initiated only a decade ago have become largely outmoded. The importance of every aspect of oceanographic investigation is receiving somewhat belated recognition, and there is evolving better nationwide coordination of oceanographic programs.

The title of this chapter implies prediction based on needs, which, in any science subject to dynamic change, can be a hazardous undertaking. Still, some attempt will be made to outline the course of future developments in survey instruments, charts, and procedures. In some cases, discussions of future events are fairly safe, involving logical extension of what are presently conceded to be worthwhile efforts. In others, the future is unclear, and any prediction is only a guess. In most instances, however, it should be possible to make some valid forecast. Of course, a forecast can be perfectly valid yet turn out to be incorrect. In predicting the future of bathymetry, every expert is his own best prophet, and it is not probable that what is written here will meet with universal acceptance. A discussion of future developments in bathymetry has firm bearing on a more complete understanding of the navigator's job.

Dynamics of Requirements

Pressing needs are not always recognized in time, but they eventually make themselves felt. It would be gratifying to be able to stipulate bathymetric chart requirements at one sitting and to know these requirements would remain unchanged for a length of time sufficient to implement means to satisfy them. But changing times bring changing requirements; this is the nature of things. Further, there are subjective factors present.

Philip M. Cohen is chief of the Marine Geophysics Group at the National Ocean Survey, Washington. He is a former member of the Marine Sciences Council in the Executive Office of the President and has worked for the Naval Oceanographic Office and Defense Intelligence Agency.

From *Bathymetric Navigation and Charting*, pp. 107-121, 1970. Reprinted with author's revisions and light editing and by permission of the author and the Naval Institute Press, Annapolis.

While many individuals could be convinced that a requirement exists for complete world coverage of bathymetric charts at a scale of 1:250,000, it could conceivably be demonstrated that no such requirement exists. Certainly requirements do exist for accurate bathymetric charts at a suitable scale, covering many areas of the world's oceans.

CHART ACCURACIES FOR VARIOUS WORK IN THE OCEANS

Most individuals engaged in work in the marine sciences would readily concur that there is need for proper bathymetric data to aid in the performance of work/tasks in the oceans. By the term *proper* is certainly meant some reasonably correct portrayal of bottom terrain features. It is probable that many things could not be done in the absence of adequate bathymetric data. But where will these data be needed? When? What are the accuracies required? Are graphical charts the only or best means suitable? These questions come quickly to mind and are rather obvious, yet are far from unimportant. There is another group of essential unknowns characterized by even greater tenuousness, e.g., to what extent is accomplishment of given tasks affected by lack of proper bathymetric information? That is, can one quantify the element of risk per kind of task under conditions in which a deficiency of these data exists? Few persons can claim to know that a ten-million dollar program of bathymetric surveys and chart production means effective support to national goals in oceanography, while a five-million dollar program means ineffectiveness. Yet, many individuals have no difficulty in spelling out the difference between both availability and lack of charts with characteristics allowing placement and maintenance of structures on the bottom. An important intangible is absent—performance criteria for work/tasks as they may be related to chart requirements or other bathymetric input.

Criteria for Satisfactory Chart

What is it that would make for a satisfactory chart? One supposes the chart would be satisfactory if it could meet its intended purpose in the hands of the user. The qualities allowing that purpose to be met might be *correctness, accuracy,* and *reliability.*

To be correct, a bathymetric chart must portray the actual terrain found in a particular area. We know that this is unlikely, or, even if possible, not able to be ascertained as such. This is attested to by the difficulties in acquiring bathymetric data. Therefore, the chart is frequently incorrect.

To be accurate, the chart must be free of mistakes and errors, i.e., it must be compiled with precision; the best-possible use of available survey data must have been made. By this token, it is not likely that very many charts are accurate. Despite this definition, in common use the accuracy of a chart is deemed a standard from which deviation is measured in linear terms. Thus, a chart would be called accurate horizontally to ±2,000 meters. Even if this definition were applied, there is no way to relate properly accuracy to given portions of the chart. Would the chart be accurate to ±2,000 meters 50 per cent of the time over 20 per cent of the area? Few charts may be considered truly accurate by whatever definition applied.

To be reliable, the chart must be counted upon to do what is expected of it, a criterion that is more useful if described in terms defined for common use of accuracy. A chart of given accuracy is, in other words, one we may assume will be reliable, the amount of accuracy being a measure of that reliability.

Work/Tasks and Associated Terrain-Accuracy Standards

Development of performance criteria by task or groups of related tasks would seem to involve no more than routine determination; however, a listing of assignments will not, for reasons which are obvious, receive general agreement for every work/task. Nevertheless, this must be attempted. Let us assume that bathymetric information is required for performance of most tasks in the oceans in the form of standard chart graphics. Table 2-1 contains a listing of work/tasks in the oceans, together with associated terrain-accuracy standards necessary for the chart product in use.

First, it must be said that no attempt is made to include all known or predictable work/tasks in the oceans. The listing is representative only. Second, some familiar aspects of work in the oceans are omitted, such as law and inspection, air-sea interaction, food from the sea, ocean engineering, etc. These efforts either are too inclusive for the purpose intended or can be divided into identifiable subtasks. The accuracies are stated in terms of terrain-accuracy standards for the chart rather than absolute accuracy for the task itself. If it is necessary, for example, to position a geodetic transponder to within ±10 meters, this may not mean (at least no one can say) that the chart used to do this must also be accurate to ±10 meters. The relation of the chart's terrain-accuracy

Table 2-1. Work/Tasks with Their Associated Terrain-Accuracy Standards

Work/Tasks	Terrain-Accuracy Standard (meters)
General navigation (long-range)	±3,000
Submersible (short-range)	± 300
Submersible (long-range)	±2,000
Farming	± 200
Fisheries research	± 400
Implantation of gear	± 25
Gravity research	± 50
Antisubmarine warfare	± 50
Mine warfare	± 50
Rescue and salvage	± 25
Mining	± 25
Deep drilling	± 25
Retrieval of treasure	± 25
Man-in-the-sea	± 100
Pollution studies	± 150
Waste disposal	± 250

standards to absolute positional accuracies necessary for the task may be direct, but may or may not be identical. It is true that this kind of rationale applies to certain groups of tasks only. It is believeable that chart accuracies of ±50 meters are superfluous if the requirement for submerged long-range navigation necessitates positional control of only ± one kilometer. But in the kind of task necessitating capability to return time and again to a given geodetic position, the greatest possible chart accuracy is desired.

Even supposing the data in Table 2-1 are valid, the biggest job still lies ahead. This is to translate the required terrain accuracy standards to appropriate chart implementation programs.

For general long-range navigation, positional achievement of ±3,000 meters is ample. The navigator desires to know his position to the greatest possible accuracy, but it is difficult to see how transit times could increase proportionately with positional accuracies greater than this value. True submersible navigation (excluding submarines) necessitates a thorough knowledge of bottom terrain features within a defined area per type of mission. If we assume this area to be at least 40 square kilometers, definition of features at a closely spaced contour interval should result in ±300 meters. In this and other examples, the nature of the operating terrain will affect ability to achieve such accuracies on the chart.

For farming and fisheries research, the accuracy standards seen are the author's best estimates. Many could dispute the values listed. However, recall that what are shown are the *chart standards* for the given tasks, and not necessarily those for the tasks themselves.

For the remainder of the work/tasks, there is direct relation with position and validity of measurement, as with gravity research; and sometimes the work/tasks can be augmented by visual sightings. Pollution and waste disposal studies are, of course, directly related. Here, again, the author falls back upon those more expert than he as to accuracies required. In certain cases, such as disposal of radiation wastes, exact position would seem to be critical.

Chart products for most of the work/tasks will invariably be of large scale, perhaps 1:10,000 and even larger. There is no need to compare such a chart with established accuracy standards for land maps. For example, it is commonly accepted that a 1:10,000-scale map (Class A) should have 90 per cent assurance of ± five meters horizontal and ± five to eight meters vertical contour accuracies. But this product is compiled with great precision from stereoscopic photographs obtained under strict conditions.

SURVEY PLATFORMS

Very few systematic bathymetric surveys have been conducted for any length of time. Prior to World War II the Navy's meager survey fleet did attempt systematic coverage of certain coastal areas. The war ended the attempt abruptly, and when hostilities ceased, the complexion of surveying changed radically. The priority of surveying was dictated by two considerations still valid today—the political nature of world affairs which involved United States peacekeeping efforts throughout the world and, second, recognition of the importance of oceanography to the nation and to naval doctrine. The necessity to increase surveys of all kinds and on a worldwide basis stemmed from these causes.

Multipurpose Collection Units

Since the magnitude of the subsequent effort exceeded available ship forces, the concept of multipurpose collection units evolved. Responsible naval officials held that, because ships were scarce and highly expensive to operate, each one should take advantage of every operation by obtaining the greatest amount of all kinds of data possible. Use of multipurpose ships did allow large amounts of all data to be collected, including bathymetric data, although the concept did not prevent assignment for collection of bathymetric data exclusively on particular missions.

Multipurpose usage led inevitably to high cost of new construction when, in 1960, Congress approved building new ships for surveying and oceanographic purposes. Before this time almost all such ships were conversions from other types. Although construction planning was based on optimum capability for a given cost, eventual costs exceeded first estimates by 10 to 30 per cent. Perhaps it is time to reappraise the concept of multimission platforms: to specialize; to build smaller, less expensive, faster, single-purpose ships for bathymetric surveys; and to have ships available that can be dispersed to many areas or assembled in a concerted effort. Such a force could complete an assignment the first time and produce results lasting for many years without the need to schedule repeat operations. More and more specialized platforms will be developed for every aspect of oceanography; this is occuring even now.

Survey and Oceanographic Ships

The distinction between survey and oceanographic ships can be defined more clearly. A survey ship may be considered one that conducts hydrographic and bathymetric surveys primarily, limiting its oceanographic work to bathythermograph collection, if done at all, and sea state and current data. An oceanographic ship is virtually the converse, with the possible exception that it may conduct survey operations in limited areas to supplement its primary oceanographic endeavors. Both ships usually operate echo sounders in transit to and from operating areas. By these definitions, only the Navy ships conduct surveys. However, the Navy also has oceanographic ships, complicated further by their designation as "oceanographic survey" ships. The present rate of bathymetric data acquisition is theoretically not limited by the number of survey ships in use alone, this can be increased by utilization of both survey and oceanographic ships.

There are far greater numbers of oceanographic ships in operation than survey ships, even including in the latter group those nonoceanographic ships that can and sometimes do perform survey work. This last category includes several types—ones whose missions are related only indirectly to charting—cable layers, NASA range ships, gravity collection and other research ships. In general, they make excellent survey platforms, and this is true of many oceanographic ships. Somewhat under 100 oceanographic ships are in operation, of which perhaps one-half are of a size and so equipped as to conduct surveys. Since the ratio of even this number to the survey ships in operation is about 5:1, conversion of a relatively few oceanographic ships would greatly increase survey capabilities. Whether this is desirable, in consideration of the number and variety of existing oceanographic requirements, is a different problem. However, if an increase in the survey fleet became necessary, it could be accomplished relatively quickly.

Ships as survey platforms have been and are today largely indispensable, but their continued use must always be examined in light of numerous factors. Some of these are cost, ability to acquire the data through alternate means, technological innovations, etc. No system that would replace survey ships can now be foreseen, and it is safe to predict that ships will never be entirely replaced for survey work. But other collection means will be developed. A single platform for complete replacement of ships, however, is doubtful.

Helicopters as Survey Platforms

In theory, the use of helicopters as survey platforms is entirely feasible. Transducers can be towed beneath the surface (but not too deeply) connected electrically to the helicopter, and geographic control would be maintained by routine means. The concept has much to commend it, particularly speed

of operation. Technologically there is no overwhelming reason why this kind of survey operation cannot be performed. But practical difficulties could offset the increased data acquistion rate. Helicopters can remain on station only a short time. Thus, extended over-water helicopter operations are impossible without ship bases. Data reduction facilities are minimal or nonexistent. And the economics of such operations, involving many helicopters and large support ships, has yet to be demonstrated. Development of longer range helicopters could change this forecast, but only if their inflight endurance was considerably increased. Speed is of slight consideration, i.e., a speed of 50 to 100 knots is such a vast improvement over present survey ship speeds that further increase to 200 knots would have little bearing. Maintenance of given altitude above the ocean surface would be accomplished by use of upward-scanning sonar so true surface-to-bottom readings could be computed. In any event, helicopter surveys cannot be considered to have great potential in bathymetric work. Their potential for hydrographic work cannot, however, be as easily dismissed. Hydrographic work in its present form cannot be efficiently continued much longer.

Satellite Platforms

It is difficult to forecast lack of success for any sensing phenomenon associated with satellites. With regard to numerous oceanographic variables it is reasonably certain that satellites will eventually perform extremely useful functions. Sea state and ice conditions, surface current data, surface temperature, and other oceanographic information can and will be observed by satellites. Even inshore hydrography, reservoir sedimentation, glacial flow, and harbor survey work will benefit through satellite photography. But neither visual nor radar photographic techniques have any great potential for deeper bathymetric work. Unless spectacular breakthroughs in photography or laser technology occur, application to bathymetry cannot be predicted in the immediate future for satellite or other airborne platforms.

Submarine Platforms

On first consideration, the ability to conduct survey operations while completely submerged offers a distinct advantage, because of the calm waters in which the platform operates. Surface waters are rough and can sometimes delay or halt survey operations. Submerged sounding is a reality now from the standpoint of putting sensors beneath the surface and through the use of completely submerged platforms, such as submarines and deep submergence vehicles (DSVs). Surveys by DSVs would probably be conducted only in limited areas. The drawback of submerged platforms for routine survey work is not lack of means to determine position, but the fact that surface ship operation is less expensive. Submerged platforms will be built to include bottom-survey ability, but none will be built for that purpose exclusively. Surveys will be conducted to augment the primary intended mission of the submerged vehicle. Positioning of vehicles close to the bottom remains a problem, but control near the surface can be obtained by towed whip antennas. Eventually, vehicles, utilizing fixed beacons for control, will conduct some surveys close to the bottom. Yet if a forecast were to be made, the probability of their use (or that of submarines) for routine bathymetric surveys is marginal.

Towed Sensors, Cable-Controlled Vehicles, and Other Platforms

Quiescence under the surface can be utilized by surface ships, towing submerged sensors, which eliminate positioning difficulties. Towed sensors will not be used for routine bathymetric collection on a large scale. Their chief use will continue to be to provide data for microbathymetric studies in relatively limited areas. Inshore hydrography may benefit from applications of paravane techniques. Remote control vehicles will be developed to transmit depth data in "real time," coupled with positioning and computerized control applications. Higher survey speeds will be available with hydrofoil and ground effect-type platforms. Man-in-the-sea programs may provide for "ground" observations of fixed-bottom control points; cable-controlled vehicles for salvage, survey work, and mining will be given increased range and utility.

COLLECTION TECHNIQUES

The data collection techniques of the future will employ the swath-sounding approach, side-looking sonar, underwater photography, lasers and holography, computer modeling, and control buoys. Some of these techniques are already in use; others may be researched and developed for the next ten years before they are used at sea.

Swath-Sounding Approach

One of the most impressive survey techniques developed during the past decade is *swath sounding.* Were it not for current limitation in its use, the concept might be termed revolutionary. Swath sounding would not have become an operational survey system without parallel advances in electronics and computer technology, and even so, development was slow and achieved only with great difficulty. Increased numbers of such systems in the future must imply a willingness to pay for the benefits derived, because of large development costs involved. This stems from miniaturization requirements for installation on smaller hulls, as well as for computer support.

It is believed that this miniaturization will occur and that additional swath-sounding units will be in use. Return in bathymetric data, whether determined by total program investment, cost per sounding mile, or cost per day at sea, dictates such a course. Procurement of individual units on a haphazard basis would be a mistake, however, without integration of hull characteristics, data-reduction techniques, computer application, and overall requirements. Further, techniques for translation of area coverage needs into survey specifications must be developed. This kind of development is particularly amenable to a typical systems approach, which could completely evolve new survey concepts and techniques.

Side-Looking Sonar

Side-looking sonar deserves technical improvement for supplemental survey purpose. This treatment places the axis of the transducer horizontally, allowing the acoustic rays to transmit and receive across, rather than from above, the ocean bottom. The advantage in this application is due primarily to the large bottom area covered. Vertical coverage is sufficiently maintained so that the bottom is "sounded" from an area very close to or even under the ship out to equipment ranges. To date, side-looking sonar can be used only in shallow waters or towed close to the bottom when in deeper areas. In either case a good definition of microbathymetry is achieved. Typical ranges depend on water conditions, target strengths, etc., but rarely exceed several miles or about five to six times the depth. The present configuration does not provide unambiguous depths for charting purposes, but it may be possible that this limitation can be overcome.

Underwater Photography

Underwater photography is invaluable for many purposes, but will lack general mapping potential except for limited bottom areas. Effective range from camera to bottom has increased since the first remote deep-sea photograph was taken by Lamont Geological Observatory in 1939, but haze still limits it to about 30 meters. Literally thousands of photographs would be necessary per square mile of ocean bottom in order to compile any but the largest scale maps even if stereoscopic overlap were not required. Other than data reduction problems caused by this volume of photographs, camera positioning poses serious difficulty. Data reduction and positioning problems could be solved, but no known lighting system will overcome the range limitation caused by haze. If haze could be eliminated (as does not seem likely), the potential for production of bathymetric charts is present.

Lasers and Holography

The laser principle was developed by the American physicist Charles H. Townes in 1960. The laser produces an intense beam of coherent light of high directivity, with the green to green-blue laser penetrating the farthest in the oceans, or rather, being absorbed the least. An invention in 1947 by Dennis Gabor of the Imperial College of Science and Technology in London was waiting for discovery of the laser: this invention was holography which freezes in film the optical wave fronts of an object and preserves original depth and parallax. The laser is used as the light source which reconstitutes on film the light waves of the object; without the laser the hologram appears as a greyish picture with a fine network of lines.

In acoustic holography, illumination of objects is accomplished by sound energy instead of laser beams as in optical holography. The resultant hologram is recorded using several different techniques and is reconstituted by lasers. One problem is that of perspective—the three-dimensional effect of the optical hologram is due to parallax achieved by the viewer. Because this kind of hologram is recorded at optical wavelengths, resolution is acceptable when viewed by the human eye. In acoustic holography, however, wavelengths are larger. Parallax effects using normal vision do not yield acceptable resolution, and other means to observe reconstituted images must be developed. However, the hologram and the laser open a new dimension in oceanography even though present difficulties in this infant science are manifold. The echo sounder has replaced the lead line; perhaps acoustic holography will someday replace the echo sounder as we know it today.

Computer Modeling

Much remains to be done in terrain simulation and statistical extension of mathematical models by computer programs. Probabilistic models should provide a means for computer simulation of a wide variety of geological processes under the oceans. In the earth sciences generally, probabilistic means aid the study of stratigraphy, sedimentation, paleontology, and geomorphology. Experimental computer modeling will increase as the basic data matrix necessary to inaugurate programs is more and more obtainable. The experimental simulation of "real-world" processes appears dependent on the computer orientation of individual oceanographers, who are motivated in part by the interests of the organization to which they are attached and the direction of their graduate training.

Computer generation of relief perspectives is feasible and should prove valuable in studies involving areas of limited horizontal extent. The importance of looking at relief in this way, other than the advantage it affords in "reading" shape, lies in the various angles from which the perspective can be displayed.

Control Buoys

Lessons learned in position maintenance from deep-drilling operations will benefit survey work through development of automatic control buoys for special search and salvage. These buoys would overcome deep-anchoring problems and would contain thrusters which correct for deviation from a radius of position error.

DATA REDUCTION AND LIBRARIES

Small-scale chart coverage, backlogged data, and repository of bathymetric data are considered under data reduction and libraries.

Small-Scale Chart Coverage

Small-scale coverage of ocean bathymetry exists in several chart series. One of these is the General Bathymetric Chart of the Oceans (GEBCO) at a scale of 1:10,000,000. The International Hydrographic Bureau (IHB) initiated this program about 25 years ago, following earlier efforts by Prince Albert of Monaco. Seventeen member nations of the IHB compile and maintain GEBCO, which is published by the Institute Geographic National in Paris. World coverage is as yet incomplete.

The Bathymetric Chart (BC) series is published by the United States Naval Oceanographic Office. Despite its rather small-scale (but much larger than GEBCO) and overly large contour interval (100 fathoms), it can be and is used by submarines for submerged position location. Positions so achieved have accuracies of 1,000 yards in exceptional cases, but this is possible on the BC chart chiefly in areas compiled from controlled surveys. As a bathymetric chart suitable for general bathymetric navigation, the series does not suffice.

The National Ocean Survey (name changed from Coast and Geodetic Survey) of the National Oceanic and Atmospheric Administration (NOAA), produces geophysical maps of many areas on the continental shelf and in deeper areas. These packages of data include a base bathymetric map, overprints of magnetic and gravity contours, and reports and data lists including seismic data. Scales are 1:250,000 and 1:1,000,000 on the margins and deeper areas respectively. NOAA was established in October 1970, and includes (besides the National Ocean Survey) the National Marine Fisheries Service, National Weather Service, National Environmental Satellite Service, as well as numerous data centers and laboratories.

Backlogged Data

Occasional statements on 20-year backlogs of unprocessed survey data, or about 50 or 100 ship-years of survey requirements that must be met, actually make little sense and upon analysis are found to be misleading. Whether very great importance can be placed on the value of bathymetric backlogs is questionable.

Most of the information in this category consists of poorly controlled and random ship tracks. Although age in itself is not a hindrance, some of the data are very old. Moreover, the majority are not entirely unprocessed, as they received some review upon receipt for possible contribution to then-current production programs. The backlogged data should not form the major basis for chart production of certain areas, except with special care, even in areas where recorded depths are sparse. Means to scan and store the backlogged data are easily developed, but elimination of the condition should more properly be attacked by solving the present rate of influx, through shipboard data reduction or procedures that store incoming data automatically and selectively.

The phrase "survey requirements" is without meaning unless qualified in use. Identification of a 50-year program of survey requirements presupposes that specific chart series have been determined to be needed. Procedures for translation of chart needs to survey times, accuracies, instruments, and even platforms are lacking at this time and are not easy to develop. Witness the use of the term ship-years; does survey time equate to ship-years? What is a ship-year? Are linear lines or square miles used as a criterion? Do different type ships have different ship-years? These questions are not meant to be facetious, and illustrate the difficulties in quantifying requirements.

Repository of Bathymetric Data

A central repository of bathymetric information ranks high in priority of need. This can be patterned after similar sorting and library organization which have proven successful—World Data Centers and the National Oceanographic Data Center (NODC), for example. NODC does not include bathymetric infor-

mation, partly because of shortages in personnel, space, and equipment, and in some measure to the belief of member organizations that other oceanographic activities should take precedence at this time.

OTHER DEVELOPMENTS

Navigational scale, control systems, positional requirements, and an ideal bathymetric navigational system that can be developed are discussed below.

Navigational Scale

Investigations are proceeding to determine the optimum chart for bathymetric navigation use. It is not difficult to forecast that an eventual product of this kind will be of larger scale than now available and will have a larger (closer) contour interval. No single scale need be decided upon covering all areas, and a 1:300,000-scale chart over certain topography might be appropriate, whereas a larger scale might be desirable over flatter areas. For ease of survey and chart construction, particular organizations may decide upon a single-scale series. An ideal scale might be supposed to be about 1:200,000, perhaps slightly larger. Scale is of less importance than utility, and this means accurate well-controlled portrayal of bathymetry.

Control Systems

Control systems will be developed which will allow accuracies of ±500 yards in all areas of the world. These may be satellites, assuming that a sufficient number can be orbited and data reduction equipment becomes less sophisticated and costly. No satellite system will replace electronic control completely. Loran-C/D will replace Loran-A in five to ten years and will be extended to cover all but the remotest areas of the earth. Fixed underwater beacons will be increasingly utilized, but will not become the general geodetic panacea once envisioned. The most urgent requirement they will meet will be for submerged vehicles. However, long-lived reliable beacons could become important aids to pinpoint positioning on continental shelves. Their use in conjunction with bathymetric methods could aid survey, farming, and mineral exploitation.

Positional Requirements

To reflect on requirements, the author contemplates that more precise positioning will be demanded in the future. However, the subject of requirements always promises to cause difficulty in achieving common viewpoints. It is not very difficult to achieve most of the positioning accuracies required by navigators today. More effort should be spent on forecasting the probable effect of future developments on the navigator's requirements many years hence. It is safe to predict such increased accuracies as being needed for many activities—farming, mineral exploitation, and so on. A real contribution to navigation would result if individuals could quantitatively determine positioning needs of the future. Systems subsequently developed could specify satisfaction of particular needs, rather than the needs accommodate themselves to the accuracies thereby available.

Ideal Bathymetric Navigational System

As one might presently envision an ideal navigational system, it would include console arrangements on or near the bridge to display the bathymetry of an area as the ship is in transit. One console would show gross bathymetry at a general navigation scale, indicating by means of a line or other CRT display mode the ship's track made good, present position, and DR advance. A large-scale console should reveal more detailed bathymetry, against the background of which the ship's position would be marked. Compact computers would correlate preprogrammed reference input with that observed in a continued series of updating

events. These would analyze, accept or reject, and supplement observed information to feed instructions to a counter-viewer arrangement which would indicate actual position by latitude and longitude in present time increments. Metric sounding data would be used.

The survey instrumentation could be a system of standard, array, directional, swath and (range and depth improvements permitting) side-scan and doppler mechanisms. The ship's course would be programmed earlier into a control computer, based upon topography, wind and sea conditions, currents, and late weather information. A decision to deviate from the planned transit would override the programmed courses, with means to return to, or determination of, a new transit plan.

What has just been described is merely one set of ideas and is not by any means the only or the best possible. This particular system is within current technical capability to develop. Costs aside, the decision to undertake such development is another matter, involving management considerations and prerogatives. With no little amount of temerity, future development of systems conceptually similar to that described is predicted.

MANAGEMENT OF BATHYMETRIC PROGRAMS

Manpower, administration, federal support, and the voice of the navigator are needed in the management of bathymetric programs.

Manpower

In the field of education and training, the urgent need for individuals trained in the ocean sciences has, of course, been recognized for some time. At present, there is also need for greater numbers of nonacademic persons—technicians—for engineering work primarily. Currently, programs in marine science are not increasing at a rate greater than can be accommodated with qualified manpower, and this situation may always be present to some extent. However, this is one problem which is receiving attention, with groups of competent individuals studying its solution.

Administration

In the administration of bathymetric collection programs, there is a danger in overspecialization. A nationwide shortage of administrative talent is conceded to be a continuing difficulty, and this is more notable in evolving disciplines. In government agencies, this shortage is acute at all levels, depending on individual organizational structures. Administration has recently been given renewed interest through the need for well-experienced personnel, whose awareness of issues cuts across technical specialties and agency lines. Many men in administrative positions today have been too long involved in specialized work, and as a result they have been given responsibilities for which their previous training has not equipped them. The administrator cannot remain a specialist if he is to be successful; rather he must broaden his abilities to consider his and other specialties as elements contributing to a whole discipline. Decision-making processes are sometimes the subject of derisive comment, with Department of Defense programs bearing the brunt of such comment. Yet proper decision-making is the keynote of the success or failure of many programs. Optimum development of programs in bathymetry will be facilitated if administrators have the perspective to make enlightened decisions.

Federal Support

Federal financial support available to the marine sciences in the broadest sense will increase in the future, but it would be a mistake with serious implications to assume that government funds are unlimited or, indeed, that the present rate of increased spending will be maintained. The need for long-range planning cannot be disparaged; yet, experience has shown the fallacy of emphasizing exotic goals which are countless in oceanographic research, at the expense of more realistic and immediate objectives. There will be ample federal spending, but no spectacular spending splurge such as seen elsewhere. This very limitation will weed out efforts that are uneconomic and inopportune and will strengthen those programs which are. Comparison of oceanography with the national space effort is sometimes overstated, but it is believed valid to state that such comparisons have shown that programs must be realistic, well-planned, capable of achievement at a pace not to be accelerated beyond a certain point without serious consequences, and subject to re-evaluation and reconsideration consistent with constant fluctuation in requirements, schedules, technology, and resource availability. These same requirements must also apply to oceanography.

The Navigator's Voice

The pace of advancement in all facets of oceanography is a rapid one. Information contained in this book may be outdated within the next few years. This is not a fault, but a situation that should be welcomed. Advances in electronics and optics alone must give material impetus to surveys, chart production, and navigational procedures which have heretofore been based on long-instituted concepts. Improvements benefiting the navigator will occur, whether he is directly involved in these developments or not and whether such improvements are considered specifically or as fallout from other primary efforts. To the extent that developments should or can proceed in response to recognizable needs, the navigator's requirements must be plainly stated and considered, and his voice heard early in the development cycle.

3

OCEAN CURRENTS,
CHEMISTRY AND WEATHER

Oceanic Surface Currents

JEROME WILLIAMS   JOHN J. HIGGINSON   JOHN D. ROHRBOUGH   1968

INTRODUCTION

Anyone having even an elementary acquaintance with the ocean knows of the great currents which exist at its surface. Some of these, such as the Gulf Stream in the North Atlantic and its counterpart in the North Pacific, the Kuroshio, have a profound effect on continental climate. This is usually considered to be beneficial since Great Britain and southwestern Alaska are both warmer than would be expected from latitudinal considerations alone. There are surface currents, however, which may produce harmful effects on the total environment.

One example of this is the Humboldt (Peru) Current which flows northward along the western coast of South America. Under normal circumstances there is a large amount of upwelling associated with the edge of this current providing nutrients for enough phytoplankton to support the largest anchovy population in the world. These fish make up one of the largest single industries in the Peruvian economy and account for the fact that Peru has led the world in fish catch for the last few years. Occasionally, however, the Humboldt Current changes its position enough to cause a cessation of the all important upwelling. Not only do the anchovies disappear, causing the loss of untold millions of dollars to the Peruvian economy, but large numbers of other marine organisms die and decay resulting in hydrogen sulfide being released into the atmosphere.

This change in position of the Humboldt Current is called *El Niño* (after the Christ child because of its usual occurrence during the Christmas season), or sometimes *Callao*

Jerome Williams is associate chairman of the Environmental Sciences Department at the United States Naval Academy. He has also been associated with the Chesapeake Bay Institute of Johns Hopkins University, Maryland. His major research interests are underwater optics, instrumentation, and physical oceanography.

Commander John J. Higginson has been an instructor of oceanography at the United States Naval Academy. He has also been associated with the Apollo Recovery Program as operations officer in Helicopter ASW Squadron Four. He is presently in receipt of orders to Helicopter ASW Squadron Two at Naval Air Station, Imperial Beach, California.

Commander John D. Rohrbough is with the United States Navy and is on the staff of the Studies, Analysis and Gaming Agency of the Office of the Joint Chiefs of Staff, the Pentagon, Washington. He has served as an instructor of oceanography and a member of the staff of the superintendent at the Naval Academy.

From *Sea and Air: The Naval Environment*, pp. 171-184, 1968. Reprinted with authors' revisions and light editing and by permission of the authors and the Naval Institute Press, Annapolis.

Painter, after the discoloring effects which are caused by the gases of decaying organisms.

Of course not all surface current systems are capable of producing a cataclysm of the order of magnitude of El Niño, but this does not make the knowledge of all surface currents any less important. Benjamin Franklin was aware of this when he produced the first chart of the Gulf Stream to aid American ships make faster crossings of the North Atlantic. But it remained for Matthew Fontaine Maury to initiate the scientific study of ocean currents about three-quarters of a century later.

Maury not only gathered together enough data to generate dependable charts, but he also tried to correlate these data in an attempt to ascertain the causes and variability of ocean currents. This work has continued to the present and is not complete, even today. However, enough information has become available to piece together a reasonably logical description of oceanic surface currents.

BASIC CAUSES

As in the case of atmospheric motion, one of the major causes of motion in the sea is uneven heating. However, the atmospheric flow pattern discussed previously is somewhat different than that in the ocean, because in addition to the direct effects of uneven heating, there are two other important factors which must be taken into account. These are (1) wind (itself produced by uneven heating) acting on the water surface and (2) the containment of the oceans within the boundaries set by land masses. Due to the interference of land masses, no currents run all the way around the world except in the Antarctic region.

In actuality there are two basic systems which must be superimposed, one upon the other. The first of these is the system produced directly by uneven heating wherein the waters at lower latitudes are heated, become less dense, and spread out over the

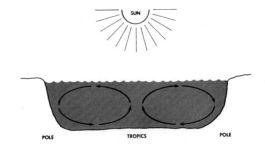

Figure 3-1. Idealized thermohaline flow in the ocean.

surface toward the poles (Figure 3-1). As they drift toward the poles these waters are cooled and finally sink. In this manner a giant convection cell is set up similar to the single cell atmospheric model, wherein surface water sinking at the poles flows toward the equator, rises in the equatorial regions, and flows away from the equator along the surface.

In addition to this basic flow poleward, the surface winds, combined with land mass placement, produce a different system. The resulting surface currents are a combination of these two flows. Since by far the greatest effect is due to winds, an attempt will be made to develop a model of ocean currents produced by wind forces and land placement alone. This will then be compared with what actually exists in nature.

AN OCEANIC CURRENT MODEL

As a start, the model assumes that the winds

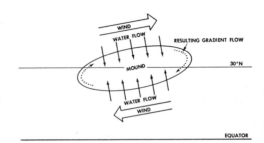

Figure 3-2. Production of an oceanic high pressure area at 30° by prevailing winds.

in existence are those in the three celled theory. This is a reasonable assumption since the tricellular model describes the *surface* winds quite well. As may be recalled, winds in the Northern Hemisphere are northeast in the latitude belt from 0° to 30°, southwest in the belt from 30° to 60°, and northeast again in the belt from 60° to the pole, with a mirror image of this system in the Southern Hemisphere.

When a wind blows there is a transport of the upper layers (about 100 meters thick) at 90° to the right of the wind in the Northern Hemisphere. The result is that with a wind blowing from the northeast, the oceanic surface layers will be caused to move toward the northwest, which is the case between the latitudes of 0° to 30° in the Northern Hemisphere.

Similarly, with a southwest wind the upper hundred meters or so of surface waters are transported to the right and a southeast* flow develops. The effect of these two currents is to pile up water within a region centered somewhere around 30° latitude, as seen in Figure 3-2.

This mound of water piled up by these two wind driven transports creates a high pressure ridge at about 30° latitude. The water, under the influence of this pressure distribution and coriolis force, will produce geostrophic flow toward the southwest between 0° and 30° and toward the northeast between 30° and 60°. Since there are land masses on each side of the ocean, the water must go somewhere. As it completes its path, it tends to produce a current gyre in a clockwise direction about this high pressure cell. Just as in the atmosphere, a clockwise rotation is found about a high.

A little farther north there is a northeast wind between the latitude of 60° and 90° which would cause the surface layer to move toward the northwest. Consequently at 60 degrees latitude water is directed toward the

*Keep in mind that winds are named by where they have been, while currents are described in terms of where they are going.

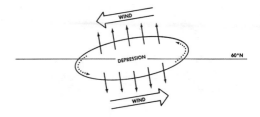

Figure 3-3. Production of an oceanic low pressure area at 60° by prevailing winds.

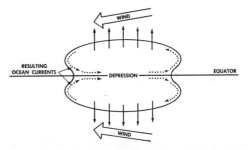

Figure 3-4. Production of an oceanic low pressure area at the equator by prevailing winds. (Note the two gyres produced from this single depression.)

southeast by lower latitude winds and at the same time toward the northwest by winds at higher latitudes (Figure 3-3). This results in a low pressure trough at a latitude of about 60 degrees. Considering both the placement of the continents on each side of the ocean and what has been learned about low pressure regions, the circulation around this low pressure area will result in a current having a counterclockwise direction.

In equatorial regions, the situation may be treated similarly. The winds north and south of the doldrums produce surface layer motion such that in both cases the water motion is away from the equator (Figure 3-4). In other words, close to the equator a low pressure system is developed in the hydrosphere. Once again, counterclockwise rotation would be expected around a low pressure system in the Northern Hemisphere, while clockwise flow would be expected in the Southern Hemisphere. Thus two gyres

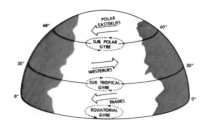

Figure 3-5. The complete model for the northern hemisphere wind driven currents.

result since flow direction about a low is opposite in the different hemispheres.

The model is now complete. Using the tricellular wind theory, it has been shown that the resulting currents in the Northern Hemisphere should consist of a counter-clockwise gyre close to the equator (*equatorial gyre*), a clockwise gyre north of that (*sub-tropical gyre*), and a counterclockwise one north of that (*sub-polar gyre*) (Figure 3-5).

Similarly, about the same condition exists in the Southern Hemisphere, except that the gyres rotate oppositely, producing a mirror image. The *equatorial gyre* is clockwise in its rotation, the *sub-tropical gyre* counter-clockwise, and the *sub-polar gyre* clockwise, due to the opposite direction of the coriolis force in the southern hemisphere.

THE MODEL VS. THE TRUE PICTURE

It is now appropriate to compare a simple model with the actual currents existing in the world's oceans. In all of the world's oceans there is a *sub-tropical gyre;* both north and south of the equator this portion of the model appears to hold fairly well. In the Pacific Ocean the *north equatorial gyre* is well established, composed of an equatorial current and an oppositely moving equatorial counter current north of the equator. There is also a *south equatorial gyre,* displaced somewhat north of the geographical equator. This is not surprising from the position of the *intertropical convergence zone* and the *oceanographic thermal*

equator, both of which are displaced north of the geographical equator.

In the northern oceans the sub-tropical gyre composed of the *Kuroshio system* in the Pacific and the *Gulf Stream system* in the Atlantic is also very well developed. In addition, the *Irminger current,* an offshoot of the Gulf Stream system,* combines with the *East Greenland current* to produce the sub-polar gyre. However, the sub-polar gyre in the Pacific is not so well developed, although the Alaska current tends to produce a flow of this type.

In both the North and South Pacific there is some evidence of the existence of the sub-polar gyre. In the South Pacific, a polar current running from east to west close to the Antarctic Continent and the West Wind Drift (Antarctic Circumpolar Current) somewhat north of this in the opposite direction, make up the larger portion of the sub-polar gyre. The southern South Atlantic also exhibits very similar properties so that it appears the model fits the southern oceans quite well.

From an unrefined point of view this crude model fits the real ocean quite well, much better than expected from the simplicity of the initial assumptions. It appears that some of the differences between the model and actual current patterns may be explained on the basis of well developed sub-surface currents. Two examples of these are the Pacific Undercurrent (Cromwell current) and the Atlantic Undercurrent, both of which flow from east to west within $1°$ of the equator. These are both well-developed currents involving transports on the order of thirty million cubic meters per second between 100 and 300 meters below the surface.

In addition the thermohaline effects have not been considered. Changes in density of surface waters produced by warming and evaporation at the lower latitudes cause a

---

* The *Gulf Stream* system is composed of the *Florida Current, Gulf Stream,* and *North Atlantic current.* See Figure 3-6.

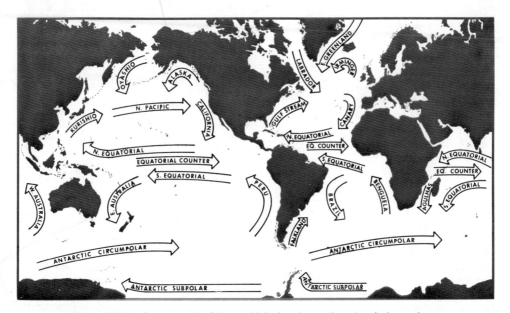

Figure 3-6. Surface currents of the world during the northern hemisphere winter.

general poleward drift at the surface. This would have the effect of strengthening such currents as the Gulf Stream while weakening those tending to oppose the drift such as the Canary current in the North Atlantic. There are, of course, a number of discrepancies in the simple model. One of these is the region of the equatorial Atlantic. Here is found a large transfer of water from the South Atlantic Ocean to the North Atlantic Ocean without the separation of the equatorial gyres that appear in the Pacific. One possible explanation for this breakdown of the equatorial gyres in the Atlantic Ocean is the closeness of the African and South American land masses. Perhaps there just is not enough expanse of water to allow the gyres to develop.

Another unusual aspect of the current systems is found in the Indian Ocean. The Indian Ocean is affected by the winds resulting from the atmospheric pressure systems present over the large Eurasian continent. These monsoon winds seasonally change direction, as do the currents associated with them. Consequently, since the Indian Ocean

current systems are very deeply influenced by the winds, the currents north of the equator will be toward the east in the summertime and toward the west during the wintertime (Figure 3-6).

SOME REPRESENTATIVE NUMBERS

It might be interesting at this point to reflect on the magnitude of some of these current systems. The *Gulf Stream* is probably the most famous of all world surface currents having speeds varying from about half a knot to in excess of three knots. The amount of water transported is somewhere around 113 sverdrups* (about 30 billion gallons per second), which is more than 65 times the amount of water moved by all the rivers of the world combined. Of course all ocean currents are not of this magnitude, but even the smaller ocean currents are involved with water transports many times larger than most rivers.

* A sverdrup (sv) is defined as a transport of one million cubic meters per second.

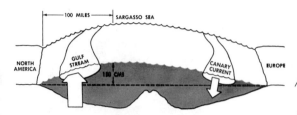

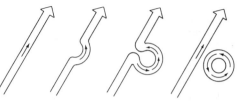

Figure 3-8. Four stages in the formation of an eddy.

Figure 3-7. The mound in the North Atlantic Ocean associated with the Gulf Stream system.

Because currents necessarily involve motion, they have the associated surface slopes. This is certainly true for the permanent currents. In the Gulf Stream system for example, the Sargasso Sea, which is the high pressure center of the sub-tropical gyre, is about 150 centimeters higher than the outside edge of the Gulf Stream itself (Figure 3-7). In other words, there is a mound of water in the center of the Atlantic Ocean corresponding to the sub-tropical gyre as there is in the center of all the high pressure gyres in the world's oceans. Similarly, there are depressions in the ocean surface on the order of magnitude of 50 centimeters, corresponding to the centers of the sub-polar and equatorial gyres which are both low pressure systems.

MATTHEW FONTAINE MAURY

As indicated previously, the first man to use large amounts of ocean data in a systematic study of surface currents, from 1841 to 1853, was Matthew Fontaine Maury, a lieutenant in the U. S. Navy. Using the data accumulated from thousands of old log books, he published the first pilot charts and sailing directions for all the world oceans. As a matter of fact, pilot charts obtained today will bear the inscription, "Founded upon researches made and data collected by Lieutenant M. F. Maury, U. S. Navy."

In addition he laid the foundation for the establishment of the U. S. Weather Bureau, did most of the work in determining the location for the first transatlantic cable, and was instrumental in the establishment of the

U. S. Naval Academy. He is also said to have urged the teaching of oceanography at the new institution, a piece of advice which was finally followed over one hundred years later.

APPLICATIONS

Ocean currents have been discussed as if they were indeed "rivers in the ocean," as Maury described them over 130 years ago. In point of fact, they may be so conceived, but if so, the bed of the river must be considered to change quite rapidly. In the Gulf Stream, for example, the path of the stream varies quite markedly from week to week. Figure 3-8 pictures schematically the Gulf Stream on four different occasions. Note that loops form in the Gulf Stream which break off after a period of time and become eddies having associated currents which may move in a direction opposite to the stream itself.

Of course it is desirable to be able to predict the formation of these eddies, since most well developed current systems appear to have eddies associated with them. However, at this time it is not possible to do this with the desired accuracy; about all that can be done in describing ocean currents is to indicate the average magnitude and direction of the motion at a particular location.

When one refers to a current atlas to determine average currents typically the information is presented in the form of a *current rose* (Figure 3-9). Probabilities of current directions are shown by indicating what percentage of the time currents have been reported in what direction, and what speed

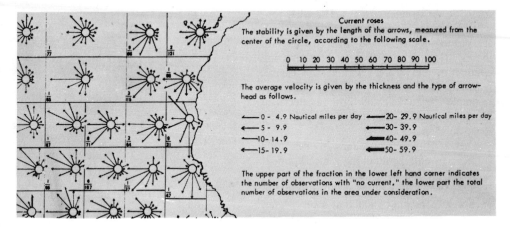

Figure 3-9. Typical current roses.

they had at the time. This allows the mariner to make a good estimate of the way the water will be moving. However, it is important to realize the surface current speed and direction cannot be predicted with absolute certainty.

Aside from the obvious effects of set and drift* on surface ships and other floating material, ocean surface currents occasionally have a fairly large effect on climate. In the main, most of the heat which is exchanged from the lower latitudes poleward is done by means of the moving atmosphere. The major exceptions to this rule are the well developed current systems such as the *Gulf Stream* in the Atlantic and the *Kuroshio*† in the Pacific. These currents involve great volumes of water capable of carrying large amounts of heat. If these current systems pass close to land areas, and if the prevailing winds are such that warmed air is carried over these land areas, the current systems will have an effect on the climate of the adjacent land areas. An example of this is the North Atlantic Current portion of the Gulf Stream system in its passage close to

the European continent. The fact that the Gulf Stream is warm and is moving into a relatively cold area, coupled with prevailing westerly winds, makes for a somewhat warmer climate in the British Isles and western Europe than normally would be expected for this latitude. This same effect occurs on the southern coast of Alaska where the effect of the Kuroshio extension (North Pacific Current) is such that this coast has a somewhat more temperate climate than would be expected.

CURRENT MEASUREMENTS

Surface currents may be measured in many different ways and they have been measured for many years. Probably the easiest and most obvious way of measuring a surface current is to put a floating object in the water and observe how far and how fast it drifts. This may be a bottle, some sort of a floating drogue, a specially designed float with a radio transmitter or radar reflector for easy tracking, or even a ship itself. In actuality most current measurements which appear on pilot charts are the result of many measured ship drifts from calculated courses. If somewhat more accurate measurements are desired, various devices may be used. One of the most esoteric is the GEK (Geo-

---

* In navigational parlance set and drift are the direction and magnitude respectively of the current velocity vector.
† So called because the *water is very clear;* Kuroshio means *Black Current* in Japanese.

magnetic electrokinetograph). The GEK consists essentially of two large electrodes which are placed in the surface water to measure the electric potential developed by a moving conductor (sea water) within the earth's magnetic field. This is basically the same principle by which a common electrical generator works, but the output is very much smaller.

One of the big problems in measuring surface currents is obtaining a measurement of water motion with respect to the earth's surface. In the deep ocean it is impossible to anchor in a manner such that a vessel does not drift, so that surface currents are not accurately measured in the deep ocean, due to lack of positioning accuracy. With the advent of better navigational systems, current measurements at sea will become more feasible.

However, if there is some method of fixing a current meter's position with respect to the earth, or if the drift of the device is known, rather conventional units may be used, the most common of which utilize some sort of a rotating vane. This may be a propeller, or a hemispherical cup as is used in Robinson's anemometer, or some other design of rotor, the speed of rotation being related to the current speed.

In recent years, instruments have been developed which measure rapidly fluctuating currents. This had not been possible in the past. With a rotating vane current meter it is difficult to measure a current which changes its magnitude or direction rapidly with time. The newer devices utilize the speed of sound in two directions to determine currents; these not only take a very small period of time to make a measurement but also have no mechanical inertia. Sound-speed is measured in one direction and compared with the sound-speed measured in the opposite direction, the difference between the two being the speed of water movement.*

Most of the devices discussed here have

* Sound energy is carried along with a moving medium.

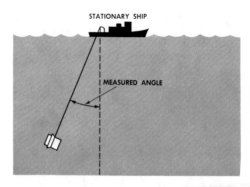

Figure 3-10. Using a current cross for current measurements.

been used with greater success in either shallow water or close to the bottom to measure bottom currents. There have been very few measurements made at sea for which great accuracies were claimed. However, a number of measurements have been made using very simple gear which have indicated the presence of currents where none had been measured before. For instance, a current cross may be lowered to the depth of interest, the angle which the line makes with the vertical is a function of the current speed (Figure 3-10). This was the case in the equatorial Pacific, for example, where the sub-surface Cromwell current was first detected in 1954 by the use of this type of current meter.

Another type of device which has been used in recent years for measuring subsurface currents is the *Swallow float*. This is a long cylinder designed to be buoyant at a particular density level, so that when it is released it will sink and remain at a particular depth. The float will then drift with the current at this level, and it may be tracked by means of acoustic gear. This has been quite successful and was utilized in affirming the previously predicted presence of a countercurrent underneath the Gulf Stream system.

Even though knowledge of currents at the present time is not complete, it is still sufficient for improving sailing times across the oceans. However for navigational purposes, especially in certain areas, it is many times woefully inadequate. A basic knowledge of surface currents is especially important for such obvious problems as determination of: personnel lifeboat tracks, paths of manmade pollutants, and the movements of plankton populations with their associated larger marine animals.

SUGGESTED READINGS

Chapin, H., and Smith, F. G. W., *The Ocean River,* Charles Scribner's Sons, 1952.

Cotter, C. H., *The Physical Geography of the Oceans,* American Elsevier Publishing Company, Inc., 1965.

Defant, A., *Physical Oceanography,* Vol. I, Pergamon Press, 1961.

Dietrich, G., *General Oceanography,* John Wiley & Sons, 1963.

Munk, Walter, "Ocean Currents," *Scientific American,* September 1955.

Neumann, G. and Pierson, W. J., *Principles of Physical Oceanography,* Prentice Hall, Inc., 1966.

Pickard, G. L., *Descriptive Physical Oceanography,* Pergamon Press, 1964.

The Kuroshio Current

RICHARD A. BARKLEY   1970

Every second the Kuroshio current carries some 50 million tons of sea water past Japan's southeast coast—a flow equal in volume to about 6000 rivers the size of the Danube or Volga. But even this massive current would take some 250 years to equal the total volume of the north Pacific. Thus, although the Kuroshio is one of the major currents in the world's oceans, and plays a vital role in the circulation of the north Pacific, it occupies only a small fraction (less than 0.1 per cent) of that ocean: a thin narrow band less than 100 km in width and about 1 km at maximum depth running for 3000 km along the western edge of the Pacific between the Philippines and the east coast of Japan (see Figure 3-11).

Asia's seamen have known the Kuroshio since ancient times. They named it Kuroshio (which means 'black stream' in the Japanese language) because of the deep ultramarine color of the warm, high salinity water which is found flowing north on the right hand side (looking downstream) of the current's axis. The heat which is carried north by this flow influences the weather throughout the northern hemisphere. The Kuroshio therefore plays an indirect but important part in the everyday life of the fishermen and farmers of eastern Asia and also, to a lesser extent, in the lives of most of the rest of mankind as well.

The first European chart to show the Kuroshio was Varenius' "Geographia Generalis" of 1650. Later, expeditions headed by Captains James Cook (1776-80) and Krusenstern (1804) added to western knowledge about the Kuroshio. Although Japanese scientists began to study the biology of the Kuroshio in 1880, it was not until 1893, when Wada started a series of drift bottle experiments, that systematic examination of its currents first began.

Today, in an attempt to learn more about the Kuroshio, scientists from China, Indonesia, Japan, Korea, Philippines, Singapore, Thailand, Hong Kong, United States, the Soviet Union and Vietnam are engaged in a major international project called the Cooperative Study of the Kuroshio (CSK). By

Dr. Richard A. Barkley is a physical oceanographer with the National Marine Fisheries Service's Southwest Fisheries Center, Honolulu Laboratory, Hawaii. He has compiled an oceanographic atlas of the Pacific Ocean, worked on the hydrodynamics of the Kuroshio-Oyashio front east of Japan, and is currently investigating boundary-layer phenomena associated with oceanic islands.

From *Science Journal*, Vol. 6, No. 3, pp. 54-60, 1970. Reprinted with author's revisions and light editing and by permission of the author and Syndication International Limited, London.

keeping a figurative finger on this pulse of the north Pacific, they hope to learn more about western boundary currents in general, the Kuroshio in particular, and the whole north Pacific Ocean as well. In the process, they should also learn more about the way in which both weather and climate respond to changes in conditions at sea, and the ways in which marine animals and plants are affected by their environment.

The influence exerted on ocean currents by the Earth's rotation was not generally appreciated until 1835, when G. de Coriolis, while studying equations of motion in a rotating frame of reference, discovered what is now called coriolis force. Coriolis showed how the effects of the Earth's rotation could be incorporated into the Newtonian equations of motion by adding two additional terms. One, the centrifugal force of the Earth's rotation, is usually absorbed into a redefined term for "gravitation," which includes gravity and centrifugal forces together and acts along a vertical defined by the direction of a plumb bob. By this definition there is no horizontal component of either force. The other term, the coriolis force, makes allowance for the effects of conservation of angular momentum on a particle which moves relative to the Earth's surface. Coriolis force acts at right angles to the Earth's axis of rotation, and thus has no effect on the energy of motion but only modifies its direction.

The vertical component of coriolis force is so small when compared to the force of gravitation that it can always be neglected in ocean current theory. In the sea, however, forces which act along the horizontal are generally weak. Indeed, frictional and inertial forces are far smaller than those we are accustomed to on land—by at least five orders of magnitude—and they have little or no effect on any but the fastest moving ocean currents. The forces exerted by wind stress are often even weaker. Thus the horizontal component of the weak coriolis force becomes an important factor when considering ocean currents.

In most ocean currents the horizontal pressure gradient is found to be in balance with coriolis force. Such currents are called geostrophic or "Earth-balanced" currents. To understand how this balance is achieved, consider a parcel of water in a region where sea level has been raised by a meter or so by convergent wind stress. This parcel of water tends to flow back toward a region of lower pressure. However, as soon as it begins to move at appreciable speeds, this motion generates coriolis force and the water is turned toward the right (in the northern hemisphere). Once it has turned 90° it cannot turn farther without flowing "uphill" and losing momentum, thus weakening the coriolis force. If it then turns slightly toward the left in response to pressure, it gains momentum, generating additional coriolis force and thus being forced to the right once again. In this way a balance is reached with horizontal pressure forces equal and opposite to horizontal coriolis forces, and the water flowing at right angles to both moving endlessly around centers of high (or low) pressure. At low latitudes the horizontal component of coriolis force is weaker than it is at high latitudes, and a relatively higher velocity is required to generate enough coriolis force to offset a given pressure gradient.

Since pressure gradients built up by wind stress are not relieved by this geostrophic flow, energy accumulates until the pressure gradient is large enough, and geostrophic flow fast enough, to generate appreciable friction in some parts of the ocean. Offshore this frictional force disrupts the geostrophic balance slightly, allowing some water to flow out of the high pressure cells back to the "lows." Along coasts friction dissipates energy by converting it to heat. Both processes counteract the effects of wind stress at the sea surface. Because of this near-equilibrium state, horizontal pressure gradients, coriolis force and geostrophic flow are closely linked; knowledge of one of these variables makes it possible to calculate the other two wherever friction can be neglected.

Thus it is possible to obtain a useful approximation (to within 15 per cent) to actual currents by making measurements of horizontal pressure gradients. To do this, a series of temperature and salinity measurements is made to depths of 1 km or more at several locations (called "stations") across a current. These observations define the field of density (which depends on temperature, salinity and pressure, or depth), from which

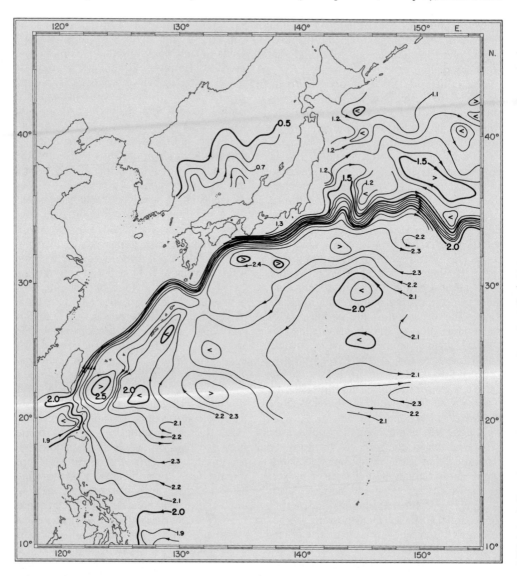

Figure 3-11. Path of the Kuroshio is shown by closely spaced contours on this chart of geostrophic flow at the sea surface. Contours show the elevation of the sea surface (pressure head in meters) or the potential energy (in dynamic meters) relative to an arbitrary horizontal reference plane. The region of narrow flow east of Japan is the Kuroshio Extension, which differs from the Kuroshio in that it has no land boundary to determine its path and absorb frictional stresses.

horizontal pressure gradients can be computed. A chart showing this field of pressure gradients is equivalent to a chart showing streamlines of geostrophic flow. This procedure is referred to as the dynamic method, and was first used by J. W. Sandstrom and B. Helland-Hansen in 1903. Unlike direct current measurements, which are more expensive and time consuming to make and more difficult to interpret, the dynamic method is little affected by transient motion due to surface waves, tides and winds, which are often much stronger than the movements of water in the permanent current system.

The theory of wind-driven ocean currents concerns itself primarily with the forces which generate, control and finally dissipate energy in the sea. Most of this energy appears in two forms: kinetic energy of motion—the currents *per se*—and potential energy of position—the "pressure head" due to slopes in the sea's surface and the internal density structure. There are also others, primarily thermal energy, which affect the distribution of density and thereby influence currents.

Those forces which carried Kuroshio water through nets set by Japanese fishermen this morning were generated weeks, months, or even years earlier by winds blowing over the entire north Pacific Ocean. Even a small gust of wind rippling the quiet waters off Baja California, for example, contributes to the oceanwide accumulation of energy which drives the inexorable flow of millions of tons of water past Asia's coasts. A major problem in ocean theory is to determine the way in which energy is transferred from the winds to the shallow wind-driven upper layers of the ocean; from there, into the geostrophic currents; and, finally, into the narrow shear zone on the left hand side of the Kuroshio, where energy is dissipated.

When V. W. Ekman provided, in 1905, a theoretical explanation for Nansen's observation that Arctic icebergs tend to drift to the right of the wind, he laid a major cornerstone for all subsequent studies of wind-driven flow. Ekman showed that a steady wind produces transport of water at right angles to the wind's direction (toward the right in the northern hemisphere, and to the left in the southern hemisphere). This movement of water is usually called Ekman transport. A balance is reached between coriolis force and wind stress at the sea surface, which is a precise analogy to the balance existing between coriolis force and the horizontal pressure gradient in geostrophic flow.

In 1947, H. U. Sverdrup, the Norwegian oceanographer and meteorologist, used Ekman's concepts to calculate the wind-driven transport in equatorial currents of the Pacific Ocean. The following year, H. Stommel showed that changes in the coriolis effect with latitude (the horizontal component of coriolis force varies as the sine of the latitude, so that it reaches a maximum value at the pole, and vanishes at the equator) were responsible for the narrow swift currents along the western boundaries.

A major advance in the theory of wind-driven circulation was made by W. H. Munk in his pioneer study of oceanwide transports. He pointed out the fundamental importance of wind shear or torque, which transmits angular momentum to the sea surface and thus generates the major circulation systems. Munk's study was based on a linear steady state mathematical model in which friction played a large part. K. Hidaka has explored such theoretical models extensively since 1949.

In more recent years the importance of large scale friction in currents such as the Kuroshio and Gulf Stream has been questioned by a number of theoreticians, who regard it as characteristic of the climatic-mean flow, but not of the instantaneous current. That is, meanders and eddies formed from time to time by the current could be considered as a form of turbulent friction of the magnitude required in Munk's and Hidaka's theories, if their effects were averaged over large areas over long periods of time. On a shorter time scale, and over smaller distances, such meanders and eddies must be treated individually, which requires the use of nonlinear time dependent mathematical models. Such models present formidable mathematical difficulties which in some cases can only be overcome by the use of numerical computer methods.

Prevailing surface winds over the north Pacific in summer describe a clockwise circulation around the mid-latitude high pressure cell, from which the air spirals outward, away from the high, towards the low pressure regions over Asia and Alaska, and towards the climatic Equator, the doldrums, near 5°N. Initially, this is cool, dry air which descends towards the sea surface and speeds up, removing both heat and moisture from the surface water layers as it goes. It then slows down as it reaches the convergent lows, where the air rises and much of the moisture returns to the sea as rain.

On streamline charts (such as Figure 3-12), constant amounts of water flow between any two streamlines (within the accuracy of the methods used), so velocity of flow is inversely proportional to the distance

between adjacent streamlines. In addition, such charts show convergences and divergences, flow of water downward out of the layer in question, or upward from below, by showing streamlines terminating (convergence) or beginning (divergence) at a coast or in midocean. In the following discussion it will be convenient to use the term "Sverdrup unit" to indicate a flow of one million cubic meters per second.

According to the map (Figure 3-12) Ekman transport removes water from the equatorial currents and the northern gyre and transports it into the central gyre. Some 45 streamlines enter the gyre at its perimeter (from the coasts, between islands and across 10° and 50°N). What happens to this massive flow? Some of it—about 4 Sverdrup units—evaporates into the cool, dry air flowing out of the atmospheric high. But Ekman transport more than makes up for this loss; the excess "piles up" in the Subtropical Convergence and sinks to depths of 100 to 200 meters. From there, much of this water

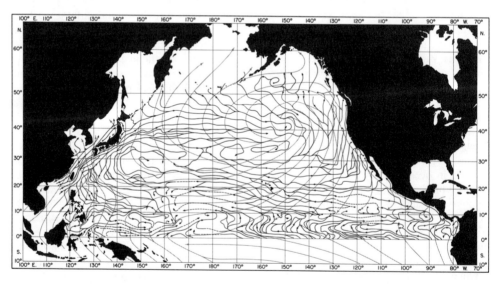

Figure 3-12. Surface current streamlines (black) and prevailing winds (heavy black) in the North Pacific during July. Off the coast of California, and the northeast coasts of China and Japan, winds blowing along the coast induce offshore Ekman transport of water. The opposite situation, where Ekman transport carried surface water toward the coast, can be seen on the northeast side of the Gulf of Alaska, and near the Gulf of Panama. Surface currents are determined primarily by the combined effects of wind-driven Ekman currents and geostrophic flow.

spreads radially outward but some of it is mixed downward, warming the underlying water. Over long periods of time, a "lens" of lower density water about one kilometer thick has accumulated in the center of the gyre. The low density layer is only one-quarter to one-tenth as thick at the perimeters of the gyre, and cold dense, deep water is found much closer to the surface there.

Density gradients due to convergence and divergence of the Ekman transport cause horizontal pressure gradients which generate the geostrophic flow shown on Figure 3-13. Such gradients are largest in the upper 100 to 200 meters and diminish to negligible values at 1 to 2 km depth, due to compensating displacements of denser, deep water.

The effectiveness of coriolis force in limiting the flow of water out of the thick, warm water lens at the convergence can be understood by comparing currents and horizontal pressure gradients in the central gyre with those at the Equator, where coriolis force vanishes. At the Equator a pressure gradient only one per cent as strong as that across the Kuroshio produces comparable current velocities. Why? Because flow at the Equator is not geostrophic and is therefore free to speed up until friction and inertia forces are large enough to balance the pressure gradient. At higher latitudes coriolis force allows much larger horizontal pressure gradients to build up before the flow reaches speeds where friction and inertia forces become important limiting factors.

The potential energy stored in the mid-latitude gyre is many hundreds of times greater than the kinetic energy of its currents, and represents an accumulation of energy five to ten times larger than the energy added by winds in a year. It is not surprising, then, that the ocean's density structure is remarkably constant, and that currents such as the Kuroshio hardly respond to changes in the local winds. Currents near the Equator, on the other hand, represent much smaller accumulations of energy and respond much more readily to changes in the wind.

After decades of effort by Japanese oceanographers, supplemented by studies made during international expeditions and by individuals from many countries, it is now possible to piece together a reasonably satisfactory description of the Kuroshio proper. Water enters the Kuroshio over a broad front 1000 km in width (13°N to about 23°N at longitude 125°E) then accelerates and narrows. Some water leaves the right hand side of the Kuroshio as soon as it begins to turn towards the east, but narrow, intense flow persists for 1500 to 2000 km after the current leaves Japan's east coast, after which there is a marked drop in velocity. This region of narrow intense flow east of Japan is called the Kuroshio extension, and it differs from the Kuroshio in that there is no land boundary on the left hand side to generate a frictional boundary layer.

Comparison of velocity profiles (plots of velocity across the surface current) of the Kuroshio and the Kuroshio extension shows that, although both have essentially identical velocity profiles on their right hand sides, the velocity gradient on the left hand side of the Kuroshio is at least six times greater (a change of 2m/s in 8 km) than that in the Kuroshio extension (2m/s in 50 km). Other things being equal, this would result in six times greater frictional stress on the left side of the Kuroshio. Such velocity profiles support the theoretical view that the Kuroshio and Kuroshio extension are the major non-geostrophic portions of the flow in the central and northern gyres, where important adjustments in the distribution of energy take place.

In the Kuroshio energy is dissipated through friction (on a small scale, since the frictional boundary layer is only about ten kilometers wide). On average, the dissipation rate must be in balance with the mean rate at which wind adds energy to the central gyre.

However, friction not only dissipates energy, it also generates counterclockwise angular momentum at a rate which more than compensates for the decrease in the

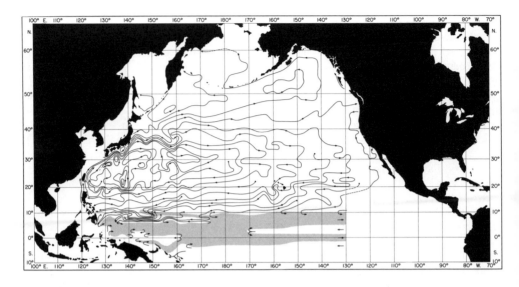

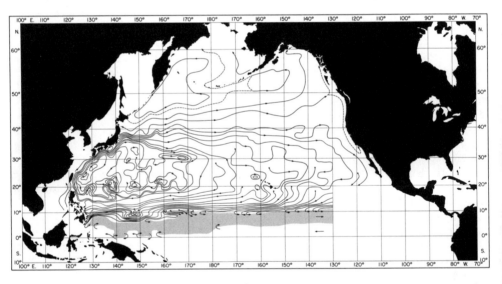

Figure 3-13. Volume transport by geostrophic currents in the North Pacific during the third quarter of the year, integrated from the sea surface down to 100 m depth (upper chart) and 1,000 m depth (lower chart). Each line represents 1 Sverdrup unit of flow on the upper chart, and 5 Sverdrup units on the lower chart. Both charts show the narrow intense flow along the western boundary, and the adjustment downstream to slower broad patterns of flow in mid-ocean. North of latitude 30°N there is little difference between the shallow and deep flows, but farther south the flow is more concentrated in the warmer upper layers, which are strongly influenced by surface winds, particularly the trade winds near 10° to 15°N. Much of the flow generated by trade winds in both hemispheres is relieved by eastward flowing currents in the countercurrent (5° to 10°N), which is caused by slack winds in the Doldrums, and by the Equatorial Undercurrent, which in the central Pacific flows along the equator at depths of 100 m or more. Note the tendency toward faster flow at lower latitudes, associated with decreased coriolis forces. The effects of Ekman transport on the surface currents can be seen by comparing these charts with the chart showing winds and surface currents.

Earth's (counterclockwise) angular momentum as the current flows north. To balance this excess, clockwise angular momentum is generated on the right hand side of the current, where the velocity decreases towards the right in what is sometimes termed an inertial boundary layer. As H. Stommel pointed out in 1948, this balance of angular momentum can only be attained on the western boundary, which accounts for the westward intensification in the current systems of the world oceans.

In the Kuroshio extension the flow adjusts to conditions in the ocean's interior, where large velocity gradients, and the concentrated angular momentum associated with such gradients, cannot persist. Friction is no longer concentrated at the boundary, nor are inertial forces restricted to the right hand side of the stream. Instead, large eddies and meanders dissipate kinetic energy throughout the path of the flow, and redistribute angular momentum at the same time. The transport decreases downstream as the flow fans out and becomes the broad, slow West Wind Drift between 155° and 160°E.

Fifty Sverdrup units of flow approach the east coast of Mindanao in the Philippines and half of this volume turns north into the Kuroshio. Most of this flow consists of warm (20°-28°C) water in the upper 200 meters; there is a layer with relatively high salinity (near 35 parts per thousand) between 100 and 200 meters with slightly more dilute water (34.5 parts per thousand) above that.

The depth of the high velocity flow increases from 200 to 400 meters, and the velocity goes up from a few tens of centimeters per second to one meter per second as the current narrows east of Taiwan. From Taiwan to Japan some dilute water from the Asian shelf, on the left of the current, is entrained by the flow, which speeds up even more, to velocities of 1.5 to 3m/s. Transport just off Japan's south coast amounts to 35 Sverdrup units, but there may be significant flow below 1000 meters since the current extends to considerable depths at that point,

and so the total transport may be as high as 45 to 50 Sverdrup units.

Just as the current reaches Japan's southeast tip, it flows over the shallow Izu-Bonin Ridge which extends due south from Honshu, Japan's main island. The Kuroshio undergoes complex and little understood fluctuations near this ridge.

Once past the Izu-Bonin Ridge, the Kuroshio may turn north along Japan's east coast for a short distance, or it may continue to flow almost due east. In either case, it joins the Oyashio current, which flows southward from the Kamchatka Peninsula. Together these two currents leave the coast and form the Kuroshio extension. Transports here amount to about 45 Sverdrup units (up to 60 on occasion) though only 25 to 35 Sverdrup units are within the high velocity core of the Kuroshio extension, where speeds of two meters per second or more are often observed.

By the time the current has reached 160°E, towards the end of the Kuroshio extension, it consists of a mosaic of water types: warm saline water from the original source off the Philippines, coastal and shelf water of lower salinity, and cold dilute water from the Oyashio, with summer temperatures of 3° to 10° C, and salinities sometimes as low as 33.8 parts per thousand.

Some mixing occurs in the core of the Kuroshio extension, but on the whole there are two distinct types of water in the current: warm saline water on the right, and cold dilute water on the left. The convoluted front separating these two types of water is often very sharp and active, with strong velocity gradients and contrasts in the properties of the water on the two sides. Rich fishing grounds are located on both sides of the frontal zone in the Kuroshio extension, so this complex feature of the western north Pacific is of particular interest to Japan's fishing fleet. A pioneer in fisheries oceanography, Professor Michitaka Uda, has studied such fronts and the fisheries associated with them since 1930. He and his colleagues in

Japan's unique system of fisheries universities have contributed much to our knowledge of the sea as an environment. They have taken particular interest in variability in the occurrence of various kinds of fish, and changes in the environment which cause much of this variability.

The path of the Kuroshio as shown in Figure 3-11 is very nearly its average position, but the current can undergo marked and fairly rapid changes in speed and in the location of its axis. Apart from changes due to tides, short term changes due to major shifts in the axis of the Kuroshio can occur as it flows past southern Japan. Meanders develop in the current which occasionally bring high velocity flow unusually close to the coastline, and part of the Kuroshio's flow may be diverted into nearby bays, where it can flush out much of the coastal water within a matter of days. These sudden and as yet unpredictable events cause widespread damage to boats and fishing gear anchored in the normally quiet waters, as offshore currents move inshore at speeds of 1-2m/s, or more.

Meanders in the Kuroshio south of Japan have been studied intensively. The axis of the current may shift onshore or offshore 100 km or more in a matter of weeks. For example, in 1959 the Kuroshio off Japan's south coast (at 133°E) began to move offshore in March or April, shifting from its initial position 20 km offshore to a distance of 140 km in about one month. This meander also drifted rather slowly downstream, reaching the central portion of the south coast (136°E) by the end of May and the eastern portion (139°E) by early August, at which time the Kuroshio's axis was back within 25 km of the coast farther upstream (133°E), where the disturbance was first observed. Such meanders may appear and disappear within a few months, but they may also remain more or less stationary for more than a decade. When the meander develops, cold water is brought up toward the surface between the Kuroshio and the coast and temperatures drop to as much as 10°C below

normal. This change has profound effects on coastal and offshore fisheries, since the area involved is fairly large—about the size of the Bay of Biscay. Familiar species of fish move away and are replaced by others, and so fishermen must either move to new grounds or market what they can catch on the old grounds.

The only nontidal changes in the Kuroshio which appear to be at all regular are annual changes in velocity and transport, which are easily obscured by the irregular variations discussed above. Japanese scientists generally agree that the speed is greatest from May until August, with a second maximum in January and February. But an analysis by Y. V. Pavlova has shown that the annual cycle is rather more complex, at least for geostrophic currents. Speed and transport not only vary with season, they vary in different ways from place to place along the Kuroshio. Off the southern tip of Japan, for example, maximum transport within the Kuroshio occurs in September and again in March or April, while maximum velocities are observed in July and January. Just east of Japan, according to Pavlova, maxima in transport occur during June and December, with velocity maxima in August and February.

What now remains to be learned about currents such as the Kuroshio? Perhaps most valuable would be information on fine structure and on fluctuations over periods of a month or less. To determine how fluctuations propagate from place to place, measurements must be made simultaneously at several points along the current, and at least a few direct measurements must be made of the currents at all depths. Many developments in ocean current theory await the results of such observations, which will also be needed for forecasts of conditions in and near the Kuroshio to serve the needs of fishermen and meteorologists. But the effort required, for a complete survey of the Kuroshio, in terms of ships, time and operating costs, is staggering and detailed rapid surveys of the Kuroshio must wait until more efficient tools, such as instrumented buoys, become commonly available.

Once routine monitoring of the marine environment becomes commonplace, we can expect marked improvements in long range weather forecasts and in catch rates of commercial fisheries, which will rapidly repay the original investment of effort and funds. Even rather minor improvements in weather forecasts can bring significant savings to farmers, public utilities, cargo ships, airlines and others who use forecasts in scheduling operations or planning routes.

Farmers in northern Japan may have good or poor harvests depending on the extent to which the Kuroshio flows north along Japan's east coast, before joining the cold Oyashio water, since water temperatures offshore strongly influence cloud cover and rainfall. Similarly, cold air from Siberia flows out over the Pacific in winter, to encounter warm water carried north by the Kuroshio; these temperature contrasts trigger formation of numerous cyclonic lows in the atmosphere over the Kuroshio. The lows carry stormy weather east to northeast across the northern Pacific Ocean towards the coasts of Alaska, Canada and the United States.

Improved knowledge of the ocean as an environment can help fishermen locate and catch protein-rich fish to feed an increasingly hungry world. Fishermen could make direct use of forecasts of the Kuroshio's flow, because the entire current system is a series of fishing grounds which move about with changes in the flow. Various species of tuna, sardine and anchovy, mackerel, squid and many other commercially important species are each found in specific zones in and near the Kuroshio. For example, fronts where coastal and offshore waters meet are often good fishing grounds. Species such as sardines are caught on the coastal side of these fronts, while mackerel and tuna occur in abundance in the warmer offshore waters. Changes in the marine environment appear to influence both the timing and the paths of fish migrations.

For more than 50 years, Kitahara, Uda and their fellow fisheries oceanographers in Japan have studied the response of fish to their environment. They have set up an extensive network for collecting and reporting temperatures at the sea surface and at various depths, movements of schools of fish near various fishing grounds, catch rates, and ocean current information for making fishing condition forecasts.

Six regional fisheries research laboratories and 38 prefectual fisheries experiment stations are responsible for collecting, analyzing and distributing information on individual fisheries, such as the salmon, sardine or albacore. Ships at sea send information on their catches and the environment to the appropriate laboratory by radio. There these reports are compiled and analyzed to produce charts of fishing conditions, conditions in the ocean, and forecasts of various kinds, which are sent to the fishing fleet by mail, radio and facsimile. Some charts and forecasts are prepared at 10 day intervals, and others are sent out once a month. Research and cargo ships also provide information on temperature, salinity, currents and other factors in the environment for use in the fishing condition broadcasts.

Forecasts are based on long term trends, the time when fishing begins or ends on various grounds, data on age and size composition of the catch, and on experience with changes in the ocean in various fishing areas and the consequences of such changes in the past. It is still too early to evaluate the system's effectiveness, except to note that the information provided to the fishermen is very much in demand.

The potential value of fishery forecasting systems can be judged from the fact that boats in some fisheries must spend 80 per cent or more of their time scouting for fish. If this time could be reduced by half, each ship in such a fishery could increase its catch as much as threefold.

We have seen that there are many reasons for undertakings such as the Cooperative Study of the Kuroshio. They range from the most abstract, through the coldly practical to the mundane—improvements in ocean current theory, better weather predictions, more efficient ways to catch fish, and improved charts of the oceans. All these and more will result from studies of the Kuroshio and other parts of the world ocean. But regardless of motive, form or content, the goal of these studies can be summarized in much simpler terms: the search for man's ultimate tool, knowledge.

SUGGESTED READINGS

Water Characteristics of the Kuroshio, J. Masuzawa in *The Oceanographical Magazine,* **17**, 37, 1965.

The seasonal variation of the Kuroshio Current (In Russian), Y. V. Pavlova, in *Okeanologia,* **4**, 625, 1964.

Description of the Kuroshio (Physical aspect), D. Shoji, in *Proceedings of Symposium on the Kuroshio, Oceanographical Society of Japan and UNESCO, Tokyo, 1965.*

On the variability of the velocity of the Kuroshio Vol 1, D. Shoji and H. Nitani, in *The Journal of the Oceanographical Society of Japan,* **22**, 192 1966.

The influence of friction on inertial models of oceanic circulation, R. W. Stewart, in *Studies on Oceanography, Hidaka Jubilee Committee, Tokyo, 1964.*

On the nature of the Kuroshio, its origin and meanders, M. Uda, in *Studies on Oceanography, Hidaka Jubilee Committee, Tokyo, 1964.*

An Introduction to Physical Oceanography, W. S. von Arx. Addison-Wesley Publishing Co., Reading, Massachusetts, 1962.

Oceanic Water Masses and Their Circulation

JEROME WILLIAMS JOHN J. HIGGINSON JOHN D. ROHRBOUGH 1968

THE UBIQUITOUS FLUIDS

All human life begins its existence enveloped in a mass of fluid. With birth, these babies are cast forth to spend the balance of their lives surrounded by other fluids. All human endeavors are partially, totally or in various combinations immersed in air or water, or in the interface region of the two. These fluids may arrange themselves in large bodies of relative homogeneity called *masses*.

WATER MASSES

A *water mass* is defined as a large homogeneous body of water which has a particular characteristic range of temperature and salinity values. The density of the water, as specified by sigma *t*, is not sufficient to identify a water mass, since a combination of various temperatures and salinities can result in the same density value.

Note that since the sigma *t* curves are not straight lines, the mixing of two water masses having the same density will result in a new mass of *greater* density. This process is known as *caballing*. For example, in Figure 3-14 water mass *a* and water mass *b* are both shown to have the same sigma *t* value. When these are mixed in equal quantities water mass *c* results wherein $T_c=(T_a + T_b)/2$, and $S_c=(S_a + S_b)/2$, but $\sigma_{tc}\neq(\sigma_{ta} + \sigma_{tb})/2$. In general, when water masses mix, resulting temperatures and salinities may be obtained by simply averaging, but resulting densities may not.

Since water masses usually gain their temperature and salinity characteristics at the

Jerome Williams is associate chairman of the Environmental Sciences Department at the United States Naval Academy. He has also been associated with the Chesapeake Bay Institute of Johns Hopkins University, Maryland. His major research interests are underwater optics, instrumentation, and physical oceanography.

Commander John J. Higginson has been an instructor of oceanography at the United States Naval Academy. He has also been associated with the Apollo Recovery Program as operations officer in Helicopter ASW Squadron Four. He is presently in receipt of orders to Helicopter ASW Squadron Two at Naval Air Station, Imperial Beach, California.

Commander John D. Rohrbough is with the United States Navy and is on the staff of the Studies, Analysis, and Gaming Agency of the Office of the Joint Chiefs of Staff, the Pentagon, Washington. He has served as an instructor of oceanography and a member of the staff of the superintendent at the Naval Academy.

From *Sea and Air: The Naval Environment*, pp. 187-195, 1968. Reprinted with authors' revisions and light editing and by permission of the authors and the Naval Institute Press, Annapolis.

Table 3-1 Characteristics of Selected Water Masses

Mass	Ocean of Origin	Location Depth (m)	Salinity (0/00) and Temp. (°C) Range
1. Antarctic Bottom	South Atlantic (Weddell Sea)	4,000 to bottom	34.66 (−)0.4[a]
2. Antarctic Circumpolar	South Atlantic	100-4.000	34.68-34.70 0.5°
3. Antarctic Intermediate	South Atlantic	500-1.000	33.8 2.2°
4. South Atlantic Central	South Atlantic	100-300	34.65-36.00 6° - 18°
5. Arctic Deep and Bottom	North Atlantic	1,300-4,000 as Deep 1,300-Bottom as Bottom	34.90-34.97 2.2° - 3.5°
6. North Atlantic Intermediate	North Atlantic	300-1,000	34.73 4° - 8°
7. North Atlantic Central	North Atlantic	100-500	35.10-36.70 8° - 19°
8. European Mediterranean	European Mediterranean	1,400-1.600	37.75 13°
9. Pacific Equatorial	Central Pacific	200-1,000	34.60-35.15 8° - 15°
10. Indian Central	Indian	100-500	34.60-35.50 8° - 15°
11. Red Sea	Red Sea	2,900-3,100	40.00-41.00 18
12. Black Sea	Black Sea	0-200	16.00 (average) various temp.

[a]This is the only negative temperature in this table.

surface and then seek their own density level by thermohaline convection, water masses in the ocean are categorized by two factors: the depth at which they reach vertical equilibrium and the geographical source region. In order of increasing depth water masses are classified as being *surface, central, intermediate, deep,* and *bottom. Surface* waters extend down to about 100 meters, *central* to the base of the main thermocline, *intermediate* from below the central waters to about 3,000 meters, and the *deep* and *bottom* waters fill the lower portions of the ocean basins (see Table 3-1).

The surface water is unique in that it does not fall into a true water mass category since the variability of parameters is so great.

In general, it would be expected that waters at greater depth are formed at the higher latitudes, while those existing closer to the surface are formed nearer the equator.

ATLANTIC OCEAN

In the immediate vicinity of the Antarctic Continent, particularly the Weddell Sea, waters reach extremely low temperatures in the winter. Due to this low temperature and high salinity resulting from ice formation, this water has the highest sigma *t* of any in the world ocean. As a consequence, having once gained these characteristics, it sinks and flows along the ocean floor in a direction toward the equator. In fact, traces of this water have been measured as far as 45° *North* latitude. This water mass is called *Antarctic Bottom Water,* obviously because of its location and formation area. The Ant-

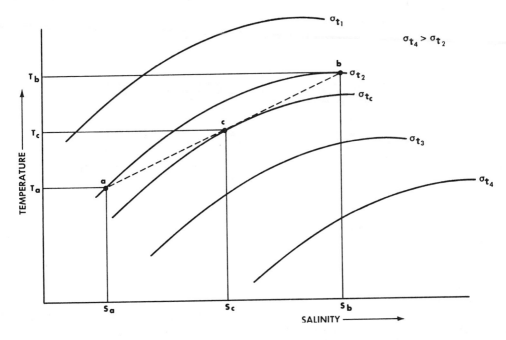

Figure 3-14. A T-S diagram showing simple mixing of two water masses.

arctic Bottom water mass also flows eastward around the Antarctic Continent due to the surprisingly deep-reaching effects of the surface West Wind Drift, mixes well below the surface with masses on its north edge, and becomes a separate, fairly homogeneous mass known as *Antarctic Circumpolar Water.* The deeper reaches of this mass, as it flows eastward, continuously provides deep water to the Indian and South Pacific Oceans. While it is true that some water circumnavigates the continent, it has been difficult to estimate the amount.

The *Arctic Deep and Bottom Water* (North Atlantic Deep and Bottom Water) is formed in relatively small areas off the coast of Greenland, one of which is the convergent region produced by the Irminger and East Greenland currents.* North Atlantic Deep

and Bottom Water, less dense than Antarctic Bottom Water overrides Antarctic Bottom Water all the way to the South Atlantic reaching the surface south of 60°S (see Figure 3-15). The North Atlantic Deep Water is continuously modified in its transit by mixing with masses yet to be discussed.

The *Antarctic Convergence Zone,* located at approximately 60°S latitude, is primarily produced by the seasonal cooling of the Antarctic Intermediate Water as it sinks to its density level. This particular convergence zone is present at nearly all longitudes of the earth; however, similar convergence zones in the North Atlantic and North Pacific are somewhat discontinuous and at times can be difficult to locate. North Atlantic Intermediate Water flows south from the Arctic Convergence to approximately 20°N where it mixes with Antarctic Intermediate Water.

North and South Atlantic Central Waters

* Periodic overflows from the North Polar Sea across the Greenland-Scotland Ridge cascade down the southern slope with relatively high velocities due to the water's very cold (-1.4C) temperature (the coldest water anywhere in the deep sea); it is

less dense, however, than North Atlantic Deep and Bottom Water because of a lower salt content.

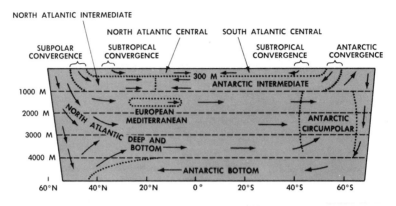

Figure 3-15. Atlantic Ocean: General subsurface movement.

form at the surface at their respective subtropical convergences during the winters. They sink and flow toward the equator losing their identities as they spread.

The one significant incursion of foreign waters is the large mass of European Mediterranean Water which finds its level at the average depth of 1,500 meters, after leaving through the Strait of Gibraltar. This water mass is continually formed in the northern area of western Mediterranean by winter cooling and evaporation by the dry air sweeping north from Africa. The cool, saline water sinks, flows south and west, and then spills out over the sill. On the surface in the Strait the less dense Atlantic waters flow in to maintain the balance, creating a two-layered stratification with each layer flowing in opposite directions (see Figure 3-16).

During World War II German submarines

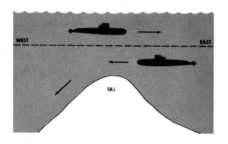

Figure 3-16. Water flow in the Straits of Gibraltar.

are said to have used the flows to transit the Strait undetected. They would dive deep or shallow depending on whether they desired to exit or enter, compensate for the required neutral buoyancy state, and then ride the flow quietly without use of their motors. This was a very ingenious use of environmental knowledge to circumvent detection.

The Mediterranean Water, with its increased salinity, has strong effects on the upper section of the North Atlantic Deep Water mass. Although its influence is felt to the west and south predominately, its telltale salinity maximum has been traced to locations up to 1,500 miles from Gibraltar.

In conclusion, the Atlantic Ocean is constantly renewing itself at all depths although at a very slow rate. Recent analyses utilizing radioactive carbon measurements indicate that it has been about 750 years since Antarctic Bottom Water in the Atlantic was at the surface. In contrast to this, 1,500 years is estimated for the age of this water mass in the Pacific Ocean.

PACIFIC OCEAN

The Pacific Ocean is noted for its generally sluggish deep water flow pattern when compared to the other oceans. However, the Antarctic Bottom Water as it flows around the Antarctic Continent, provides a fairly

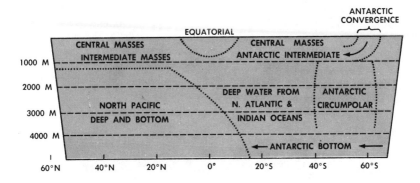

Figure 3-17. Pacific Ocean: General subsurface structure.

continuous input to the South Pacific Basin. On the other hand, Antarctic Circumpolar Water, which has been partially mixed with waters of the Atlantic and Indian Oceans, enters from the west to slowly but continuously push into the South Pacific deep layer (below 1,000 m).

The intermediate and central layers of the entire Pacific Ocean are diffuse and not well defined. The various convergence zones which would be analogous to those in the Atlantic are discontinuous and misplaced. Several masses existing at the same depth in different areas make a cross-sectional depiction difficult to construct. It is important to note that there is near surface mixing of masses from distant regions at the equator forming the Pacific Equatorial Water Mass—the only major water mass which does not receive any characteristics from the surface near the formation area.

The North Pacific is unique because no extremely dense water masses form in its most northerly reaches. The Deep and Bottom Water of the North Pacific experiences little interchange with other areas. Its origin is in doubt both in time and space, and it is characterized by an oxygen minimum due to the sluggish flow present.

Because of the slow movement of the subsurface mass, the surface current motion reaches deeper and has greater effect on the subsurface characteristics in the North Pacific than do the surface currents in other oceans. This is probably due to the general absence of vital thermohaline convective activity in the North Pacific Ocean (see Figure 3-17).

INDIAN OCEAN

Of the three major oceans only the Indian Ocean does not extend into the North Hemisphere. There is no cold water sinking along its northern edge causing the deep water mass to have a lesser movement than that in the North Atlantic Ocean. However, there is a well defined bottom flow in the South and, oceanographically speaking, is like the South Atlantic south of the Subtropical Convergence Zone at about 40°S latitude.

The Antarctic Bottom Water is present at all latitudes of the Indian Ocean. The deep layer is that which is led around the south tip of Africa from the Atlantic; it is reasonably well oxygenated, especially considering the distance from its source region in the North Atlantic. The Antarctic Intermediate Water forms at the Antarctic Convergence Zone and spreads to the north. The Indian Central Water sinks at the Subtropical Convergence and flows north toward equatorial regions.

Bottom water from the Red Sea flows over the sill and on through the entrance at the Strait of Bab el Mandeb to spread and mix with deep layers of the Indian Ocean. Red Sea Water is characterized by its very

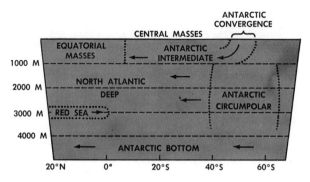

Figure 3-18. Indian Ocean: General subsurface structure.

high salinity of about $40^{\circ}/_{oo}$ to $41^{\circ}/_{oo}$. This water mass is formed within the Sea by constant evaporation by dry air from Africa and by winter cooling periods in much the same manner as European Mediterranean Water. High salinity causes its density to be such that it spreads out in the Indian Ocean at depths near 3,000 meters. Traces have been identified as far as 1,250 miles south of the Gulf of Aden. Red Sea Water provides the only significant modifying effect in the entire deep Indian Ocean. As an aside, hot spots have been discovered recently at great depths (2,040 m) in the Red Sea. The anomalous temperatures measured thus far range from $22^{\circ}C$ to $56^{\circ}C$; salinities have been determined to be in excess of $250^{\circ}/_{oo}$. Their causes remain unexplained but future concerted investigations are planned to develop answers as to how the spots have been formed and how they continue to exist.

The Equatorial shallow layers of the Indian Ocean are not clearly defined. This is partially due to seasonal monsoon changes of surface currents. The water is being constantly overturned by changing winds and does not have significant characteristics. Little distinguishable subsurface flow is present (see Figure 3-18).

BLACK SEA—A SEA APART

The Black Sea, with its complete lack of thermohaline convection, has a complete oxygen disappearance at all depths below 200 meters. Precipitation and runoff far exceed evaporation. The flow into the sea through the Bosphorus from the European Mediterranean is so meager that it would renew the waters below 30 meters only once in 500 years. Consequently, the deep waters have become stagnant; hydrogen sulfide is present; and only anaerobic bacteria can live in the blackened waters.

CONCLUSION

The preceding discussion, at best, is only a very cursory qualitative treatment of deep water ocean circulation. Although the mechanisms producing subsurface flows were discussed, bottom topographic effects have been neglected. But to ignore the latter in a detailed study would be a serious omission. Only the most prominent of the marginal sea effects have been introduced.

Air can be subdivided into masses displaying similar identifiable characteristics just as the water masses of the oceans. In addition, there can be identified clear boundaries between these masses, both in the sea and in the air. Those in the sea have not achieved the importance that those in the atmosphere enjoy; however, the wall of the Gulf Stream has been clearly identified as a "front" in the ocean.

SUGGESTED READINGS

Bailey, H. S., Jr., "The Voyage of the *Challenger*," *Scientific American*, May, 1953.

Dietrich, Gunter, *General Oceanography*, John Wiley and Sons, New York, 1963.

King, Cuchlaine, A. M., *An Introduction to Oceanography*, McGraw-Hill, Inc., 1963.

Kort, V. G., "The Antarctic Ocean," *Scientific American*, September, 1962.

Stommel, Henry, "The Anatomy of the Atlantic," *Scientific American*, January, 1955.

Sverdrup, H. V., Johnson, Martin, W., and Fleming, Richard, H., *The Oceans,* Prentice-Hall, Inc., 1942.

Williams, Jerome, *Oceanography*, Little, Brown and Company, Inc., Boston, 1962.

Yasso, Warren E., *Oceanography*, Holt, Rinehart and Winston, Inc., New York, 1965.

Special Issue: "Deep Ocean Engineering," *Naval Ship Systems Command Technical News*, January, 1967.

Upwelling

ROBERT L. SMITH   1968

INTRODUCTION TO
UPWELLING IN 1972

Since the writing of the following review article in 1967, interest in the study of upwelling has increased greatly. The importance of upwelling, and especially coastal upwelling, has been discussed frequently. Ryther (1969) has suggested that perhaps 50% of the world's fish supply is produced in upwelling regions. A major report on the future of marine sciences, *An Oceanic Quest* (National Academy of Sciences, 1969), gives priority to the study of coastal upwelling. The Office of the International Decade of Ocean Exploration, National Science Foundation, has placed emphasis on "studies of the physical, chemical, and biological dynamics of convective systems such as regions of upwelling. . . ." The motivation for much of the interest in upwelling is the hope that with an increased understanding of the dynamics of the upwelling system and by relating the physical dynamics to the biological and chemical dynamics, it may eventually be possible to predict commercially important

nekton stocks from a knowledge of a few oceanographic and meteorological variables.

In the next decade one may expect many papers appearing in the literature as a result of studies measuring upwelling synoptically and modeling it numerically with computers. Hopefully not all the works will be strictly specialized to just physics or just biology— but rather interdisciplinary and directed toward understanding the "upwelling ecosystem." With this in mind I should like to call the reader's attention to two recent publications: a review article on upwelling and fish production (Cushing, 1969) and a collection of papers resulting from a working conference on the analysis of upwelling systems held in Barcelona (Instituto de Investigaciones Pesqueras, 1971).

Vertical motions are an integral part of oceanic circulation. Particularly important are the ascending motions that result in an exchange between near-surface and deeper waters. Here we shall use upwelling to mean only an ascending motion, of some mini-

Dr. Robert L. Smith is an associate professor of oceanography at Oregon State University. Most recently he has served as scientific officer for the Physical Oceanography Program with the Office of Naval Research, Virginia. He has written over fifteen technical articles.

From *Oceanography and Marine Biology Annual Review*, Vol. 6, pp. 11-46, 1968. Reprinted with author's notes. Editing by permission of the author and George Allen and Unwin Ltd., London.

mum duration and extent, by which water from subsurface layers is brought into the surface layer and is removed from the area of upwelling by horizontal flow. This use of the term upwelling follows Sverdrup (1938) and Wyrtki (1963), who points out that although upwelling is a widely used term in oceanography, a precise definition has not yet been given. In general, upwelling is the result of horizontal divergence in the surface layer, and usually the water comes from depths not exceeding a few hundred meters. Upwelling may occur anywhere, but it is a particularly conspicuous phenomenon along the western coasts of the continents (the eastern boundary current region) where prevailing winds carry the surface water away from the coast.

Because upwelling brings subsurface water into the surface layer, it induces horizontal anomalies in the distribution of physical and chemical properties that normally have marked vertical gradients. Such anomalies are often useful indicators of upwelling (Park, Pattullo and Wyatt, 1962), but the effects of upwelling and the physical process of upwelling should not be confused. Effects similar to those produced by upwelling can be caused by wind-induced mixing or by the baroclinic adjustment of the density field associated with an increase in the geostrophic transport of a current; however, the *persistence* of these effects is probably possible only with upwelling. The interrelations among currents, internal waves, mixing, and upwelling are far from simple and are not well understood.

In the last decade interest in upwelling, the process and the effects, has increased. This is perhaps not unrelated to the quest for more food from the sea to support the burgeoning world population, and to the increasing awareness that upwelling affects the economy of man as well as the economy of the sea.

The emphasis in this paper is on the physical process of upwelling. We must not ignore, however, the effects of upwelling. The effect on climate is well known to those who live on the western coasts of continents. That upwelling has important effects on the ecology of marine life is readily apparent: a map showing the areas of upwelling in the world ocean also rather adequately serves as a map of the areas of high organic productivity. Gordon J. F. MacDonald (1967), chairman of the Panel on Oceanography of the President's Science Advisory Committee, has said: "The yield of fish, whether in the natural environment or in scientific farming of restricted areas of the sea—agriculture—will depend on the nutrients supplied by upwelling. Technical opportunities exist here for schemes to utilize natural hydrodynamic or atmospheric energy sources to bring to the surface nutrient-rich deep water, or to fertilize selected marine habitats, such as bays, coral lagoons, and fenced in areas."

Since the importance of upwelling is widespread, so is the literature. Important papers are to be found in the journals of the fisheries biologist, the meteorologist, the oceanographer, and the applied mathematician; but to this author's knowledge no general review article on upwelling exists. The papers of Gunther (1936) and Hart and Currie (1960), the article on eastern boundary currents by Wooster and Reid (1963) and, for the mathematically inclined, the theoretical discussion by Yoshida (1967), contain valuable discussions on upwelling and have considerable breadth. Symposia on upwelling have been held at the Ninth Pacific Science Congress, in Bangkok in 1957, and at the Second International Oceanographic Congress, in Moscow in 1966.

EARLY WORK

Some of the first recorded observations of the conspicuously cold water associated with upwelling were made by early explorers and conquerers along the west coasts of Africa and the Americas. At Callao (Peru), the Conquistadores made good use of the cold water, hanging flagons of wine in the sea to cool (Acosta, 1604). A variety of theories were put forward to explain the anomalously cold water. By the early nineteenth century it was generally held that the cold coastal water observed off the coasts of Peru, California, and South West Africa was simply the result of water advected from higher latitudes. By the middle of the nineteenth century, however, enough observations had been accumulated to make it apparent that temperatures did not increase monotonically with decreasing latitude in these regions. Indeed, the coldest coastal water might be found far downstream (towards the equator). De Tessan, as early as 1844, had explained the cool water off Peru as a result of upwelling (Gunther, 1936). Witte (1880) in a theoretical discussion concluded that upwelling could come about either by the effect of the Earth's rotation on a meridional current or by offshore winds driving the surface water away from the coast. Buchan (1895), summarizing the results of the Challenger Expedition, held that offshore winds drive the surface waters offshore, causing upwelling which brings cold water of low depths to the surface. Thus he explained the low water temperatures at the surface off the south-east Arabian coast during the monsoon and off Peru. He did not seem to be aware that the winds along those coasts were predominately longshore.

The work of Ekman (1905) provided a basis for understanding the effect of wind stress on ocean circulation. He showed that due to the effect of the Earth's rotation and frictional forces, the net transport of water due to the wind stress is directed 90° to the right of the wind in the Northern Hemisphere (90° to the left in the Southern Hemisphere); an appreciation of this result led to the first application of Ekman's theory to an upwelling situation, explained the upwelling off the Pacific coast of the United States. He showed that the upwelling was a direct effect of the coastal winds and that a northerly wind blowing parallel to the coast could be sufficient to induce offshore transport of surface water, necessitating a compensatory replacement of water from deeper layers. McEwen (1912) extended the study

and was able to show that the temperature anomalies observed off the California coast were in satisfactory agreement with those he predicted by an application of Ekman's theory.

DYNAMICS OF UPWELLING— GREATLY SIMPLIFIED

In the original paper the dynamics of upwelling were discussed in some detail and with considerable reference to theoretical works through 1967. Since some acquaintance with mathematics beyond elementary calculus was assumed, the section on dynamics of upwelling (pp. 13-27 in the original paper) has been deleted from this edition. It will, however, be both helpful and useful for the reader to have some idea of the most basic dynamics involved in upwelling and the following very brief and greatly simplified discussion has been prepared for this edition.

As discussed briefly in the section on early work, the Ekman theory provides the basic "intuitive" understanding for the mechanism behind most upwelling. An excellent and clear discussion of the effects of wind on ocean water in general and Ekman theory in particular can be found in Stewart (1969). Ekman theory states that in a homogeneous, infinite ocean, a uniform steady wind would provide a net transport of water in a direction 90° *cum sole* to the wind. A brief, heuristic explanation follows:

(1) The Coriolis acceleration, an effect of the earth's rotation, affects any object moving on the earth and is directed at right angles to the motion of the object (90° to the right in the Northern Hemisphere). The magnitude of the force on a unit mass (the acceleration) is proportional to the speed of the object (v) and the latitude or, more precisely, the Coriolis parameter f ($f = 2\Omega \sin \phi$: where Ω is the earth's angular velocity and ϕ the latitude). The Coriolis force per unit mass is fv.

(2) When the wind blows over the water, it exerts a force on the surface in the direction of the wind. The force is "transmitted" (and successively reduced) to the water be-

neath the surface by a frictional mechanism (turbulence).

(3) At some depth beneath the surface, the water velocity and frictional forces associated with it become negligibly small. The layer above, in which frictional effects are significant and which "responds" to the wind, is known as the Ekman layer—and, in the idealized case, it moves over the underlying water with negligible friction. The Ekman layer is tens of meters thick.

(4) The Ekman layer, viewed as a slab of water moving frictionlessly over the underlying water, is under the influence of two forces: the wind stress on the surface acting in the direction of the wind and the Coriolis force on the water in the slab acting at right angles to the water motion. (The individual parcels of water within the Ekman layer, or the slab, are moving in a complex manner and in varying directions—but we are concerned with the *net* or average motion of all the water parcels in the slab. The Coriolis force on the slab is then at right angles, $90°$ to the right in the Northern Hemisphere, to the net or average motion of the water in the Ekman layer.)

(5) In the steady state equilibrium case the two forces on the slab, the wind stress and the Coriolis force, must balance. The net motion of the Ekman layer, or slab, must be such that the resultant Coriolis force on the slab is directed opposite to the wind—which is the case if the net motion in the Ekman layer is at right angles to the wind ($90°$ to the right in the Northern Hemisphere).

As Stewart (1969) comments: "This Ekman-layer flow has some important fairly direct effects in several parts of the world. For example, along the coasts of California and Peru the presence of coastal mountains tends to deflect the low-level winds so that they blow parallel to the coast. Typically, in each case, they blow equatorward, and so the average Ekman flow—to the right off California and to the left off Peru—is offshore. As the surface water is swept away deeper water wells up to replace it. The upwelling water is significantly colder than

the sun-warmed surface waters, somewhat to the discomfort of swimmers (and, since it is also well fertilized compared with the surface water, to the advantage of fishermen and birds)."

Quantitatively, the Ekman transport (in, say, grams of water moved at right angles to the wind in 1 second through a strip 1 cm wide and extending from the surface to the Ekman depth) is equal to the wind stress divided by the Coriolis parameter:

$$M = \tau/f.$$

The wind stress in dynes/cm^2 can be estimated from $\tau = 10^{-3}\rho_a w^2$, where ρ_a is density of air in gr/cm^3 and w is wind speed in cm/sec.

Although the dynamics of upwelling are far more complicated than simple Ekman theory can account for, the component of the Ekman transport directed offshore does give a rather good prediction of where and what season coastal upwelling occurs (see section on Regions of Upwelling). Ekman theory has also been used quantitatively in studying coastal upwelling and, although the Ekman transport was derived for the equilibrium state of the ocean under a constant wind stress in the region away from the effects of a coastal boundary, the results have been surprisingly good. We discuss two examples:

In applying Ekman's theory quantitatively to upwelling off southern California, Sverdrup (1938) and Sverdrup and Fleming (1941) arrived at a dynamical interpretation of coastal upwelling. Following Yoshida (1967) we shall refer to this as the Ekman-Sverdrup model. The motivation for the model was a series of three hydrographic surveys off the southern California coast in the spring and early summer of 1937. A "remarkable" change occurred between late March and early May in the section running southwest from the coast at $35°N$. The change in the distribution of properties indicated that the surface water was carried away from the coast and replaced by subsurface water. Since the wind during the six weeks between cruises had consistently a northerly component, which would transport water away from the coast, there was qualitative agreement with Ekman theory. The transport computed from the winds using the equation given above was in very close agreement with the transport deduced from changes in the distribution of properties.

It was observed by Sverdrup that there was a relatively well-defined boundary offshore between the upwelled water and the lighter water. At this outer boundary, dynamic computations showed a strong current flowing parallel to the coast, and hence with the wind. This current, called a "convection" current by Sverdrup, was the result of the strong horizontal density gradient at this boundary. In this "convection" current the velocity decreased rapidly with depth. This led Sverdrup to suggest intense frictional stress associated with the large vertical shear in the "convection" current. Sverdrup argued that the stress in the top layer of the "convection" current would partly balance the wind stress, reducing the Ekman transport in this region. The boundary region would move outward at a velocity less than the water closer inshore. The boundary region acts as a slowly moving wall and a cellular vertical circulation will develop between the coast and the boundary region, beyond which upwelling may again take place. Figure 3-19, from Sverdrup (1938), shows the computed flow and the "convection" current.

An investigation by Smith, Pattullo and Lane (1966) of an early stage of coastal upwelling supports the use of Ekman transport as an estimate even in a non-equilibrium situation: They reported on a study of the early stage of upwelling based on oceanographic observations made in May 1963 off the southern Oregon coast. A line of hydrographic stations was run perpendicular to the coast at 42°N and repeated three days later on the same phase of the tide. The change in the temperature is shown in Figure 3-20.

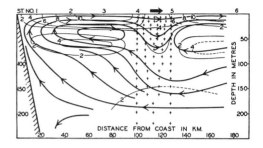

Figure 3-19. Computed circulation in section normal to coast during upwelling off southern California; heavy lines are streamlines, light lines show areas of equal horizontal velocity (cm sec^{-1}), and crosses indicate region of strong "convection" current directed away from reader (from Sverdrup, 1938).

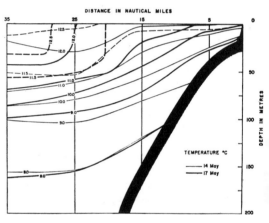

Figure 3-20. The distribution of temperature off southern Oregon during the early stages of upwelling (from Smith, Pattullo and Lane, 1966).

The offshore transport associated with the upwelling was computed from changes in the distribution of temperature and salinity. These computations, similar in principle to those of Sverdrup (1938) off southern California, are shown in Figure 3-21. The offshore transport was also computed from heat budget considerations and the Ekman transport was computed from the observed winds. All estimates fell between 4.4 and 8.8 $\times 10^9$ gr cm^{-1} for the amount transported

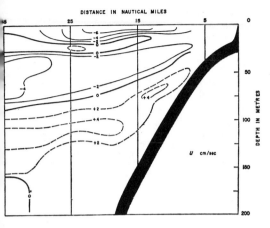

Figure 3-21. Onshore velocity (cm sec^{-1}) during early stage of upwelling computed from hydrographic data off southern Oregon (from Smith, Pattullo and Lane, 1966).

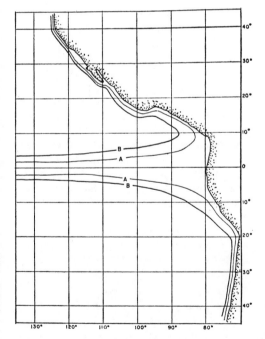

Figure 3-22. Potential upwelling region in eastern Pacific; amplitude of boundary disturbances, viz. upwelling, at A and B are, respectively, e^{-1} and $e^{-\pi}$ of those along the equator and coastal boundaries (from Yoshida, 1967).

offshore during the 76-hour period. The vertical velocities at the base of the Ekman layer, inferred from the displacement of the isolines of the physical properties, decreased from 7.0×10^{-3} cm sec^{-1} inshore to 0.2×10^{-3} cm sec^{-1}, 35 nautical miles from the coast.

The reader is cautioned that the above discussion has oversimplified the dynamics of upwelling. The reader interested in the theoretical development is, of course, urged to read the original section on dynamics of upwelling. However, considerable progress toward a better theoretical understanding of upwelling is being made and hopefully the original section will become largely obsolete within the next few years. The reader should watch for the articles now beginning to appear in the published literature (see especially the new *Journal of Physical Oceanography*). Particularly promising are the results from numerical model studies. The advent of high speed electronic computers has allowed the numerical solutions to complicated sets of equations and these techniques are now being applied to the upwelling problem.

REGIONS OF UPWELLING

Qualitatively, we expect upwelling where there is a divergence of the surface flow, usually induced by the wind. From the theoretical discussion by Yoshida (1967) comes the concept of a potential upwelling region which extends from a narrow coastal belt at high latitudes along the eastern boundaries of the oceans, i.e., the west coasts of the continents, to an equatorial zone (see Figure 3-22). Upwelling occurs in this "boundary region" whenever the winds are favorable, but outside this region no marked upwelling or undercurrents will occur. The effects of upwelling, *e.g.* anomalously low temperatures and high productivity, are particularly conspicuous along the eastern boundaries of the oceans and in the eastern equatorial regions. The most intense upwelling is in the coastal regions off the west coasts of continents where a one-sided divergence of the surface layer is induced by a wind stress parallel to the coastal boundary. Many of the descriptive studies of upwelling are included in the more extensive studies of the major eastern boundary currents. Thus it is convenient to refer to the regions by the name of the current.

Whatever deficiencies there may be in the simple Ekman theory, the Ekman transport away from the coastal boundary is a useful index of coastal upwelling. Wooster and Reid (1963), using Hidaka's (1958) seasonal mean wind stress charts of the world ocean, plotted the Ekman transport away from the coast for $5°$ latitudinal intervals along the eastern boundaries of the oceans for each season. They found that the "seasonal and geographical variations in the index do appear to bear some relation to corresponding variations in coastal upwelling." In partial summary:

(1) The maximum values are usually observed in the spring or summer, with the exception of the distinct winter maximum for the Peru Current region north of $30°$ S.

(2) The maximum values of the index migrate from south to north, from spring to summer, in the California, Benguela, and the Canary Current regions.

(3) The maximum values for the index are in the Southern Hemisphere. The smallest values are off the west coast of North America.

(4) Negative values are observed in high latitudes ($\phi \tilde{>} 50°$).

The index would predict upwelling off the west coast of Australia, but, as will be discussed later, such does not seem to occur in any significant amount. Indeed, the Indian Ocean is anomalous. The upwelling is most intense off the east coast, i.e. along the southeast Arabian coast, where the index would predict upwelling during the southwest monsoon, and off the Somali coast during the monsoon, in a western boundary current!

Peru Current System

There are a number of early observations, reports and discussions of the Peru Current region, but most are now of historic rather than scientific interest. Schott (1931) discussed the region and classified the areas of upwelling, using surface temperature, salinity and current observations. The first major survey of the Pacific coastal waters off South America was made by the *William Scoresby* in 1931.

Gunther (1936) discussed in detail the results of the expedition, with emphasis on the upwelling phenomenon. He distinguishes the Peru Coastal Current as that part of the Peru Current System which represents the narrow belt of cold, upwelled water running from about Valparaiso to the Gulf of Guayaquil. As a measure of upwelling, Gunther compares the inshore, upwelled water temperature with the mean of the temperature at 150 m outside the upwelling zone and the surface temperature offshore ($100°W$). From the temperature and salinity structure, Gunther estimates the depth affected by upwelling for 12 latitudes (from $35°S$ to $2°S$) and concludes that upwelling takes place from about 130 mi, with a minimum estimate of 40 m and a maximum depth of 360 m. He found that while the appearance of upwelling was somewhat irregular in time and space, it had a very definite relation to the wind, in general agreement with the Ekman theory. No relation to the bottom topography was apparent. Gunther also found evidence of the existence of a subsurface countercurrent, carrying high salinity water south, and observed an eddy-like structure in the surface isotherms. These features appear common to all coastal upwelling regions.

Posner (1957) has reported on the Yale Expedition to Peruvian waters in 1953. Some observations were made during the El Niño of that year and although conditions were probably never normal, upwelling was observed. The emphasis of the study was on the nutrients and plankton; quantitative estimates of productivity and transport of nutrients by upwelling were made. Schweigger (1958) gives a detailed description of the upwelling along the Peruvian coast, and distinguishes five regions.

In October and November 1960, the Step-I Expedition of the Scripps Institution of Oceanography made a survey of the entire region off Peru. Wooster and Gilmartin (1961) confirmed the existence of subsurface flow to the south (Peru-Chile Undercurrent) both by direct measurement and by geostrophic calculations. Wyrtki (1963) used the hydrographical data, with wind stress values for the same reason, to obtain the horizontal and vertical field of motion in the Peru Current. The calculations indicate that the upwelling along the coast is restricted to depths of less than 100 m but that ascending motion at greater depths and further offshore is of the same magnitude as the coastal upwelling and presumably is important in the upwelling process. Wyrtki found the flow near the coast to be highly convergent between 100 m and 400 m. From the calculations, vertical velocities averaged over large areas at 100 m depth, are of the order of 10^{-5} or 10^{-4} cm sec^{-1}. The maximum vertical values are likely at shallow depth and are, presumably, greater. The computations also imply that south of 15°S upwelling is supplied by Subantarctic Water flowing north with the Peru Coastal Current, and north of 15°S by Equatorial Subsurface Water flowing south with the Peru Countercurrent. The countercurrent flows south between the Peru Coastal Current and the Peru Oceanic Current, chiefly as a subsurface current, but is distinct from the undercurrent inshore.

One significant result, in view of Yoshida's (1967) theory, is the extension of the upwelling region, shown in Wyrtki's calculations, towards the northwest and away from the coast as the equator is approached. As Wyrtki notes, this area of ascending motion probably extends toward the equator and leads over into equatorial upwelling. This would provide a better explanation of the low temperatures in the region than advection from the south.

El Niño

The antithesis of upwelling along the coast of Peru is El Niño. During an abnormal Southern Hemisphere summer, coastal upwelling ceases and the strip of cool coastal water vanishes. The consequences can be catastrophic to the climate and the ecology. El Niño is characterized by unusually high surface temperatures, northerly winds, heavy rains, the disappearance of the anchovies, and the resulting mass mortality of the 'guano' birds. El Niño has been known to occur in 1891, 1925, 1941, 1953, and 1957-58 (Wyrtki, 1966). It appears that 1965-66 were also El Niño years (Bjerknes, 1966).

El Niño is related to a weakening of the atmospheric system, but the exact mechanism is not yet understood. Schott (1931) attributes El Niño to exceptional weakness of the Southeast Trades and a displacement of the Intertropical Convergence southward beyond the equator. Posner (1957) defines El Niño as a tongue of surface water with higher temperatures and low salinity extending southward over the northward Peru Current along the coast. He suggests as a cause, a substantial shift of the atmospheric pressure and wind system. Wooster (1960) attributes the cause of El Niño to a general weakening of atmospheric circulation with a reduction of the wind component parallel to the coast for an extended period, causing a weakening or cessation of upwelling. Under this hypothesis, El Niño does not essentially differ from the normal cessation of upwelling in such upwelling regions as the California Current, the Benguela Current, and the southeast Arabian coast, where marked seasonal changes occur in the atmospheric circulation; El Niño is abnormal because of its relative rarity, unpredictability, and the severity of its effects on the climate, ecology, and the economy of the Peru coastal region.

Bjerknes (1961) concluded that the meteorological control over the occurrence of El Niño lies in the slightly fluctuating strength of the Pacific Trade Winds. A necessary condition, however, is an accumulation of large masses of warm water in the tropical eastern Pacific. The disappearance of southerly winds for any appreciable time results in the overflow of the warm tropical water over the cooler water of the Peru Current, causing El Niño. Schell (1965) studied the sea surface temperature anomalies along the west coast of South America in relation to the atmospheric pressure distribution. It appeared to him that weaker southerlies and south-easterlies during the March to November period are associated with higher sea surface temperatures during the following December to February quarter—El Niño season. Bjerknes (1966) states that El Niño is a summer phenomenon of the Southern Hemisphere and the high sea temperature is a by-product of the annual insolation maximum. El Niño summers are rare and differ from an average summer by anomalously weak trade winds. The Peru Countercurrent, flowing southward from the equatorial region between the Peru Ocean Current and the Peru Coastal Current, beneath a shallow drift current driven by the southeast trade winds, would allow water of equatorial temperature to appear rather suddenly at the surface off Peru during an unusual summer when the southeast trade winds become too weak to drive the drift current.

California Current System

There are many papers concerned with the various aspects of oceanography in the California Current region. Most are rather specialized, geographically and/or topically, and only touch upon the upwelling. Smith (1964, 1966) has investigated upwelling along the Oregon coast. The papers of Sverdrup (1938) and Sverdrup and Fleming (1941) concerning upwelling off southern California have been previously referred to, as has that of Smith, Pattullo and Lane

(1966) on the upwelling along the Oregon coast. The most comprehensive discussion of the California Current system, including a substantial discussion of the upwelling, is that of Reid, Roden and Wyllie (1958). Much concerning the system and the upwelling is found in the reports of the California Cooperative Oceanic Fisheries Investigations, Scripps Institution of Oceanography, the Departments of Oceanography at Oregon State University and at the University of Washington, and the Pacific Oceanographic Group of Canada.

Upwelling is markedly seasonal in the California Current region. The North Pacific High dominates the atmospheric circulation during the spring and summer; the predominating winds are northerly or north-westerly, blowing equatorward parallel to the coast and driving the surface waters offshore. The winds are strongest off Baja California in April and May, off southern California in May and June, off northern California in June and July, and off Oregon and Washington in July and August. Since upwelling is the result of the northerly and north-westerly winds, the most intense upwelling shifts up the coast as spring and summer progress. In the late fall and winter the North Pacific High has weakened and moved southwards. The region north of Point Conception (California) experiences westerly or south-westerly winds under the dominance of the deepening Aleutian Lows moving across the North Pacific, and upwelling ceases. There are non-seasonal occurrences of upwelling and, occasionally, subsidence of upwelling during the summer. These variations in the strength and occurrence of upwelling also appear to be closely related to the winds (Reid, 1960; Lane, 1965a).

A deep countercurrent, below 200 m, flows northwest along the coast from Baja California to Cape Mendocino (Reid, Roden and Wyllie, 1958). Reid (1962) has measured the countercurrent at 250 m off central California. Stevenson (1966) does not find evidence of it 45 nautical miles off Oregon at about 45°N, although Collins

(1968) finds northward subsurface flow on the Oregon shelf. Sverdrup, Johnson and Fleming (1942) speculate that when the north winds are weak or absent in the late fall and early winter, this countercurrent flows to the north from the tip of Baja California to north of Point Conception, where it is known as the Davidson Current. The analogy to the explanation of El Niño by Bjerknes (1966) is to be noted.

Benguela Current

Hart and Currie (1960) use the term Benguela Current to denote the easternmost periphery of the anticyclonic gyre of the south Atlantic, the region of cool upwelled water between 15°S and 34°S. The distinction between the coastal current (the Benguela Current) and offshore flow is analogous to the distinction between the Peru Coastal Current and the Peru Oceanic Current.

The first major investigation was done by the *Meteor* during the Deutsche Atlantische Expedition of 1925-27. Defant (1936a) used the *Meteor* data in a discussion of the oceanic and upwelling conditions. He found the region of strongest temperature anomaly, the coldest water, to be in the coastal current between 23°S and 31°S.

In 1950 the *William Scoresby* made a more localized and detailed study of the Benguela Current upwelling region. Curie (1953) summarized the findings, and Hart and Currie (1960) gave a detailed report of the investigation. Two surveys were made, the first during March (southern autumn) when upwelling was minimal, and the second during September-October (southern spring) when intense upwelling was encountered. The cool, low salinity water moving offshore did not appear to originate uniformly along the coast. It seemed to be more a function of the local winds than bottom topography. From a comparison of T-S relationships they concluded that the upwelled water originates at depths between 200 and 300 m. At this depth there appears to be a subsurface coun-

tercurrent flowing southward along the edge of the shelf, which they termed a "compensation current." Eddy-like structure was also observed in the surface temperatures. The continental shelf in the Benguela Current upwelling region is of the order of 100 nautical miles wide. This is considerably wider than in the California Current region, where it is generally less than 25 nautical miles, or the Peru Current region, where the shelf width may be almost negligible.

Hart and Currie give a diagrammatic and idealized version of the water movements observed off the coast, shown in Figure 3-23. It is, as they state, probably a fair representation of the circulation in the major upwelling regions associated with the eastern boundary currents. They also discuss the diversity of opinion as to seasons of maximum upwelling. Böhnecke's (1936) surface temperature anomaly charts suggest upwelling at all seasons but most intense in summer and fall, consistent with the Ekman

Figure 3-23. Schematic of circulation during upwelling in Benguela Current region; features shown appear to be common to all major coastal upwelling regions (mirror image in Northern Hemisphere) (from Hart and Currie, 1960). Note the sinuous boundary indicative of eddies between upwelled coastal water and oceanic surface water, the circulation in vertical section (cf. Figure 3-19), and subsurface countercurrent or "compensation" current.

transport index. From the surface temperature anomaly charts there also appears to be a northward migration of maximum upwelling as the seasons progress from spring to summer.

Upwelling along the coasts of Southern Africa, including the Benguela Current region, has also been discussed by Darbyshire (1966, 1967), and Orren and Shannon (1967), with some diversity of opinion.

Canary Current

Associated with the Canary Current is the upwelling region along the northwest coast of Africa extending from the Canary Islands to Cape Verde. I would repeat the comments of Sverdrup, Johnson and Fleming (1942) and, more recently, Wooster and Reid (1963) as to the lack of a detailed study of the upwelling in the area. This region is not oceanographically unexplored; indeed, the *Meteor* made six profiles at right angles to the coast in 1937 with the object of studying the upwelling (Defant, 1961), but apparently no interpretation or detailed discussion has yet been published. From the surface temperature anomaly charts of Böhnecke (1936), Wooster and Reid conclude that the upwelling occurs throughout the year, but is most intense in spring and summer and migrates northward from winter to summer. Defant (1961) gives January to May as the period of upwelling with temperature anomalies of almost 7°C.

Indian Ocean

On the basis of geographical analogy one might expect an eastern boundary current and upwelling along the west coast of Australia. It is only in this decade that the region has received much oceanographical attention although Schott (1933) had speculated on the existence of upwelling along the west coast of Australia. Temperature anomaly charts give no indication of upwelling at any time, even though the offshore Ekman transport inferred from mean wind stress

charts would indicate strong summer coastal upwelling from 35°S to 15°S (Wooster and Reid, 1963). Surface temperature charts did indicate the possibility of upwelling along the northwest Australian shelf and led Schott (1935) to postulate upwelling there. But an examination by Wyrtki (1962) of the oceanographical conditions in the region between Java and Australia during the southeast monsoon season shows that the main upwelling in this region is situated along the coast of Java and Sumbawa, and not the northwest Australian shelf. Indeed the amount of upwelling south of Java is estimated to contribute 2.4 million m^3 sec^{-1} to the South Equatorial Current and the upwelling velocity is of the order of 0.5×10^{-3} cm sec^{-1}. Preliminary assessment of the results of several recent cruises along the coast of Australia between 35°S and 25°S shows no clear evidence of upwelling nor of an appreciable north-flowing West Australian Current (Hamon, pers. comm.).

On the western side of the Indian Ocean, upwelling develops along the south Arabian coast during the southwest monsoon (Bobzin, 1922). During the International Indian Ocean Expedition more detailed investigation of the probable upwelling regions was made (Currie, 1966). The first cruise of the R.R.S. *Discovery* during the Indian Ocean Expedition, was made to study the southeast Arabian upwelling region. The survey was made during the southwest monsoon of 1963 and five lines of stations were occupied between the end of June and end of August, which indicated the general distribution of upwelled water along the Arabian coast (Royal Society, 1963). Upwelling was present over some 1000 km of the coast and was the result of the wind stress parallel to the coast. From shipboard wind observations, the Ekman transport directed offshore was estimated to be $10 \times 10^6 m^3$ sec^{-1}. Observations made the following year in March and in May by the R.R.S. *Discovery* during the inter-monsoon period showed little or no evidence of upwelling on the coast (Royal Society, 1965). Duing (1966) discusses the vertical motion along the coasts of the Arabian Sea during the northeast monsoon.

Another cold water area is in the vicinity of the Somali coast where the western boundary current, which develops during the southwest monsoon, flows from the Southern Hemisphere up the east coast of Africa into the southern part of the Arabian Sea. The current leaves the coast at about 10°N and it is here that the coldest water is observed. This unique seasonal western boundary current is discussed in papers by Stommel and Wooster (1965), by Swallow and Bruce (1966), and by Warren, Stommel and Swallow (1966). The cold upwelling area is apparently related to baroclinic adjustment in an accelerating current. The winds in the region are consistent, however, with an off-shore Ekman transport.

Equatorial Regions

The relatively low temperatures in the equatorial regions of the Atlantic and Pacific, apparent on charts of surface temperature (Böhnecke, 1936; Schott, 1935) are not simply the results of advection from the Benguela and Peru Currents. Defant (1936b) and Sverdrup, Johnson and Fleming (1942) recognized two zones of divergences in the equatorial regions, one at the equator and the other at the northward edge of the Equatorial Countercurrent. These divergences bring water of low temperature, low oxygen content, and high nutrient content into the upper layers. Like the upwelling regions on the eastern boundaries of the ocean, the equatorial region is characterized by green water and an abundance of fish which is in contrast to the cobalt blue and desert-like regions of the trade wind currents (Dietrich, 1957).

Since World War II, the eastern equatorial region of the Pacific has been rather intensively studied, both because of the economic importance of the fisheries and, more recently, because of the interesting oceanographical features, e.g., the Cromwell Undercurrent. A comprehensive review of the region

is given by Wyrtki (1966). It is known generally that the equatorial upwelling is related to the Undercurrent. In the pronounced shear and mixing associated with the Undercurrent, the mixed water ascends into the surface layer and the "upwelling" effects are intensified by the wind-induced divergence at the surface. Cromwell's (1953) discussion of the divergence at the equator and resulting upwelling in the Pacific east of about 180° meridian, is still qualitatively correct although it was made before the discovery of the Undercurrent. The upwelling, with the associated low-temperature water, low oxygen content and rich nutrients, extends west along the equator from the Galapagos Islands to about 180° (Austin, 1958; Wooster and Cromwell, 1958). Observations indicate that the effects of this upwelling extend to 50 m in the eastern Pacific and 100-150 m in the central Pacific (Wooster and Jennings, 1955), consistent with the mean depth of the thermocline increasing westward (Wooster and Cromwell, 1958). Austin and Rinkel (1958) discuss the variation in upwelling in the equatorial Pacific. They state that the most persistent and pronounced upwelling is in the eastern and central Pacific, decreasing or being absent in the western Pacific. They see no evidence of cooler water at the surface or even the doming of subsurface isotherms in sections at 164°E and 167°E. Tsuchiya (1961) also found no strong evidence of upwelling at the equator in the western Pacific. Noting that the winds in the far eastern Pacific are more southerly than easterly, Austin and Rinkel point out that with a very shallow thermocline, wind-mixing may play a more important role in bringing cool water to the surface than does the wind divergence and upwelling.

Cromwell (1958) attempted to make a distinction between upwelling and ridging. The ascending motion at the intense divergence at the equator is upwelling, and is usually strong enough for the isotherms to intersect the sea below. Cromwell does not call the divergence along the southern boundary of the North Equatorial Current, i.e., the northern edge of the countercurrent, up-welling but rather ridging, because of the low phosphate. Wyrtki (1966) feels, how-ever, that the distinction between upwelling and ridging reflects only the intensity of divergence rather than the type of associated circulation.

In the Atlantic recent work has concen-trated on the Equatorial Undercurrent and the upwelling associated is noted only in passing (Neumann, 1960; Knauss, 1963). Dietrich (1957) associated the cold water which is sometimes present along the Gold Coast of Africa with the upwelling at the northern edge of the Equatorial Counter-current (Guinea Current). Upwelling here has been discussed by Howat (1945) and mentioned by Lawson (1966).

The winds along the equator in the Indian Ocean change markedly with the seasons, and for this reason the region was of particu-lar interest during the International Indian Ocean Expedition. The first papers (Knauss and Taft, 1963; Knauss, 1963) reported strong but not steady currents in the region of the thermocline and no evidence of up-welling, vertical mixing, or spreading of the thermocline, as observed in the Atlantic and Pacific. After further sections were made, Knauss and Taft (1964) concluded that "even in the Indian Ocean where the under-current is not as well developed or as steady as in the other oceans, these indications of upwelling of water from the thermocline may be found."

There is also upwelling in the eastern tropical Pacific which is not directly asso-ciated with either the eastern boundary cur-rents or the Equatorial Undercurrent. The upwelling is of sufficient strength and dura-tion to appear clearly in the monthly charts of sea surface temperature recently pro-duced by Wyrtki (1964a). The effects of upwelling may be seen in the Gulf of Tehuantepec from November to March, and in the Gulf of Panama from February to April. Upwelling in the Costa Rica Dome, so called because the thermocline rises to with-in 10 m of the surface, is indicated from December to May.

The Gulf of Tehuantepec (about 15°N, 95°W) lies in the region of the warmest surface water in the eastern Pacific, but dur-ing the winter months this water is at least 1°C colder than the rest. The region has been described by Holmes and Blackburn (1960) and Blackburn (1962). The relation between winds and sea surface temperature was investigated by Roden (1961). The low winter temperatures are found to be related to the wind-induced mass transport diver-gence. The vertical velocities computed from the curl of the wind stress are of the order of 10 m day^{-1} during northerly gales.

The upwelling in the Gulf of Panama has been recently discussed by Schaefer, Bishop and Howard (1958), Roden (1962) and Forsbergh (1963). Schaefer, Bishop and Howard found upwelling in the Gulf of Pan-ama during northerly winds (November to March). There is the usual influence on the distribution of properties, and associated with the upwelling there is a seasonal de-crease in sea level and surface temperature at Balboa. Roden (1962) carried out a statis-tical analysis of the sea level and wind re-cords in the Gulf of Panama and showed the temperature and sea level are related inversely to the winds, as is typical of up-welling regions. The power spectra showed most of the energy at low frequencies and no significant periodicities.

The Costa Rica Dome is a large cyclonic eddy located in the region where part of the countercurrent turns into the North Equa-torial Current off Central America. Wyrtki (1964b) has explained the upwelling in the center of the Costa Rica Dome by the trans-verse circulation with the cyclonic flow around the dome. This upwelling is not directly wind-induced but is nevertheless in-tense, the vertical transport adding 7×10^{10} cm^3 sec^{-1} to the surface circulation of 20×10^{12} cm^3 sec^{-1}. The dome appears to be in thermal balance and the nutrients added by upwelling agree with the observed high pro-ductivity of the area.

Antarctic

In Antarctic waters, between about 40°S and 60°S, strong westerly winds drive the West Wind Drift, the eastward flowing circumpolar current. Due to the effect of the Coriolis force there is, of course, some northward component to the surface flow. To the south, near the Antarctic Continent, easterly winds blow causing some southward as well as westward movement. The region between the West and East Wind Drifts is thus a region of diverging flow in which deep water tends to upwell (Deacon, 1963). This divergence shows up close to the Continent by the high temperature and salinity of the water at 100 m (Sverdrup, 1943). Owing to the ascending motion, nutrient-enriched deep water is brought into the surface layer making possible the extraordinary blooms of plankton. The plankton forms the food for "Krill" (*Euphausia superba*) which is eaten by the blue and the finback whales. This is why the Antarctic divergence zones are the main whaling grounds (Dietrich, 1957). There is another significant divergence zone in the Atlantic Antarctic, the Bouvet Divergence, between the Antarctic Divergence and the Antarctic Convergence to the north (Dietrich, 1957). The evidence for a Subantarctic Divergence between the Antarctic Convergence and the Subtropical Convergence is discussed by Houtman (1967).

Other Regions

In the regions discussed above, the major atmospheric circulation systems are favorable to upwelling over extended areas and periods. Upwelling of more limited extent can be found elsewhere, when and where the local winds transport water in the surface layers offshore.

Such is the case of the upwelling in the Bay of Bengal on the coast of India (LaFond, 1958a, b; Ramasatry and Balaramamurty, 1958). LaFond (1958a) attributes the upwelling to the persistent southwest monsoon, which blows nearly parallel to the coast from February to September. Jayaraman (1965) doubts the large scale upwelling off the east coast and claims there is no evidence of marked upwelling along the east coast of continents in the Northern Hemisphere with the exception of the Somali-Arabian coast. He does argue that significant upwelling occurs along the southwest coast of India.

Localized upwelling does occur along the east coast of North America. A correlation with wind has been obtained with upwelling off Nova Scotia (Hachey, 1937; Longard and Banks, 1952). Upwelling has been reported along the north-eastern coast of Florida (Green, 1944; Taylor and Stewart, 1959), North Carolina (Wells and Gray, 1960), and in the eastern end of the Gulf of Carico on the Caribbean coast of Venezuela (Richards, 1960; Fukuoka, Ballester and Cervigon, 1964). Upwelling is also found in smaller bodies of water, e.g., Lake Michigan (Moffett, 1962).

EFFECTS OF UPWELLING

Temperature

Upwelling, through vertical advection and the accompanying horizontal advection, will markedly alter the distribution of physical and chemical properties. Much to the annoyance of bathers in regions of coastal upwelling, the surface water temperature is much below normal for the latitude! The vertical sections of temperature in an upwelling region show the shallow isotherms ascending steeply. In the eastern boundary currents there is the general rise of the thermocline (and isotherms) toward the coast reflecting the distribution of mass associated with the equatorward geostrophic flow. With upwelling the thermocline rises steeply near the coast and may even intersect the surface; thus the vertical temperature gradient may be decreased inshore. There are relatively large horizontal gradients in the surface temperature. For example in July and August, the months of most intense upwelling along the Oregon coast, the water at 10 m depth 5 nautical miles off Newport, Oregon, is characteristically 8°C cooler than the water at the depth seaward of the shelf edge. In the winter the difference is seldom even 1°C.

Equatorial upwelling is apparent in the relative minimum in the surface temperature along the equator in the eastern and central Pacific. East of 180° there is cooler water at the equator than for several degrees of latitude poleward (Austin and Rinkel, 1958). The meridional gradient in the eastern equatorial Pacific is of the order of 2°C in 100 miles.

Where there is summer upwelling, the annual range of surface water temperature will be suppressed. Along the west coast of the United States the annual range for the coastal water is less than 3°C, but offshore the range is in excess of 7°C. Winter upwelling, as along the coast of Peru, accentuates the annual range of the surface temperature. The observed annual inshore and offshore temperature ranges in coastal upwelling regions are listed by Wooster and Reid (1963). A qualitative comparison with the Ekman transport for the regions shows general agreement.

Salinity

Only in the upwelling region of the California Current does the salinity generally increase with depth. In other regions, as in most of the world ocean, the salinity decreases with depth to a minimum value at several hundred meters. The usual effect of upwelling is thus to decrease the surface salinity, and anomalously low salinity and temperature are thus indicative of upwelling. Off the west coast of North America, however, upwelling increases the surface salinity.

Density

The density structure follows the temperature structure, with isopycnals rising in upwelling regions. Off the west coast of North America, the combined effect of the thermocline and halocline is to give a very strong pycnocline which may intersect the surface as a front during intense upwelling.

The vertical density gradient below 100 meters or so, off many of the upwelling coasts, decreases because of the descent of the isopycnals toward the coast at depths below 100 meters. This is apparently associated with the poleward undercurrent indicated in the theoretical and observational studies of upwelling regions. The eastward subsurface countercurrent along the equator may be an analogous case for equatorial upwelling.

Oxygen

In general, the oxygen content of the surface waters of the ocean is near saturation. The oxygen content decreases with increasing depth to a minimum at intermediate depths. During upwelling subsurface water of relatively low oxygen content ascends to the surface, and the surface waters may become markedly undersaturated. This is particularly common in the major upwelling regions off the west coasts of the continents. Posner (1957) has reported a surface water oxygen content in the Peru Current region as low as 35% of saturation. Hart and Currie (1960) report a phenomenally low 6% off the southwest coast of Africa. Values of 60-70% saturation are common off Oregon during upwelling (Park, Pattullo and Wyatt, 1962).

Pearson and Holt (1960) have found that upwelling may negate conclusions in the pollution analysis of estuaries which are based on the dissolved oxygen-content, unless the oxygen content of the water at the ocean source is determined. In an analysis of the estuary at Grays Harbor, Washington, the oxygen deficiency due to upwelling, as compared with the normally assumed saturation levels for ocean water, was the equivalent of the oxygen demand associated with the domestic sewage discharge of 20 million people.

Phosphate

Phosphate, or dissolved inorganic phosphorous, generally increases with depth from very low values at the surface to maximum values at some intermediate depth. It is one of the principal nutrients and therefore important to life in the sea. High values occur in upwelling regions where subsurface water comes to the surface. The usual mid-latitude oceanic surface concentration is of the order of 0.2 μg-atom/1. In the upwelling regions of Peru (Posner, 1957) and the Benguela Current (Hart and Currie, 1960) the surface concentration is between 1 and 2 g-atom/1. Values of over 2 μg-atom/1 were found at the surface of the south-east Arabian coast (Royal Society, 1963), and Laurs (1967) has reported values of 2.5 μg-atom/1 near the coast during intensive upwelling off Oregon.

Sea Level

Upwelling can be expected to have an effect on sea level through steric variation, i.e., the change in height of the sea surface arising from isostatic adjustments due to changes in the specific volume of the water column (Pattullo, Munk, Revelle and Strong, 1955). Upwelling, by replacing warm, less dense surface water with cooler, more dense water, increases the mean density of the water column; in adjusting isostatically the mean sea surface is lowered. Schaefer, Bishop and Howard (1958), in the study of upwelling in the Gulf of Panama, reported that a decrease in the surface temperature and the sea level at Balboa was associated with upwelling. Taylor and Stewart (1959) observed that periods of colder water, along the northeastern coast of Florida during the summer, were accompanied by a lowering of sea level. The phenomenon was associated with winds from directions most conducive to producing upwelling. Stewart (1960) has discussed the relation of coastal water temperature and sea level along the Pacific coast of the United States and attributes some of the variation to upwelling.

Panshin (1967) investigated the relation of sea level winds and upwelling along the Oregon coast, using data taken by the U.S. Coast and Geodetic Survey in 1933 and 1934. The monthly mean sea level, adjusted for the effects of atmospheric pressure variations, was lowest in July, generally a month of intense upwelling. The July monthly mean at Newport, Oregon, was 15 cm below the annual mean. No hydrographical data were available, but regression analysis was used to establish the relation between daily mean sea level and winds. The relation was significant and the correlation was most strong for longshore winds averaged over four preceding days; mean sea level was lowered by northerly winds, consistent with the idea that upwelling induced variations in sea level.

Climate

Upwelling has a marked effect on the climate along the western coasts of the continents. It is true that along these coasts the eastern boundary currents flow from higher latitudes, and the temperature of the water lags behind the new surroundings. But upwelling makes the west coasts of the continents at these latitudes still cooler than they would be otherwise. The surface water is coldest at San Francisco in April and May, the months of the most intense upwelling there (Sverdrup, 1943). The air temperature is close to the water temperature but generally somewhat higher. The mean air temperature at San Francisco does not reach its maximum until September, after upwelling has ceased, while at Sacramento, 100 miles inland and unaffected by upwelling, the maximum mean air temperature is reached in July.

The presence of cold upwelled water along the coast cools the air and increases the relative humidity. Low stratus clouds or fog are therefore common and in general the layer of moist air is shallow, and there is warm air aloft. The clouds are seldom deep enough, nor is there sufficient atmospheric moisture to yield much precipitation, though the fog may be so thick as to be almost a drizzle. The prediction of fog and the relation of fog and upwelling on the southern California coast have been studied by Petterssen (1938) and, more recently, by Cohen (1966).

Where tropical arid climates are located along the west coasts of the continents, upwelling modifies the climate and fog is common. Such regions are described climatologically as cool and humid coastal deserts (Haurwitz and Austin, 1944). The principal examples are coastal southern California and western Baja California (Mexico), the Peru and Atacama deserts of coastal Peru and Chile, the Sahara along the northwest African coast (Canary Current region) and the Namib desert in southeast Africa (Benguela Current region). The thick mist or fog typical of such upwelling regions is also encountered off the southeast Arabian coast during the upwelling associated with the southwest monsoon (Royal Society, 1963).

Upwelling contributes to the diurnal sea breeze, which is a particularly common phenomenon along coasts affected by upwelling. The pressure gradient from sea to land, established by the differential heating of sea and land, results in onshore winds which bring cool moist air as much as 50 miles inland. Since the sea breeze is, in effect, driven by the difference in the temperature over the land and over the sea (Haurwitz, 1947) upwelling increases the strength of the sea breeze. Lowry (1962) has studied the sea breeze along the Oregon coast.

Lane (1965b) has carried out a study of the climate and heat exchange in the upwelling region off Oregon using the classical empirical equations. In the summer, during upwelling, the temperature of the water, air, and wet bulb are lower in the coastal upwelling region than further to sea. The converse is true during the winter (non-upwelling period). Upwelling in summer affects the heat budget by slightly reducing the back

radiation, greatly reducing the conduction from the sea to the atmosphere (conduction to the sea may occur frequently), and greatly reducing the heat loss due to evaporation. Despite the reduced evaporation there is an increase in the relative humidity. These conclusions, found by Lane, appear to hold true in most coastal upwelling regions.

Biota

The high productivity of upwelling regions is due to the increased supply of nutrients brought into the euphotic zone from below. The production of organic matter is limited to the surface layers where the light is sufficient for photosynthesis. Nutrient supplies are generally depleted in the surface layers, but where there is an upward transport of nutrients from the higher concentrations below, productivity can be high and continuing.

It should be pointed out that the shallowness of the mixed layer and the velocity shear associated with upwelling are conducive to the turbulent transport of nutrients. Thus, nutrients can be brought into the surface layer from below by turbulent processes (mixing) which, while not upwelling *per se,* are part of the upwelling dynamics. Posner (1957) attributed half of the phosphate brought into the surface layer off Peru to vertical advection (upwelling) and half to vertical eddy diffusion.

Primary productivity can be measured with ^{14}C techniques; the quantitative measure is the mass of inorganic carbon converted to organic carbon by photosynthesis per unit time and per unit volume or surface area of ocean. An estimate of an average for the world ocean might be 0.2 g C m^{-2}day^{-1} (Steemann Nielsen and Jensen, 1957). Values of over 1 g C m^{-2}day^{-1}, reported by Holmes, Schaefer and Shimada (1957) for a station in the Peru Current region and by Anderson (1964) during upwelling off the Oregon coast, are perhaps typical of coastal upwelling regions. The standing crop of phytoplankton, measured in terms of the chlorophyll a concentration, for the above regions were 2.0 mg m^{-3} and up to 8.0 mg m^{-3} respectively. During upwelling off the southeast Arabian coast the concentration of chlorophyll a was 4 to 5 mg m^{-3} (Royal Society, 1963). Other recent work involving the relation of primary productivity and coastal upwelling is to be found in papers by Curl and Small (1965), Bolin and Abbott (1963), and Smayda (1966). The high primary production of equatorial regions is discussed by Doty and Oguri (1958) and by Vinogradov and Voronina (1962).

It is not always easy to obtain quantitative correlations between upwelling and the standing crop at the various trophic levels. As Blackburn (1965) has pointed out, the production of phytoplankton in an upwelling region may be temporarily limited by the coldness of the water, some time will elapse before the phytoplankton is transformed into the zooplankton, and still more time will elapse before the zooplankton are eaten by the primary carnivores, etc., during which time the biota involved may be transported away from the site of the upwelling and initial high primary productivity.

Qualitatively the relation between upwelling and at least the lower trophic levels is readily apparent. The standing crops of zooplankton are high in the coastal regions of the eastern boundary currents (Wooster and Reid, 1963); the anchovy fishery and the guano industry of Peru are well known.

Forsbergh (1963) in a study of the relationship of meteorological, hydrographical, and biological variables in the Gulf of Panama showed the productivity and standing crop of phytoplankton to be significantly greater during upwelling. He also obtained an "almost" [sic] significant correlation of the high zooplankton crop during the upwelling period (January to April) and the catch per unit effort of skipjack tuna from April to June, with data from five successive years. Smayda (1966) also working in the Gulf of Panama, found a correlation among coastal upwelling, winds, phytoplankton, zooplankton, and anchovies.

Laurs (1967) has studied the relation of the ecology of the lower trophic levels (phytoplankton, herbivores, and primary carnivores) to the coastal upwelling off Oregon. The standing stock of phytoplankton was highest inshore during upwelling, and the highest standing stocks of primary carnivores tended to be found successively seaward as the upwelling season progressed. The highest standing stock of herbivores and primary carnivores occurred in the fall, after upwelling began to subside, in the region where the front had intersected the surface.

Studies of the relation of zooplankton, foraging organisms, and tuna are found in the papers of King (1958) and Murphy and Shomura (1958), while Blackburn (1965) discusses the ecology of tuna and upwelling; further references to that subject can be found in his paper.

More detailed studies of the biological implications of upwelling in the Peru Current region can be found in the work of Gunther (1936) and Posner (1957). Wyrtki (1966), in his review of eastern equatorial Pacific oceanography, cites more references to the Peru Current region and to the equatorial region of the Pacific. The work of Hart and Currie (1960) is a comprehensive study of the biology in the Benguela Current upwelling region. For more information on the biological effects in the California Current upwelling region the reader is referred to the Progress Reports of the California Cooperative Oceanic Fisheries Investigations.

In summary, a quotation from Wooster and Reid (1963) is appropriate: "If one estimates that at least four million tons of anchovy can be removed in a year from the inshore water of Peru, by birds and man, this is equivalent to about one-seventh of the annual world landings of fish. Yet these fish are caught in a coastal strip less than 800 miles long and 30 miles wide, or only about 0.02% of the surface of the world ocean. Similar figures could be cited for other eastern boundary current regions to emphasize the great fertility of their waters."

EPILOGUE

Defant (1961) in summarizing the upwelling problem has stated: ". . . there is no doubt that the upwelling and sinking of oceanic waters is primarily connected with convergence and divergence regions, occurring at the sea surface. The cause of these divergences and convergences in most cases lies in the distribution of wind stress exerted by the prevailing wind on the sea surface. A totally satisfying explanation of upwelling at continental coasts has not yet been given, and is probably not possible at all since the total process is composed of a number of substages each of which is always controlled by other factors."

To understand the upwelling process and its interactions with the other energetic phenomena is certainly not an easy task, but with the increasing ability of the oceanographer to make *in situ* time-series measurements from buoys, advances in mathematical and statistical analyses, and the increasing facilities for analysis afforded by the computer, our understanding will increase. Indeed it must, for there are two obvious and immediate practical concerns whose solutions may lie in a better understanding of upwelling: food from the sea and waste disposal. To be sure, the answer is not just in understanding upwelling. With regard to the utilization of upwelling in aquiculture, MacDonald (1967) has said: "the problems involved are biological as well as technological, and their solution requires the marriage of engineering and marine biology on a scale not previously attempted." Similar statements can be made by those who see among the wonders of the sea its availability to receive waste products—like some great hole in the ground (Fϕyn, 1965). But if the waste is to be pumped into the sea off coasts where upwelling is prevalent, it is apparent that an adequate scientific knowledge of the upwelling process is needed.

Perhaps the central oceanographical problem can be stated as prediction. One cannot expect a "totally satisfying" analytical solution to the equations of motion so that one can "plug in" the wind stress and "get out" the oceanic response. But numerical solutions of the equations with time-varying forcing functions, in principle possible with modern computers and being done in meteorology with some success, can "predict" the oceanic response. Another approach to the prediction problem, using the principles of time-series analysis, is through Wiener's prediction theory (1949). The oceanic region of interest could be viewed as a dynamic physical system with meteorological and oceanographical inputs and outputs. By studying the input and output time series the various transfer functions (the "hydrodynamic impedances") can be obtained. A mathematical or electronic prediction system could be made using the prior knowledge of transfer functions and real time data acquisition.

A start in the study of upwelling by these means has been made by O'Brien (1967) in his computerized study of time-dependent solutions of the upwelling induced by a hurricane, and in the studies by Mooers and Smith (1967) and by Collins (1968) with time-series data from a coastal upwelling region.

REFERENCES

Acosta, J. de, 1604. *The Naturall and Morall Historie of the East and West Indies,* London. (Reprinted: 1880, Hakluyt Society, London, No. 60, 295 pp.)

Anderson, G. C., 1964. *Limnol. Oceanogr.* **9,** 284-302.

Austin, T. S., 1958. *Trans. Am. geophys. Un.* **39,** 1055-1063.

Austin, T. S. and Rinkel, M. O., 1958. *Proc. Ninth Pacif. Sci. Congr.* **16,** 67-71.

Bjerknes, J., 1961. *Bull. inter-Am. trop. Tuna Commn.* **5,** 217-303.

Bjerknes, J., 1966. *Bull. inter-Am. Trop. Tuna Commn.* **12,** 25-86.

Blackburn, M., 1962. *Spec. scient. Rep. U.S. Fish. Wildl. Serv. Fish.* No. 404, 28 pp.

Blackburn, M., 1965. In, *Oceanogr. Mar. Biol. Ann. Rev.*, edited by H. Barnes, George Allen and Unwin Ltd., London, 3, 299-322.

Bobzin, E., 1922. *Dt. übersee. met Beob.* 23, H 1-18.

Böhnecke, G., 1936. *Wiss. Ergebn. dt. atlant. Exped. 'Meteor'*, 5, Part 2, 74 charts.

Bolin, R. L. and Abbott, D. P., 1963. *Prog. Rep. Calif. coop. ocean. Fish. Invest.* 9, 23-45.

Buchan, A., 1895. *Rep. scient. Res. Voy. 'Challenger'*, Physics and Chemistry, Part 8, Appendix, 33 pp.

Cohen, S. H., 1966. *Bull. Am. met. Soc.* 47, 486 only (abstract).

Collins, C. A., 1968. Ph. D. Thesis, Oregon State University, Corvallis, 154 pp.

Cromwell, T., 1953. *J. mar. Res.* 12, 196-213.

Cromwell, T., 1958. *Bull. inter-Am. trop. Tuna Commn.* 3, 135-164.

Curl, H. and Small, L. F., 1965. *Limnol. Oceanogr.* 10, R 67-73.

Currie, R. I., 1953. *Nature, Lond.* 171, 497-500.

Currie, R. I., 1966. In, *Oceanogr. Mar. Biol. Ann. Rev.*, edited by H. Barnes, George Allen and Unwin Ltd., London, 4, 69-78.

Cushing, D. H. 1969. "Upwelling and fish production." *FAO Fish. Tech. Pap.*, 84, 40 pp.

Darbyshire, M., 1966. *Deep Sea Res.* 13, 57-81.

Darbyshire, M., 1967. *Deep Sea Res.* 14, 279 only.

Deacon, G. E. R., 1963. In, *The Sea*, Vol. 2, edited by M. N. Hill, Interscience Publishers, New York, 281-296.

Defant, A., 1936a. *Landerkdl. Forsch., Festschr. N. Krebs* 52-66.

Defant, A., 1936b. *Wiss. Ergebn. dt. atlant. Exped. 'Meteor'* 6, Pt. 1, 289-411.

Defant, A., 1961. *Physical Oceanography*, Vol. 1. Pergamon Press, New York, 729 pp.

Dietrich, G., 1957. *Allgemeine Meereskunde*. Borntraeger, Berlin, 492 pp.

Doty, M. S. and Oguri, M., 1958. *Proc. Ninth Pacif. Sci. Congr.* 16, 94-97.

Duing, W., 1966. *Second Int. Oceanogr. Congr.* (abstracts), 103 only.

Ekman, V. W., 1905. *Ark. Mat. Astr. Fys.* 2, 11, 1-52.

Forsbergh, E. D., 1963. *Bull. inter-Am. trop. Tuna Commn.* 7, 1-109.

Føyn, E., 1965. In, *Oceanogr. Mar. Biol. Ann. Rev.*, edited by H. Barnes, George Allen and Unwin Ltd., London, 3, 95-114.

Fukuoka, J., Ballester, A. and Cervigon, F., 1964. In, *Studies on Oceanography*, edited by K. Yoshida, Univ. of Tokyo, Tokyo, 145-149.

Green, C. K., 1944. *Science, N.Y.* 100, 546-547.

Gunther, E. R., 1936. *Discovery' Rep.* 13, 107-276.

Hachey, H. B., 1937. *Proc. Trans. Nova Scotian Inst. Sci.* 19, 264-276.

Hart, T. J. and Currie, R. I., 1960. *'Discovery' Rep.* 31, 123-298.

Haurwitz, B., 1947. *J. Met.* 4, 1-8.

Haurwitz, G. and Austin, J. M., 1944. *Climatology.* McGraw-Hill, New York, 410 pp.

Hidaka, K., 1958. *Rec. oceanogr. Wks Japan*, 4, 2, 77-123.

Holmes, R. W. and Blackburn, M., 1960. *Spec. scient. Rep. U.S. Fish Wildl. Serv.*, Fish. No. 345, 106 pp.

Holmes, R. W., Schaefer, M. B. and Shimada, B. M., 1957. *Bull. inter-Am. trop. Tuna Commn.* 2, 129-156.

Houtman, T. J., 1967. *N.Z. Dep. sci. industr. Res. Bull.* No. 174, 40 pp.

Howat, G. W., 1945. *Nature, Lond.* 155, 415-417.

Instituto de Investigaciones Pesqueras. 1971. Analysis of upwelling systems. *Inv. Pesq.* 35(1), 1-362.

Jayaraman, R., 1965. *Curr. Sci.* 34, 121-122.

King, J. E., 1958. *Proc. Ninth Pacif. Sci. Congr.* 16, 98-107.

Knauss, J. A., 1963. *Trans. Am. geophys. Un.* 44, 447-478.

Knauss, J. A. and Taft, B. A., 1963. *Nature, Lond.* 198, 376-377.

Knauss, J. A. and Taft, B. A., 1964. *Science, N.Y.* 143, 354-356.

LaFond, E. C., 1958a. *Proc. Ninth Pacif. Sci. Congr.* 16, 80 only.

LaFond, E. C., 1958b. *Proc. Indian Acad. Sci.* (B) 46, 1-46.

Lane, R. K., 1965a. *Res. Briefs Fish Commn. Ore.* 11, 25-28.

Lane, R. K., 1965b. Ph.D. Thesis, Oregon State University, Corvallis, 115 pp.

Laurs, R. M., 1967. Ph.D. Thesis, Oregon State University, Corvallis, 121 pp.

Lawson, G. W., 1966. In, *Oceanogr. Mar. Biol. Ann. Rev.,* edited by H. Barnes, George Allen and Unwin Ltd., London, **4,** 405-448.

Longard, J. R. and Banks, R. E., 1952. *Trans. Am. geophys. Un.* **33,** 377-380.

Lowry, W. P., 1962. Ph.D. Thesis, Oregon State University, Corvallis, 216 pp.

MacDonald, G. J. F., 1967. *Int. Sci. & Technol.* April 1967, 38-48.

McEwen, G. F., 1912. *Int. Revue ges. Hydrobiol. Hydrogr.* **5,** 243-286.

Moffett, J. W., 1962. *Proc. Fifth Conf. Gt Lakes Res.,* 1962, 126 only.

Mooers, C. N. K. and Smith, R. L., 1967. *Symposium on Mean Sea Level* (abstracts), 7 only.

Murphy, G. I. and Shomura, R. S., 1958. *Proc. Ninth Pacif. Sci. Congr.* **16,** 108-113.

National Academy of Sciences. 1969. *An Oceanic Quest.* Washington, D. C. 115 pp.

Neumann, G., 1960. *Deep Sea Res.* **6,** 328-334.

O'Brien, J. J., 1967. *J. atmos. Sci.* **24,** 208-215.

Orren, M. J. and Shannon, L. V., 1967. *Deep Sea Res.* **14,** 279 only.

Panshin, D. A., 1967. M.S. Thesis, Oregon State University, Coravallis, 71 pp.

Park, K., Pattullo, J. G. and Wyatt, B., 1962. *Limnol. Oceanogr.* **7,** 435-437.

Pattullo, J. G., Munk, W., Revelle, R. and Strong, E., 1955. *J. mar. Res.* **14,** 88-155.

Pearson, E. A. and Holt, G. A., 1960. *Limnol. Oceanogr.* **5,** 48-56.

Petterssen, S., 1938. *Bull. Am. met. Soc.* **19,** 49-55.

Posner, G. S., 1957. *Bull. Bingham oceanogr. Coll.* **16,** 106-155.

Ramasastry, A. A. and Balaramamurty, C., 1958. *Proc. Indian Acad. Sci.* (B) **46,** 293-323.

Reid, J. L., 1960. *Prog. Rep. Calif. coop. ocean. Fish. Invest.* **7,** 77-90.

Reid, J. L., 1962. *J. mar. Res.* **20,** 134-137.

Reid, J. L., Roden, G. I. and Wyllie, J. G., 1958. *Prog. Rep. Calif. coop. ocean. Fish. Invest.* 28-57.

Richards, F. A., 1960. *Deep Sea Res.* **7,** 163-182.

Roden, G. I., 1961. *Geofisica Internacional.* **1,** 55-76.

Roden, G. I., 1962. *Geofisica Internacional.* **2,** 77-92.

Royal Society, 1963. International Indian Ocean Expedition R.R.S. *Discovery* Cruise 1. Cruise Report, London, 24 pp.

Royal Society, 1965. International Indian Ocean Expedition R.R.S. *Discovery* Cruise 3. Cruise Report, London, 55 pp.

Ryther, J. H. 1969. Photosynthesis and fish production in the sea. *Science.* **166,** 72-76.

Schaefer, M. B., Bishop, Y. M. and Howard, G. V., 1958. *Bull. inter-Am. trop. Tuna Commn.* **3,** 79-132.

Schell, I. I., 1965. *J. geophys. Res.* **70,** 5529-5540.

Schott, G., 1931. *Annln Hydrogr. Berl.* **59,** 161-169, 200-213, 240-252.

Schott, G., 1933. *Annln Hydrogr. Berl.* **61,** 225-233.

Schott, G., 1935. *Geographie des Indischen und Stillen Ozeans.* Boysen, Hamburg, 413 pp.

Schweigger, E., 1958. *J. oceanogr. Soc. Japan.* **14,** 87-91.

Smayda, T., 1966. *Bull. inter-Am. trop. Tuna Commn.* **11,** 355-612.

Smith, R. L., 1964. Ph.D. Thesis, Oregon State University, Corvallis, 83 pp.

Smith, R. L., 1966. *Rep. Challenger Soc.* **3,** 18, 43 only.

Smith, R. L., Pattullo, J. G. and Lane, R. K., 1966. *J. geophys. Res.* **71,** 1135-1140.

Steemann, Nielsen, E. and Jensen, E. A., 1957. *Galathea Rep.* **1,** 49-136.

Stevenson, M. R., 1966. Ph.D. Thesis, Oregon State University, Corvallis, 140 pp.

Stewart, H. B., 1960. *Prog. Rep. Calif. coop. ocean. Fish. Invest.* **7,** 97-102.

Stewart, R. W. 1969. The atmosphere and the ocean. *Scientific American,* **221**(September), 76-86.

Stommel, H. and Wooster, W. S., 1965. *Proc. natn. Acad. Sci. U.S.A.* **54,** 8-13.

Sverdrup, H. U., 1938. *J. mar. Res.* **1,** 155-164.

Sverdrup, H. U., 1943. *Oceanography for Meteorologists.* Prentice-Hall, New York, 246 pp.

Sverdrup, H. U. and Fleming, R. H., 1941. *Bull. Scripps Instn Oceanogr.* **4**, 261-378.

Sverdrup, H. U., Johnson, M. W. and Fleming, R. H., 1942. *The Oceans.* Prentice-Hall, New York, 1087 pp.

Swallow, J. C. and Bruce, J. G., 1966. *Deep Sea Res.* **13**, 861-888.

Taylor, C. B. and Stewart, H. B., 1959. *J. geophys. Res.* **64**, 33-40.

Thorade, H., 1909. *Annln Hydrogr. Berl.* **37**, 17-34, 63-76.

Tsuchiya, M., 1961. *Oceanogrl Mag.* **13**, 1-30.

Vinogradov, M. and Voronina, N., 1962. *Rapp. P.-v. Réun. Cons. perm. int. Explor. Mer,* **153**, 200-204.

Warren, B., Stommel, H. and Swallow, J. C., 1966. *Deep Sea Res.* **13**, 825-860.

Wells, H. W. and Gray, I. E., 1960. *Limnol. Oceanogr.* **5**, 108-109.

Wiener, N., 1949. *Extrapolation, Interpolation, and Smoothing of Stationary Time Series.* Massachusetts Institute of Technology, Cambridge, 163 pp.

Witte, E., 1880. *Annln Hydrogr. Berl.* **8**, 192-193.

Wooster, W. S., 1960. *Prog. Rep. Calif. coop. ocean. Fish. Invest.* **7**, 43-45.

Wooster, W. S. and Cromwell, T., 1958. *Bull. Scripps Instn Oceanogr.* **7**, 169-282.

Wooster, W. S. and Gilmartin, M., 1961. *J. mar. Res.* **19**, 97-122.

Wooster, W. S. and Jennings, F., 1955. *Calif. Fish Game.* **41**, 79-90.

Wooster, W. S. and Reid, J. L., 1963. In, *The Sea,* Vol. 2, edited by M. N. Hill. Interscience Publishers, New York, 253-280.

Wyrtki, K., 1962. *Aust. J. mar. Freshwat. Res.* **13**, 217-225.

Wyrtki, K., 1963. *Bull. Scripps Instn Oceanogr.* **8**, 313-346.

Wyrtki, K., 1964a. *Dt. hydrogr. Z.,* Ergänz-Hft. (A), Nr. 6, 84 pp.

Wyrtki, K., 1964b. *Fishery Bull. Fish Wildl. Serv. U.S.,* **63**, 355-372.

Wyrtki, K., 1966. In, *Oceanogr. Mar. Biol. Ann. Rev.,* edited by H. Barnes, George Allen and Unwin Ltd., London, **4**, 33-68.

Yoshida, K., 1967. *Jap. J. Geophys.* **4**, 2, 1-75.

ADDENDUM

Several papers, which have appeared since the writing of this review, should be noted: Armstrong, Stearns and Strickland (1967) have reported on the measurement of the upwelling and subsequent biological processes using a Tecnicon Autoanalyzer to record continuously the surface properties while the ship was under way. Banse (1968) has given a thorough discussion of upwelling along the west coast (Arabian Sea) of India and Pakistan. Hidaka (1967) has computed the vertical velocity associated with equatorial upwelling from the equation of continuity using directly measured horizontal velocities. Hidaka's results indicate that the quality of the directly measured current data at present available are not accurate enough to permit reliable computations of vertical motion using the equation of continuity.

REFERENCES

Armstrong, F. A. J., Stearns, C. R. and Strickland, J. D. H., 1967. *Deep Sea Res.* **14**, 381-398.

Banse, K., 1968. *Deep Sea Res.* **15**, 45-79.

Hidaka, K., 1967. *La mer* **5**, 2, 23-31.

Deep Sea Tides

DAVID E. CARTWRIGHT   1969

The tides are by far the most predictable of all movements of the ocean. Yet their apparent regularity conceals mysteries which have taxed mathematicians for more than a century, and which demand the most advanced research tools available to oceanography for their solution.

The basic reason for this paradox is that one can predict tides, if necessary very accurately, without really understanding much about their mechanism. Fishermen have for ages known how to tell the times and heights of high water at their particular harbor from the phase of the Moon. Every seaside holiday maker learns that tomorrow's tide will be much the same as that observed today but about an hour later. Such predictions serve their humble purposes well enough, even though made in total ignorance of the laws of gravity and hydrodynamics.

The calculations made to produce the official tide tables are of course much more elaborate in execution but not so very different in principle. A year's records from a tide gauge at the place in question are analyzed for certain periodicities which occur in the motions of the Moon and the Sun. The analysis produces a series of constant tidal amplitudes and phases appropriate to the place, from which the tide there can be predicted almost indefinitely into the future. The fact that the tide tables are remarkably good—errors of 30 cm in height or 15 minutes in time are rare, except in bad weather—confirms the popular opinion that there is nothing more to be learnt about tides. One just consults the "Oracle." But these predictions tell us nothing about the behavior of the tide outside certain chosen ports nor, indeed, how the oceans respond as a whole to the Moon's and Sun's disturbing forces.

The real problems cannot be appreciated without understanding what is known about the origin of tides in the deep ocean. The upper of the two world maps (Figure 3-24) shows contours of the gravitational potential produced by a disturbing body such as the Moon which, for the moment, is overhead at the equator on the Greenwich meridian (0° longitude). A mass gains gravitational poten-

Dr. David E. Cartwright is a senior principal scientific officer at the National Institute of Oceanography, Surrey, Great Britain. His early research was on measurements and statistical theory of sea waves and ship motions. In his later studies on mean sea level and storm surges, he became increasingly involved with tides, and, since about 1963, his research has been largely on tides.

From *Science Journal*, Vol. 5, No. 1, pp. 60-67, 1969. Reprinted with author's revisions and light editing by permission of the author and Syndication International Limited, London.

tial on the earth normally by going uphill, so these contours are exactly similar to height contours on an ordinary map, although the heights involved are generally less than one meter. The water "feels" as if the local surface of the globe were tilted according to the

gradient of the contours and so it tends to move downhill towards the nearest "basin." There are two such basins where no potential gradient or tilt is felt. One is immediately under the Moon, where its attraction is greatest; the other is on the opposite side of

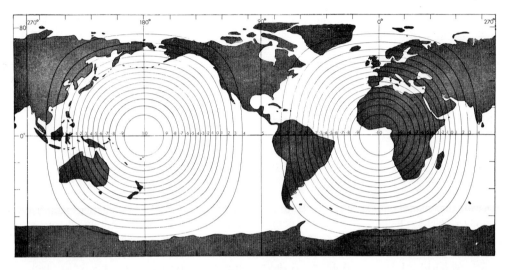

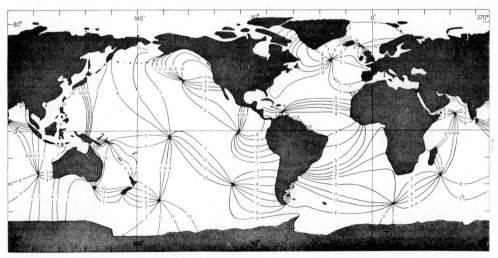

Figure 3-24. Contours of gravitational potential produced by a celestial body when it is above the equator at the Greenwich meridian (0° longitude) are shown in the top map. On a sphere these contours are actually circles. For the Moon 10 units represent 358 mm water; for the Sun they represent 162 mm. Contours in bottom map show positions of high tide at various hours after the situation represented in the top map, as estimated by G. Dietrich. Low tides are obtained by adding or subtracting six hours. At the nodal points, or amphidromes, some of which are conjectural, the rise and fall of the tide is zero.

the Earth (in the Pacific Ocean) where it is least. Here the local value of g is effectively reduced slightly, but this in itself causes no movement. Another zone where the water is undisturbed by the Moon is round the circular ridge halfway between the basins where the Moon's attraction is exactly balanced by the centrifugal effect of the orbit. The water movements which build up the tidal waves are caused only by the effective tilting where the contours are close, and this is greatest $45°$ away from the centers of the basins. However, the magnitude of the greatest tilt is only about 1 in 10^7, equivalent to, say, a rise of 100 mm of water across the 1000 km width of the Black Sea. So tides are very small in lakes and enclosed seas of that order of size. But across the major oceans of the Earth the gravitational potential difference becomes quite large. Between the center and the eastern edge of the Pacific Ocean, for example, it varies from -10 to $+5$ units of height, which for the Moon is equivalent to more than a half meter change in sea level.

If the Earth rotated at about one thirtieth of its present speed, so that the Moon stayed overhead for a long time, then sea level would adjust itself to match this potential almost exactly. Its slope would balance the potential gradient; bulges 358 mm high would build up at the basin centers and troughs 179 mm low would form around the $90°$ meridian. The hypothetical egg shape is known as the equilibrium tide. However, the pattern of disturbance sweeps around the Earth at more than 3000 kilometers per hour near the equator, so the equilibrium tide never gets a chance to establish itself. Water particles are accelerated at the potential gradients but, as they acquire velocity, they are deflected to the right north of the equator and to the left south of it because of the gyroscopic effect of the Earth's rotation. They are deflected horizontally at the coastlines and vertically at the underwater mountain ranges. The resulting "storm in a teacup" is sketched in the lower world map (Figure 3-25).

This tidal map, drawn by G. Dietrich in 1944, is largely guesswork, helped by measurements at coastlines and islands and some general knowledge of how tides behave in mathematically described basins. The curves delineate the positions of the tidal wave crests at successive lunar hours—a lunar hour is about 62 ordinary minutes—0 (or 12) corresponding to the time of the upper diagram. Heights vary along the wave crests, but height contours are not given because they are unknown and virtually impossible to extrapolate from coastal measurements. A striking feature of the map is the system of nodal points known as amphidromes around which the crest lines rotate. The vertical rise and fall of the tide is permanently zero at the amphidromes. The existence of such points is well established by theory but in practice very few of the amphidromes shown here have been substantiated by measurement. In fact, the French hydrographer C. Villain later produced a modified version of Dietrich's map in which at least one of Dietrich's amphidromes is omitted entirely and large areas are left blank for lack of evidence.

Nevertheless, some system of tidal waves, not wholly different from Dietrich's sketch, must be set up in the large oceans by the action of the Moon's gravity. Where these waves impinge on the shallow seas surrounding the continents, two things happen. The associated tidal currents are funnelled up from some 5000 meters depth to some 100 meters and hence strengthen by a factor of about 50; they cause friction on the bottom and magnify the effects of the Earth's rotation. For similar reasons the waves have to travel more slowly, so the amplitude increases to preserve the power transmission, in this case by $\sqrt{50}$ or about 7. Both these effects tend to make the tidal waves in coastal waters larger and more complicated than in the open ocean and more sensitive to small coastal irregularities. Notable features are the enormous local magnifications produced by France's Cotentin peninsula south of Cherbourg, where the range is more than

14 meters, and by Canada's Bay of Fundy, where the range is sometimes as much as 18 meters. The complexity of the tidal pattern in a small area such as the North Sea can be computed with some precision. However, such computations are not based upon the Moon's gravitational field at all but upon the known characteristics of the waves entering the area from the open sea. Thus, only a subsidiary technical problem is solved.

In broad outline, then, there are two connected but distinct zones of tidal activity: the wide deep ocean basins which respond directly to the external, gravitational disturbances; and the shallow seas surrounding the continents, in which the tides are driven by the system generated in the deep ocean basins but tend to have much larger amplitudes, especially in their associated tidal currents. Unfortunately, the vast majority of tidal measurements are crowded into certain parts of the shallow seas surrounding the continents. Practical requirements also tend to direct research activity into this zone. The problems of principal interest to tidal science, however, are mostly concerned with the deep oceans.

One problem area centers on the evaluation of tidal friction. The tides over the whole globe contain a large quantity of ener-

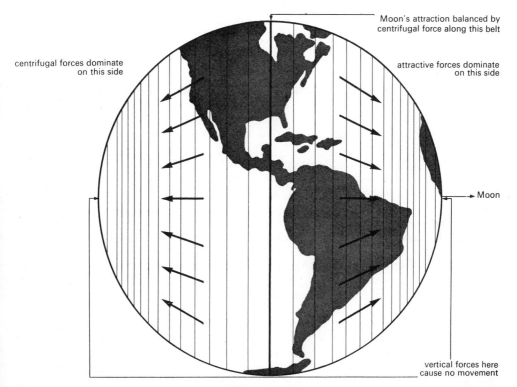

Figure 3-25. Tides are caused by the disturbances in gravitational potential produced by a celestial body. Contours of this potential shown here correspond to those in the top map on page 149; they are exactly similar to height contours on an ordinary map. The effect is as if the Earth's surface was being tilted locally according to the gradient indicated by the contours and so the water tends to move downhill towards the nearest "basin" where no tilt is felt. These basins occur directly under the Moon, where its attraction is greatest, and on the opposite side of the Earth, where the centrifugal force resulting from the Earth's rotation dominates. Movement of water towards these two points would produce an egg shape if it were not for the Earth's rotation.

gy, partly potential, partly kinetic. We do not know quite how much, but a rough calculation gives an order of magnitude of 10^{17} J or ten million million ton-meters. Much of this energy is conservative; that is, only a little of the motion does work against the external forces. It is like a child's swing which, once a certain amplitude has been reached, requires only moderate exertion to keep it going. But some dissipation does occur and an obvious mechanism is through the friction of tidal currents on the sea bed. Such energy dissipation is proportional to the cube of the velocity, so it is very largely concentrated in the shallow seas. Several people, following H. Jeffrey's work on tidal friction in 1920, have calculated the total dissipation rate. Their results vary, depending on the quantity and reliability of the tidal current data used, but 10^{12} Js^{-1} seems to be an authentic figure. This is one million megawatts, equivalent to the output of more than 2000 large power stations! It also implies that the ocean dissipates all its tidal energy through friction in about 10^5 seconds, or the order of one day.

This loss of energy is of course made good by the generating forces, which are mainly effective over the large zone of the deep ocean basins. It is ultimately drawn from the rotational energy of the Earth itself. The deceleration of the Earth, an increase in length of the day by some milliseconds per century, is well authenticated from astronomical measurements, at least in modern times. It is affected by other factors besides friction; they can be allowed for but I will not discuss them here. The deceleration calls for a total dissipation in the tides of 2.7×10^{12} Js^{-1} which is at least twice the figure calculated from bottom friction in shallow seas.

There is thus a distinct gap in the energy budget, and nobody yet knows how to account for it. There is firstly the question of whether the loss occurs in the ocean tides at all or whether perhaps in the bodily tides of the Earth's interior. Geophysicists think the latter unlikely. The question would be answered if we knew the ocean tide map accurately enough, for we could then calculate the total transfer of energy from the gravitational forces into the ocean. If this does amount to 2.7×10^{12} Js^{-1}, then the oceanographers must find a mechanism for dissipating about half of this in the deep ocean basins. Here, bottom friction is certainly ruled out; but there is growing evidence for the importance of the role of internal tides. These are waves of tidal period occurring at sharp changes of density within the ocean. They can be many meters in amplitude but are soon dissipated because they are somewhat unstable and tend to curl up and break into turbulent eddies. Theorists have shown that the ordinary "surface" tides certainly lose some energy by creating internal tides but the magnitude and extent of the conversion rate is as yet but vaguely known.

So far I have scarcely mentioned the tides caused by the Sun. Most accounts describe them as being caused in an exactly similar way to the Moon's tides. However, the solar tides do present some anomalous features which concern modern researchers. To start with, the gravitational potential of the Sun is exactly similar to that sketched on the upper world map (Figure 3-26), except that 10 units represent not 358 mm of water as for the Moon but only about 162 mm. The Sun's mass is of course enormously greater than that of the Moon, about 27 million times, but the mean distance is 389 times as great and, since the tidal force involves the inverse cube of the distance, the effective solar forces are only about 45 per cent of the lunar forces.

The solar and lunar tides arising from these forces are easily separated by analyzing long records, and their amplitudes might reasonably be expected also to bear a ratio of 0.45. In fact, this is not generally so although, until about 1900, measurement of tides and comparison of the solar and lunar amplitudes was regarded by astronomers as being one of the most reliable means of determining the ratio of the Sun's and Moon's masses! It was P. G. D. de Pontécoulant who, in his *Système du Monde* of 1846, reviewed the current astronomical estimates of the Earth: Moon mass ratio (the Earth: Sun ratio was known then with some accuracy) and, in deploring their lack of agreement, first suggested that the figure adopted should be that derived by P. S. Laplace from a long series of tidal measurements, namely 75.2. The modern value, based on observations of close approaches of the asteroid Eros and confirmed by recent satellite measurements, is 81.30, correct to the last figure. This alone is a clear indication that the tides are not in simple proportion to their generating forces.

We now know that, although the Sun's and Moon's gravitational forces are exactly similar in spatial distribution, their different periods, 12 hours and 12.4 hours, cause the ocean to react differently in the tides it produces. At Bermuda, for example, the solar tide is about 0.23 of the lunar tide; at Mauritius the ratio is 0.77; both are very different from the "ideal" 0.45. The general pattern of tidal waves from the Sun over the oceans is rather similar to that of the lunar tides but has significant differences of detail. For example, there are islands where the Sun's tides predominate, so that high water occurs at about the same time every day, all the year round. This applies at certain of the Solomon Islands to the east of New Guinea. The islands happen to be near an amphidrome of the lunar tide but some distance from the corresponding solar amphidrome. There are also some islands on solar amphidromes where the tide lacks the familiar fortnightly beating at new and full Moon but varies principally in a monthly cycle because of the varying lunar distance. Thus, the deep sea tide map, complicated though it is, really requires a third dimension to cover the full spectrum of frequencies.

A closely related problem, which has been appreciated for decades but never satisfactorily solved, is that of tidal "age." At full and new Moons the Sun and Moon are in line with the Earth, so their gravitational potentials reinforce each other. We then get the so-called spring tides, which have greater range than usual. These spring tides nearly always occur later than the astronomical alignment of the Sun's and Moon's forces, by one or two days. "Age" is an old term for the delay in time between when this alignment occurs and when the tides reinforce each other at a particular place. A positive age means that the solar tide lags its generating force in phase by more than the corresponding lunar phase lag. This may be in some way another result of frictional forces, being stronger at the higher frequency of the Sun, but advanced mathematical studies with idealized ocean basins have so far failed to confirm the phenomenon as a general rule. To add to the mystery, there are a very few places, not unusually situated, where the observed "age" is negative. At Honolulu, for example, the highest tides occur before full Moon. This then is another matter which we should like to clear up.*

These properties are not really peculiarities of the Sun itself but of the response of the ocean to oscillations of slightly higher frequency. The Sun does however possess an inherent peculiarity because of its heat radiation. The pattern of radiation over the Earth differs fundamentally from that of the Sun's gravitation in that it occurs on the daylit hemisphere only. It is only half an egg shape and has a 24 hour periodicity rather than a 12 hour one, but a 12 hour and an 8 hour component are also effectively present as second and third "harmonics." The atmosphere, which is in a sense an extension of the ocean, is much more responsive to the heating effect of the Sun than to its gravitational effect and, further, its dynamic structure makes it more responsive to 12 hourly than to 24 hourly heating. As a result the atmosphere has a predominantly 12 hourly "tide" not caused by gravitation at all. This tide is in turn transmitted to the ocean through the atmospheric pressure and causes

* For a modern analysis of the problem of age, see: C. J. Garrett and W. H. Munk (*Deep-Sea Research* *18*, 5, 493-503, 1971).

an anomalous addition to the 12 hourly tide caused by gravity. (There are other means by which the Sun's radiation may affect the ocean tides, but I will not discuss them here.)

The concept of a "radiational" tide in the ocean is at least implicit in the writings of Sir George Darwin in 1895, but it is only recently that a direct estimate of such a tide has been made. Walter Munk of the University of California and I examined the spectrum of sea levels in relation to the separate spectra of the gravitational potentials of the Sun and the Moon and also of an analogous "potential" matched to the Sun's radiation. (In this context, a spectrum analysis is just like an analysis of a complex sound or other vibration, in that one examines how the energy in the vibration is distributed over a range of frequencies. But whereas one analyses vibration by direct mechanical sensors, in this case one computes the spectrum by numerical operations on long series of tide gauge readings.) If the tide consisted merely of a 12.4 hourly and a 12 hourly oscillation, the spectrum would contain only two "lines." In fact it contains a whole hierarchy of lines of considerable complexity over a wide band of frequencies because of the irregularities in the Moon's and Sun's orbits.

In the spectrum of the Sun's gravity at the top of the diagram on page 156 (Figure 3-27), 2.0 cycles per day (c/d) line, known as the S line, indeed dominates. The two adjacent lines, T and R, differ from S by one cycle per year (c/y) and so reinforce the solar tide at yearly intervals. These lines are produced by the annual change in distance of the Earth from the Sun as the Earth goes round its elliptical orbit. A more important line is K, 2 c/y above S. K is caused by the changing declination of the Sun—its elevation above the equator. Small lines outside the range T-K are unimportant. The spectrum of radiations also occupies the lines T-K, although their proportions are slightly different.†

† Strictly, these symbols should be written S_2, K_2, etc. to distinguish them from diurnal tides S_1, K_1, etc.

Somewhat surprisingly, the Moon also contributes to the solar frequencies. The principal lunar line, at 1.93 c/d, is well off the scale of the diagram on page 156 (Figure 3-27) but it also has a strong K line produced by the Moon's declination, which changes in a monthly cycle, and even a very small S line, for more complicated reasons. The lunar K line has strong side lines, the adjacent side lines differing from K by about one nineteenth c/y because of the 18.6 year irregularity of the Moon's orbit. The spectrum of sea level reveals contributions from all the above potentials and, in addition, has a "continuum," or background noise level caused by random oscillations in the sea.

Having sorted out the coherent spectrum from the noise continuum, our procedure was first to isolate the side lines to K. Since these, unlike K itself, are uniquely caused by the Moon, they pinpoint the purely gravitational effect, and so we could evaluate it. From this, we deduced the ocean's response to the Moon's K line itself and, further, the purely gravitational part of its response to the Sun. Finally, by subtraction, we isolated the radiational anomaly in the ocean tide. At Honolulu, this radiational anomaly turned out to have an amplitude of 1.8 cm but at Newlyn, Cornwall, it has an amplitude of 19 cm. The latter seems at first sight a very considerable anomaly, but the solar gravitational tide at Newlyn is as large as 81 cm because of the peculiar magnification of the half daily tide as a whole in the North Atlantic Ocean. Analysis of tides at other localities suggests that the ratio of radiational: gravitational tide of 0.2 is fairly typical.

The diurnal tides, with spectra in the range 0.8-1.1 c/d, are an entirely separate species from the half daily tides, being caused entirely by the motion of the Moon and Sun above and below the equator. They are most pronounced in the Pacific Ocean and, because of a local weakening of the half daily tides, in the Gulf of Mexico. In most of the Atlantic they are rather dwarfed by the magnified half daily tides but here they are especially interesting in having a pronounced local "age." Their greatest amplitude follows new and full Moon by up to six days in the NE Atlantic but nowhere else. Nobody has yet explained this phenomenon.

There are also tidal variations of very low frequency, with periods from 2 weeks to 19 years. They are difficult to estimate from records because they have amplitudes of only a few centimeters, scarcely distinguishable from the ocean's noise continuum which is high at this end of the spectrum. Analysis by C. Wunsch of MIT of new records from Pacific islands taken during the International Geophysical Year, 1957-8, has recently exploded an old notion that the low frequency tides are smoothly distributed round the globe and has demanded a new approach to their mathematical theory.

There are also high frequency tides, chiefly caused by distortion of the wave profile in shallow water such as the double high tide at Southampton. But, apart from their frictional aspect that I mentioned earlier, these have little bearing on the main problems of the deep sea tides.

What effort is being applied to these problems today? The main reason that so many long standing fundamental issues are still unsolved is that powerful researches by mathematicians in the 19th and early 20th centuries have shown these problems to be very difficult indeed. The flood of interest set off by the discoveries of Laplace and W. T. Kelvin slowly dwindled to a trickle of activity, mainly concerned with practical details of tide prediction. However, the coming of the computer and the rapid advances in scientific technology stimulated by space research promise a new and exciting lease of life.

One approach which has been transformed by the computer is the mathematical solution of the world tide map. Before 1950, valiant efforts had been directed towards solving the tidal equations in hypothetical

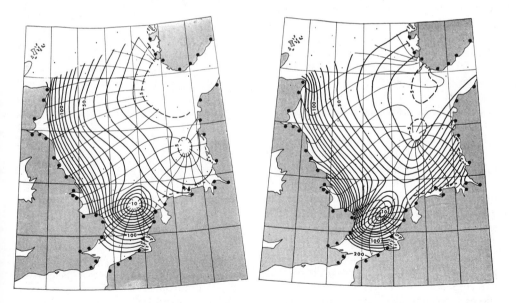

Figure 3-26. Computer simulation of tides in a small area like the North Sea is possible provided the tides are correctly specified at the open sea boundaries. Diagram on right was deduced from tidal measurements; diagram on left was computed by S. Ishiguro on an analogue computer at the National Institute of Oceanography although similar results have been obtained using digital computers. Only the principal lunar tide is represented here. Radial lines are time in lunar hours; heavy lines are contours of equal amplitude (cm).

basins with uniform depth and convenient geometrical boundaries. During the 1930s J. Proudman and A. T. Doodson of the Liverpool Tidal Institute had calculated a notable series of tidal charts for oceans bounded by meridional segments of various sizes. They show an impressive complement of amphidromes and wave crest contours but bear little resemblance to what is known of the pattern of tides in the real oceans. Realistic solutions are possible only if the full complexity of boundaries and depth contours of the real ocean can be allowed for. The ocean basins interact, so ultimately the whole world map must be covered. The mathematical labor required was beyond the wildest dreams of the pre-computer age but now the problem can at least be tackled with some hope of success.

So far, the problem has stretched even modern resources. C. L. Pekeris of Israel's Weiszmann Institute, who has been pioneering this work for the past ten years, reported his latest progress to the IUGG assembly in Berne in 1967.* A one degree mesh covering the world's oceans has taken the *Golem* computer 81 hours continuous running time to reach a solution for one frequency only. This, Pekeris rightly observed, is as long as anyone would care to run a computer and hope to have confidence in the results. Thankfully, the results do resemble what we know about the ocean tidal pattern but there are signs that even the one degree mesh is too coarse for accuracy. Amplitudes are rather high and phases are up to three hours early. Perhaps the results could be improved by adjusting the value of friction in the computer model but reduction of the mesh size to half a degree would require the computer to run for two months.

W. Hansen of Hamburg University has proposed a compromise between the roles of the computer and direct measurement. He suggests dividing the oceans into some 20

zones of convenient shape. The computer would handle each zone separately but, since there are inevitably open sea boundaries, the mutual influence of adjacent zones would have to be supplied empirically from tidal measurements at sea. This principle, which can be highly successful when applied to such a relatively small area as the North Sea, was in fact used by Hansen to compute the tidal pattern for the whole Atlantic Ocean with surprising accuracy as long ago as 1952. In view of Pekeris's difficulties, it looks very likely that some method akin to Hansen's will have to be accepted as the most realistic compromise bewteen computing the whole ocean and measuring the whole ocean.

The other principal area of research is techniques of measurement. For well over a

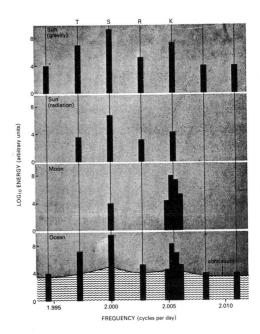

Figure 3-27. Spectrum of tides (at bottom of chart) with a period very close to 12 hours reveals separate contributions from the three sources of disturbance shown above it, as explained in the text on page 154. The strong side bands of the Moon's K line, caused by the 19 year irregularity in the Moon's orbit, are of particular interest. Continuum in ocean spectrum is caused by random effects of weather and currents.

century, all records have been derived from the conventional harbor tide gauge, essentially a float linked mechanically to a pen which records on a clockwork driven drum. Such gauges have the virtues of simplicity and robustness, requiring only minimal attention by a harbor master to keep going satisfactorily for years. The data thus accumulated over decades form an extremely valuable record, indispensable for high resolution tidal spectroscopy, for example. Accordingly, some effort is being made to collect together the material which is at present scattered round the world and in some cases merely accumulating dust on obscure shelves. W. H. Munk is building up a library of some ten million numbers, for the most part hourly values of sea level at various ports, carefully examined for errors and stored in duplicate on magnetic tapes at the Scripps Institution of Oceanography at La Jolla in California.

But we are still desperately in need of measurements from the deep sea itself. The oceanic islands at which conventional tide records can be made are too few and scattered to define the intricate detail of the tidal map and are in any case useless in providing data along clean-cut boundaries as required for Hansen's computational method. The solution is to design instruments which can be left on the ocean bed to record the pressure variations caused by the tides and which can be recovered after a suitable period. There are several difficulties, however. The chief one is to record accurately variations of a few hundred millimeters of water above a total ambient pressure of some kilometers. In other words, an accuracy of the order of one in a million is required. Most pressure sensors are also sensitive to temperature changes, so an accurate thermometer which will work at pressures of the order of 500 kg cm^2 also has to be provided.

Work has so far progressed on the basis of a sensor which in America is known under the trade name of Vibrotron. This is a taut wire a few centimeters long held in a thick steel casing. The natural frequency of the wire varies with pressure on the casing, and can be measured by a counting technique. The first *bona fide* records from a depth of more than 1000 meters have come from the French AFEGPO organization working on the advice of the Chief Hydrographer M. Eyriès, with instruments designed principally by H. Erdelyi. The AFEGPO principle is to back off the large ambient pressure by means of a subsidiary chamber which is filled with compressed nitrogen automatically on reaching the bottom of the ocean. The vibrating wire sensor (of French design) measures the pressure difference between the sea and the compressed gas, which never exceeds a meter or so of water. Nitrogen is chosen because of its low thermal coefficient of expansion but the instrument is still very sensitive to small changes in gas temperature. The temperature of the gas is monitored by means of a second vibrating wire cased in Invar, an iron-nickel alloy with a very small coefficient of thermal expansion. The recording medium is electro-mechanical counters photographed on 16 mm film. The instrument is recovered by hauling in a nylon cord linking the bottom unit with a surface buoy.

At La Jolla, California, Frank Snodgrass has obtained several good records from a more elaborate deep sea tide recorder, which has been described as the nearest thing to a space satellite ever put on the sea bed. Snodgrass uses the Vibrotron to measure the *absolute* pressure, thus avoiding some difficulties associated with the French backing gas pressure. Besides the temperature sensor, the equipment includes current meters. Recording is by magnetic tape. There is also an impressive system of transmitted coded acoustic signals to which the parent ship on the surface can listen to check that the various parts of the instrumentation are functioning correctly. There is no buoyed line to the surface but, on receiving a coded signal from the parent ship, the whole recording unit releases ballast (including most of its batteries) and floats upwards. A radio transmitter switches on at the surface to aid finding and recovery. Records are usually made for one month, which is the shortest useful period for tidal information, but Snodgrass and Munk hope eventually to leave at least some tide capsules at the bottom of the ocean for a whole year.

However, gathering tidal data from large cross sections of the oceans, pinpointing amphidromes and evaluating tidal friction require more than the development of complex instrumentation. They also require many long ocean voyages and a lot of time, work and money. The task is clearly one for international cooperation, and so the International Association of Physical Sciences of the Ocean has set up a Working Group on Deep Sea Tides, consisting of most of the theoreticians, instrument engineers and general oceanographers interested in the subject. So far the group has been too concerned with instrumental teething troubles to plan a hard-and-fast plan for concerted action. It appears moreover that before we can "run" in the deep ocean we shall have to learn to "walk" on the shallow continental shelf. In-

strumental problems are easier and most of the countries represented have instruments which can work usefully at up to 200 meters depth. There are in fact very large areas of shelf seas which are virtually unexplored tidally. These form interesting boundary zones which look inwards to the shoreline ports with their tide gauges, and outwards to the deep ocean where the fundamental problems lie. Altogether, there is enough work to keep a handful of scientists from various countries busy for a good many years to come.

SUGGESTED READINGS

Physical Oceanography, volume 2 by A. Defant (*Pergamon Press, Oxford,* 1961).

Tides by M. Hendershott and W. H. Munk (*Annual Review of Fluid Mechanics* 2, 205-224, 1970).

Maregraphes de Grandes Profondeurs by M. Eyriès (in *Cahiers Oceanographiques, 20,* 5, May 1968).

Tidal Spectroscopy and Prediction by W. H. Munk and D. E. Cartwright (*Philosophical Transactions of the Royal Society,* A259, 533-581).

Marine Chemistry

EDWARD D. GOLDBERG   WALLACE S. BROECKER   M. GRANT GROSS
KARL K. TUREKIAN   1971

INTRODUCTION

Chemical species introduced into the oceans will initially be partitioned among three phases: the living biosphere, sea water, and inorganic and organic particles. On different time scales and in various sites, the ultimate fate of all such elements is removal to the sea floor or discharge into the atmosphere. Although it is not possible to treat in detail the paths of all chemical species in the various types of marine environments, we can attempt to systematize available chemical data for radionuclides in sea water and in sediments and to direct attention to those areas where information is lacking.

Marine environments can be conveniently categorized into two major domains: the coastal ocean and the open ocean. The coastal ocean includes estuaries, lagoons, the water over the continental shelves, and many marginal seas. The open ocean is that part not significantly affected by its boundaries with the continents or by the shallow-ocean bottom.

Obviously, there is no clear-cut boundary between the coastal and open ocean; therefore, for this report, we will consider water deeper than 1,000 m to be part of the open ocean.

The different physical and biological conditions of these two oceanic domains, relevant to problems of elemental distribution, are summarized in Table 3-2. The remainder of the chapter elaborates on these concepts.

CHEMICAL SYSTEMATICS AND ELEMENTAL REACTIVITIES IN SEA WATER

The formulation of models to study the dispersion of radioactive species introduced into the marine environment can be ap-

Dr. Edward D. Goldberg is a chemical oceanographer with Scripps Institution of Oceanography, University of California.

Dr. Wallace S. Broecker is a geochemical oceanographer with the Lamont-Doherty Geological Observatory, Columbia University, New York.

Dr. M. Grant Gross is a geological oceanographer with the Marine Sciences Research Center, State University of New York, Stony Brook.

Dr. Karl K. Turekian is a geochemical oceanographer with the Department of Geology, Yale University, Connecticut.

From *Radioactivity in the Marine Environment*, Publication ISBN 0-309-01865-X, Committee on Oceanography, National Academy of Science—National Research Council, Washington, D.C., 1971, pp. 137-146. Reprinted with light editing and by permission of the authors and the National Academy of Science.

Table 3-2. Primary Factors That Can Alter the Chemical Composition of Sea Water.

Factor	Effect upon Elemental Behavior	Coastal Ocean	Open Ocean
Primary productivity	Fixation of elements in biomass, with subsequent transfer to deeper waters or to sediment	High, with local variability	Generally low, except in areas of divergence and high latitude
Particle input by river runoff, or resuspension from bottom	Provide surfaces for reactions of dissolved species and sites for bacterial activity	High	Low
Reservoirs for element accumulation	Storage of elements for various periods of time	Exist in sediments of ocean bottom	Exist in deep water and sediments of ocean bottom
Water circulation	Dispersion or retention of introduced species	Partial retention near coast	Dispersion dominant

proached within the framework of chemical characteristics and the behaviors of their stable counterparts in sea water. The assumption is made that the dispersion paths of radionuclides introduced in soluble forms will be the same as those of stable nuclides introduced or existing in the marine environment. This implies that the chemical speciation is the same for both stable and radioactive nuclides in sea water.* Hence, it is possible to utilize existing knowledge of oceanic chemistry in formulating models. The complexities of sea water as an electrolyte solution and the inadequacies in our quantitative description of the oceanic system are well known. Substantial advances

have been made in the chemical description of sea water over the past years, especially in regard to the speciation of the elements and to their relative reactivities. This new information permits meaningful statements about the expected behavior of at least some of the nuclides.

Table 3-3 lists average values for the concentration of elements in the ocean, together with what appears to be, from thermodynamic considerations, the most important chemical form, or forms, in solution. Column 4 divides the variability in the abundance of the elements into the following three categories.

A. Concentration is directly proportional to the salinity (such elements are referred to as "conservative").

B. There is a well-developed and readily described variability in concentration, as a function of depth, ocean basin, or both.

C. Reported variations in abundance are independent of salinity and not clearly dependent on depth or oceanic basin.

The elements in Class A are, in general, unreactive and display a remarkable stability in solution, while elements in Class B are usually involved in biological cycles, perhaps in inorganic processes that may result in an

* There are evidences in the literature of exceptions to this assumption. For example, stable zinc appears to exist in a number of complex species, some of which exchange slowly with the uncomplexed forms. Zinc-65, introduced into the marine environment from nuclear installations, may not attain the same chemical speciation in short time intervals as its stable isotopic counterparts. Some of the reported differences of specific activities between organisms and the water may result from the preferential uptake of a specific dissolved species of zinc. Another explanation may be that organisms may have picked up these elements from water masses where the specific activities were different but the speciation was similar. At present, resolution of this difficulty is not evident.

inhomogeneous oceanic distribution. The behavior of Class C elements in the ocean is not well understood.

There are several approaches to the description of the reactivity of an element in the oceanic chemical system, the most useful of which are residence times and the degree of undersaturation or supersaturation.

Residence times is defined as the average time an element spends in ocean water between introduction and incorporation into the sediments. Reactive elements generally have relatively short residence times in the oceans, while chemically inert species generally have much longer residence times, assuming a steady-state system in which the amount of an element entering the marine environment is compensated by the transfer of an equivalent amount from sea water to the sediments. The residence time, T, is defined by the relationship

$$T = A/(dA/dt),$$

where A is the total amount of the element in solution in the oceans and dA/dt is the amount introduced, and therefore precipitating, per unit time. Table 3-3 gives values of the residence times based upon values of dA/dt calculated from stream-input data.

Because of the oversimplified nature of the model, the absolute values of these residence times should be taken as a measure of reactivity rather than as a meaningful chronological number. The alkali metals and the alkaline-earth metals, for example, with long residence times, are characterized by the lack of reactivity of their ions in solution, while those elements intimately involved in biological cycles—phosphorus and silicon—have short residence times (10^3 to 10^4 years). The shortest residence times are calculated for elements primarily associated with lithogenous particles, such as aluminum, titanium, and thorium.

A second measure of reactivity derives from the degrees of undersaturation of ions with respect to their least soluble compound and their most stable dissolved species. It has been noted that, for reactive elements whose expected concentrations are calculated on the basis of their least soluble salts (Table 3-4) or on the basis of stable complexes (Table 3-5), and that are classified according to residence times (except for the rare earths and thorium), the observed concentrations are much lower than the limiting ones for oxygenated sea water. This indicates that phenomena other than solubility equilibria are determining elemental concentrations.

For lanthanum, cerium, and thorium, the solubility of the phosphate appears to govern their concentrations in sea water, keeping them at a remarkably low level (Table 3-4). Their removal from sea water, or their transfer from surface to deeper waters in which phosphate is regenerated by the decomposition of organic detritus descending through the water column.

COASTAL OCEAN

Several characteristics of the coastal ocean appear to be especially significant in determining the behavior of elements. They are (a) rapid mixing of substances injected into the ocean; (b) circulation patterns that tend to favor retention near the coast of substances introduced into the coastal ocean; (c) relatively intense biological activity; and (d) the abundance of particles (both biogenous and lithogenous) suspended in the water.

The discharge of a river quickly mixes with a volume of sea water several times as large to form a low-salinity surface layer that flows into the coastal ocean, mixing continuously. Such discharges can be identified on the basis of salinity and other parameters. The distribution of dissolved substances discharged by rivers usually replicates the patterns shown by these identifying parameters. Most soluble substances introduced along the shore will be mixed fairly rapidly, even when not clearly associated with a major river discharge.

The circulation of the coastal ocean tends to favor the retention of dissolved sub-

Table 3-3. Geochemical Characteristics of the Elements[a]

Element	Sea Water Concentration (μg/liter)	Principal Dissolved Species	Category[b]	Concentration Dissolved in Stream Waters (μg/liter)	Residence Time in Ocean (yr)
H	1.1×10^8	H_2O	A	—	—
He	7×10^{-3}	He (gas)	A	—	—
Li	1.7×10^2	Li^+	A	3	2.3×10^6
Be	6×10^{-4}	—	—	—	—
B	4.5×10^3	$B(OH)_3$, $B(OH)_4^-$	A	10	1.8×10^7
C	2.8×10^4	HCO_3^-, CO_3^{-2}	A	—	—
C (org.)	1×10^2	—	A	—	—
N	1.5×10^4	N_2 (gas)	A	—	—
N	6.7×10^2	NO_3^-	B	—	—
O	8.8×10^8	H_2O	A	—	—
O	6×10^3	O_2	B	—	—
O	1.8×10^6	SO_4^{-2}	A	—	—
F	1.3×10^3	F^-	A	100	5.2×10^5
Ne	0.12	Ne (gas)	A	—	—
Na	1.1×10^7	Na^+	A	6.300	6.8×10^7
Mg	1.3×10^6	Mg^{+2}	A	4.100	1.2×10^7
Al	1	—	—	400	1.0×10^2
Si	3×10^3	$Si(OH)_4$, $SiO(OH)_3^-$	B	6,500	1.8×10^4
P	90	HPO_{-2}, $H_2PO_4^-$, PO_4^{-3}	B	20	1.8×10^5
S	9.0×10^5	SO_4^{-2}	A	—	—
Cl	1.9×10^7	Cl^-	A	1,800	1×10^8
Ar	4.5×10^2	Ar (gas)	A	—	—
K	3.9×10^5	K^+	A	2,300	7×10^6
Ca	4.1×10^5	Ca^{+2}	A	15,000	1.0×10^6
Sc	$<4 \times 10^{-3}$	$Sc(OH)_3^0$	—	0.004	$<4 \times 10^4$
Ti	1	$Ti(OH)_4^0$	—	3	1.3×10^4
V	2	$VO_2(OH)_3^{-2}$	—	0.9	8.0×10^4
Cr	0.5	CrO_4^{-2}, Cr^{+3}	—	1	2.0×10^4
Mn	2	Mn^{+2}	C	7	1.0×10^4
Fe	3	—	—	670	2.0×10^2
Co	0.4	Co^{+2}	C	0.1	1.6×10^5
Ni	7	Ni^{+2}	C	0.3	9.0×10^4
Cu	3	Cu^{+2}	C	7	2×10^4
Zn	10	Zn^{+2}	C	20	2×10^4
Ga	3×10^{-2}	—	—	0.09	1×10^4
Ge	7×10^{-2}	$Ge(OH)_4$	—	—	—
As	2.6	$HAsO_4^{-2}$, $H_2AsO_4^-$	—	2	5×10^4
Se	9×10^{-2}	SeO_4^{-2}	C	0.2	2×10^4
Br	6.7×10^4	Br^-	A	20	1×10^8
Kr	0.2	Kr (gas)	A	—	—
Rb	1.2×10^2	Rb^+	A	1	5×10^6
Sr	8×10^3	Sr^{+2}	A	70	4×10^6
Y	1×10^{-3}	$Y(OH)_3^0$	C	—	—
Zr	3×10^{-2}	—	—	—	—
Nb	0.01	—	—	—	—
Mo	10	MoO_4^{-2}	A	0.6	7×10^5
Ru	—	—	—	—	—
Rh	—	—	—	—	—

Table 3-3. (Continued)

Element	Sea Water Concentration ($\mu g/liter$)	Principal Dissolved Species	Category[b]	Concentration Dissolved in Stream Waters ($\mu g/liter$)	Residence Time in Ocean (yr)
Pd	–	–	–	–	–
Ag	0.3	$AgCl^2$	C	0.3	4×10^4
Cd	0.1	Cd^{+2}	–	–	–
In	<20	H_2O	–	–	–
Sn	0.8	–	–	–	–
Sb	0.3	–	C	2	7,000
Te	–	–	–	–	–
I	60	IO_3^-, I	A	7	4×10^5
Xe	5×10^{-2}	Xe (gas)	A	–	–
Cs	0.3	Cs^+	A	0.02	6×10^5
Ba	20	Ba^{+2}	C	20	4×10^4
La	3×10^{-3}	$La(OH)_3^0$	C	0.2	6×10^2
Ce	1×10^{-3}	$Ce(OH)_3^0$	C	–	–
Pr	0.6×10^{-3}	$Pr(OH)_3^0$	C	–	–
Nd	3×10^{-3}	$Nd(OH)_3^0$	C	–	–
Sm	0.5×10^{-3}	$Sm(OH)_3^0$	C	–	–
Eu	0.1×10^{-3}	$Eu(OH)_3^0$	C	–	–
Gd	0.7×10^{-3}	$Gd(OH)_3^0$	C	–	–
Tb	1.4×10^{-3}	$Tb(OH)_3^0$	C	–	–
Dy	0.9×10^{-3}	$Dy(OH)_3^0$	C	–	–
Ho	0.2×10^{-3}	$Ho(OH)_3^0$	C	–	–
Er	0.9×10^{-3}	$Er(OH)_3^0$	C	–	–
Tm	0.2×10^{-3}	$Tm(OH)_3^0$	C	–	–
Yb	0.8×10^{-3}	$Yb(OH)_3^0$	C	–	–
Lu	0.1×10^{-3}	$Lu(OH)_3^0$	C	–	–
Hf	$<8 \times 10^{-3}$	–	–	–	–
Ta	$<3 \times 10^{-3}$	–	–	–	–
W	0.1	WO_4^{-2}	–	0.03	1.2×10^5
Re	0.008	–	–	–	–
Os	–	–	–	–	–
Ir	–	–	–	–	–
Pt	–	–	–	–	–
Au	1×10^{-2}	$AuCl_2^-$	C	0.002	2×10^5
Hg	0.2	$HgCl_4^{-2}$, $HgCl_2^0$	C	0.07	8×10^4
Tl	<0.1	Tl^+	–	–	–
Pb	0.03	$PbCl_3^-$, $PbCl^+$, Pb^{+2}	C	3	4×10^2
Bi	0.02	–	–	–	–
Po	–	–	–	–	–
At	–	–	–	–	–
Rn	6×10^{-13}	Rn (gas)	–	–	–
Ra	1×10^{-7}	Ra^{+2}	C	–	–
Ac	–	–	–	–	–
Th	$<5 \times 10^{-4}$	$Th(OH)_4^0$	–	0.1	<200
Pa	2.0×10^{-6}	–	–	–	–
U	3	$UO_2(CO_3)_3^{-4}$	A	0.04	3×10^6

[a] Compiled from Goldberg (1965) and Turekian (1969).

[b] See text for explanation of letters.

Table 3-4. Expected Equilibrium Concentrations for Some Elements, Based on Insoluble Salts of Phosphate, Carbonate, Hydroxide, and Sulfide (Concentrations in log moles/liter)[a]

Element	PO_4^{-3} (log a = -9.3)	CO_3^{-2} (log a = -5.3)	OH^- (log a = -6)	S^{-2} (log a = -9)	Observed in Sea Water
La^{+3}	-11.1	-	0	-	-10.7
Ce^{+3}	-10.0	-	-	-	-10.2
Th^{+4}	-11.8	-	-	-	-11.7
Cr^{+3}	-11.3	-	-	-	-8.0[b]
UO_2^{+2}	9.2	-	-	-	-7.8[c]
Fe^{+3}	-10.6	-	-	-	-7.3[d]
Fe^{+2}	-	-	-	-6.4	-7.3[d]
Mn^{+2}	-	-3.1	+0.2	-2.6	-7.4
Co^{+2}	-4.4	-6.5	-2.2	-12.1	-8.2
Ni^{+2}	-2.9	-0.6	-3.2	-10.7	-6.9
Cu^{+2}	-5.1	-3.5	-5.8	-26.0	-7.3
Ag^{+1}	-2.0	-2.7	-1.5	-19.8	-8.5[e]
Zn^{+2}	-3.5	-3.7	-3.5	-14.1	-6.8
Cd^{+2}	-3.7	-5.0	-0.5	-16.2	-9.0
Hg^{+2}	-	-	-12.5	-43.7	-9.1[e]
Al^{+3}	-	-	-12.0	-	-8.3
Ga^{+3}	-	-	-16.0	-	-9.3
Sn^{+2}	-	-	-15.0	-16.0	-8.2
Pb^{+2}	-6.8	-6.8	-2.0	-16.6	-9.8[e]

[a] Calculations made with the following activity coefficients: monovalent ions, 0.7; divalent ions, 0.1; trivalent ions, 0.01.
[b] Occurs primarily as CrO_4^{-2},
[c] Occurs primarily as $UO_2(CO_3)_3^{-4}$.
[d] May occur as particulate phases.
[e] Occurs primarily as chloride complexes.

stances near the coast. The thin plume of low-salinity water formed by the discharge of a river moves along the coastline for many kilometers, carrying with it many of the substances injected into it. The subsurface circulation also tends to favor, near the

coast, the retention of most substances injected into the coastal ocean. Where precipitation and runoff exceed evaporation, an estuarine-like circulation results in which fresh water added to the ocean surface mixes with salt water from below and moves generally seaward (Figure 3-28). A shoreward subsurface flow replaces the salt water that has

Table 3-5. Complexes Formed in Sea Water and Their Expected and Observed Equilibrium Concentrations (in moles/liter).

Element	Complex	Expected[a]	Observed
Silver	$AgCl_2^-$	-4.2	-8.5
Mercury	$HgCl_4^=$ $HgCl_2^0$	+1.9 } -0.3 }	-9.1
Lead	$PbCl_3^-$ $PbCl^+$	-5.6 } -5.8 }	-9

[a] Using least soluble salt, from Table 3-4.

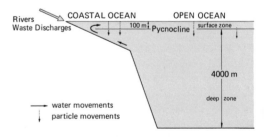

Figure 3-28. Schematization of the coastal and open ocean.

moved upward into the surface layers. In areas where evaporation predominates, the coastal ocean does not exhibit estuarine-like circulation, but such areas are relatively uncommon. The increased nutrient supply due to this vertical circulation and the nutrients supplied by rivers are major causes of the relatively large primary production in the coastal ocean, because subsurface waters brought into the surface layers supply nutrients to the photic zone.

Certain substances in the surface waters become associated with the descending organic debris. Chemical species released by the decomposition of particles sinking out of the surface layer will tend to be moved landward in the subsurface flow, to return eventually to the surface layer. Although materials are lost to the sediments or to the surface layers of the open ocean, the circulation of the surface and subsurface waters tends to retain some chemical species in the coastal ocean.

Dissolved oxygen in the near-bottom waters or in the sediment may be depleted or even completely exhausted, depending on the supply rate of dissolved oxygen relative to the rate of consumption in the decomposition of organic matter formed in the photic zone. Where exhaustion of dissolved oxygen occurs, sulfate-reducing bacteria are involved in the production of H_2S. Complete exhaustion of the dissolved oxygen in the water column is not common in the coastal ocean except where water circulation is greatly restricted, as in certain fjords, or where primary productivity is extremely high. In areas of large primary productivity, where the supply rate of dissolved oxygen is large enough to prevent oxygen depletion in the near-bottom waters, H_2S may nevertheless form in the sediments because the rate of oxygen diffusion into the sediment is slow.

The presence of hydrogen sulfide may act as a control on the concentration of many metallic ions in sea water, as can be seen in Table 3-4. A sulfide ion activity of 10^{-9} moles/liter, corresponding to that of a "stinking mud," results in an environment favorable to the removal of metals that form highly insoluble sulfides. Certain metals will be depleted even in moderately sulfide-rich water, i.e., $a_s^{-2} = 10^{-13}$ moles/liter.

Manganese and iron appear to be enriched in sulfide-rich waters inasmuch as the sulfides of the reduced forms of these elements are relatively soluble (Table 3-4). Higher concentrations of manganous ions in the anoxic waters of the Black Sea, relative to the oxygenated surface waters, have been reported, apparently confirming the existence of this process of enrichment.

Reactive elements brought into the ocean may become associated with land-derived solid materials. Such materials derived from organisms incorporate certain elements into their tissues or skeletons, often concentrating them many-fold relative to their concentrations in seawater. After an organism's death, some of these elements are incorporated in the bottom sediment, along with undecomposed organic remains. Nutrient elements, such as phosphorus and nitrogen, may be released as soluble species during decomposition of the organic remains in the water.

Particle-associated chemical species and reactive elements that were taken up by the organisms and not quickly released by decomposition are eventually incorporated in the sediment, which is a major reservoir for such elements.

OPEN OCEAN

The open ocean differs significantly from the coastal ocean; thus, there appear to be major differences in elemental behavior (Table 3-2). Among the significant differences between the coastal and open ocean are the general nutrient deficiencies and relative scarcities of organisms and particles in surface layers of the latter. The well-developed density stratification in the open ocean inhibits large-scale vertical mixing and the upward movement of nutrients from deep waters, except at divergences or in high latitudes. Unreactive elements generally pass through the coastal ocean into the open ocean. Reactive elements tend to become associated with particles or to be utilized by organisms and thus are removed from seawater in the coastal ocean. Elements with short residence times tend to accumulate in the sediment deposited in the coastal ocean; elements with long residence times tend to accumulate in the open ocean. Uranium, which can be reduced to a tetravalent state in coastal deposits, is an exception—it has a long residence time in the ocean but tends to accumulate in inshore deposits.

In the open ocean, some chemical species removed from the surface layers by organisms are released in adjacent deeper waters as the particles decompose before reaching the bottom. Among these are elements such as phosphorus and nitrogen, which are retained by the partially decomposed organic matter and deposited with sediments in the coastal ocean. Consequently, open ocean sediments are relatively less important as a reservoir for reactive elements than are the coastal bottom deposits.

Downward transport and subsequent decomposition of the particulate remains of marine organisms are the major cause of chemical inhomogeneities within the ocean. Various elements fixed by marine organisms in the surface waters of the ocean are transported to, and released in, depths where solution or oxidation of the debris takes place. The net result is a depletion of these elements in surface waters and an enrichment in deep waters.

The ratio of the concentration of an element (normalized to a standard salinity) in the deep waters to that in the surface waters of the Pacific Ocean provides a convenient index of the effectiveness of this process. Values are given in Table 3-6 for several elements.

Nitrogen, phosphorus, and silicon are removed very efficiently, the last element primarily as the opaline tests of diatoms and radiolaria, and the first two as basic constituents of the organic material itself. The

Table 3-6. Ratios of Elemental Concentrations in Deep Waters C_{deep} and in Surface Waters C_{surf} of the Pacific Ocean.

Element	C_{deep}/C_{surf}
Li	1.0
Na	1.0
K	1.0
Rb	1.0
Cs	1.0
Mg	1.0
Ca	1.02
Sr	<1.1
Ba	4
Ra	4
CO_2	1.2
S as SO_4^{-2}	1.0
P as phosphate ions	>10.0
Si as $Si(OH)_4^0$	>10
N as NO_3^-	>10.0
Mn	~1
Mo	~1
U	~1
Br	~1

fact that the nitrogen/phosphorus ratio (16/1) is nearly constant in all the major open ocean water masses and in all major types of organisms attests to the dominance of organic activity in the generation of inhomogeneities in the concentrations of these two elements. Carbon is depleted not only by its incorporation in organic molecules but also by its fixation and removal as $CaCO_3$ (in foraminifera, coccoliths, pteropods). Calcium owes its depletion to $CaCO_3$ and possibly to calcium phosphate precipitation. The absence of significant concentration variations for relatively unreactive elements such as cesium strongly suggests that transport by organisms is unimportant compared to transport by physical mixing in the water column.

Nonsystematic variations in the strontium/chlorinity and cesium/chlorinity ratios have been reported, but the results are difficult to interpret. Possibly, the strontium concentration of sea water is altered by the dissolution of large numbers of celestite (strontium sulfate) radiolaria. For cesium no mechanism is evident.

The depletion factors given in Table 3-2 can be easily converted to residence times for the element in surface water, T_S, relative to transfer to the deep sea. Assuming that such elements as Na, K, Mg, Cl, and S are transported only by physical mixing, we can use their residence times as references. From simple material-balance considerations, it can be shown that

$$T_S \text{ element} = \frac{C_{surf}}{C_{deep}} \; (T_S \text{ chlorine})$$

where C_{surf} and C_{deep} are the concentrations of the element in surface and in deep water, respectively. Radium, for example, because of its fourfold enrichment in deep water, apparently resides in surface water only one quarter the time that chlorine resides there. The silicon residence time is on the order of one tenth the chlorine residence time.

The residence time of a relatively nonreactive element is about 20 years in the mixed layer for the world ocean—this estimate is based upon physical mixing processes solely. Thus, barium or radium would have passage times of about 5 years in the mixed layer, while silicon would be removed in about 2 years.

The variations with depth of many trace elements in sea water are poorly known, as indicated in Table 3-3. For such elements, another approach is needed to delimit their reactivities in biological cycles. One possibility is to use the chemical composition of marine plankton. The degree to which a given element is enriched by marine organisms should be related to the importance of its transport by particles. Three types of solid phases must be considered: organic detritus, calcareous exoskeletal materials, and opaline exoskeletal materials. The total ocean-wide productions can be estimated from the deep-water excesses (concentration in deep water minus concentration in surface water) of silicate, of total dissolved inorganic carbon or carbonate alkalinity, or of nitrate and phosphate.

The deep Pacific excesses are as follows:

P	3×10^{-6} moles/liter
N	5×10^{-6} moles/liter
C	4×10^{-4} moles/liter
Si	2×10^{-4} moles/liter
Carbonate alkalinity.	2×10^{-4} equivalence/liter

For the total dissolved inorganic carbon, 3×10^{-4} moles/liter result from the combustion of organic carbon, and 1×10^{-4} moles/liter from the fallout of $CaCO_3$. The carbonate alkalinity, nitrogen, phosphorus, and carbon data are internally consistent, and show that for each mole of $CaCO_3$ dissolving in the deep sea, organic material containing 3 moles of organic carbon must be oxidized. The silicon data suggest that for each mole of $CaCO_3$, there must be 2 moles of opaline silica precipitated. The corresponding weight ratios would be—dry organic material: $CaCO_3$: $SiO_2 = 0.7 : 1.0 : 1.3$.

The elemental concentration factors for marine organisms, with respect to sea water, are given in Table 3-7. Although there is an uncertainty of at least an order of magnitude in these values for many of the elements, they provide an entry to the problem of the dissemination of elements through biological activity. Where reliable concentrations are known, we can estimate the deep-water excesses of trace elements. If we assume, for example, that any element incorporated into plant tissue is released in a manner similar to that of phosphorus without fractionation, then the enrichment values in Table 3-7 assume an importance. Those elements that are reported to be enriched to the same level as phosphorus, or higher (Al, Sc, Pb, Fe, and Cr in Table 3-7), would be expected to have significant deep-water enrichment. Such elements are removed from surface water by sinking particles. Since phosphorus can attain enrichments of an order of magnitude or so in the deep ocean relative to the surface ocean, elements that are enriched in marine plants by even one order of magnitude less than phosphorus would be expected to have detectable nonhomogeneous vertical distributions, whereas, if the enrichment of an element relative to phosphorus is down by more than one order of magnitude,

Table 3-7. Elemental Concentration Factors for Marine Organisms (Expressed as the Logarithm)[a].

Element	Plants	Animals
Al	4	4.5
As	3	2
Ba	–	1
Be	3	3
B	0.1	0.5
Cd	2	5
Ca	0.5	1
Ce	2.5	2
Cs	0.5	1
Cl	–	-1
Cr	4.5	5
Co	3	2
Cu	2	3
F	0.1	–
Ga	1.5	1.5
Au	3.5	1.5
I	3.5	–
Fe	4.5	3.5
Pb	4.0	4.0
Li	–	-0.3
Mg	-0.2	-0.2
Mn	3.5	3.5
Mo	1.5	1.5
Ni	2.0	3.0
Nb	3.0	5.5
P	3.7	4
Pu	3	3
K	-0.1	0.5
Ra	–	–
Ru	2.0	1.0
Sc	4.5	3.5
Si	2.3	2.0
Ag	3.0	3.0
Na	-1.5	-0.8
Sr	1.0	0.5
S	-0.5	-0.5
Sn	2.5	2.0
Ti	3.5	3.5
W	3.0	2.0
U	–	–
V	2.5	2.0
Zn	3.0	4.0
Zr	3.0	3.2

[a]Concentration factors for organisms from the open and coastal areas are included. Data are insufficient to ascertain differences that might result from systemic variations in elemental concentrations in these two marine domains.

no significant vertical-concentration changes are to be expected.

Concentration factors based upon direct measurements of elemental contents in ashed organisms are given in Table 3-8 in the arbitrary units of liters of sea water per gram of ash. On this scale, the concentration factor for phosphorus is about 200 (or more, if we use the lower phosphorus concentration of surface waters). These more recently determined values provide an additional frame of reference for considerations of vertical distributions of elements in the marine environment.

CONCLUSION

The chemical factors governing the dispersion of species introduced into the marine environment are in need of better definition, especially with regard to coastal environments. The involvement of many elements in primary plant productivity is not known. The comparative chemistries of oxic and anoxic waters warrant much additional work. The natures and roles of suspended inorganic and organic phases are poorly understood. An elaboration of such problems would provide firmer bases from which to systematize the general dispersion problem. In particular, we see need for chemical investigations along the following lines:

1. Elemental concentrations should be sought for the marine plants that are most important in the fixation of carbon in both coastal and open ocean areas. The relative importance in the coastal oceans of attached plants to planktonic species with regard to the uptake of specific elements warrants attention.

2. For the coastal ocean, the relative effects of interactions of introduced species with nonliving particulate organic and inorganic phases and with viable organic phases should be ascertained.

3. The distribution of heavy metals in the open ocean and their distribution among dissolved, colloidal, and particulate states should be determined.

Table 3-8. Concentration Factors between Sea Water and Plankton Ash for Some Trace Elements[a]

Element	Sea Water (μg/liter)	Plankton Ash (μg/g): Plants (Sargassum)	Animals	Concentration Factors (liters sea water per grams plankton ash): Plants (Sargassum)	Animals
P	88	20,000	20,000	230	230
Ag	0.3	0.3	0.3	1	1
Al	1	65	300	65	300
B	4,450	1,200	140	0.27	0.031
Ba	20	120	52	6.0	2.5
Cd	0.11	8	13	72	120
Co	0.4	3	3	7.5	7.5
Cr	0.5	9	7	18	14
Cu	2	270	270	135	135
Li	170	6	40	0.04	0.2
Ni	6.6	27	12	4	2
Sr	8,100	8,500	930	1	0.1
Ti	1	26	120	26	120

[a]Based on unpublished data for marine plankton species from G. Thompson and V. T. Bowen (Woods Hole), H. Curl (Oregon State), G. Nicholls (Manchester), and K. K. Turekian (Yale).

SUMMARY

Although initially incorporated in the biosphere, or dissolved in sea water or associated with particles, materials introduced into the ocean are eventually removed from sea water and deposited on the ocean bottom. The available data are systematized for certain important radionuclides in sea water and sediments.

Marine environments may be divided into coastal ocean (above the continental shelf) and open ocean, where the water is deeper than 1,000 m. Physical and biological conditions are different in these ocean areas, and these in turn influence the behavior of the elements in the ocean. Elements are classified as (A) conservative elements, (B) those exhibiting well-developed and regular variability in concentration with depth or ocean basin or both, and (C) those whose concentration is independent of depth or ocean basin. Elements in class A are generally unreactive in sea water. Elements in class B are usually involved in biological cycles. Elements in class C are not well understood. Elements have also been classified according to their residence time and reactivity in sea water.

The coastal ocean is characterized by rapid mixing of substances, partial retention close to coast of solids and certain reactive elements, relatively intense biological activity, and abundance of particles. Many reactive elements become associated with particles and are deposited near the continents.

Open ocean waters are characterized by relative deficiency of nutrients in surface waters and relative scarcity of organisms and particles. Elements with long residence times tend to accumulate in open ocean waters. Sediments depositing in deep ocean areas are less important as a reservoir for reactive elements than are sediment deposits in coastal areas. Downward transport and subsequent decomposition of particles is the dominant cause of depletion of the element in surface waters and enrichment in subsurface waters.

SELECTED REFERENCES

These references were selected to provide background information about marine chemistry and an introduction to the literature.

Barnes, C. A., and M. G. Gross. 1966. Distribution at sea of Columbia River water and its load of radionuclides, p. 291-301. In *Disposal of Radioactive Wastes into Seas, Oceans, and Surface Waters.* IAEA, Vienna.

Burton, J. D. 1965. Radioactive nuclides in sea water, marine sediments, and marine organisms, p. 425-476. In J. P. Riley and G. Skirrow [ed.] *Chemical Oceanography.* Vol. 2. Academic Press, New York.

Chester, R. 1965. Elemental geochemistry of marine sediments, p. 23-80. In J. P. Riley and G. Skirrow [ed.] *Chemical Oceanography.* Vol. 2. Academic Press, New York.

Culkin, F. 1965. The major constituents of sea water, p. 121-162. In J. P. Riley and G. Skirrow [ed.] *Chemical Oceanography.* Vol. 1. Academic Press, New York.

Dietrich, G. 1963. *General Oceanography— An Introduction.* Wiley, New York, 588 p.

Duursma, E. K. 1966. Molecular diffusion of radioisotopes in interstitial water of sediments, p. 355-369. In *Disposal of Radioactive Wastes into Seas, Oceans, and Surface Waters.* IAEA, Vienna.

Goldberg, E. D. 1965. Minor constituents in sea water, p. 163-196. In J. P. Riley and G. Skirrow [ed.] *Chemical Oceanography.* Vol. 2. Academic Press, New York.

Harvey, H. W. 1960. *The Chemistry and Fertility of Sea Waters.* Cambridge Univ. Press, London.

Horne, R. A. 1969. *Marine Chemistry.* Wiley (Interscience), New York. 568 p.

International Atomic Energy Agency. 1966. *Disposal of radioactive Wastes into Seas, Oceans, and Surface Waters.* Vienna. 898 p.

Kautsky, H. 1966. Possible accumulation of discrete radioactive elements in river mouths, p. 163-173. In *Disposal of Radioactive Wastes into Seas, Oceans, and Surface Waters.* IAEA, Vienna.

Lowton, R. J., J. A. Martin, and J. W. Talbot. 1966. Dilution, dispersion and sedimentation in some British estuaries, p. 189-204. In *Disposal of Radioactive Wastes into Seas, Oceans, and Surface Waters.* IAEA, Vienna.

Osterberg, C. L., N. Cutshall, V. Johnson, J. Cronin, D. Jennings, and L. Frederick. 1966. Some non-biological aspects of Columbia River radioactivity, p. 321-333. In *Disposal of Radioactive Wastes into Seas, Oceans, and Surface Waters.* IAEA, Vienna.

Riley, J. P. and G. Skirrow [ed.] 1965. *Chemical Oceanography.* Academic Press, New York. 2 vol.

Templeton, W. L., and A. Preston. 1966. Transport and distribution of radioactive effluents in coastal and estuarine waters of the United Kingdom, p. 267-288. In *Disposal of Radioactive Wastes into Seas, Oceans, and Surface Waters.* IAEA, Vienna.

Turekian, K. K. 1965. Some aspects of the geochemistry of marine sediments, p. 81-126. In J. P. Riley and G. Skirrow [ed.] *Chemical Oceanography.* Vol. 2. Academic Press, New York.

Turekian, K. K. 1969. The ocean, streams, and atmosphere, p. 297-323. In K. H. Wedepohl [executive ed.] *Handbook of Geochemistry.* Vol. 1. Springer-Verlag, Berlin.

Wedepohl, K. H. [executive ed.]. 1969. *Handbook of Geochemistry.* Springer-Verlag, Berlin. 2 vol.

Understanding the Weather

ROBERT H. SIMPSON 1968

Breakthroughs in the science of weather forecasting in the last half century have stemmed from two technological developments.

First was the radiosonde, a balloon sounding device for measuring pressure, temperature and humidity. From this advance came air-mass analysis, the concepts of warm and cold fronts, of the jet-stream, and of large planetary waves that generate and steer temperate latitude storms (Figure 3-29).

The second was the development of high-speed digital computers, which enabled the meteorologists to make real-time predictions of changes in atmospheric properties.

A third technological triumph is the weather satellite, which may play a significant role in future breakthroughs.

However, the important missing link in the forecaster's computations is the link between the atmosphere and the oceans—the important action that occurs at the sea-air interface, especially the exchanges of energy and of heat.

The systematic flow of heat to the air exerts a controlling influence on the development of most severe ocean storms. Even in the absence of severe storms, the oceans' powerful influence on air circulation causes prediction models to go astray after a few days of extrapolation.

It is not an oversight that most operational models essentially ignore the action at the sea-air interface. There is neither the basic understanding nor sufficient observational data to deal with this problem in real time. Since it is easier to study and understand the impact of the action at this interface on the weather regimes of the tropic oceans than in the temperate latitudes, let us first consider the nature of tropic ocean weather and its problems.

Study of the ocean weather has brought closer interdisciplinary ties between meteorology and oceanography, and the pooling of manpower and resources—in such projects as BOMEX (Barbados Oceanographic and Meteorological Experiment) and the Barbados experiment of 1968.

Significantly, the emphasis in these projects is upon the weather of the tropic oceans, rather than the more frequently

Dr. Robert H. Simpson is the former director of the National Hurricane Center of the National Oceanic and Atmospheric Administration's National Weather Service. He was named the first director of the National Hurricane Research Project in 1955. He has written and helped write over forty articles on hurricanes.

From *Oceanology International*, Vol. 3, No. 7, pp. 42-45, 1968. Reprinted with light editing and by permission of the author and *Oceanology International*.

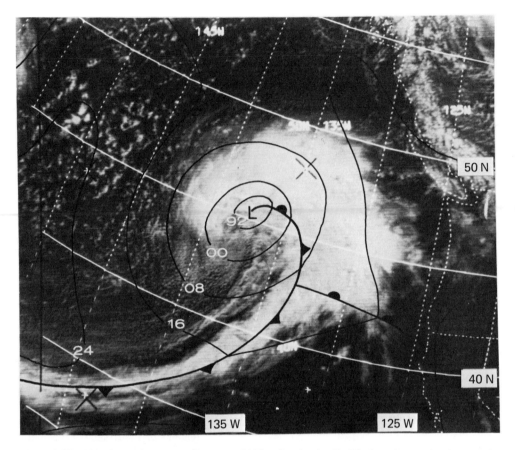

Figure 3-29. Gigantic cyclone extending over 2000 miles in the Pacific in a temperate storm system (NOAA National Environmental Satellite Service).

storm-ridden oceans of higher latitudes. Day-to-day weather hazards in temperature latitudes are far more frequent, and have a much greater impact on commerce and the economy, than those in the tropics. While the tropical hurricane (or typhoon) generally is recognized as the "greatest storm on earth," it is comparatively rare. Only about 10 major tropical cyclones occur each year in the entire area of the Atlantic Ocean.

Of these, only about seven reach full hurricane intensity, and an average of only two cross the United States coastline and have any impact on the economy and welfare of the nation. On the other hand, temperate latitudes suffer weather handicaps, although less severe, almost weekly.

Therefore, the initial emphasis on using the computer to test and apply numerical prediction models has centered on the weather regimes of middle latitudes. Physical numerical models, effective in temperate latitudes, do not measure up well in lower latitudes, tending to "lose their way" in the iteration of weather trends when the computations extend beyond 72 hours.

One important reason for these failures is that present models are unable to simulate the input of the oceans to the energy budget of the atmosphere, or to model the rapid heat flow from the sea to the air during development of extreme storm systems, including hurricanes. More information is available from mid-latitude oceans than tropic oceans, but the role of the oceans in higher latitudes usually is masked by the other kinds of energy releases.

Operational models have been engineered to consider the conservative, reversible processes occurring within the free atmosphere alone. Complex, irreversible processes involving condensation and precipitation, radiation, transfers of heat and momentum, and frictional losses near the earth's surface often are ignored.

Investigations of the sea-air interface can be conducted more effectively in the warm tropic oceans where the exchanges are largest and are not masked or confused with other large-scale energy processes. Here, small disturbances can be traced over thousands of miles of ocean by satellite and investigated regionally by reconnaissance aircraft. The sea-air interface action can be studied from oceanographic vessels and buoys.

These techniques were used last summer by scientists from Florida State University, the National Oceanic and Atmospheric Administration (NOAA), and the National Center for Atmospheric Research working from the West Indian Island of Barbados. This program is a forerunner to the BOMEX experiment, which will begin in 1969.

Tropic oceans are swept continually by steady and persistent east winds (tradewinds) generated by large, persistent mounds of high pressure that extend in east-west ridges across the sub-tropical "horse latitudes."

These easterlies prevail through a depth of 5 to 9 km during summer, and, in the North Atlantic, reach their greatest persistence in August. In winter, they often are quite shallow, and give way to the occasional intrusion of west winds from temperate latitudes as far equatorward as the 12th or 15th parallels.

In their transit of the tropic oceans, tradewinds gradually are drawn into the doldrums, an east-west low-pressure trough that migrates seasonally with the sun. When air movements are steady, the tradewinds from both hemispheres approach this trough and merge along an east-west line known as the intertropical confluence (ITC). Under other circumstances, they feed into a lethargic system of large eddies.

The ITC is the most important single feature of tropic ocean weather, the most prevalent producer of weather, and the means of generating the great cloud systems that must export heat to higher latitudes to maintain the global heat balance. Weather activity along the ITC varies widely in time and space, mainly in response to the migration of the tradewind sources.

At times, the confluence of trade winds about the ITC may develop low pressure centers which deform it and concentrate the bad weather about an eddy several hundred kilometers wide, moving slowly westward. Sometimes the ITC is transformed into a continuous system of eddies parading westward across the tropical scene.

Such eddies generally remain benign, although they bring a succession of light to heavy rain showers and shifting winds. However, when they move away from the ITC and become imbedded in the tradewinds, they tend to grow "fangs" and develop vicious winds.

The growth and development of eddies into severe storms depend mainly upon an expanse of warm ocean water, and a cluster of the rain storms, which permits the latent heat released by rain to be concentrated in a deep vertical column, allowing surface pressures to fall.

At the top of the rain-storm cluster there must be some means of conducting the heat away to a colder environment. In the absence of this factor, the would-be storm stifles in its own hot air. These conditions for development or intensification more frequently are met after an eddy becomes deeply imbedded in the tradewinds than while it is still a part of the ITC.

There are other sources of tradewind disturbance, which are independent of the ITC, and probably are responsible for more than half of the "seedling" circulations that later become hurricanes on the Atlantic. These seedlings migrate westward across the tropic oceans, many of them visible only to the satellite.

Those that do deform the normal flow of the trades are known as tropical waves. These waves are essentially north-south low-pressure troughs, imbedded in the trade-winds. They have no low-pressure centers initially, and are associated with squally weather and heavy rains from the axis of the wave eastward for several hundred kilometers. The meteorological satellite has thrown much new light on these waves and their origin, at least in the Atlantic area.

While no one has been able to identify the precise manner in which these waves form, many are detected first near the Abyssinian plateau in Africa. After migrating westward at a speed of 22 to 27 km/hour they emerge in the Atlantic as benign rain storms, visible to the satellite as cumulus clouds, in an inverted "V."

As it progresses westward across the warm Atlantic, moisture is swept up from the ocean and distributed through a deep layer of the atmosphere. Sailing vessels caught in these waves often experience heavy rain accumulating to more than 5 cm an hour, with squally winds reaching 70 to 90 km/hour for relatively short periods of time.

Another source of tropical waves comes as a spin-off from the jetstream region of temperate latitudes. At the jetstream level (about 12,000 m), whorls or eddies several hundred kilometers in diameter sometimes spin away equatorward from the prevailing west winds. These migrate southwestward and westward as low-pressure systems, superimposing themselves on the trade winds at lower levels. The low-pressure center of these eddies often is reflected into the trade winds, giving rise to a tropical wave with

weather characteristics similar to those that originate over Africa.

In the course of a hurricane season in the Atlantic, as many as 60 potential storms or hurricanes are formed, but less than 10% of them actually develop into full-blown hurricanes. Why, with so many good opportunities, do so few tropical disturbances reach hurricane intensity?

If the hurricane is the "greatest storm on earth," it also has been considered one of the simplest, most uncomplicated of atmospheric storms—uncomplicated because it has near-circular symmetry and is surrounded by a homogeneous environment with no fronts or contrasting air masses. It derives its energy almost solely from the latent heat released by the tall cumulus clouds near its center.

However, the process by which a disturbance grows into a hurricane is *not* uncomplicated. It is here that the dramatic role of the oceans enters in the struggle between the stormy winds and the ruffled ocean surface.

The tradewinds normally are laden with moisture, acquired gradually by a long trajectory over water. But, if you should condense all the moisture normally carried by these winds as they converge on the low-pressure center of a disturbance, the heat of condensation liberated would be sufficient to lighten the air column only about 1%. This might reduce pressure at the center of the disturbance to about 1,000 mbs (or about 29.50 in.). This reduction would support a sustained wind of about 85 km/hr in an ordinary tropic ocean setting.

To become a major hurricane, the disturbance would have to acquire a central pressure of 950 mbs or lower. The heat required to further lighten the air at the storm's center and lower the surface pressure so drastically has to come from the ocean through more complicated, rapid transfer processes.

In part, the heat is transferred by conduction as the rain-cooled air near the surface comes in vigorous contact with the warm ocean water. And, in part, it comes in latent form as the moisture content of the air is increased by evaporation from the ocean surface and from the spray carried into the air by the stormy winds.

These processes are of enormous interest to the meteorologist and the oceanographer alike, although for different reasons. Investigations of the oceans' role in storm development have led both oceanographers and meteorologists to suggest that a chemical film spread artificially over the water surface might inhibit the vital flow of latent heat to a hurricane, depriving it of the energy needed to sustain damaging winds. This presumably would cause a hurricane to "shrivel on the vine."

However, to significantly limit the flow of moisture from the water surface to the air, it would be necessary to maintain a film the thickness of a single molecule. Obviously, in a turbulent sea with 12 to 18 m waves generated by the hurricane it would be difficult, if not impossible, to maintain the integrity of such a film.

Regardless of weather modification potentials, machine models for weather prediction must consider the energy processes at the sea-air interface. This is important in tropical latitudes, and in temperate and higher latitudes, especially for models that may successfully simulate weather developments for periods longer than 72 hours.

The weather satellite provides meteorologists with a daily view of large-scale cloud systems exceeding anything that could be achieved with ground-based or airborne observing platforms. However, its utilization will remain restricted until its observations are calibrated in terms of the dynamics of wind systems, which generate the clouds it sees and the action occurring at the sea-air interface. These can be observed only by airborne or earthbound probe systems.

The satellite looks down from on high, mainly upon the exhaust products of a storm system. It sees clouds generated by cold and warm fronts that may extend in lines of thousands of kilometers. It views clusters of patchy small clouds that accompany the fair-weather regimes typical of high pressure and of the undisturbed tradewinds. It sees the cirrus, or ice-crystal, clouds that characterize the jet stream, the rain showers generated by the sea breezes in Florida and on other coasts, and numerous other benign weather exigencies, which can be classified and tracked.

But when the satellite looks down on a disturbance that is struggling to become a storm or hurricane, it primarily sees the products that are expelled at the top of the taller clouds, an amorphous outflow of cloud matter that often masks the number, the sizes, and the organization of the participating convective clouds.

Nevertheless, experience has shown that certain characteristic cloud patterns viewed by the satellite can be correlated with the growth and intensity of a storm system. The National Environmental Satellite Center at Suitland, Md., now issues bulletins regularly on all tropical disturbances that exceed a certain intensity. They estimate the maximum winds in terms of the brightness and the diameter of the exhaust product or outflow from the cluster of tall cumulus clouds.

The weather satellite is a diagnostic tool of unparalleled value. It provides positive information on the existence of cloud and storm systems over areas where there is rarely any other kind of observation, provides useful information on the intensity of storm systems and is helpful in tracking their movement.

Unfortunately, the satellite cannot furnish explicit information for predicting the changes of intensity or movement of a storm or an incipient disturbance. However, coupled with earthbound probe systems, it becomes a powerful tool of investigation, and ultimately might become an indirect tool for prediction.

The use of ocean buoys, oceanographic vessels, and research aircraft, together with the weather satellite, may succeed someday in cracking the problems of predicting storm development in all oceans.

Hopefully, a practicable means will be found to observe and assess action at the sea-air interface, and incorporate this intelligence into machine prediction models.

BOMEX and GATE experiments in the tropical Atlantic comprise a brave start in this direction. Even if nothing else is accomplished, the closer bond that has resulted between experimental meteorology and experimental oceanography is in the interests of both disciplines.

Atmospheric-Oceanic Observations in the Tropics

MICHAEL GARSTANG   NOEL E. LaSEUR   KENNETH L. WARSH
RONALD HADLOCK   JOSEPH R. PETERSEN   1970

In the early part of this century L. F. Richardson enunciated what may be one of the most fundamental concepts in meteorology (Shaw 1931, p. 338). In describing how kinetic energy might be transferred between eddies of various sizes in the atmosphere, he wrote:

Big whirls have little whirls that feed on their velocity.
And little whirls have lesser whirls and so on to viscosity.

Such energy-transfer processes are inherent in turbulent fluid behavior and apply with equal validity to the ocean and to interaction between ocean and atmosphere. The direction of transport from large to small

scale implied by the above parody is, however, not unique. J. S. Malkus (1961, p. 84), describing such fluid action and interaction, provides a vivid picture of "the coexistence in each (sea and air) of many interacting scales of motion, from tiny eddy to planetary gyre, supplying and removing energy from one another, coupled in loops within loops of stable and unstable interaction, inseparable and non-linear, where the whole is frequently spectacularly different from the sum of its separate parts."

It is not immediately obvious why a fluid system deriving its energy from a stable source (meteorologically speaking)—the sun—should respond in this fashion. In part the reason must lie in the fact that the

These five authors were associated with Florida State University when this article was first published:

Dr. Michael Garstang is a professor of environmental science at the University of Virginia. He is studying the energy exchange and transport at the air-sea interface.

Dr. Noel E. LaSeur is a professor of meteorology with the Department of Oceanography at F.S.U. He studies hurricanes and tropical storms.

Dr. Kenneth L. Warsh is an assistant professor in the Department of Oceanography at F.S.U. He studies ocean waves and near-surface circulation.

Dr. Ronald Hadlock is an assistant professor of oceanography and meteorology with the Department of Oceanography at F.S.U. His models of hurricane experiments in the laboratory played an important role in the design of the Barbados Experiment.

Joseph R. Petersen is a research associate with the Department of Oceanography at F.S.U. and managed the field operations for the Barbados Experiment.

From *The American Scientist*, Vol. 58, No. 5, pp. 482-495, 1970. Reprinted with light editing and by permission of the authors and the *American Scientist's* Board of Editors.

atmosphere absorbs a relatively small amount (less than 20 per cent) of the solar radiation. The greater part of the solar energy entering the earth-atmosphere system is absorbed at the surface where it is used primarily to evaporate water. This energy—latent in water vapor, terrestrial radiation, and sensible heat—is returned across the air-surface interface to the atmosphere. The latent heat of the water vapor is typically reconverted to heat by condensation far from its original source.

An examination of the planetary radiative balance between incoming solar and outgoing terrestrial radiation reveals additional complications. On an annual basis there is an excess of incoming over outgoing radiation between the latitudes of 38 degrees north and south and a deficit poleward of these latitudes. Since the average temperature does not change much over the years, the fluid parts of the system act in such a way as to remove the excess heat from the tropical latitudes to provide for the radiation deficit in middle and high latitudes. Yet this simple solution does not appear valid, since it is established that the greater part of the energy input to the atmosphere must occur across the air-surface interface where in the tropical regions of excess the mean surface wind field is directed toward the equator. But this paradox is resolved if in the equatorial regions the latent and sensible heat accumulated in the lower levels of the atmosphere can be transported to great heights. At these heights the mean flow is no longer directed toward the equator, but unsteady eddies are present that can perform poleward transport of the energy originally injected at the surface.

The fluid system is therefore recognizably complicated. To understand it, the oceanic and atmospheric scientist must be able to make quantitative observations with sufficient spatial and temporal resolution to describe its behavior. Given adequate description, he is in a position to analyze and quantify these physical processes. Once the physical processes have been adequately represented in quantitative form, he can predict and perhaps even modify and control the behavior of the system.

In fact, the atmospheric scientist has been unable to accomplish all of these tasks equally well, if at all. Early successes were achieved in describing the large-scale features of the planetary circulations that could be presented in terms of balances between major forces acting in the fluid. Computer technology has permitted treatment of increasing complexity, but in no instance has the meteorologist been adequately able to include energy sources and sinks or to specify explicitly how such energy is converted and utilized by atmospheric systems. The understanding of the workings of the atmospheric fluid-heat engine is far from complete. Little is known about the precise manner in which the engine receives its fuel and perhaps less about how this fuel is converted into motion. It is no small wonder, therefore, that we are unable to provide accurate estimates of the behavior of the engine at some future point, even though the time steps may be only a matter of days. Part of the difficulty stems from inadequate observation of "what is there."

The field programs that form the basis for this paper were conceived in an attempt to remove part of the observational deficiencies. The fundamental philosophy underlying the experiments was to attempt a multiscale observational program in which simultaneous measurements in the atmospheric boundary layer, the region of dry convection between the boundary and cloud bases, and in and above the cloud layer would be made. Such a scheme not only would permit quantitative description of scales of motion in each of these atmospheric layers, but would lead to insight into their interaction and role in atmospheric energy transfer. The tropical ocean, a prime source of atmospheric energy, was chosen for the measurements. These experiments represent only the first steps being taken by atmospheric scientists to make major integrated field measurements designed to provide data for numerical models that will simulate the behavior of the atmosphere. The Global Atmospheric Research Program (National Academy of Sciences 1969) stands as a culmination of these efforts. GARP has as its objectives "to advance the state of atmospheric sciences and technology so as to provide greater knowledge of basic physical forces affecting climate . . . ; to develop existing weather forecasting capabilities"; and "to develop an expanded program of atmospheric science research."

FIELD EXPERIMENTS

The vast tropical and equatorial oceans represent the source region for a very significant fraction of the energy required to power the motions of the sea and air. Investigations into planetary energetics must logically place considerable emphasis on these tropical ocean areas, from which the absorbed solar energy is returned to the atmosphere across the air-sea boundary surface. While the initial transfer process is molecular in scale, turbulent eddy transport rapidly takes over as the primary mechanism of upward flux of heat and water vapor. Within the first few decameters of the sea surface these fluxes are nearly constant with height and are largely governed by mechanical turbulence produced by friction as the wind blows over the surface of the earth. Yet this mechanical mixing is neither constant with time nor independent of other atmospheric scales of motion. Processes taking place primarily on one scale may have an important effect, direct or cumulative, on another. Perhaps the outstanding example is the hurricane; it is now established that energy transferred on a convective cloud scale to create and support the storm on a space scale many times that of the original turbulent scale at the boundary.

The experimental design must therefore provide for the sampling of a hierarchy of scales of motion, commencing at the air-sea boundary and terminating in motions of a planetary scale. A possible approach lies in the intensive observation of the small-scale processes over the entire range of variation in the synoptic and larger scales. Such observations might permit the establishment of parametric relationships between the small- and large-scale systems, allowing the implicit inclusion in atmospheric models of the small-scale effects upon the larger-scale processes.

Although variations in atmospheric structure can be measured over the entire observable range of space and time, certain scales are recognized as having particular significance. The smallest scales are referred to as turbulent, involving eddies in the range from 1 to 100 meters with lifetimes of 1 to 10 seconds. These motions are of particular importance in the lowest few hundreds of meters above the surface. At a height of about 600 meters over the tropical oceans condensation occurs in rising air, and cumulus clouds form. These convective scale clouds are the dominant cloud type of the tropics and range in size from small trade-wind cumuli a few hundred meters in diameter to giant cumulonimbus (thunderstorms) a few kilometers in diameter. The height of these clouds is of the same order as their diameter. Tens or a few hundred of these convective clouds are frequently found in organized lines or groups on the meso-scale covering areas with at least one dimension of the order of a hundred kilometers. Individual cloud lifetimes range from 10 to about 100 minutes. These zones of organized convection are associated with various disturbances of the tropical atmosphere on the synoptic scale, such as waves and vortices, characterized by spatial dimensions of 100-1,000 km and lifetimes of days. Finally, these synoptic systems occur with greater frequencies in preferred larger-scale patterns, which on the seasonal averages yield the planetary scales of motion.

To execute the plan based upon these concepts, a specific tropical ocean area was selected. This area had to be one in which the range of atmospheric systems described above occurred with some predictable regu-

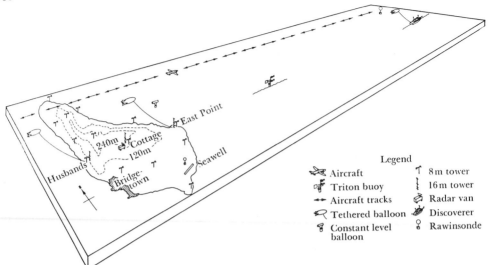

Figure 3-30. Schematic representation of the observational network east of and on Barbados (13°N, 59.5°W). The island, the easternmost of the Lesser Antilles, is 20 by 15 miles, with maximum elevation just over 1100 ft. The E-W line of towers, buoy, and ship is traced by the aircraft track. The single-level towers are on a grid of about 4 by 4 miles.

larity yet be free of complicating effects such as might be induced by the presence of large land masses. The sea area immediately to the east of Barbados (Figure 3-30) was chosen as the most practical location. The time selected was the period July through August, when a variety of meso- and synoptic scale systems develop and move through the Barbados region, but when there is less danger of hurricanes, which typically occur later in the season.

Barbados, which has moderate relief, provided a base for the measurements which were taken in July-August 1968 and repeated in May-August 1969 as part of the Barbados Oceanographic and Meteorological Experiment (BOMEX) (Garstang and LaSeur 1968; Davidson 1968; Kuettner and Holland 1969). The island served as a base of operations and also formed part of the experiment. For example, because of the difference in absorption (radiation) characteristics of land and sea, the island represents a heat source by day and a heat sink by night. It is also an obstacle to atmospheric flow. The measurement of the perturbations of the

temperature and velocity field produced by these island effects was one important objective of the observational program.

Sensing systems were designed to make measurements at the air-sea and air-land interface, in the region of dry convection below cloud base, in the cloud layer, and up to approximately 100,000 ft. The experiment was designed to coincide with the presence of the ATS-3 satellite positioned at synchronous altitude over the Amazon mouth. Additional satellite observations were available in 1969.

The spatial deployment of these sensing systems in 1968 is illustrated in Figure 3-30. In 1969 the stable buoy *Triton* was positioned in the center of the BOMEX array to the northeast of the island (Kuettner and Holland 1969). The field experiments were concluded in August 1969; it will be at least two years from that date before the most fundamental scientific results of these efforts will emerge. The measuring systems, measurements, and results described below illustrate what has been achieved and provide experience useful in the design and execution of the even larger-scale programs now under consideration.

BOUNDARY LAYERS OF THE OCEAN AND ATMOSPHERE

The stable spar buoy *Triton,* illustrated in Figure 3-31 and 3-32 and described in some detail by Warsh, Garstang, and Grose (1970), was designed at the Florida State University as a platform from which a variety of critical measurements could be made in the deep ocean and under both fair and foul conditions. A digital recording current meter attached at a depth of 25 m below mean sea level measured a time series of current speed and direction for a period of 60 days in 1968 (Warsh, Echternacht, and Garstang 1970a, 1970b).

An analysis of the kinetic energy of the current variations shows (Fig. 3-33) large values at three periods: 12 hours, about 52 hours, and the longest periods (lowest frequencies) of 30-60 days. The first two maxima have clear physical explanations in terms of the semi-diurnal tidal oscillations and inertial oscillations, respectively. The long-period oscillations may be associated with effects on the scale of the ocean basin itself; a longer period of record will be required to answer that question. The inertial oscillations arise from interactions between accelerations of the current and the deflecting influence of the earth's rotation. To test the qualitative impression that variations in the magnitude of these inertial motions is related to atmospheric disturbances, the current variations were divided into two portions: the shorter periods equal to or less than the inertial period, and the longer periods. Variations in the magnitude of these two parts during the period are shown by the lower curves in Figure 3-34. The long-period oscillations exhibit two maxima separated by about 30-35 days; the shorter periods (dominated by inertial motion) show many more maxima with an average period of 3-5 days. The magnitude of the shorter-period variations is generally greater during

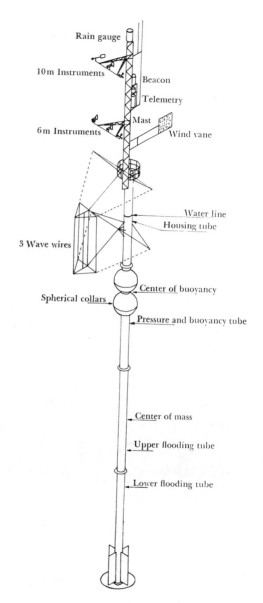

Figure 3-31. Assembly diagram of *Triton* in the configuration used in 1968.

the period of least amplitude in the longer period motions. The upper curve in Figure 3-34 is a measure of the degree of atmospheric disturbance on the organized convective meso-scale and synoptic scale calculated

Figure 3-32. *Triton* on station in 1969 in water over 5000 m deep. Instruments in the atmosphere are at 2 and 10 meters above mean sea level. Instrumentation at 6 m is not complete in this illustration. The large vane (emblazoned with symbol and name) serves to wind-orientate the buoy.

in a manner such that minima represent a greater degree of such disturbances.

Associated with such atmospheric systems are variations from the prevailing steady trade winds; such variations could produce the current accelerations necessary to induce the inertial oscillations. Minima in this curve occur with about the same periods as maxima in the shorter-period oscillations of the ocean current, with a tendency for the maxima in the latter to coincide with or slightly lag minima in the former. We suggest that this is evidence for coupling between atmosphere and ocean of the kind postulated by Malkus (1962) and identified by Webster (1968) at higher latitudes. Extensive previous work (Palmer 1951; Riehl 1954; Yanai *et al.* 1968; and Chang 1970) has documented the tendency for atmospheric disturbances in the tropics to occur with an average period of about four days.

As is seen in Figure 3-32 *Triton* carried a

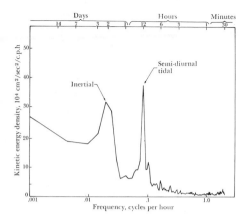

Figure 3-33. The relative distribution (variance spectrum analysis) of the kinetic energy of the variations in the ocean current as measured by the current meter on *Triton* during the period 2 July - 13 August 1968.

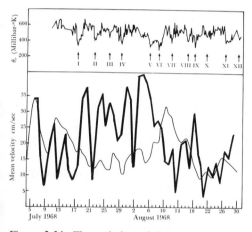

Figure 3-34. The variation of the relative magnitudes of the long-period (light line) and shorter-period (primarily inertial motions): (heavy line) variations in the ocean current data of Figure 3-33. For a description of the upper curve, see pages 181-182.

multiple (3 to 5) resistance-wire wave system. Horizontal and vertical accelerometers were used to monitor the relative small buoy motions (10 cm, heave; 35 cm, surge; and less than $1°$ pitch in seas ranging from one to two meters in height). Simultaneous meteorological data were recorded from thirteen sensors located at two levels (6 m and 10 m in 1968, 2 m and 10 m in 1969 above mean sea level). This system therefore permits three-dimensional computations of the wave fields with a concurrent display of meteorological data.

Such a display is presented in Figure 3-35 (Grose, Garstang, and Warsh 1970) for a five-day period in 1968. On 18 August 1968, starting at 14:00 LST, wave data were collected continuously. Prior to this time, wave data were collected over an 11-min period every two hours. The contoured portion of Figure 3-35 depicts the smoothed variance density per band for the first twenty-three of one hundred bands (each 0.02 Hz wide) into which the total variance was partitioned. Directly above the contours of variance density is shown the net wave direction, which was obtained by multiplying the principal wave direction by the variance density for each band and then finding the vectorial average of all the one hundred frequency bands. The principal directions were calculated from cross-spectral analysis of the sea surface slopes from the three wave staffs. The net directions are shown every six hours for the two-hour data, unless major deviations occurred, and every hour for the continuous-wave data. The total variance is shown in units of centimeters squared on the lower graph of Figure 3-35, which was constructed using the average of the total variance measured by each of the three wave staffs.

From this short record a series of interactions between the ocean and the atmosphere can be detected. On 14 August the variance density shows the greatest energy to be concentrated in swells with periods near 8 sec (frequency of 0.14 Hz). Commencing early on the afternoon of 14 August the total variance drops rapidly by 50 per cent. Wave and wind directions are congruent. Wind speed shows an initial drop coincident with the pronounced drop in total variance. An increase from 8 to 10 m sec^{-1} in wind speed is accompanied by a 25 per cent increase in total variance, but variance con-

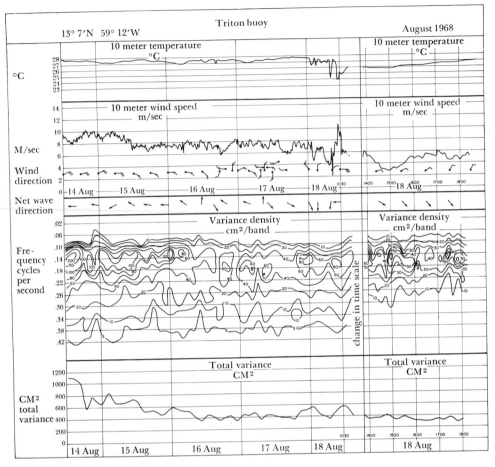

Figure 3-35. Meteorological data and characteristics of ocean waves as measured on *Triton*, 14-18 August 1968. The lower curve is a measure of the total kinetic energy of wave motion; the distribution of this kinetic energy in waves of different frequency is shown in the contoured section immediately above.

tinues to decline while wind speed remains high until midday on 15 August. As long as the wind does not have to reorient the wave direction, similar linear relationships between wind speed and wave variance can be found. When reorientation is necessary, redistribution in the wave spectrum occurs. This phenomenon is most evident on 18 August in the continuous data where an increase in wind speed shows no accompanying increase in total variance. There is, however, a redistribution of the variance spectrum resulting first in a smoothing of

the spectrum when the wind is nearly opposite to that of the waves and then in an increase in peakedness as the wind and wave directions become more nearly congruent.

Such interaction is equally evident on 17 August, when a well-organized atmospheric system passed over the buoy location. With the arrival of the initial squalls (also documented by island-based radar), there is a 33 per cent increase in total wave variance strongly concentrated at 0.16 Hz. As the center of the disturbance passes, wind speeds drop below 4 meters per second, and the sea surface responds with a rapid decrease in total variance and a leveling out of the spectral peaks.

Three mechanisms contribute to this decrease and redistribution of variance. First, the decrease in wind speed leads to a decrease in the driving force. Second, energy is dissipated in reorienting wave directions. Third, the effect of nearly one inch of rainfall cannot be neglected, both in terms of oceanic and atmospheric surface stability and in terms of actual impact.

These data permit the examination of the coupling between the atmosphere and the ocean, in particular the effect of the non-steady atmospheric scales of motion. The meterological sensors carried on *Triton* and on the U.S.C.G.S.S. *Discoverer* permitted the estimation of momentum, sensible heat, and latent heat flux between the ocean and the atmosphere. While *Triton* represents a nearly ideal platform from which to make this type of measurement, the 300-ft *Discoverer* generates problems of interference and contamination of the measurements. Careful comparisons between the buoy and the ship measurements are not yet complete. Figure 3-36 (Garstang and Warsh 1970), however, presents four days of data from *Triton* during August 1968. The flux computations were made using the bulk aerodynamic equations adjusted for departure from neutral stratification. The use of these equations is justified on the grounds that the stratification varied between -0.001 and -0.01 as computed from the Richardson bulk number (Garstang and Warsh 1970).

Except for a brief period at the start of the sequence, when sea and air temperatures appear to be unusually high, all variables are consistent with expected values. The early discrepancy may be the result of buoy servicing on the afternoon of 19 August. Both sea and air temperatures exhibit a weak diurnal oscillation. Air temperature is affected strongly by rain and related to cold downdrafts associated with the precipitation cycle of convective clouds. That this effect is real is confirmed by the fact that it is observed at

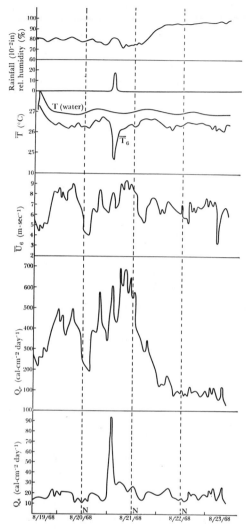

Figure 3-36. Curves of relative humidity, rainfall amount, surface temperature of the ocean, air temperature at a height of 6 meters above the ocean, wind speed at a height of 6 meters, and the vertical fluxes of latent heat (Q_e) and sensible heat (Q_s) by turbulent eddies, at a site 25 km east of Barbados.

night, when a reduction in insolation caused by increased cloudiness cannot be called upon to explain the drop in temperature; the fact that the cooling commences before the start of the rain and ends some time after rain has ceased to fall adds credence to the data. Surface wind speeds also exhibit a fairly pronounced diurnal oscillation. A complete explanation of the diurnal variation in surface wind fields over the open ocean is not yet available (see for example Shibata 1964). Additional insight into this question should emerge from a detailed analysis of the time series of surface wind speed and direction now available from *Triton*.

Relative humidity, in agreement with previous results (Garstang 1967), does not show any obvious diurnal changes. No clear explanation is offered for the increase in relative humidity after midday of the second day. This increase has an obvious impact upon the latent heat flux, which decreases to low values after this time. Tests are being made to ascertain that the indicated high values of relative humidity are not a spurious result caused by either drying or contamination (sea salt deposition) of the wet-bulb wick.

Momentum flux proves to be fairly stable, ranging between 0.37 and 2.16 dynes cm^{-2}. Highest values coincide with the highest average surface-wind speeds some 44 hours after the start of the data train shown in Figure 3-36. Lowest values occur 27 hours after the start, coincident with the second-to-lowest average wind-speed period. The low values of stress at this point, rather than during the lowest wind-speed period, are related to small differences in stability prevailing during each period.

Despite some of the observational difficulties, significant inferences can be drawn from the computed values of latent and sensible heat flux shown in Figure 3-36. Petterssen, Bradbury, and Pedersen (1962), Garstang (1967), and others have all recognized the fact that organized systems of a meso- or synoptic-scale in the atmosphere may significantly amplify the flux of latent and sensible heat between the ocean and the atmosphere. It has been further pointed out that only in regions of organized atmospheric systems do clouds grow to significant heights (Holle 1968).

Figure 3-36 indicates a period of rain in the first hour of 21 August. Other data (satellite pictures, conventional meteorological data) indicate that a poorly organized but extended east-west line of cloudiness was present over Barbados and *Triton* during the daylight hours of 21 August. Immediately prior to, during, and subsequent to this rain, latent heat flux increased by 100 per cent while sensible heat flux increased by nearly one order of magnitude when compared to the average fluxes prior to the rain. These observations made from *Triton* confirm the ship measurements made by Garstang (1967) and the speculation of other authors. It is, therefore, expected that a complete analysis of the *Triton* data will provide a reliable measure of the interdiurnal fluctuations of energy flux over an open tropical ocean site. This information is of vital importance to the understanding of individual synoptic-scale systems and may provide an essential link in the energy-exchange chain.

THE BOUNDARY LAYER AND SUBCLOUD LAYER

Three multilevel towers (see Figure 3-37) and 14 single-level towers, together with one special purpose tower, were erected on the island. The single-level towers, located on approximately a 4 × 4 mile grid, as shown in Figure 3-30, provided near-surface velocity divergence as a function of time. Such information is important to the study of the island effect upon the atmospheric circulation. Computations from these data reveal that there is a consistent diurnal and interdiurnal near-surface divergence.

These results, together with others described below, suggest that a heated island does not function in a simple, direct thermal sense. That is, at times of maximum heating, air does not simply rise over the island, leading to inflow from the sea at the boundaries. Instead, what happens over the heated island is critically dependent on the vertical structure of the wind field. If, as is usually the case in the tropics, the wind speed increases to a maximum near 600 m above the sur-

Figure 3-37. Sixteen-meter tower with sensors for temperature, humidity, horizontal wind speed and direction, and vertical wind speed, at 4 m and 16 m. At 2 m and 8 m, the vertical wind speed is omitted.

face, heating will result in increased buoyant mixing over the island. The stronger winds will be mixed downward to the surface at times of maximum heating. The ocean surrounding the island warms only slightly during the day, and no additional mixing occurs.

The result is stronger near-surface velocities over the island compared to those over the ocean. More low-level air leaves the is-

land than can be replaced by the horizontal wind field, and net downward motion must set in to compensate for this divergence. Thus, at times of maximum heating there are actually descending currents over the island instead of the classical picture of rising of hot air. The measurements obtained from the single-level towers confirm this hypothesis. The island-induced circulation therefore not only becomes a diurnal phenomenon but is strongly tied to interdiurnal changes in the structure of the larger-scale wind field, reflecting once again the interaction between various scales of motion.

The multiple-level towers provide time series of temperature, humidity, and three components of the wind field at three locations on an east-west line across the island. The time series are expected to yield information on the diurnal and interdiurnal turbulent structure of the boundary layer. At the central site, Cottage, detailed solar and terrestrial radiation measurements were made, which will be combined with the tower measurements to establish heating-rate models for the atmospheric boundary layer. The tower observations will also permit the computation of momentum, water vapor, and heat flux through the boundary layer at each location. Since vertical motion of the atmosphere was measured explicitly at each tower, more direct techniques of computation of momentum and heat can be employed. In addition, a fourth mobile tower carrying complex instrumentation was used as a check on the fixed tower. By moving this tower with respect to the fixed tower, a measure of terrain effects was obtained.

Three techniques were employed to sample the region of dry convection between the surface and the base of the convective clouds near 600 m. Tethered balloons (see Figure 3-38) were flown at altitudes up to 600 m from East Point and Husbands, stations on the extremities of the east and west coasts respectively. The balloons were tethered by 1,000-lb breaking-strain nylon line and controlled by a specially designed winch. The sensor-modulator-transmitter package was suspended from the tethering line in an adjustable but semirigid frame. A drogue chute damped high-frequency oscillations of the package, while a horizontally mounted vane maintained the package in a near-horizontal position offsetting variations caused by wind pressure on the package.

Temperature (thermistor), humidity (carbon strip hygristor), horizontal wind speed (cup anemometer), and vertical wind speed

Figure 3-38. Fifteen-hundred-cubic-foot helium-filled tethered balloon ascending from storage pit. The balloon is 26 ft long, with a maximum diameter of 10 ft.

(propeller anemometer) were transmitted in a pulse-delay fashion by means of a 403 MHz carrier frequency. The pulse delay technique permitted sampling at a rate of about three times per second without rapid depletion of the battery power supply, making possible continuous flights as long as 12 hours, after which deterioration of the humidity element necessitated replacement of that sensor.

More than one instrument package could be flown at a time if the carrier frequency was offset so that signals could be discriminated. The signals were received at the ground, translated into voltages by a demodulator, and recorded on a digital magnetic tape recorder. Figure 3-39 shows a short section of record with two instruments at 125 and 250 m, respectively. The balloon height, which was not monitored continuously, is also shown on this record. Long-period oscillations of the balloon must be removed from the record to obtain true values for a given height, but in many applications of these data the balloon oscillations are relatively unimportant compared to the shorter-period turbulent fluctuations.

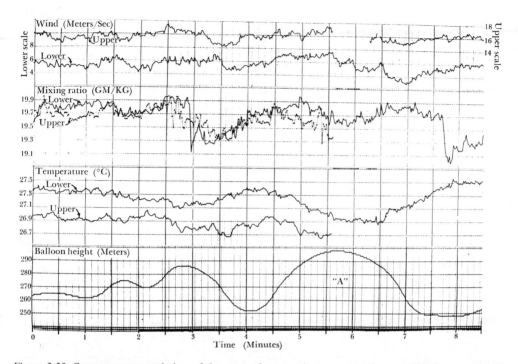

Figure 3-39. Computer-generated plots of the output from two instrument packages at 125 (lower) and 250 m (upper) respectively. The vertical oscillations of the balloon are shown on the lower graph. ("Mixing ratio" is grams of water vapor per kilogram of air.)

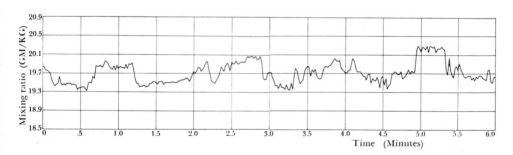

Figure 3-40. Expanded humidity record (mixing ratio) at 124 m, showing the discrete increases in water vapor.

Examination of the temperature records for the east-coast station show short-period random fluctuations. There is little evidence in the temperature record of "bubbles" of hot air that would carry energy through the subcloud layer and initiate convective clouds at the level of condensation. However, the humidity record (displayed in greater detail in Figure 3-40) shows distinct parcels of moist and dry air. Since moist air is lighter than dry air of the same temperature, the moist air parcels are unquestionably buoyant and rising.

Clear evidence for this phenomenon is

Figure 3-41. Constant-volume balloon of clear mylar plastic carrying a cylindrical aluminum-foil radar target and an instrumented (temperature and humidity) transponder, seen in the right hand of the scientist. The tow balloon is in the foreground, nearly obscuring the fuse timed to release the CV-balloon at flight altitude.

seen in Figure 3-39 in the major balloon oscillation marked A. Although humidity decreases with height, it can be seen that, as the balloon rises, humidity at both levels increases, the increase commencing at the sensor levels before the balloon responds. These observations suggest that we are dealing with a rising parcel of moist air; parcel sizes are estimated from these data to be in the order of 100-300 m. The results strongly imply that energy transport through the subcloud layer over the open ocean is closely allied to water vapor concentrations, in agreement with findings of Woodcock (1960).

Further analysis may lead to a better understanding of the mechanisms for initial concentration and the subsequent relationship between the moist parcels and individual convective clouds.

Further information on the structure of the subcloud layer is being derived from trajectories of constant-volume balloons (CV-balloon). Cylindrical balloons, which have a volume of 0.3 m^3 and are constructed from 1-mil mylar plastic (see Figure 3-41), were inflated with helium and ballasted so that the balloon displaced exactly its own weight plus that of its payload at a predetermined altitude above ground. The CV-balloon was filled to an overpressure (well in excess of atmospheric pressure) at the

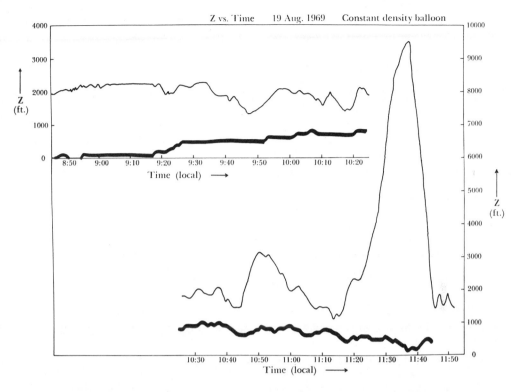

Figure 3-42. The height (Z) of the island (heavy curve) and of the constant volume balloon (light curve) along its trajectory over Barbados 19 August 1969. The balloon was launched at East Point at 8:30 local time and fell back to the ground in a rain shower over two hours later. The continuous flight is shown in two parts here for convenience.

ground, ensuring that no further changes in volume could take place with changes in pressure or temperature. The CV-balloon was then launched by aircraft or towed to the predetermined altitude by a conventional weather balloon, which was then cut loose by a fuse. At the altitude where the CV-balloon exactly displaces its own weight it is neutrally buoyant and is assumed to behave as an air parcel.

The CV-balloon also carried a passive radar reflector and a radar-triggered transponder. The transponder utilized the same pulse-delay techniques as were employed on the tethered balloon to transmit temperature and humidity sensed by a thermistor and carbon-strip hygristor. The position data obtained from the 3 cm tracking radar, to-

gether with the temperature, humidity, and suitable date/time marks, were recorded about three times per second in digital form on magnetic tape.

Results from one such flight are shown in Figure 3-42. During the first 100 minutes the balloon which leveled off at slightly above 2,000 ft, gradually lost altitude. It is likely that this phenomenon reflects the decrease in air density as daytime heating progresses. There is no doubt, however, that the balloon is behaving as a CV-balloon with neutral buoyancy. The series of displacements ranging between 200 and 500 ft are reflections of the eddy turbulence in the subcloud layer. Vertical velocities associated with these displacements approach 1.5 m sec^{-1}. If mean values of the velocity components are calculated over intervals of 10 to 100 seconds along the trajectory, departures from the mean velocity can be obtained, and from these the momentum flux can be computed. Values range from 1 to 5 dynes cm^{-2}. In cases where the balloon crossed the coastline from land to sea, momentum flux drops dramatically—at least 50 per cent.

In the flight presented in Figure 3-42, the CV-balloon is entrained into the circulation of a cumulus congestus and swept from less than 400 ft above the terrain to 9,600 ft in 26 minutes. When corrections for negative buoyancy and liquid water on the surface of the CV-balloon are applied, calculated vertical motion in the cloud exceeds 10 m sec^{-1}. The vertical distribution of the velocity field in the cloud agrees remarkably well with an adiabatic steady-state jet model for convection. The horizontal trajectory of the balloon, which is not shown, reveals that the balloon underwent a complete spiral as it rose in the cloud.

The final method used to sample the subcloud layer was by means of an instrumented aircraft—a twin-engined Queen Air outfitted and operated by the National Center for Atmospheric Research, in Boulder, Colorado. The aircraft made flight-level measurements of horizontal wind speed, temperature, and humidity. An infrared radiometer was used to measure the surface radiative temperature, and motion-picture cameras were used to photograph the cloud fields. With the exception of the cloud pictures, all this information, together with relevant data on aircraft heading, speed, time, and the like, were recorded on magnetic tape. The reduction of these data has only just commenced.

The combination of these three methods of monitoring the behavior of the atmosphere in the layer between the surface and the convective cloud base will produce far more information than has ever been available for this layer at any other time or place in the tropics. From these data the links between the surface and the clouds will be sought.

THE CLOUD LAYER AND MIDDLE AND UPPER TROPOSPHERE

Measurements in this region were carried out in a most intensive fashion during BOMEX in 1969 (Kuettner and Holland 1969), by the largest fleet of scientific aircraft and surface vessels ever assembled for a single meteorological experiment. The reduction of the enormous amount of data collected is still in progress. The results presented below on scales of motion in and above the cloud layer are drawn primarily from the observations made during the 1968 experiment.

Intensive use was made of conventional sounding techniques, using balloon-borne radio transmitters carrying temperature, humidity, and pressure sensors whose signals were modulated and transmitted to ground. Soundings made in 1968, together with other measurements, were made in two modes. Days of intensive experiments were separated by days of routine observations. A five-day cycle consisted of 72 hrs of intensive observations followed by 48 hrs of routine observations. During intensive periods soundings were made at 3-hour intervals from the island and from the *Discoverer* 60 mi east of Barbados. On routine days soundings were made every 6 hours.

The tropical troposphere is, on the average, in a state of convective and conditional unstable equilibrium. This instability, however, is not released at random but rather through organized systems in the tropical atmosphere. Such organized systems are of the meso- to synoptic-scale and are most frequently centers of intense convection, under which large quantities of sensible and latent heat (which is released in condensation-precipitation) are conveyed to high altitudes, creating in these very regions of activity a more stable atmosphere.

An effective measure of the total process—and hence the existence of an organized tropical atmospheric system—is the vertical distribution of equivalent potential temperature, θ_e, obtained from a radiosonde. Such vertical distributions of θ_e, are

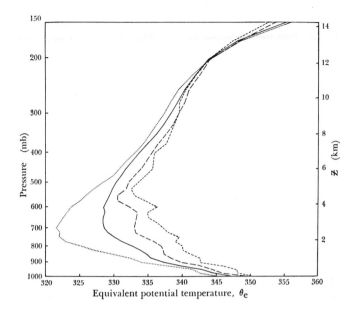

Figure 3-43. The distribution of θ_e as a function of pressure (P) in millibars or height (Z) in kilometers for days with four characteristic weather types: average (solid); suppressed convection (dotted); moderately enhanced convection (long dashes); and strongly enhanced convection (short dashes). The differences in the distributions result primarily from the vertical transport of water vapor by the convective clouds.

shown in Figure 3-43. In this figure soundings have been classified according to the vertical distribution of θ_e, yielding mean curves for low, moderate, and high values of θ_e in the column.

Similarly, these curves correspond to organized states of weather that can be classified as undisturbed, average, moderately disturbed, and disturbed. When days for which any of these soundings might be typical are compared with such measures of the weather as rainfall and cloudiness (from island observations and satellite pictures), good agreement is obtained.

Particularly good results are obtained if an integrated value of θ_e is derived from individual soundings by summing up the area below the curve to the left of a line arbitrarily located at $\theta_e > 350°$K in Figure 3-43. (Such results are also presented in the upper part of Figure 3-34.) While this method is diagnostic rather than prognostic, it is a valuable means of quantitative classification of

the tropical atmosphere and forms a basis for linking the surface, subcloud layer, and cloud layer to the synoptic-scale processes.

THE FUTURE

The above discussion gives some indication of the magnitude and complexity of the problem even when it is viewed only from the standpoint of a single experiment. We have made no reference to the cost in terms of human effort and money. Two and one half years were spent on preparatory work by the authors and a nucleus of some ten graduate students and technicians before the execution of the 1968 experiment. When major facilities—such as aircraft, buoys, ships, and radar—must be assembled, together with the necessary people and expendable supplies, logistics alone dictates that dates and times must be fixed well in advance.

Once such a timetable is established,

system development and testing must proceed on schedule. Schedules and delivery times which appeared entirely reasonable during the planning stages could not be adhered to. As a result, the final phase of testing and evaluation prior to deployment could not be adequately accomplished. More than forty scientists, students, and technicians were in the field in 1968.

About $750,000 in direct funding was expended on this experiment over a period of four years. If the services provided at nominal or no charge by the Air National Guard, Environmental Science Services Administration, National Center for Atmospheric Research, Caribbean Meteorological Institute, and the Barbados Meteorological Service had had to be paid for, the cost would have been more than double.

In 1969, for BOMEX the total experiment was at least one order of magnitude greater in complexity and cost.

For each scale of motion, links with other scales in the atmosphere were indicated. Thus the regular periodicities noted in the ocean-current measurements were tied to organized synoptic-scale disturbances in the atmosphere, whose existence was summarized by variations of a particular parameter. Similarly, the behavior of the surface wave fields and the fluxes of sensible and latent heat could be linked to synoptic systems in the atmosphere. In the case of the heat fluxes, a fundamental link in the energy transport of the system was identified. In the subcloud layer the vertical transport of heat and water vapor was tied to eddies, measuring hundreds of meters across, which received their buoyancy not from temperature differences but from moisture differences. The task of unifying the components into a whole remains to be done, but many of the parts have been identified and their roles clarified in this first major attempt at an integrated field program. Many more pieces will be identified as analysis proceeds.

At this stage we may summarize the experience gained from the 1968 program and participation in the BOMEX, 1969, in the following conclusions: This type of field program is essential to better understanding of the atmosphere and its interactions with land and ocean. Specialized measuring systems must be developed and tested if the required data are to be obtained. In spite of attempts to formulate objectives of the program in advance, unanticipated problems arise in the execution or are revealed by analysis of the data; conversely, serendipity typically provides unexpected results.

Much more extensive field programs are now in the planning stages as part of the Global Atmospheric Research Program. Considerable effort has been expended in formulating the scientific objectives of the first GARP tropical experiment, now tentatively scheduled for the tropical Atlantic region in 1974. Much additional effort will be needed to develop and select the technological tools and experimental design to fulfill these objectives. In addition, an administrative and scientific management structure must be established to provide the framework for cooperative efforts of many university and government groups on both the national and international level. Given optimum success, this field program is not likely to answer all questions; rather, succeeding experiments will have to be built upon its results. The potential benefits fully justify the effort needed.

REFERENCES

Chang, Chih-Pei. 1970. Westward propagating cloud patterns in the tropical Pacific as seen from time-composite satellite photographs, *J. Atmos. Sci.* **27**, 133-38.

Davidson, B. 1968. The Barbados Oceanographic and Meteorological Experiment, *Bull. Amer. Meteor. Soc.* **49**, 928-34.

Garstang, M. 1965. Some meteorological aspects of the low-latitude western Atlantic: Results of Crawford Cruise #15. Woods Hole Oceanographic Inst., Ref. No. 58-72, 97 pp.

Garstang, M. 1967. Sensible and latent heat exchange in low latitude synoptic-scale systems, *Tellus* **19**, 492-508.

Garstang, M., and N. E. LaSeur. 1968. The 1968 Barbados Experiment, *Bull. Amer. Meteor. Soc.,* **49**, 627-35.

Garstang, M., and K. L. Warsh. 1970. Energy flux measurements at the sea-air interface, *Marine Tech. Soc. J.* in press.

Grose, P., M. Garstang, and K. L. Warsh. 1970. Surface wave spectra in the deep ocean, EOS, *Trans. Am. Geophys. Un.* **51**, 302.

Holle, R. 1968. Some aspects of tropical oceanic cloud populations, *J. Appl. Met.* **7**, 173-83.

Kuettner, J. P., and J. Holland. 1969. The BOMEX Project, *Bull. Am. Meteor. Soc.* **50**, 394-402.

Malkus, J. S. 1962. Large-scale interactions, pp. 88-294 in M. N. Hill, ed., *The Sea,* 1. New York: Wiley (Interscience).

National Academy of Sciences. 1969. Plan for U.S. Participation in the Global Atmospheric Research Program. Washington, D.C.: National Academy of Sciences, 79 pp.

Palmer, C. E. 1951. Tropical Meteorology, pp. 859-80 in T. Malone, ed., *Compendium of Meteorology.* Boston: Am. Meteor. Soc.

Petterssen, S., D. L. Bradbury, and K. Pedersen. 1962. The Norwegian cyclone models in relation to heat and cold sources, *Geofys. Publ. Geophysica Norvegica* **24**, 243-50.

Riehl, H. 1954. *Tropical Meteorology.* New York: McGraw-Hill, 392 pp.

Shaw, Sir William Napier. 1931. *Manual of Meteorology,* **4**. Cambridge: Cambridge University Press, 359 pp.

Shibata, E. 1964. The atmospheric tide hypothesis on the diurnal variation of the cloudiness in the tropics. M. A. dissertation, Dept. of Meteorology, University of California (Los Angeles), 82 pp.

Warsh, K. L., M. Garstang, and P. Grose, 1970. A sea-air interaction deep-ocean buoy, *J. Marine Res.* **28**, 99-111.

Warsh, K. L., K. L. Echternacht and M. Garstang. 1970a. An observation of an inertial flow at low latitude, *J. Geophys. Res.* **75**, 2207.

Warsh, K. L., K. L. Echternacht, and M. Garstang. 1970b. Structure of near-surface currents east of Barbados, submitted for publication to *J. Physical Ocn.*

Webster, F. 1968. Observations of inertial-period motions in the deep-sea, *Rev. of Geophysics* **6**, 473-90.

Woodcock, A. H. 1960. The origin of the trade-wind orographic shower rains. *Tellus,* **12**, 315-26.

World Meteorological Organization, 1970. The Planning of GARP Tropical experiments, GARP Pub. Series No. 4.

Yanai, M., T. Maruyama, Tsuyoshi Nitta, and Y. Hayashi. 1968. Power spectra of large scale disturbances over the tropical Pacific, *J. Met. Soc. Japan* **46**, 308-23.

4

LIFE AT SEA

Some Food Chains

A. P. ORR AND S. M. MARSHALL   1969

Little has been said about the food links between the phyto- and the zooplankton and the food chains which link the nutrient salts in the water with the fish and whales which are caught for man's use. We shall describe a few examples of these in detail.

Food-web is perhaps a better description than food-chain for the complicated relationships between plants and animals in the diet of the herring as shown by Hardy. The herring is our most valuable plankton-feeding fish. It lays its eggs on a suitable surface on the sea bottom and they hatch in a week or two (depending on the temperature). After the young have consumed the yolk in the yolk sac, which lasts them only a few days, they have to fend for themselves and this is a critical period in their life. They can take in both plants and animals but these have to be very small because of the size of the larval gullet. If too large an animal is eaten it sticks in the gullet and the young larva dies.

Each organism is taken by a definite act of capture and not by chance encounter. The larva, therefore, cannot feed in the dark, and in the sea maximum feeding is in the daylight hours especially in the early morning and in the evening. The small larva when it first tries to feed is very inefficient and in only about one out of twenty darts does it succeed in capturing food. As it grows older it learns (or the inefficient die off) and when it is seven or eight weeks old about 85 per cent of its efforts are successful.

First the larvae take diatoms (although it is not certain that they can digest these) and the nauplii of the small copepods. As they grow larger they can take the copepodites and adults of the small copepods. When they are about three months old their silvery scales develop and they then look like herring and swim more actively. In the Clyde when the baby herring are nearly three months old the *Calanus* of the first generation lay their eggs and hundreds of these eggs can be seen in the gut of each young fish. As they grow they progress to larger and larger food.

Some races of herring spawn in winter and some in summer. The first have larger eggs and a bigger supply of yolk which en-

Dr. A. P. Orr and Dr. Sheina M. Marshall were on the staff of the Marine Station, Millport, Isle of Cumbre, Great Britain as biochemist and biologist, respectively. Dr. Marshall specializes in plankton research and has co-authored numerous articles with Dr. Orr. Dr. Orr died in 1962.

From *The Fertile Sea*, pp. 97-106, 1969. Reprinted with light editing and by permission of the family of the late Dr. Orr, Dr. Marshall, and The Buckland Foundation, Welling, Kent.

ables them to grow larger and more capable before they begin to feed. At that time of year there is not much food in the water but there are few predators. The summer larvae with a smaller yolk supply hatch at a less efficient stage but their food is much more abundant. Predators however are more common.

It is difficult to keep herring alive in an aquarium but it has been done and feeding experiments have been made with them. When they were given a good supply of copepods (about 9 per liter) they broke shoal formation and began to feed. They continued to feed for about two hours and then the shoal reformed. The rate of digestion depended on size and temperature. For a fish of six inches complete digestion took 24 hours. Even at this size they still cannot feed in darkness and do best in a moderately good light. Near the surface even moonlight is bright enough—"the herring it loves the merry moonlight" has a basis of truth. When they are feeding actively they break their normal shoaling behavior and dart about rapidly after the food.

A careful study of the food of the herring has been made in the North Sea. Nearly half was crustacean; in this fraction copepods and amphipods were equally abundant and there were also a few euphausiids. Most of the other fraction was made up of fish (sand eels) but there were a few *Spiratella*, arrow-worms, and appendicularians. The herring on the west coast of the British Isles eat a much higher proportion of crustaceans, mainly *Calanus* and euphausiids. Most of these food organisms feed on the phytoplankton, but the arrow-worms and some euphausiids are themselves carnivorous and will feed on copepods, and the first even on small herring larvae.

The plaice is an example of a bottom-living fish. Its eggs however are laid freely in the sea and the larvae spend some weeks in the plankton. At first they look just like round fish, but soon one eye begins to move over to the other side, the body flattens from side to side and the little fish settles on one side on a sandy bottom. While it is in the plankton it feeds on plankton organisms suitable for its size, usually copepod nauplii, small copepods, larval molluscs, and worms. In the southern North Sea, however, it depends almost entirely on the appendicularian *Oikopleura*, which itself feeds on the very minute flagellates of the plankton. Once on the bottom the plaice changes over to feeding on bottom-living animals such as amphipods, worms and bivalves, which in turn have a variety of ways of feeding. Many bivalves live buried in the sand, connected with the surface by two tubes, the siphons, one of which takes in water and the other discharges it. The tips of these extended siphons are often nipped off by newly settled plaice and in fact, for a short time, they may form an important part of the young fish's diet. Siphons may be cropped more than once from the same animal but they can be regenerated fairly rapidly. In the bivalve *Scrobicularia* one of the siphons is extended into a long thin tube which moves actively about over the surface (Figure 4-1) like a vacuum cleaner, sucking up the detritus. In this detritus there are many little bits of organic matter, some living, such as bottom diatoms, small nematode worms or bacteria, some the dead remains of plants and animals which have sunk from the surface layers, for in shallow coastal water these

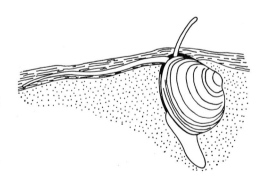

Figure 4-1. Mud-living, deposit-feeding, bivalve *Scrobicularia plana,* showing the mode of feeding. (From Yonge.)

will reach the bottom before dissolving. *Scrobicularia* is a deposit feeder, but there are many burrowers both molluscan and crustacean which are suspension feeders and live on the small planktonic plants and animals.

Apart from being commercially valuable the plaice is a hardy and easily handled fish and so a good deal of work has been done on its feeding and growth. A number of plaice were kept in boxes floating in the sea and were fed with known quantities of mussel flesh. Some were given just enough food to keep them at a steady weight—a maintenance ration; some were given more than this—an intermediate ration, and some were given as much as they would eat—a maximum ration. To keep a plaice of 40 g (about an ounce and a half) at a steady weight without growth needed between 4 and 10 g of food a week, or between 1½ and 3½% of its own weight per day. If the water is warm it needs more, if cold, less. Bigger fish need proportionately less to maintain themselves. With a larger food supply they will grow, but with twice the maintenance ration they grow almost as fast as with a maximum ration. With fish fed on the maximum ration only a quarter of a fifth of the food was used for growth.

Whalebone whales depend on plankton for food, although some species take small fish as well. They filter off the plankton through the fringes of whalebone which hang inside their mouths and, since they need enormous quantities to maintain themselves, they are found most commonly where zooplankton is richest, in the Arctic and Antarctic. The commercially valuable whales of the Arctic have been destroyed by overfishing, and the populations of the Antarctic are rapidly going the same way. The food of the northern whalebone whales is largely *Calanus* and *Spiratella,* that of those of the southern ocean mostly *Euphausia superba.* As we have seen these foods themselves exist on phytoplankton.

There are not many herbivorous fish but among them are some of the herring family,

sardines and anchovies. One curious feature about their feeding is that the larval stages eat more copepods and other zooplankton than the adults; both eat phytoplankton but some of the adults eat little else. Off the coasts of California, Chile and Peru where there is up-welling there are, or were, enormous shoals of fish—anchovies and menhaden off the South American coast, sardines off the Californian. There was a great fishery and canning industry in California (immortalized in Steinbeck's *Cannery Row*) until some years ago when the sardines disappeared. In Peru nobody seems to have thought of using the fish until recently and innumerable sea birds, the guayanes, fed on them. A hundred years ago the chief source of nitrate, or salt-peter as it used to be called, was in Chile where, in this rainless area, huge beds of "guano" have been formed over the centuries from the excretions of these sea birds. Until the mid-1950s there was only a small fish-canning industry in Peru, and a small fishmeal industry depending on its residues. With an increasing demand for fishmeal the latter industry was expanded, processing machinery was brought south from the derelict Californian factories, and a small-meshed purse seine was introduced. This caught the "anchoveta," a small fish of the sardine family, 10-17 cm long, which occurs in countless millions off the Peruvian coast and is the chief food of the guayanes.

These fish are not canned; they are cooked, dried, and ground to a meal which is used as a high-grade supplement to foods for pigs and poultry. Besides meal, oil is recovered and used in the manufacture of paint and linoleum. A third product is a liquid residue containing nutrients and this is also sold as a stock-food supplement or as liquid fertilizer.

The industry rocketed. In 1948, 1,000 metric tons of fish were landed; in 1965, 138,500; in 1962 and 1963, more than 6½ million tons. With calm weather and the fish supply close to the shore, even unskilled fishermen can fill their holds once or twice a day. The resulting meal can therefore under-sell European fishmeals.

Every few years however the nutrient-rich Peruvian current water is overlaid by a warm sterile water mass flowing from north to south, a current which has been given the name "El Nino." When this happens the fish are no longer found near the surface, the guayanes are no longer able to catch their usual food and many starve to death. Such bad years have been recorded ten times in the last half-century; the last was in 1957 when the guayanes are supposed to have been reduced from 35 to 10 million. Some-thing of the same sort happened in 1963 (although it was not caused by El Nino) and although the total catch remained high there had been an increase in fishing effort. After this comparatively poor year the catch con-sisted largely of small (8-11 cm) and im-mature fish. It seems probable that the stock is now being overfished.

The production of guano has decreased steadily since the anchoveta fishery began, and although El Nino has always caused fluc-tuations in the number of guayanes, the guano producers now blame the decrease entirely on the anchoveta fishermen. It has been estimated that the birds have to eat 11-16 tons of fish to produce one ton of guano, and since a great deal of their excre-tions must be lost at sea (thus increasing surface fertility) a conservative estimate would be 22 tons of fish for one ton of guano, whereas 5-7 tons of fish produce one ton of fishmeal besides the oil and fish solu-bles.

This is one of the biggest fisheries in the world, taking about 18% of the total world catch, and conservation measures are ob-viously called for.

There is a commercial herring fishery in the Barents Sea and recent work has shown an unexpected and intricate relationship among the fish and zooplankton. In winter, when *Calanus* is in deep water, the herring are in deep water too and they follow the *Calanus* up to the surface in spring when these spawn, feeding on them actively. In some years however the ctenophore *Bolinopsis* is so abundant that it destroys the *Calanus* and spoils the herring fishery. Ctenophores eat herring larvae and those of other fishes was well as copepods. Russian scientists have found that one ctenophore 2½-3 cm long can eat 10-11 copepods in two hours, digesting half of them in one day and throwing the rest out dead. With *Calanus* at 300/cubic meter, and two *Bolinopsis* the *Calanus* would all be eaten in a month. How-ever there is a larger ctenophore *Beroe* which does not eat crustacean food but lives on other ctenophores, often *Bolinopsis; Beroe* in turn is eaten by the cod, which sometimes gorges on it. Haddock eat *Beroe* too al-though not so intensively as the cod. In years when cod are abundant in the Nur-mansk coast fishery they reduce the popula-tion of *Beroe,* thus increasing the population of *Bolinopsis,* and so the stocks of *Calanus,* and then of herring, suffer. A good herring fishing may therefore depend on a bad cod fishery, and this through a whole series of inconspicuous links.

Some of the herbivorous zooplankton, if themselves useless as food, may be in direct competition for food with the useful species and so have an injurious effect on fish popu-lations. Salps may be an example of this but they are not often common in northern waters.

Breakdown and regeneration are part of the cycle of life and this is where bacteria are important. When a plankton animal dies its substance begins to decompose at once and within a day or two the body has lost almost all its nitrogen and phosphorus. This does not seem to be by bacterial action; probably the enzymes present in the body can account for it. It is important for the multiplication of the phytoplankton in sum-mer that dissolved nutrients should be re-turned to the water quickly.

The remains of the animal bodies and of the fecal pellets, which are not dissolved quickly, sink to the bottom (if they are not eaten on the way) and are there decomposed by bacteria and gradually returned in solution to the sea water.

Bacteria are found throughout the oceans, but they are much more common on the bottom than elsewhere, particularly on the surface of the bottom sand and mud deposits. Here they break down dead animal matter and here they themselves are a rich food for the bottom-living detritus feeders. In the water itself they are much scarcer and become rare as one goes toward the open ocean. Many bacteria are washed into the sea from the land and fresh waters, but these do not survive for long. Even the dangerous coliform bacteria and typhoid bacilli are killed within a day or two. If, however, they are eaten by shellfish such as mussels or oysters they can survive in the body fluids of their hosts, and this is why there are such stringent regulations about the cleansing of those shellfish taken for human food. Marine bacteria are however adapted to salt water and can even withstand the pressure at great depths. Although in a sample of water taken from the open sea very few bacteria can be counted, yet if this same sample is put in a bottle the numbers rise immediately. The smaller the bottle the greater the number that develops. They are living upon the very dilute solution of organic matter in the sea water but they also seem to need a surface to settle on, and the greater the surface the higher the number. Putting glass beads in the bottle increases the surface available and raises the numbers.

The reason why these bacteria do not multiply while they are in the sea is not fully understood. Possibly it has something to do with the extreme dilution of their nutrient medium, the sea, for if nutrients are added to the sample bottles the difference (in numbers developing) between large and small bottles disappears. It has been suggested that strong light at the surface is injurious to them, but this effect could not go very deep. It may be that in open water the bacteria are all attached, either to plankton organisms or to particles of detritus.

It is for these reasons that most bacterial decomposition takes place on the sea bottom and there the dead material is eventually broken down into nitrate, phosphate and carbon dioxide which are again available for phytoplankton growth. If the sea is shallow these nutrients will be returned to the surface waters with the temperature overturn in winter. If it is very deep they may be lost to the productive cycle unless they are carried along in some deep bottom-flowing current, to well up to the surface in places far from their origin.

Bacteria are of very many types and have different functions. There are some which are able to decompose tough substances like chitin, cellulose or lignin. There are others which oxidize ammonium compounds to nitrite and nitrite to nitrate. This is the usual sequence, but in the sea there are also bacteria which do the opposite and reduce nitrate to nitrite or even to nitrogen gas. They can use all these substances for building up their own bodies and in addition they produce enzymes and even vitamins which are of vital importance to the growth of the phytoplankton. It is much more difficult to grow cultures of phytoplankton without bacteria than with them, but the substances they supply have not always been identified.

Photosynthesis
and Fish Production in the Sea

JOHN H. RYTHER   1969

INTRODUCTION

Numerous attempts have been made to estimate the production in the sea of fish and other organisms of existing or potential food value to man.[1-4] These exercises, for the most part, are based on estimates of primary (photosynthetic) organic production rates in the ocean[5] and various assumed trophic-dynamic relationships between the photosynthetic producers and the organisms of

interest to man. Included in the latter are the number of steps or links in the food chains and the efficiency of conversion of organic matter from each trophic level or link in the food chain to the next. Different estimates result from different choices in the number of trophic levels and in the efficiencies, as illustrated in Table 4-1.[2]

Implicit in the above approach is the concept of the ocean as a single ecosystem in which the same food chains involving the same number of links and efficiencies apply throughout. However, the rate of primary production is known to be highly variable, differing by at least two full orders of magnitude from the richest to the most impoverished regions. This in itself would be expected to result in a highly irregular pattern of food production. In addition, the ecological conditions which determine the trophic dynamics of marine food chains also vary widely and in direct relationship to the absolute level of primary organic production. As is shown below, the two sets of

1. H. W. Graham and R. L. Edwards, in *Fish and Nutrition* (Fishing News, London, 1962), pp. 3-8; W. K. Schmidt, *Ann. N. Y. Acad. Sci.* 118, 645 (1965).
2. M. B. Schaeffer, *Trans. Amer. Fish. Soc.* 94, 123 (1965).
3. H. Kasahara, in Proceedings, 7th International Congress of Nutrition, Hamburg (Pergamon, New York, 1966), vol. 4, p. 958.
4. W. M. Chapman, "Potential Resources of the Ocean" (Serial Publication 89-21, 89th Congress, first session, 1965) (Government Printing Office, Washington, D. C., 1965), pp. 132-156.
5. E. Steemann Nielsen and E. A. Jensen, *Galathea Report*, F. Bruun *et al.*, eds. (Allen & Unwin, London, 1957), vol. 1, p. 49.

Dr. John H. Ryther is a marine biologist and senior scientist with Woods Hole Oceanographic Institution, Massachusetts. His research interests include most aspects of biological oceanography, particularly plankton ecology and physiology and aquaculture. He has written or helped write over sixty-five technical articles.

Table 4-1. Estimates of potential yields (per year) at various trophic levels, in metric tons. (After Schaeffer.[2])

| | Ecological Efficiency Factor | | | | | |
| | 10 per cent | | 15 per cent | | 20 per cent | |
Trophic Level	Carbon (tons)	Total Weight (tons)	Carbon (tons)	Total Weight (tons)	Carbon (tons)	Total Weight (tons)
0. Phytoplankton (net particulate production)	1.9×10^{10}		1.9×10^{10}		1.9×10^{10}	
1. Herbivores	1.9×10^{9}	1.9×10^{10}	2.8×10^{9}	2.8×10^{10}	3.8×10^{9}	3.8×10^{10}
2. 1st stage carnivores	1.9×10^{8}	1.9×10^{9}	4.2×10^{8}	4.2×10^{9}	7.6×10^{8}	7.6×10^{9}
3. 2nd stage carnivores	1.9×10^{7}	1.9×10^{8}	6.4×10^{7}	6.4×10^{8}	15.2×10^{7}	15.2×10^{8}
4. 3rd stage carnivores	1.9×10^{6}	1.9×10^{7}	9.6×10^{6}	9.6×10^{7}	30.4×10^{6}	30.4×10^{7}

variables—primary production and the associated food chain dynamics—may act additively to produce differences in fish production which are far more pronounced and dramatic than the observed variability of the individual causative factors.

PRIMARY PRODUCTIVITY

Our knowledge of the primary organic productivity of the ocean began with the development of the ^{14}C-tracer technique for *in situ* measurement of photosynthesis by marine plankton algae[6] and the application of the method on the 1950-52 *Galathea* expedition around the world.[5] Despite obvious deficiencies in the coverage of the ocean by *Galathea* (the expedition made 194 observations, or an average of about one every 2 million square kilometers, most of which were made in the tropics or semitropics), our concept of the total productivity of the world ocean has changed little in the intervening years.

While there have been no more expeditions comparable to the *Galathea,* there have been numerous local or regional studies of productivity in many parts of the world. Most of these have been brought together by a group of Soviet scientists to provide up-to-date world coverage consisting of over 7000 productivity observations.[7] The result has been modification of the estimate of primary production in the world ocean from 1.2 to 1.5×10^{10} tons of carbon fixed per year[5] to a new figure, 1.5 to 1.8×10^{10} tons.

Attempts have also been made by Steemann Nielsen and Jensen,[5] Ryther,[8] and Koblentz-Mishke *et al.*[7] to assign specific levels or ranges of productivity to different parts of the ocean. Although the approach was somewhat different in each case, in general the agreement between the three was good and, with appropriate condensation and combination, permit the following conclusions.

(1) Annual primary production in the open sea varies, for the most part, between 25 and 75 grams of carbon fixed per square meter and averages about 50 grams of car-

6. E. Steemann Nielsen, *J. Cons. Cons. Perma. Int. Explor. Mer.* 18, 117 (1952).

7. O. I. Koblentz-Mishke, V. V. Volkovinsky, and J. G. Kobanova, in *Scientific Exploration of the South Pacific,* W. Wooster, ed. (National Academy of Sciences, Washington, D. C., in press).
8. J. H. Ryther, in *The Sea,* M. N. Hill, ed. (Interscience, London, 1963), pp. 347-380.

bon per square meter per year. This is true for roughly 90 per cent of the ocean, an area of 326 × 10⁶ square kilometers.

(2) Higher levels of primary production occur in shallow coastal waters, defined here as the area within the 100-fathom (180-meter) depth contour. The mean value for this region may be considered to be 100 grams of carbon fixed per square meter per year, and the area, according to Menard and Smith,[9] is 7.5 per cent of the total world ocean. In addition, certain offshore waters are influenced by divergences, fronts, and other hydrographic features which bring nutrient-rich subsurface water into the euphotic zone. The equatorial divergences are examples of such regions. The productivity of these offshore areas is comparable to that of the coastal zone. Their total area is difficult to assess, but is considered here to be 2.5 per cent of the total ocean. Thus, the coastal zone and the offshore regions of comparably high productivity together represent 10 per cent of the total area of the oceans, or 36 × 10⁶ square kilometers.

(3) In a few restricted areas of the world, particularly along the west coasts of continents at subtropical latitudes where there are prevailing offshore winds and strong eastern boundary currents, surface waters are diverted offshore and are replaced by nutrient-rich deeper water. Such areas of coastal upwelling are biologically the richest parts of the ocean. They exist off Peru, California, northwest and southwest Africa, Somalia, and the Arabian coast, and in other more localized situations. Extensive coastal upwelling also is known to occur in various places around the continent of Antarctica, although its exact location and extent have not been well documented. During periods of active upwelling, primary production normally exceeds 1.0 and may exceed 10.0 grams of carbon per square meter per day. Some of the high values which have been reported from these locations are 3.9 grams

for the southwest coast of Africa,[5] 6.4 for the Arabian Sea,[10] and 11.2 off Peru.[11] However, the upwelling of subsurface water does not persist throughout the year in many of these places—for example, in the Arabian Sea, where the process is seasonal and related to the monsoon winds. In the Antarctic, high production is limited by solar radiation during half the year. For all these areas of coastal upwelling throughout the year, it is probably safe, if somewhat conservative, to assign an annual value of 300 grams of carbon per square meter. Their total area in the world is again difficult to assess. On the assumption that their total cumulative area is no greater than 10 times the well-documented upwelling area off Peru, this would amount to some 3.6 × 10⁵ square kilometers, or 0.1 per cent of the world ocean. These conclusions are summarized in Table 4-2.

FOOD CHAINS

Let us next examine the three provinces of the ocean which have been designated according to their differing levels of primary productivity from the standpoint of other possible major differences. These will include, in particular, differences which relate to the food chains and to trophic efficiencies involved in the transfer of organic matter from the photosynthetic organisms to fish and invertebrate species large and abundant enough to be of importance to man.

The first factor to be considered in this context is the size of the photosynthetic or producer organisms. It is generally agreed that, as one moves from coastal to offshore oceanic waters, the character of these organisms changes from large "microplankton" (100 microns or more in diameter) to the

9. H. W. Mernard and S. M. Smith, *J. Geophys. Res.* 71, 4305 (1966).

10. J. H. Ryther and D. W. Menzel, *Deep-Sea Res.* 12, 199 (1965).
11. J. H. Ryther, E. M. Holbert, C. J. Lorenzen, and N. Corwin, "The Production and Utilization of Organic Matter" in the *Peru Coastal Current* (Texas A & M Univ. Press, College Station, in press).

Table 4-2. Division of the Ocean into Provinces according to Their Level of Primary Organic Production.

Province	Percentage of Ocean	Area (km²)	Mean Productivity (grams of carbon/m²/yr)	Total Productivity (10⁹ tons of carbon/yr)
Open ocean	90	326 × 10⁶	50	16.3
Coastal zone[a]	9.9	36 × 10⁶	100	3.6
Upwelling areas	0.1	3.6 × 10⁵	300	0.1
Total				20.0

[a]Includes offshore areas of high productivity.

much smaller "nannoplankton" cells 5 to 25 microns in their largest dimensions.[12,13]

Since the size of an organism is an essential criterion of its potential usefulness to man, we have the following relationship. the larger the plant cells at the beginning of the food chain, the fewer the trophic levels that are required to convert the organic matter to a useful form. The oceanic nannoplankton cannot be effectively filtered from the water by most of the common zooplankton crustacea. For example, the euphausid *Euphausia pacifica,* which may function as a herbivore in the rich subarctic coastal waters of the Pacific, must turn to a carnivorous habit in the offshore waters where the phytoplankton become too small to be captured.[13]

Intermediate between the nannoplankton and the carnivorous zooplankton are a group of herbivores, the microzooplankton, whose ecological significance is a subject of considerable current interest.[14,15] Representatives of this group include protozoans such as Radiolaria, Foraminifera, and Tintinnidae, and larval nuplii of microcrustaceans. These organisms, which may occur in concentrations of tens of thousands per cubic meter, are the primary herbivores of the open sea.

Feeding upon these tiny animals is a great host of carnivorous zooplankton, many of which have long been thought of as herbivores. Only by careful study of the mouthparts and feeding habits were Anraku and Omori[16] able to show that many common copepods are facultative if not obligate carnivores. Some of these predatory copepods may be no more than a millimeter or two in length.

Again, it is in the offshore environment that these small carnivorous zooplankton predominate. Grice and Hart[17] showed that the percentage of carnivorous species in the zooplankton increased from 16 to 39 per cent in a transect from the coastal waters of the northeastern United States to the Sargasso Sea. Of very considerable importance in this group are the Chaetognatha. In terms of biomass, this group of animals, predominantly carnivorous, represents, on the average, 30 per cent of the weight of copepods in the open sea.[17] With such a distribution, it is clear that virtually all the copepods, many of

12. C. D. McAllister, T. R. Parson, and J. D. H. Strickland, *J. Cons. Cons. Perma. Int. Explor. Mer* 25, 240 (1960); G. C. Anderson, *Limnol. Oceanogr.* 10, 477 (1965).
13. T. R. Parsons and R. J. Le Brasseur, in "Symposium Marine Food Chains, Aarhus (1968)."
14. E. Steemann Nielsen, *J. Cons. Cons. Perma. Int. Explor. Mer* 23, 178 (1958).
15. J. R. Beers and G. L. Stewart, *J. Fish. Res. Board Can.* 24, 2053 (1967).

16. M. Anraku and M. Omori, *Limnol. Oceanogr.* 8, 116 (1963).
17. G. D. Grice and H. D. Hart, *Ecol. Monogr.* 32, 287 (1962).

which are themselves carnivores, must be preyed upon by chaetognaths.

The oceanic food chain thus far described involves three to four trophic levels from the photosynthetic nannoplankton to animals no more than 1 to 2 centimeters long. How many additional steps may be required to produce organisms of conceivable use to man is difficult to say, largely because there are so few known oceanic species large enough and (through schooling habits) abundant enough to fit this category. Familiar species such as the tunas, dolphins, and squid are all top carnivores which feed on fishes or invertebrates at least one, and probably two, trophic levels beyond such zooplankton as the chaetognaths. A food chain consisting of five trophic levels between photosynthetic organisms and man would therefore seem reasonable for the oceanic province.

As for the coastal zone, it has already been pointed out that the phytoplankton are quite commonly large enough to be filtered and consumed directly by the common crustacean zooplankton such as copepods and euphausids. However, the presence, in coastal waters, of protozoans and other microzooplankton in larger numbers and of greater biomass than those found in offshore waters [15] attests to the fact that much of the primary production here, too, passes through several steps of a microscopic food chain before reaching the macrozooplankton.

The larger animals of the coastal province (that is, those directly useful to man) are certainly the most diverse with respect to feeding type. Some (mollusks and some fishes) are herbivores. Many others, including most of the pelagic clupeoid fishes, feed on zooplankton. Another large group, the demersal fishes, feed on bottom fauna which may be anywhere from one to several steps removed from the phytoplankton.

If the herbivorous clupeoid fishes are excluded (since these occur predominantly in the upwelling provinces and are therefore considered separately), it is probably safe to assume that the average food organism from coastal waters represents the end of at least a three-step food chain between phytoplankton and man.

It is in the upwelling areas of the world that food chains are the shortest, or—to put it another way—that the organisms are large enough to be directly utilizable by man from trophic levels very near the primary producers. This, again, is due to the large size of the phytoplankton, but it is due also to the fact that many of these species are colonial in habit, forming large gelatinous masses or long filaments. The eight most abundant species of phytoplankton in the upwelling region off Peru, in the spring of 1966, were *Chaetoceros socialis, C. debilis, C. lorenzianus, Skeletonema costatum, Nitzschia seriata, N. delicatissima, Schroederella delicatula,* and *Asterionella japonica.* [11, 18] The first in this list, *C. socialis,* forms large gelatinous masses. The others all form long filamentous chains. *Thalossiosira subtilis,* another gelatinous colonial form like *Chaetoceros socialis,* occurs commonly off southwest Africa[19] and close to shore off the Azores.[20] Hart[21] makes special mention of the colonial habit of all the most abundant species of phytoplankton in the Antarctic—*Fragiloriopsis antarctica, Encampia balaustrium, Rhizosalenia alata, R. antarctica, R. chunii, Thallosiothrix antarctica,* and *Phaeocystis brucei.*

Many of the above-mentioned species of phytoplankton form colonies several millimeters and, in some cases, several centimeters in diameter. Such aggregates of plant material can be readily eaten by large fishes without special feeding adaptation. In addition, however, many of the clupeoid fishes (sardines, anchovies, pilchards, menhaden,

18. M. R. Reeve, in "Symposium Marine Food Chains, Aarhus (1968)."
19. Personal observation; T. J. Hart and R. I. Currie, *Discovery Rep.* 31, 123 (1960).
20. K. R. Gaarder, Report on the Scientific Results of the "Michael Sars" North Atlantic Deep-Sea Expedition 1910 (Univ. of Bergen, Bergen, Norway).
21. T. J. Hart, *Discovery Rep.* 21, 261 (1942).

and so on) that are found most abundantly in upwelling areas and that make up the largest single component of the world's commercial fish landings, do have specially modified gill rakers for removing the larger species of phytoplankton from the water.

There seems little doubt that many of the fishes indigenous to upwelling regions are direct herbivores for at least most of their lives. There is some evidence that juveniles of the Peruvian anchovy (*Engraulis ringens*) may feed on zooplankton, but the adult is predominantly if not exclusively a herbivore.[22] Small gobies (*Gobius bibarbatus*) found at mid-water in the coastal waters off southwest Africa had their stomachs filled with a large, chain-forming diatom of the genus *Fragilaria*[23] There is considerable interest at present in the possible commercial utilization of the large Antarctic krill, *Euphausia superba,* which feeds primarily on the colonial diatom *Fragilariopsis antarctica.*[24]

In some of the upwelling regions of the world, such as the Arabian sea, the species of fish are not well known, so it is not surprising that knowledge of their feeding habits and food chains is fragmentary. From what is known, however, the evidence would appear to be overwhelming that a one- or two-step food chain between phytoplankton and man is the rule. As a working compromise, let us assign the upwelling province a 1½-step food chain.

EFFICIENCY

The growth (that is, the net organic production) of an organism is a function of the food assimilated less metabolic losses or respiration. This efficiency of growth or food utilization (the ratio of growth to assimilation) has been found, by a large number of investigators and with a great variety of organisms, to be about 30 per cent in young, actively growing animals. The efficiency decreases as animals approach their full growth, and reaches zero in fully mature or senescent individuals.[25] Thus a figure of 30 per cent can be considered a biological potential which may be approached in nature, although the growth efficiency of a population of animals of mixed ages under steady-state conditions must be lower.

Since there must obviously be a "maintenance ration" which is just sufficient to accommodate an organism's basal metabolic requirement,[26] it must also be true that growth efficiency is a function of the absolute rate of assimilation. The effects of this factor will be most pronounced at low feeding rates, near the "maintenance ration" and will tend to become negligible at high feeding rates. Food conversion (that is, growth efficiency) will therefore obviously be related to food availability, or to the concentration of prey organisms when the latter are sparsely distributed.

In addition, the more available the food and the greater the quantity consumed, the greater the amount of "internal work" the animal must perform to digest, assimilate, convert, and store the food. Conversely, the less available the food, the greater the amount of "external work" the animal must perform to hunt, locate, and capture its prey. These concepts are discussed in some detail by Ivlev[27] and reviewed by Ricker.[28] The two metabolic costs thus work in opposite ways with respect to food availability, tending thereby toward a constant total effect. However, when food availability is low, the added costs of basal metabolism and external work relative to assimilation may

22. R. J. E. Sanchez, in Proceedings of the 18th Annual Session, Gulf and Caribbean Fisheries Institute, University of Miami Institute of Marine Science, 1966, J. B. Higman, Ed. (Univ. of Miami Press, Coral Gables, Fla., 1966), pp. 84-93.
23. R. T. Barber and R. L. Haedrich, *Deep-Sea Res.* 16, 415 (1952).
24. J. W. S. Marr, *Discovery Rep.* 32, 34 (1962).

25. S. D. Gerking, *Physiol. Zool.* 25, 358 (1952).
26. B. Dawes, *J. Mar. Biol. Ass. U. K.* 17, 102 (1930-31); *J. Mar. Biol. Ass. U. K.* 17, 877 (1930-31).
27. V. S. Ivlev, *Zool. Zh.* 18, 303 (1939).
28. W. E. Ricker, *Ecology* 16, 373 (1946).

have a pronounced effect on growth efficiency.

When one turns from consideration of the individual and its physiological growth efficiency to the "ecological efficiency" of food conversion from one trophic level to the next,[2,29] there are additional losses to be taken into account. Any of the food consumed but not assimilated would be included here, though it is possible that undigested organic matter may be reassimilated by members of the same trophic level.[2] Any other nonassimilatory losses, such as losses due to natural death, sedimentation, and emigration, will, if not otherwise accounted for, appear as a loss in trophic efficiency. In addition, when one considers a specific or selected part of a trophic level, such as a population of fish of use to man, the consumption of food by any other hidden member of the same trophic level will appear as a loss in efficiency. For example, the role of such animals as salps, medusae, and ctenophores in marine food chains is not well understood and is seldom even considered. Yet these animals may occur sporadically or periodically in swarms so dense that they dominate the plankton completely. Whether they represent a dead end or side branch in the normal food chain of the sea is not known, but their effect can hardly be negligible when they occur in abundance.

Finally, a further loss which may occur at any trophic level but is, again, of unknown or unpredictable magnitude is that of dissolved organic matter lost through excretion or other physiological processes by plants and animals. This has received particular attention at the level of primary production, some investigators concluding that 50 per cent or more of the photoassimilated carbon may be released by phytoplankton into the water as dissolved compounds.[30] There appears to be general agreement that the loss of dissolved organic matter is indirectly proportional to the absolute rate of organic production and is therefore most serious in the oligotrophic regions of the open sea.[11,31]

All of the various factors discussed above will affect the efficiency or apparent efficiency of the transfer of organic matter between trophic levels. Since they cannot, in most cases, be quantitatively estimated individually, their total effect cannot be assessed. It is known only that the maximum potential growth efficiency is about 30 per cent and that at least some of the factors which reduce this further are more pronounced in oligotrophic, low-productivity waters than in highly productive situations. Slobodkin[29] concludes that an ecological efficiency of about 10 per cent is possible, and Schaeffer feels that the figure may be as high as 20 per cent. Here, therefore, I assign efficiencies of 10, 15, and 20 per cent, respectively, to the oceanic, the coastal, and the upwelling provinces, though it is quite possible that the actual values are considerably lower.

CONCLUSIONS AND DISCUSSION

With values assigned to the three marine provinces for primary productivity (Table 4-2), number of trophic levels, and efficiencies, it is now possible to calculate fish production in the three regions. The results are summarized in Table 4-3.

These calculations reveal several interesting features. The open sea—90 per cent of the ocean and nearly three-fourths of the earth's surface—is essentially a biological desert. It produces a negligible fraction of the world's fish catch at present and has little or no potential for yielding more in the future.

Upwelling regions, totaling no more than about one-tenth of 1 per cent of the ocean

29. L. B. Slobodkin, *Growth and Regulation of Animal Populations* (Holt, Rinehart & Winston, New York, 1961), Chap. 12.
30. G. E. Fogg, C. Nalewajko, W. D. Watt, *Proc. Roy. Soc. Ser B Biol. Sci.* 162, 517 (1965).

31. G. E. Fogg and W. D. Watt, Mem. *Inst. Ital. Idrobiol. Dott. Marco de Marshi Pallanze Italy* 18, *suppl.*, 165 (1965).

Table 4-3. Estimated Fish Production in the Three Ocean Provinces Defined in Table 4-2.

Province	Primary Production [tons (organic carbon)]	Trophic Levels	Efficiency (%)	Fish Production [tons (fresh wt.)]
Oceanic	16.3×10^9	5	10	16×10^5
Coastal	3.6×10^9	3	15	12×10^7
Upwelling	0.1×10^9	1½	20	12×10^7
Total				24×10^7

surface (an area roughly the size of California) produce about half the world's fish supply. The other half is produced in coastal waters and the few offshore regions of comparable high fertility.

One of the major uncertainties and possible sources of error in the calculation is the estimation of the areas of high, intermediate, and low productivity. This is particularly true of the upwelling area off the continent of Antarctica, an area which has never been well described or defined.

A figure of 360,000 square kilometers has been used for the total area of upwelling regions in the world (Table 4-2). If the upwelling regions off California, northwest and southwest Africa, and the Arabian Sea are of roughly the same area as that off the coast of Peru, these semitropical regions would total some 200,000 square kilometers. The remaining 160,000 square kilometers would represent about one-fourth the circumference of Antarctica seaward for a distance of 30 kilometers. This seems a not unreasonable inference. Certainly, the entire ocean south of the Antarctic Convergence is not highly productive, contrary to the estimates of El-Sayed.[32] Extensive observations in this region by Saijo and Kawashima[33] yielded primary productivity values of 0.01 to 0.15 gram of carbon per square meter per day—a

value no higher than the values used here for the open sea. Presumably, the discrepancy is the result of highly irregular, discontinuous, or "patchy" distribution of biological activity. In other words, the occurrence of extremely high productivity associated with upwelling conditions appears to be confined, in the Antarctic, as elsewhere, to restricted areas close to shore.

An area of 160,000 square kilometers of upwelling conditions with an annual productivity of 300 grams of carbon per square meter would result in the production of about 50×10^6 of "fish," if we follow the ground rules established above in making the estimate. Presumably these "fish" would consist for the most part of the Antarctic krill, which feeds directly upon phytoplankton, as noted above, and which is known to be extremely abundant in Antarctic waters. There have been numerous attempts to estimate the annual production of krill in the Antarctic, from the known number of whales at their peak of abundance and from various assumptions concerning their daily ration of krill. The evidence upon which such estimates are based is so tenuous that they are hardly worth discussing. It is interesting to note, however, that the more conservative of these estimates are rather close to figures derived independently by the method discussed here. For example, Moiseev[34] calculated krill production for

32. S. Z. El-Sayed, in *Biology of the Antarctic Seas* III, G. Llano and W. Schmitt, eds. (American Geophysical Union, Washington, D. C., 1968), pp. 15-47.
33. Y. Saijo and T. Kawashima, *J. Oceanogr. Soc. Japan* 19, 190 (1964).

34. P. A. Moiseev, paper presented at the 2nd Symposium on Antarctic Ecology, Cambridge, England, 1968.

1967 to be 60.5×10^6 tons, while Kasahara[3] considered a range of 24 to 36 $\times$ 10^6 tons to be a minimal figure. I consider the figure 50×10^6 tons to be on the high side, as the estimated area of upwelling is probably generous, the average productivity value of 300 grams of carbon per square meter per year is high for a region where photosynthesis can occur during only half the year, and much of the primary production is probably diverted into smaller crustacean herbivores.[35] Clearly, the Antarctic must receive much more intensive study before its productive capacity can be assessed with any accuracy.

In all, I estimate that some 240 million tons (fresh weight) of fish are produced annually in the sea. As this figure is rough and subject to numerous sources of error, it should not be considered significantly different from Schaeffer's[2] figure of 200 million tons.

Production, however, is not equivalent to potential harvest. In the first place, man must share the production with other top-level carnivores. It has been estimated, for example, that guano birds alone eat some 4 million tons of anchovies annually off the coast of Peru, while tunas, squid, sea lions, and other predators probably consume an equivalent amount. [22, 36] This is nearly equal to the amount taken by man from this one highly productive fishery. In addition, man must take care to leave a large enough fraction of the annual production of fish to permit utilization of the resource at something close to its maximum sustainable yield, both to protect the fishery and to provide a sound economic basis for the industry.

When these various factors are taken into consideration, it seems unlikely that the potential sustained yield of fish to man is appreciably greater than 100 million tons. The total world fish landings for 1967 were just

over 60 million tons,[37] and this figure has been increasing at an average rate of about 8 per cent per year for the past 25 years. It is clear that, while the yield can be still further increased, the resource is not vast. At the present rate, the industry can continue to expand for no more than a decade.

Most of the existing fisheries of the world are probably incapable of contributing significantly to this expansion. Many are already overexploited, and most of the rest are utilized at or near their maximum sustainable yield. Evidence of fishing pressure is usually determined directly from fishery statistics, but it is of some interest, in connection with the present discussions, to compare landings with fish production as estimated by the methods developed in this article. I will make this comparison for two quite dissimilar fisheries, that of the continental shelf of the northwest Atlantic and that of the Peruvian coastal region.

According to Edwards,[38] the continental shelf between Hudson Canyon and the southern end of the Nova Scotian shelf includes an area of 110,000 square miles (2.9 $\times 10^{11}$ square meters). From the information in Tables 4-2 and 4-3, it may be calculated that approximately 1 million tons of fish are produced annually in this region. Commercial landings from the same area were slightly in excess of 1 million tons per year for the 3-year period 1963 to 1965 before going into a decline. The decline has become more serious each year, until it is now proposed to regulate the landings of at least the more valuable species such as cod and haddock, now clearly overexploited.

The coastal upwelling associated with the Peru Coastal Current gives rise to the world's most productive fishery, an annual harvest of some 10^7 metric tons of anchovies. The maximum sustainable yield is estimated at,

---

35. T. L. Hopkins, unpublished manuscript.
36. W. S. Wooster and J. L. Reid, Jr., in *The Sea*, M. N. Hill, ed. (Interscience, London, 1963), vol. 2, p. 253.

37. FAO Yearb. *Fish. Statistics* 25 (1967).
38. R. L. Edwards, *Univ. Wash. Publ. Fish.* 4, 52 (1968).

or slightly below, this figure,[39] and the fishery is carefully regulated. As mentioned above, mortality from other causes (such as predation from guano birds, bonito, squid, and so on) probably accounts for an additional 10^7 tons. This prodigious fishery is concentrated in an area no larger than about 800 × 30 miles,[36] or 6 × 10^{10} square meters. By the methods developed in this article, it is estimated that such an upwelling area can be expected to produce 2 × 10^7 tons of fish, almost precisely the commercial yield as now regulated plus the amount attributed to natural mortality.

These are but two of the many recognized examples of well-developed commercial fisheries now being utilized at or above their levels of maximum sustainable yield. Any appreciable continued increase in the world's fish landings must clearly come from unexploited species and, for the most part, from undeveloped new fishing areas. Much of the potential expansion must consist of new products from remote regions, such as the Antarctic krill, for which no harvesting technology and no market yet exist.

39. R. J. E. Sanchez, in Proceedings, 18th Annual Session, Gulf and Caribbean Fisheries Institute, University of Miami Institute of Marine Science (Univ. of Miami Press, Coral Gables, 1966), p. 84.
40. The work discussed here was supported by the Atomic Energy Commission, contract No. AT(30-1)-3862, Ref. No. NYO-3862-20. This article is contribution No. 2327 from the Woods Hole Oceanographic Institution.

Minerals from Sea Animals

CHARLES E. LANE   1968

Every cubic mile of sea water contains about 160-million tons of solids. It has been estimated that these solids are worth more than $6 million.

Most of the solid material consists of sodium chloride—common table salt. But precious metals are there, too—including gold, silver, and platinum. As far as we know, in fact, every element known on earth is found in the oceans. Some elements exist in such dilute quantities, however, that they are undetectable by any means known to science.

In some cases, we know that certain minerals are in sea water only because we find these minerals greatly concentrated in the bodies of marine organisms.

The way that marine plants and animals extract some of these minerals is a complete mystery to biologists. If we can learn how they do it, we might eventually reap a rich harvest of metals that now are extremely rare and extremely valuable.

It is a biological truism that all cells differ in ionic composition from the medium that bathes them, whether this be sea water or human blood plasma, because of the characteristic activities of their boundary membranes. Cellular organization as we know it today probably never could have evolved unless there had occurred an earlier development that gave these membranes the ability to regulate the transport of solute particles and ions against concentration gradients.

Without this capacity for active transport, the enormous functional diversity of animal and plant cells could not be sustained. Active transport requires an external source of energy, because in thermodynamic terms the membrane pushes a load uphill—in this case, up a concentration gradient. Phosphate bond energy derived from adenosine triphosphate, creatine phosphate, and other phosphogens provides the power for most active transport processes.

All cells contain relatively more potassium and less sodium than their medium. This is true of nerve, muscle, and blood cells of man as well as the epithelial membranes enclosing the simplest marine jellyfish. Other

Dr. Charles E. Lane is a professor of functional biology, Institute of Marine and Atmospheric Sciences, Rosenstiel School of Marine and Atmospheric Sciences, University of Miami, Florida. He studies the general biology of marine organisms, more specifically, the physiology of internal secretions and the metabolism of protein, carbohydrates, and fats.

From *Oceanology International*, Vol. 3, No. 2, pp. 27-30, 1968. Reprinted with light editing and by permission of the author and *Oceanology International*.

ions commonly differing in concentration between the inside and outside of cells include magnesium, calcium, sulfate, and chloride.

In addition to this difference between intracellular and extracellular concentration of the relatively common ions, various marine organisms have acquired a special capacity with regard to certain ions or elements.

Among the halogens, chlorine and iodine tend to be concentrated in animals more than bromine or fluorine. Chlorine, the most abundant single element in sea water, is the most ubiquitous anion in animals. Iodine—although occurring in sea water at less than .001 the concentration of bromine—is used perferentially by many marine animals.

Iodine is concentrated by sponges and by many marine plants—primarily the algae, or seaweeds. Some seaweeds have provided the raw materials for the commercial production of iodine. Interestingly, although many invertebrates combine iodine with proteins, none appear to employ these iodinated compounds for metabolic regulation.

This capability was apparently first achieved by ancestral vertebrates and has persisted throughout vertebrate evolution as a principal function of the thyroid gland. This organ, found only in vertebrates, combines iodine and the amino acid tyrosine to form the active metabolic regulator thyroxine. Fluorine is concentrated by some sponges and occurs as fluorite in mollusk shells.

Although silicon occurs in sea water in only trace concentrations (0.02 to 4.0 mg/kg of water) the siliceous sponges extract this element efficiently from the sea water ceaselessly circulating through their bodies. The sponge constructs a complex siliceous skeleton, which may comprise a significant portion of its total dry substance (Figure 4-2). Some protozoans and a few bryozoans also concentrate significant amounts of silicon. The mechanisms by which this element is extracted and deposited are unknown.

Calcium is abundant in sea water (400 mg/kg of water) and is widely used as skeletal material by marine organisms. Calcium carbonate, either in the form of aragonite or calcite, is the most common compound, although calcium phosphate may also be employed.

Corals generally concentrate calcium from sea water and combine it with bicarbonate derived from carbon dioxide to form the carbonate skeleton of individual polyps. Collectively these structures form massive coral reefs that are such conspicuous features of both ancient and modern tropical seas.

The one-celled animals known as foraminiferans, or forams, deposit calcium carbonate in their tests. Massive deposits of these microscopic "shells" comprise the chalk cliffs seen in various coastal locations around the world.

The shells of marine mollusks consist chiefly of calcium carbonate deposited by specialized epithelial cells in the mantle. Calcium salts are widely used to reinforce the exoskeleton of crustaceans. These animals grow only by replacing their relatively inflexible "shells" with larger ones at ecdysis.

To minimize loss, deposited calcium salts may be removed from a crustacean's "shell" before the molt and stored in some other regions of the body until the new, larger exoskeleton is formed. In the crayfish and some of its relatives, this material is stored in calcified patches in the wall of the stomach. It is released to the newly formed larger shell immediately after the molt.

The wood-boring crustacean *Limnoria* molts in two stages. First, the posterior segment of the exoskeleton is shed. Calcium salts then are mobilized from the anterior region of the body and deposited in the new exoskeleton covering the posterior part of the body. Then the anterior part of the animal molts. By this procedure, calcium loss is reduced by at least half.

Marine organisms concentrate various metals. These are important in the normal physiology of all animals, because they participate in various ways in enzymatic activ-

Figure 4-2. Skeleton of Euplectella, the Venus flower basket shows the tendency to concentrate silicon. (Courtesy of Charles E. Lane, University of Miami.)

ities, facilitate gas transport, or are essential ingredients in vitamins.

Prominent among the metals of biological importance is iron, since it accounts for the oxygen-transport capability of the blood pigment, hemoglobin, that occurs in some members of every phylum of marine animals. The iron-containing portion of the molecule, the heme moiety, is apparently the same throughout the series, variations in the composition affecting only the protein or globin part of the molecule. Sea water generally contains from 2 to 20 μg of dissolved iron per kg of water.

The marine snails are efficient concentrators of iron, the metal appearing in the hard radular teeth or in the shell pigment. The mechanism of extraction of dissolved iron from sea water by these animals is unknown. It is generally thought that most animals acquire much of their iron with their food, and the original extraction and concentration of the iron is accomplished by smaller organisms in the food chain.

Copper occurs in sea water in concentrations of from 1 to 10 mg/kg of water. In the blood of some mollusks and arthropods, copper occurs in concentrations of 10,000 mg/kg of blood. In these forms, the copper occurs as an essential component of the respiratory pigment, hemocyanin.

This oxygen-transporting pigment occurs

in crustaceans, in mollusks (both in bivalves and gastropods), and in arachnids—such as the horseshoe crab, *Limulus.* Unlike the iron-containing pigment hemoglobin, which may be concentrated in special cells, hemocyanin always occurs in solution in blood plasma. Perhaps it remains in solution and remains within the vascular compartment of the animal because of its large molecular size, the molecular weight of hemocyanin being over 100 times that of hemoglobin.

Zinc occurs in sea water in concentrations of about 5 mg/kg of water. It is found widely in much higher concentration in animal tissues. Zinc is important in the normal biology of animals because it forms an essential component of the enzyme carbonic anhydrase that catalyzes the combination of carbon dioxide and water in many tissues, such as gills, red blood cells, and the vertebrate kidney.

Bivalves and other mollusks contain large amounts of zinc. Octopus urine, for example, contains nearly 200 times as much zinc as does the sea water in which the animal lives. Zinc participates in many other biological reactions in ways that remain obscure. Secretions of the prostate gland of mammals, including man for example, contain significant quantities of zinc.

Zinc is complexed with the hormone insulin by commercial producers to prolong its effect. This metal is also abundant in the eye of carnivores such as dogs, cats, seals, and foxes, where it is concentrated in the reflecting layer, called the tapetum, that causes the eyes of these animals to reflect light at night.

Here zinc occurs as a complex with the sulfur-containing amino acid, cysteine. Zinc constitutes 16% of the dry weight of the cells of the tapetum of the silver fox. It is reported that this animal is blinded if the zinc be complexed with an injected chelating agent, suggesting that zinc participates in some fundamental but still unknown way in the processes of vision in this animal.

Molybdenum, vanadium, silver, and nickel all appear in the sea in fractions of mg/kg of water. Of this group, nickel has been reported to be concentrated by sponges and some mollusks. If silver is concentrated by any marine organism, it is in such small amounts as to be questionable in ordinary, even microanalytic, procedures. Frequently the level of silver contamination of analytic reagents is higher than the amount of silver detected in the sample being analyzed.

Although vanadium is very rare in the sea and is not found in most animals, it is highly concentrated in primitive protochordate animals, the tunicates. Vanadium is restricted to specialized cells suspended in the blood, called vanadocytes. These are various colors, depending upon the particular oxide of vanadium that is carried.

A respiratory function was at one time attributed to these cells, but it has been shown recently that under conditions that exist within the animal it is highly unlikely that they could participate in any oxygen transport function. The functional significance of large concentrations of vanadium in these cells remains unknown.

The tunicates once were thought to extract vanadium from the sea water they pump over their gills. The measured pumping rate and length of life of the animal, however, make it unlikely that observed amounts of the metal could be explained by this mechanism. It seems more likely that vanadium somehow is ingested along with the tunicate's food.

This brief survey suggests that many, perhaps most, marine organisms have boundary membranes capable of either retaining or excluding a number of components of their environment.

It is quite clear that the largely passive "semipermeable membrane" that biologists once relied on is no longer adequate to explain compositional differences between organism and environment. Boundary membranes are living metabolizing structures with complex functions, the explanation of which will challenge the best efforts of biologists.

Light in the Sea

GEORGE L. CLARKE 1967

The penetration of light into the sea provides the energy for the growth of all green plants in the ocean. These plants serve, directly or indirectly, as the food for all marine animals, large and small.

In addition, light is necessary for the vision of fishes, other animals, and human divers. Light also controls the migration, breeding, and other responses of many species. A thorough knowledge of the conditions of light in the sea, therefore, is of paramount importance.

The minimum intensity of light that permits photosynthesis to take place is roughly 1% of sunlight, but the minimum intensity to which animals can respond is many orders of magnitude lower.

Early studies of light penetration and transparency were carried out using photovoltaic cells placed in a watertight case with an upward-directed receiving window and an insulated cable carrying the electrical output to a meter aboard ship.

Since photometers of this type are sensitive to about 0.1% of sunlight, they can be used to investigate light conditions in the upper layers of the ocean and the relationship of these conditions to the growth of green plants. But they cannot be used to measure the weaker levels of illumination, which are the most significant in the photic responses of animals.

With the development of the photomultiplier tube, a new dimension was added to light studies in the sea. Our photometers, provided with tubes of this type and lowered on steel-armored conducting cables, now are capable of responding to intensities as low as 10^{-7} $\mu w/cm^2$ or about 10^{-12} of full sunlight.

Using the new instruments, we can trace sunlight to a depth of more than 800 meters in the clearest water. We also can measure the penetration of daylight into turbid water; and we can investigate the light available in the water from the moon, from the stars, and even from the heavily clouded night sky.

When we began using these sensitive photometers, not only were we pleased with the extension of our working range, but we

Dr. George L. Clarke is a professor of biology at Harvard University and an associate in marine biology with Woods Hole Oceanographic Institution, Massachusetts. In 1931 he took part in the maiden voyage of the research vessel *Atlantis*. He developed the Clarke-Bumpus plankton sampler and the bathyphotometer, an instrument that responds to minimum visible daylight and flashes of bioluminescence deep in the ocean.

From *Oceanology International* Vol. 2, No. 7, pp. 40-42, 1967. Reprinted with light editing and by permission of the author and Oceanology International.

also had a new surprise. At great depths during the day, and at shallower depths at night, the recorder frequently showed sudden strong deflections lasting from a fraction of a second to several seconds. These were found to result from the flashing of millions of luminescent organisms.

Further investigation showed that, at certain depths in many localities, this bioluminescence, or living light, often is stronger than the light penetrating from the surface. It therefore has a significant effect on the light conditions in the sea.

As the sun's radiation enters the sea surface, only a small percentage is lost by reflection and back-scattering, but about half of the energy—the infrared portion—is absorbed completely in the upper few meters. The far ultraviolet radiation also is absorbed rapidly. The near ultraviolet and the visible components of the spectrum are absorbed and scattered at rates that vary according to the water purity.

In the clearest sea water (just as in pure fresh water) blue light, of wavelength about 470 mμ, is the most penetrating, and is the predominant color scattered back to the eye as one looks down into the clear, tropical ocean.

In other regions, the dissolved and particulate matter present (including living organisms) may be sufficiently abundant to exert a secondary selective action on the light, with the result that the water may appear green, gray, brown, or occasionally yellow or reddish.

Thus, in the upper layers of the sea, animals and plants live in light of widely differing spectral composition, but with increasing depth the spectrum is narrowed. In the open ocean the light is reduced to the blue region, and no further significant change in color takes place below about 30 m. The all-pervading blueness of the illumination below the surface layers is familiar to divers.

The transparency of the upper strata of the ocean varies greatly from the clearest water, found chiefly in tropical seas (and which sometimes is almost as transparent as distilled water), through intermediate conditions in the temperate, polar, and most coastal regions, to the highly turbid waters found off the mouths of rivers, in estuaries, and in many shallow areas.

In any one locality, transparency may fluctuate widely as populations of animals and plants—particularly microorganisms—flourish and then die; or as storms, currents, or runoff affect the dissolved and suspended matter in the water. However, there probably is little seasonal change below a depth of 200 m. The transparency of deeper strata has been found to be uniformly high in all regions where measurements have been made.

Since we know the approximate light intensities necessary for important photic reactions of plants and animals, we can determine the depths at which these responses take place if we know the rate of light penetration.

To understand the extent to which light exerts control throughout the oceanic regions of the world, many more measurements of the seasonal ranges of transparency should be made in all types of situations.

The maximum depth at which photosynthesis can take place is about 150 m in the clearest ocean water, but only about 50 m in clear coastal water. In turbid inshore water, this vital plant reaction is limited to still shallower depths—sometimes as little as 1 m or less. At deeper levels, green plants simply cannot grow even though other conditions, such as the supply of nutrients, may be favorable.

Seaweeds attached to the bottom can extend outward from the coasts only to the depth limit imposed by light. Therefore, by far the greatest bulk of plant material, on which all other forms of life in the sea ultimately depend, is composed of floating plant life, particularly the microscopic phytoplankton.

Since the photic responses of marine animals have a much lower threshold than that of plants, animal responses can be elicited by the weaker sunlight found at much greater depths. A light intensity sufficient to attract crustaceans would be found at over 600 m in the clearest water and about 150 m in coastal water.

Many planktonic crustaceans and certain kinds of fishes display diurnal vertical migration in response to changes in daylight. Animals in the deep scattering layers (DSL) of the ocean, as recorded by echo-sounders on ships, have been found in certain studies to follow individual isolumes ranging from 10^{-1} to 10^{-5} μw/cm^2 and to move vertically from about 75 m at night to as deep as 460 m in the day. Other diurnal migrations may extend to 800 m or more.

Of particular interest and importance is the maximum depth at which animals can see—and we can include human divers in our consideration.

Under ideal conditions, the human eye can detect light of about 10^{-3} μw/cm^2 from a broad source and of about 10^{-10} μw/cm^2 from a small source. A man probably could detect the presence of sunlight penetrating from the surface at a depth of 900 m in the clearest ocean water and at 200 m in clear coastal water.

Deep-sea fishes are believed to have more sensitive eyes than man and may be able to sense the differences between day and night at slightly greater depths.

When human divers, fishes, or other animals are looking at objects below them or to one side, they are seeing them by reflected light, requiring higher incident illumination, and hence this type of vision could not extend so deep. Nevertheless, the evidence suggests that objects can be detected visually during the middle of the day at depths much greater than previously supposed.

The distance away that an object could be recognized depends not only on the intensity of light, but also on the amount of scattering by the water and suspended material.

In a nighttime test in clear Mediterranean waters, it was found that an observer in Cousteau's diving saucer at a depth of 25 m could see a 500-watt lamp submerged to the same depth at a horizontal distance of 220 m. At a depth of 200 m, the lamp could be seen 275 m away. But at depth of 50 m, where the water stratum contained some detritus, the visibility was only 150 m.

The foregoing considerations are further complicated by the existence of bioluminescence produced in the water by a great variety of marine organisms ranging from unicellular dinoflagellates to crustaceans and fishes.

The growth of plants is not affected, however, because although the luminescence of the sea often seems dazzlingly bright to our dark-adapted eyes, measurements have shown that its maximum intensity (10^{-2} μw/cm^2) falls far short of what is necessary in order for green plants to photosynthesize.

Quantitative studies of bioluminescence have been carried out from the equator to high latitudes at a variety of stations in the Atlantic, Pacific, and Indian oceans, and in the Mediterranean Sea.

Contrary to the earlier impression that luminescence was a sporadic phenomenon, we have found that when suitable instrumentation is used, at least some flashing can be detected at every locality investigated and at almost every level down to the maximum depth tested (3,750 m).

Calculations indicate that in very clear water the brightest flashes probably can be seen by a fish (or a diver) at distances somewhat greater than 50 m.

During the day in the upper strata of the ocean, the high ambient light level obscures luminescence. Also, the flashing response tends to be inhibited by daylight. Studies with photometers shielded from the ambient light have shown, however, that many luminescent organisms are present in these layers during daytime.

When many flashing organisms are close together, their luminescence may raise the general light level for an appreciable time. This change may increase the extent of vision for some of the animals present or may modify the vertical migration of others.

The surface strata of most regions contain

many luminescent dinoflagellates, which flash when the water is agitated. If these organisms are especially abundant on a dark night, an observer may see a brilliant display of fireworks in the wake of a ship, and around each breaking wave or other moving object.

At deeper levels dinoflagellates tend to be less numerous, but populations of other luminescent forms take their place at successive depths, each flashing in its own way and often with a characteristic pattern of lights. More than 150 flashes per minute have been recorded at depths below 100 m when the water was agitated.

One of the more intriguing current problems of investigation is to find the reason for the existence of bioluminescence and to determine the extent to which it has survival value for organisms that possess it.

In some cases, it may be a by-product of metabolism and of no importance. On the other hand, many animals in the sea undoubtedly use their internally generated light as a protection from enemies and for other purposes. The light organs may form obliterative counter-shading, may cause a fright reaction, or may produce a luminous discharge that serves as a decoy.

Other species may use their luminescence as a search light, as a lure, or as a means of recognition. The utility of luminescence is particularly hard to explain for microorganisms, but the raised light level produced by dense populations of these forms may cause predatory animals to turn away or to remain at deeper levels.

We now have a general understanding of the penetration of light into the sea and of differences in transparency in various regions. We have information as to the probable depths at which vision and other photic responses are possible under certain circumstances. We know that at night and at great depths the sea is not inky black but is frequently lit up by a variety of luminous displays.

All of these lines of investigation must be extended and broadened. We need to determine to what extent and in what manner light controls the growth of plants in the sea and guides the migration of animals in different types of water.

We wish to ascertain which organisms produce the various types of luminescence, to learn what role this phenomenon plays in the interrelations of inhabitants of the sea, and to understand the interplay between light from the surface and light from luminescent organisms in controlling the reactions of marine creatures.

More knowledge is sought on how the optical characteristics of the water limit the vision of marine animals and of human divers, and how we can improve underwater vision, photography, and television. There is the intriguing possibility that measurements from airplanes or space craft can gather useful information rapidly on the transparency, color, and luminescence of water masses.

Investigations are being launched to explore these problems and to interrelate more clearly the various aspects of the light conditions in the sea—many of them of critical interest to man.

Conditions for the Existence of Life on the Abyssal Sea Floor

ROBERT J. MENZIES   1965

Investigations of deep-sea biological life have been increasing ever since the Danish *Galathea* Expedition. The recent expeditions of the *Vitiaz* (U.S.S.R.), *Ob* (U.S.S.R.), *Vema* (U.S.A.), and *Eltanin* (U.S.A.) have resulted in the accumulation of large collections of animals from depths in excess of 2000 m. Systematic studies on the fauna collected by these expeditions have appeared at an increasing rate (Foraminifera, Schjedrina, 1959; Bandy and Echols 1964: Coelenterata, Pasternak, 1961; Squires, 1961: Mollusca, Filatova, 1958; Clarke, 1961, 1962a, b: Monoplacophora, Clarke and Menzies, 1959; Lemche and Wingstrand, 1959; Menzies and Layton, 1963; Bryozoa, Menzies, 1963c: Isopoda, Menzies, 1962; Wolff, 1962; Birstein, 1963: Amphipoda, Barnard, 1961, 1962, 1964: Tanaidasea, Wolff, 1956: Cumacea, Băcescu, 1962: Echiuroidea, Zenkevitch, 1958: Polychaeta, Levenstein, 1961: Pogonophora, Ivanov, 1963: Echinodermata, Madsen 1961) and these studies have provided us with some data for new views regarding the abyssal environment and its biological components.

Although old concepts have been both substantiated and questioned a great deal more needs to be done.

Almost 90% of the surface of the ocean lies above depths in excess of 2000 m and in geographical extent, therefore, the abyssal realm of the marine environment represents the largest single biotic unit in the world. Besides being of enormous extent, the abyssal realm has some characteristic conditions for life which are not met with elsewhere on the earth. It is proposed to give a brief review of the more salient features of the biota and its environment as derived from the results of more recent investigations.

Deep-sea biological investigations started many years before the epoch making voyage of the *Challenger* (1873-1876). The real stimulus to deep-sea biological exploration appears to have come from efforts designed to test the concept of an azoic zone below 300 fathoms which had been proposed by Edward Forbes (1844). A little earlier Sir John Ross (1818) (see Murray and Hjort, 1912, p. 5) had recovered the basket star, *Astrophyton*, from a sounding line recovered

Dr. Robert J. Menzies is a professor of oceanography with the Department of Oceanography at Florida State University. His research interest is biological oceanography and especially deep-sea biology. He has written and helped write over one-hundred books and technical articles.

From *Oceanography and Marine Biology Annual Review*, Vol. 3, pp. 195-210, 1965. Reprinted with light editing and by permission of the author and George Allen and Unwin Ltd., London.

from a depth of 800 fathoms and the great marine scientist, Michael Sars (1868) (*loc. cit.*) listed 19 species of invertebrates living at depths greater than 300 fathoms. In 1863 G. O. Sars recorded 92 species of animal life between 200 and 300 fathoms off Norway. This Norwegian work lead Thomson (1873) and Carpenter to the world's first deep-sea biological investigations from the *Lightning* (1868), *Porcupine* (1869), and *Valorous* (1875). These expeditions took samples from depths greater than 1000 fathoms, proved the fallacy of Forbes' suggestion, and were the direct precursors of the *Challenger.* The work of the Danish *Galathea* and the Russian *Vitiaz* finally demonstrated the existence of life in the greatest depths of the sea, that is, in the deep trenches of the Pacific.

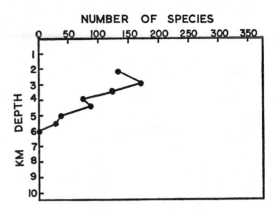

Figure 4-3. Numerical distribution of abyssal species plotted against depth (from Vinogradova, 1962a).

THE ABYSSAL ENVIRONMENT

A universally accepted definition of the abyssal zone of the sea remains to be developed. Nomenclatural variation in depth limits of the abyssal zone are given by Hedgpeth (1957) and by Vinogradova, Birstein, and Vinogradov (1959). There is a common agreement that this abyssal zone includes the sea floor at and below the average depth of the oceans and excludes the trench depths. The upper limit needs to be agreed upon; the present upper limits which have been used, namely, 200 m, 1000 m, 2000 m, and 3200 m, are all to a great extent limits of convenience for the treatment of distributional data. Vinogradova (1959) has attempted to define the upper limit of the abyssal zone on the basis of faunal zonation (Figure 4-3). She has demonstrated an abrupt faunal change in the abyss at 3800 m—and because her study was based upon a large amount of biological data it seems best to accept her definition. Essentially Vinogradova excludes the entire continental slope from consideration as abyssal. Perhaps a new term is required to define the slope fauna and to distinguish it from the abyssal fauna. Perhaps also it is not possible to place an upper limit to the abyssal zone

which will apply to all parts of the ocean. Temperature, sediment type, food supply, and predators, are all factors which vary almost independently of depth and with similar values at different depths in different parts of the world. Because these factors may have a greater influence than depth or pressure on the vertical zonation of marine life and because they show regional variations with depth it may be that an upper limit to an abyssal zone is a variable regional phenomenon for which we need not and should not seek a universal scheme; for example, the temperature below 2000 m in the Mediterranean is 14°C, in the Antarctic and Arctic it is −1.2°C and in the Caribbean 6°C, but over much of the world ocean it ranges between 1 and 2°C. The vertical variation of temperature regionally with depth is seldom considered in articles about the abyss and it would be most useful if writers would define the region they are considering as well as their definition of abyssal.

DISTRIBUTION OF PHYSICAL AND CHEMICAL CHARACTERISTICS

Because the lives of abyssal organisms are influenced by the physical characteristics of the environment it is useful to have a broad view of the distribution and variation of the

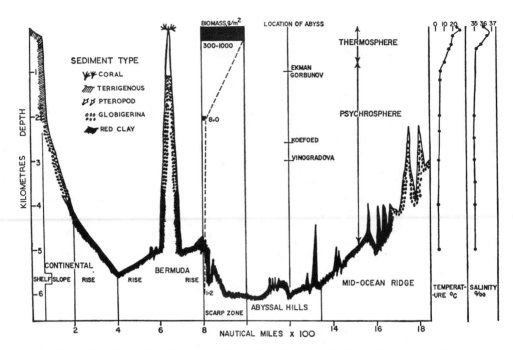

Figure 4-4. A schematic section of the ocean from Cape Hatteras, North Carolina, to the mid-ocean ridge; vertical exaggeration about 1:400; biomass data shown here are derived from Pacific values because required Atlantic data are not yet available. (Compiled from various sources.)

major physical and chemical features in the deeper parts of the oceans.

Temperature (Figure 4-4)

Excluding the accessory seas, the temperature variation below 2000 m is rather small (Sverdrup, Johnson, and Fleming, 1942), and from 50°N to 58°S the maximum range is only 3.6°C to −0.6°C. Bruun (1957) applied the term *psychral* to the fauna living in the sea at temperatures below 10°C. Recent studies suggest added thermal zonation at abyssal depths and Menzies (1963a) has proposed the term *hypopsyschral* for the fauna inhabiting water of temperatures less than 0°C. Verification through added collecting and taxonomic studies is required to substantiate this suggestion.

Within the abyss it might be expected that low temperature would have significant bearing on the physiology and growth of

marine poikilotherms. The relations established for shallow-water life may not, however, apply to animals living under high pressures. One would expect abyssal Arctic and Antarctic species (hypopsychral) to have an antifreeze substance in the body fluids as shown by Gordon, Amdur, and Scholander (1959) for Arctic organisms, or an increase in blood osmo-concentration (see Kinne, 1963). A temperature-salinity correlation will probably not be found to exist among abyssal organisms because of the absence of any known physiologically significant salinity variation (Figure 4-4) at abyssal depths. Animals will probably be found to be strict stenotherms; since one can distinguish hypopsychral (less than 0°C) from psychral (greater than 0°C) populations on a taxonomic basis, it seems reasonable to believe that a physiological explanation relating this distinction to temperature will be discovered. It is also reasonable to believe that the degree of stenothermy exhibited by abyssal forms will greatly exceed that of other known groups of poikilotherms and one might expect a certain degree of genetic and hereditary adjustment to a seasonally, and possible even epochally, uniform and constant temperature. It is perhaps of significance to the phenomenon of abyssal gigantism or overgrowth (Birstein, 1957; Wolff, 1960) that there is an attainment of a larger final size at low temperatures. This may be correlated with reduced rates of metabolism and growth both of which tend to result in a postponement of sexual maturity and a prolongation of life. This is, however, speculative and based on *a priori* reasoning from data on shallow-water organisms (Ray, 1960).

Postponement of sexual maturity appears to occur with a high frequency in abyssal isopods and in many cases the larval legs have been found with developing sexual characteristics in animals of moderate size, for example, *Haploniscus helgei,* referred to a pre-hermaphroditic stage by Wolff (1962), *Haploniscus tridens* and *Haploniscus percavix* called intersexes by Menzies (1962a), as well as *Echinothambemia ophiuroides* (Menzies, 1962a). Wolf (1962) had difficulty determining the stages of development in the two former and it is quite possible that comparison with known shallow-water stages may not be possible or even desirable. Metabolic studies on abyssal organisms are needed and even simple determinations of high and low lethal temperatures should be of considerable interest, especially if they are conducted over the vertical and horizontal ranges of temperature distribution in the sea.

Salinity (Figure 4-4).

From north to south in the Atlantic the salinity of water below 2000 m is everywhere near to 34°/oo with a variation of little more than 0.3°/oo (Sverdrup, Johnson, and Fleming, 1942). Seasonal and other variations are negligible or possibly absent.

Light

Solar light does not penetrate to water below 1000 m, and the only light present is that from bioluminescence.

Pressure

The pressure at any point in the sea is a function of water density and depth. Generally speaking it increases one bar for each 10 m of depth so that the pressure at 10,000 m is equal to 1000 bars.* Although a diversified deep-sea fauna is known to exist to the greatest depths of the sea (Wolff, 1960) where pressures may be in excess of 1000 atmos., very little is known regarding the influence of high hydrostatic pressure on the physiology of marine life in nature. Publications concerned with the ecology of the deep-sea fauna (Bruun, 1957) have generally minimized the role of pressure as a factor influencing the penetration of life into the deeps and the failure to recover living *animals* from the deep sea has usually been ascribed to temperature changes encountered by the animals and not to pressure change or decompression. Bacteria have, however, been recovered in a living state from great depths (ZoBell and Morita, 1959) and when cultured at *in situ* pressures and temperatures they showed growth (Table 4-4) and physiological activity. Bacteria which grow best at high pressure have been termed barophyllic (ZoBell and Johnson, 1949). These have been maintained in culture for as long as 30 months at *in situ* pressures and temperature (ZoBell and Morita, 1959). In contrast, Oppenheimer and ZoBell (1952) demonstrated the inability of shallow-water bacteria to survive high pressures over an extended period of time. Clearly the data from bacterial studies suggest a special adaptation of bacteria to high hydrostatic pressure and it is highly probable that multicel-

* Except at the very greatest depths the difference between 1 bar and 1 atmosphere is negligible for the present discussion.

Table 4-4. Numbers of Bacteria in Sediment Samples from the Philippine Trench.[a]

Depth (m)	Incubation Pressure:	
	1 atmos. MPN[b]	1000 atmos. MPN[b]
10,060	2,300	760,000
10,190	930	3,500,000
10,210	680	210,000
10,160	8,400	920,000
1,000	540,000	0
1,960	2,300,000	0
10,120	5,900	2,800,000

[a] From ZoBell and Morita (1959).

[b] MPN = most probable number per gram wet weight; estimated by the minimum dilution method in nutrient medium; incubated in refrigerator at different pressures.

lular organisms will demonstrate even more strictly pressure dependent physiology.

Sano (1959) in an interesting study of invertebrates attached to the French bathyscaphe reported that fouling species of the green algae *Enteromorpha linza* and *Ulva,* the bryozoans *Bugula neritina* and *Menepea occidentales,* the bivalve *Mytilus edulis* and the crustacean *Balanas amphitrite hawaiiensis* survived a dive from the surface to 200-300 m. Menzies and Wilson (1961) in an experiment to separate temperature from pressure effects showed that *Mytilus edulis* was killed when submerged to depth of 3480 m but survived a 2227 m depth submergence. In contrast they reported that the brachyuran crab, *Pachygrapsus crassipes,* does not survive submergence to depths in excess of 1000 m. Individuals of each species when kept at the equivalent *in situ* temperature in the absence of pressure change (compression-decompression) survived. Laboratory experiments by Regnard (1891), Fontaine (1930), and Ebbecke (1935) have suggested an organic complexity-pressure sensitivity series in animals, the more complex organisms appearing to suffer most from increasing pressure and the less complex least. The data on which this generalization is based are far from complete since the following animal groups of various degrees of organic complexity remain yet to be investigated: Porifera, Anthozoa, Platyhelminthes, Nemertina, Kinorhyncha, Gephyrea, Ostracoda, Copepoda, Amphipoda, Isopoda, Cumacea, Amphineura, Scaphopoda, Pteropoda, Cephalopoda, Phoronida, Brachiopoda, Crinoidea, Holothuroidea, Chaetognatha. These groups are all represented by species which live in the deep sea and it would be of considerable interest to see more work done along these lines to verify the organic complexity-pressure sensitivity hypothesis and to have many more ecological investigations concerning the effects of pressure on living plants and animals.

From the work of ZoBell and his associates on bacteria it is logical to believe that marine invertebrates will show special physiological adaptations to high pressure (Knight-Jones and Qasim, 1955); even quite small pressure change elicits behavioral response in some marine invertebrates and not in others (Knight-Jones and Qasim, 1955; Enright, 1962, Rice, 1964) but the anatomical basis for this response remains unknown.

Experiments have also shown a variation in the response of cells of different species

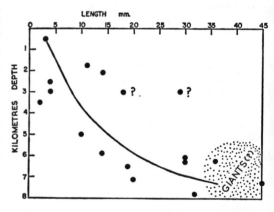

Figure 4-5. Size-depth distribution of species of the abyssal isopod genus *Storthyngura.* Coefficient of correlation between depth and increasing length in plus 0.62 ± 0.15: possible giants indicated (from Birstein, 1957).

to hydrostatic pressure at room temperature (Johnson, Eyring, and Polissar, 1954); the pressure required to block cell division or to cause rounding of certain cells is 400 atmos. in amoeba, 300-400 in echinoderms, more than 800 in *Ascaris,* and between 220-270 in certain molluscs and annelids.

There appears to be a tendency for certain species of isopod Crustacea belonging to certain genera to show unusual size in association with depth, a phenomenon which Birstein (1957) has termed abyssal gigantism (see Figure 4-5). Wolff (1960) initially associated gigantism with the effect of hydrostatic pressure on metabolism, but has since (Wolff, 1962) associated unusual size increase with increased longevity or "overgrowth" and hence does not consider it to represent true gigantism. Wolff (1962) considers that low temperature, a large supply of food in restricted areas, and the effect of pressure, may all contribute to this effect either by giving an increased metabolic rate, or by retarding the attainment of sexual maturity, together with greater longevity. In contrast, Birstein (1963) believes that the causes of deep-sea gigantism are to be sought in the peculiarities of metabolism at increased hydrostatic pressure. I consider that a large supply of food as a cause of gigantism, as postulated by Wolff, is unlikely at most abyssal depths in the sea since at such depths food supply must surely be limited. "Overgrowth" to produce giants is a possibility but it would, in such cases, probably be shown in a gradual increase in size. This does not seem to be the case in all genera that show this phenomenon. The possibility of radiation damage should not be ignored.

Recently some effort has been directed toward the construction of apparatus for study of marine animals under moderate hydrostatic pressure (Sundnes, 1962).

Sediment Types

The sediment cover of the sea floor of the abyss is derived from three main sources; the land masses contribute terrigenous sediment;

the pelagic fauna contributes the skeletal material to most of the oceanic area; and everywhere outer space contributes cosmic matter. Terrigenous sediments are confined to the borders of land masses and islands. Pelagic sediments cover the balance of the sea floor. The distribution of these reflects the pelagic source (Figure 4-4) and the solubility of the skeletons at various bottom temperatures and pressures. Pelagic sediments include pteropod, globigerina, radiolarian, and diatom material, the first two of which are calcareous and the second two calcareous or siliceous. Red clays appear to result from very low carbonate deposition due to the solution of the carbonates in the deep, cold water of the abyss. Table 4-5 shows the percentage distribution of the various types of sediments mentioned above (Kuenen, 1950), and Figure 4-4 their average distribution over the eastern North Atlantic sea floor generally without taking into account minor variations.

Oxygen

The abundance of oxygen at all levels in the abyssal zone results in sediments that are in an oxidized state over most of the abyssal sea floor. For this reason sulfides are rare in current abyssal sediments and are restricted

Table 4-5. Percentage of Sea Floor covered by Marine Sediment and Percentage of Lime.[a]

	Coverage (%)	Mean Depth (m)	% Lime
1. Shelf	8	<100	Great variation
2. Terrigenous (hemipelagic)	18	±2300	Great variation
3. Pelagic	74	4300	—
(a) Globigerina	35	3600	65 (av.)
(b) Pteropod	1	2000	80 (av.)
(c) Red Clay	28	5400	0–30
(d) Diatom	9	3900	2–40
(e) Radiolarian	2	5300	20 (max.)

[a]From Kuenen (1950).

to enclosed or semi-enclosed basins, such as the Cariaco Trench.

Topographical Features

The recent work of Heezen, Tharp, and Ewing (1959) has provided us with a more detailed and modern view of the sea floor, but biologists have yet to determine the extent to which their data fit this new topographic picture. It is apparent, however, that the older view of flatness of the sea floor must be abandoned and further biological sampling should be designed to investigate the provinces now known to exist. Statistical grid sampling should be designed to coincide with topographic features if we are to learn or discern the importance of submarine topography on the distribution of abyssal benthos.

Radiation

Cosmic radiation is absorbed in the surface layers of the sea and because of this the planktonic, the pelagic and the bathypelagic populations are effectively shielded from the radiation provided by cosmic rays (Day, 1963). Marine sediments do, however, contain radium, uranium, and other radioactive elements. The general features of radium distribution in the sea have been known for many years (Sverdrup, Johnson, and Fleming, 1942; Pettersson, 1953). Various oozes show differing average concentrations of radium with radiolarian oozes the highest, thus,

Radiolarian ooze	14.1×10^{-12} g/g
Red clay	8.7×10^{-12} g/g
Globigerina ooze (siliceous)	7.2×10^{-12} g/g
Diatom ooze	5×10^{-12} g/g
Globigerina ooze (lime)	3.7×10^{-12} g/g
Terrigenous mud	2.5×10^{-12} g/g

Nearshore sediments average in contrast only 0.3×10^{-12} g/g (grams/gram) and gran-

ites only 1.6×10^{-12}. The conclusion is inescapable hat the abyssal environment has one of the highest radium concentrations known in the sea. Its effect on animal life is unknown and unexplored. For reasons which are not well understood (Petterson, 1953) the quantity of radium in sediments may be higher 10-20 cm beneath the surface. It seems probable therefore that abyssal benthic life (especially the infauna) has been subject to radiation from radium and its products for many thousands of years. It is intriguing to consider the abyssal benthic fauna as a long-term natural experiment in low level radiation effects.

Uniformity of Characteristics

The *relatively* uniform nature of the distribution of temperature, salinity, and dissolved oxygen in the abyssal realm, a uniformity which increases with depth (below 2000 m) is a dominant feature of the sea floor. Thus at $50°N$ in the Atlantic the temperature at 2000 m is $3.32°C$ and only less than $1°C$ lower, namely, $2.38°C$ at 4000 m. The salinity ranges from $34.92°/oo$ at 2000 m to only $34.95°/oo$ at 4000 m and the dissolved oxygen varies between 6.30 and 6.34 ml/l between the same depths. In contrast, however, sediment type changes from pteropod ooze to red clay between the same depths at the same latitude (Sverdrup, Johnson, and Fleming, 1942; Kuenen, 1950). The picture of uniformity of physical characteristics led early oceanographers and some modern ones (Zenkevitch and Birstein, 1960) to a view of constancy (in time) of the characteristics. This view does not appear to fit the geological evidence (Menzies, Imbrie, and Heezen, 1961) which suggests changes in abyssal characteristics with geologic time. The old view of widespread uniformity in abyssal characteristics of topography, life, sediment, and water character is, in spite of the above statements, giving way to a new concept (still emerging) of distinct niches in the abyssal realm. Submarine photography has been a help in showing the

Figure 4-6. Fauna at floor of Milne-Edwards Trench, USARP photograph, *Eltanin* Cruise 3, Camera Sta. 1, frame 2, 8 June 1962, A, stalked ascidian; H, holothurian *Peniagone* sp.; P, polychaete worm tube; T, track of holothurian (?); C, crinoid, Depth 3301 fathoms, lat. 08° 18'S, long. 81° 05'W (after Menzies, 1963b). (Courtesy of the Smithsonian Institution.)

existence of micro-variations in sediment. Zenkevitch (1961) makes this point more strongly than other modern scientists. There can be little doubt that the environments shown in the series of two photographs are significantly different from one another and that the animals (species) of each are probably distinct (Figures 4-6 and 4-7). If we now add to this macroscopic picture of physically distinct niches a biotic factor which concerns the interrelationships between species as proposed by Day (1963), we can arrive at a large number of distinct abyssal "total" environments even in a limted area, and this is in agreement with the fact that diversity among abyssal species appears to be the rule rather than the exception, and with the recent work of Sanders (1963) who has shown that the distribution of lamellibranchs at abyssal depths is not uniform throughout a given basin but shows restricted and circumscribed patterns.

CHARACTERS OF ABYSSAL ANIMALS

Color

For the most part abyssal animals are pale in color, or at least quite drab when compared with shallow-water species. Over 80% of the abyssal isopods are white and a small percentage are cream. In the isopod crustaceans such dominant pigments as exist are not cuticular but are present in the "liver" or digestive glands which are often green. These colored internal organs are concealed by the cuticle. By contrast, the bathypelagic fauna has a high proportion of red or black animals and the pelagic realm many green and multi-colored animals. The absence of skeletal pigment seems to be a dominant feature of abyssal species.

Size

The size of abyssal animals is a matter of

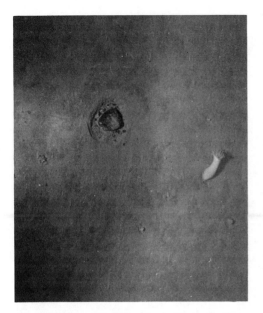

Figure 4-7. Abyssal sea floor in the Antarctic showing actiniarian or sea cucumber and rock. Depth 2247 fathoms (sonic), lat. 56° 03′ S, long. 60° 48′ W; *Eltanin* Cruise 4, Camera Sta. 3, frame 19, 20 July 1962 (U.S. Antarctic Research Program). (Courtesy of the Smithsonian Institution.)

considerable interest to zoologists. Generally speaking the average size of all members of an abyssal population appears to be smaller by an order of magnitude than the average size of an equivalent population from the shelf of intertidal zone (personal observations). Underwater photographs from abyssal depths seldom indicate an animal whose size is greater than 10 cm and generally indicate very little in the way of living animals which may suggest that most abyssal animals are very small or concealed from view. The fact that the average size of the individuals in any abyssal population is smaller than those of the shallow-water fauna is presumably in part genetically determined, but it may also be dependent upon the limited amount of food available to abyssal animals; much more information is required.

Food and Feeding

The primary source of food for abyssal organisms comes from plants at the sun-lit surface of the earth. Transport (Menzies, 1962b) of this nutritive material into the abyss is accomplished through the agency of submarine turbidity currents (including slumps, submarine landslides, and so on), rafted floating marine and terrestrial plant debris, a rain of dead phytoplankton and zooplankton and their products; the vertical migrations of zooplankton also perhaps provide a secondary food source. All of these phenomena are best developed near the margins of land-masses where plant nutrients are most effectively re-cycled and available in sufficient quantity for a high rate of primary organic production. The recent Russian studies (Zenkevitch, Barsonov, and Belyaev, 1960) have shown that the quantity of benthic biomass is related more closely to distance from land than to depth. In the open ocean far from land the source of food for the abyssal fauna is derived mostly from water masses above it which are relatively poor in primary food.

Filter feeding organisms generally dominate in shallow waters and Ockelmann (1958) has shown that the proportion of filter feeding bivalves diminishes with increasing depth. In the main, abyssal benthonts are deposit feeders. Among the benthic Isopoda over 90% feed in whole or part on bottom deposits (Menzies, 1962b). Some species of isopods tend to show particle size selectivity while others do not.

The abyssal zooplankton appears to subsist mainly on other plankton organisms as a zoophagous population (Vinogradov, 1962a), and in this respect contrasts markedly with the deposit-feeding abyssal benthic fauna. A regional study such as has been carried out by Sokolova (1959) on the distribution of animal feeding types with depth should be rewarding. Such studies should take into account not only depth but also water turbidity and substrate type; the anatomy of the feeding apparatus of the species, and the actual gut contents should be investigated.

DISTRIBUTION

The quantitative and qualitative aspects of distribution of the abyssal fauna have been reviewed by Vinogradova (1962b). The total biomass of the world ocean has been provisionally estimated at 10 billion tons. Eighty-two per cent of this biomass is assigned to shallow coastal waters and only 0.8% to abyssal depths greater than 3000 m (Zenkevitch, Barsanov, and Belyaev, 1960). This small biomass is, however, not uniformly spread over the oceanic abyss. The highest abyssal biomass values are found nearest to the shore and the lowest farthest from shore. Tropical offshore regions of the open ocean are poorest in biomass which in these regions amounts to only 0.08-0.03 g/m^2. The decrease of biomass with depth is shown in Figure 4-4. The data on biomass in Figure 4-4 are the Pacific values (Filatova and Levenstein, 1961). In the Antarctic the abyssal biomass is greater than elsewhere by a factor of 3-4 at 2000 m.

Contrary to popular belief, the abyss contains a highly diversified fauna and every known major marine taxon (phylum, class, order) has been found in the deep sea. Curiously the deep sea contains the Monoplacophora representing one class of the Mollusca which is not yet known from shallow water (Menzies, 1963b). If any single group of organism may be said to be numerically dominant at abyssal depths, it is the benthic Foraminifera and this is followed by the Polychaeta (Menzies, 1963b). Unfortunately, neither of these animal groups is particularly well known taxonomically. A high degree of species diversity within a limited area seems characteristic of abyssal populations and from a single small trawling at 4960 m depth Menzies (1962a) reported five species of the isopod genus *Haploniscus*. To illustrate the significance of this diversity one need only realize that a collection of shallow-water isopods seldom exceeds more than three species belonging to a single genus even when a great variety of habitats are sampled. The significance of abyssal species diversity requires more attention than it has been given in the past. In an examination of 1031 species of abyssal organisms Vinogradova (1962b) found only 15% occurring in more than one ocean and only 4% in all three oceans. These figures are probably conservative (they include eurybathial species) because Clarke (1962a) gives only one species of non-cephalopod mollusc out of 1152 species of abyssal non-cephalopod molluscs as cosmopolitan while he reports one-fourth of the abyssal cephalopods as cosmopolitan. A comparison of abyssal isopods (58 species) (data from Menzies, 1962a) in the adjacent Argentine and Cape Basins in the South Atlantic shows only 14% of the species in common, while none of these species is as yet known from other oceans. In contrast all of the genera (22 genera) are known also from the Pacific abyss (Menzies, unpublished data). These data are in agreement with those presented by Vinogradova (1962a) and the early suspicion of Ekman (1953) that abyssal genera were cosmopolitan. Not all abyssal genera are, however, cosmopolitan and *Neopilina* is a well-known exception. Among the isopods *Mesidothea* appears restricted in its abyssal distribution to the Arctic whereas *Serolis* is restricted in its abyssal distribution to Antarctic waters (Menzies, 1962a). If other examples from other animal groups are discovered to have a distribution parallel to that of *Neopilina, Mesidothea,* and *Serolis* then the broad outlines of abyssal divisions presented by Vinogradova (1959) will have to be modified. The Arctic abyss (Menzies, 1962a) may show a species endemism in excess of 50% (excluding eurybaths) and should not be looked upon as an appendage to the Atlantic.

Even with the changes in zoogeographical limits that must come as the fauna of the abyss becomes better known, it is very likely

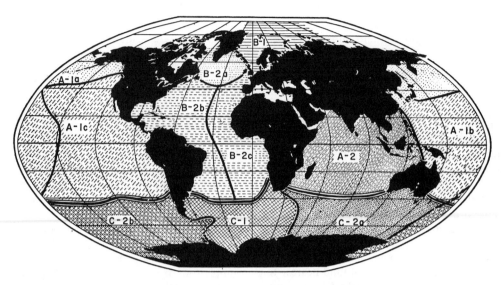

Figure 4-8. Deep-sea faunal areas and provinces established by Vinogradova, see text pp. 000-000 for symbols (after Vinogradova, 1959).

that the broad concepts of abyssal zoogeography outlined by Vinogradova (1959) will prove valid. She distinguishes Pacific, Atlantic, and Antarctic regions, within the seas, each with subdivisions and provinces. Those given below are shown in Figure 4-8.

Pacific-North Indian deep-water area (A)
    Pacific sub-area (A-1)
        North Pacific Province (A-1a)
        West Pacific Province (A-1b)
        East Pacific Province (A-1c)
    North Indian sub-area (A-2)
Atlantic deep-water area (B)
    Arctic sub-area (B-1)
    Atlantic sub-area (B-2)
        North Atlantic Province (B-2a)
        West Atlantic Province (B-2b)
        East Atlantic Province (B-2c)
Antarctic deep-water area (C)
    Antarctic-Atlantic sub-area (C-1)
        Indian Province (C-2a)
        Pacific Province (C-2b)

Her division of the abyss of the world oceans into areas, provinces and sub-provinces sets the stage for a refreshing and new con-

cept of restricted distribution among abyssal species which has been generally confirmed by the more recent investigations (Barnard, 1962; Clarke, 1962a; Menzies, 1962a) and contrasts boldly with the old view of wide distribution among abyssal species (Bruun, 1957).

EVOLUTION

At present we have no solid evidence on the rate of evolution of abyssal species and even the data for marine species as a whole are poor. Detailed studies on sediment cores selected from special regions of the sea where checks on time and sedimentary rate are possible could give much needed data. Day (1963) gives a minimum estimate of 2.7 million years per species in the shallow-water genus *Micraster* in support of Zeuner's (1958) contention that the rate of speciation in the sea is much slower than that on land. Day (1963) suggests that a slower mutation rate for marine species may be due, among other factors, to absorption of cosmic rays in the surface layers, but he did not give the significant radium concentration of abyssal muds. High concentration of radium in radiolarian oozes and red clays may have had an effect on abyssal species in such environments. One might suspect a significantly higher proportion of mutants from radium-rich sediments than from radium-poor sediments, and the possibility that abyssal giants may be mutant forms needs to be examined.

Speciation in the abyss may be at a significantly higher rate than elsewhere in the sea. If accelerated mutation rate due to radiation is a factor involved in evolutionary rate than the rate of evolution in the sea should be most rapid in abyssal radiolarian environments and least in planktonic and bathypelagic environments. This entire problem is in need of basic data which will yield information on evolutionary rates in the sea.

REFERENCES

Băcescu, M. C., 1962, *Vema Res. Ser.* 1, 207-223.

Bandy, O. L. and Echols, R. J., 1964, *Antarctic Res. Ser.* 1 (in press).

Barnard, J. L., 1961, *Galathea Rept.* 5, 23-128.

Barnard, J. L., 1962, *Vema Res. Ser.* 1, 1-76.

Barnard, J. L., 1964, *Bull. Amer. Mus. Nat. Hist.* 127, 1-46.

Birstein, J. A., 1957, *Zool. Zh.* 36, 961-985.

Birstein, J. A., 1963, *S.S.S.R.Akad. Nauk, Inst. Okeanol.*, 213 pp.

Bruun, A. F., 1957, *Geol. Soc. Amer., Mem.* 67, 641-672.

Clarke, A. H., Jr., 1961, *Bull. Mus. Comp. Zool. Harvard* 125, 345-387.

Clarke, A. H., Jr., 1962a, *Bull. Nat. Mus. Can.*, No. 181, 114 pp.

Clarke, A. H., Jr., 1962b, *Deep-Sea Res.* 9, 291-306.

Clarke, A. H., Jr., and Menzies, R. J., 1959, *Science* 127, 1026-1027.

Day, J. H., 1963, Systematics Assoc. Publ. 5, Speciation in the Sea, 31-49.

Ebbecke, U., 1935, *Pflüg. Arch. ges. Physiol.* 236, 648-657.

Ekman, S., 1953, *Zoogeography of the Sea*, Sidgwick and Jackson Ltd., London, 417 pp.

Enright, J. T., 1962, *Comp. Biochem. Physiol.* 7, 131-145.

Filatova, Z. A., 1958, *Trud. Inst. Okeanol.* 27, 204-218.

Filatova, Z. A. and Levenstein, R. J., 1961, *Trud. Inst. Okeanol. Akad. Nauk S.S.S.R.* 45, 190-213.

Fontaine, M., 1930, *Ann. Inst. Oceanogr.* 8, 1-97.

Forbes, E., 1844, Advance. Sci. Lond., Rept. 13th Meeting, 130-193.

Gordon, M. S., Amdur. B. H. and Scholander, P. F., 1959, Intern. Oceanogr. Congr. Preprints, edited by M. Sears, Amer. Ass. Advanc. Sci., Washington, 234-236.

Hedgpeth, J. W., 1957, *Geol. Soc. Amer., Mem.* 67, 1-27.

Heezen, B. C., Tharp, M. and Ewing, M., 1959, *Geol. Soc. Amer. Spec.* Pap. 65, 112 pp.

Ivanov, A. V., 1963, *Pogonophora*, Academic Press, London, 479 pp.

Johnson, F. H., Eyring, H. and Polissar, M. J., 1954, *The Kinetic Basis of Molecular Biology*, John Wiley & Sons, Inc., New York, 874 pp.

Kinne, O., 1963, *Oceanogr. Mar. Biol. Ann. Rev.* 1, 301-340.

Knight-Jones, E. W. and Qasim, S. Z., 1955, *Nature, Lond.* 175, 941-943.

Kuenen, Ph. H., 1950, *Marine Geology*, John Wiley and Sons, Inc., New York, 568 pp.

Lemche, H. and Wingstrand, K. G., 1959, *Galathea Rept.* 3, 9-71.

Levenstein, R. J., 1961, *Trud. Inst. Okeanol. Acad. Nauk S.S.S.R.* 45, 214-222.

Madsen, F. J., 1961, *Galathea Rept.* 4, 174 pp.

Menzies, R. J., 1962a, *Vema Res. Ser.* 1, 79-206.

Menzies, R. J., 1962b, *Int. Rev. ges. Hydrobiol.* 47, 339-358.

Menzies, R. J., 1963a, Proc. Arctic Basin Sympos., Tidewater Publ. Co., Md., 46-66.

Menzies, R. J., 1963b, *Int. Rev. ges. Hydrobiol.* 48, 185-200.

Menzies, R. J., 1963c, *Amer. Mus. Novit.* No. 2130, 1-8.

Menzies, R. J. and Layton, W., 1963, *Ann. Mag. nat. Hist., Ser. 13*, 5, 401-406.

Menzies, R. J. and Wilson, J. B., 1961, *Oikos* 12, 302-309.

Menzies, R. J., Imbrie, J. and Heezen, B. C., 1961, *Deep-Sea Res.* 8, 79-94.

Murina, W. and Starologatov, L., 1961, *Trud. Inst. Okeanol. Akad. Nauk S.S.S.R.* **46**, 179-200.

Murray, J. and Hjort, J., 1912, *Depths of the Ocean,* Macmillan & Co., Ltd., London, 821 pp.

Ockelmann, W. K., 1958, *Medd. Grønland* **122** (4), 1-256.

Oppenheimer, C. H. and ZoBell, C. E., 1952, *J. mar. Res.* **11**, 10-18.

Pasternak, F. A., 1961, *Trud. Inst. Okeanol.* **45**, 240-258.

Pettersson, H., 1953, *Amer. Scient.* **41**, 245-255.

Ray, C., 1960, *J. Morph.* **106**, 85-108.

Regnard, P., 1891, Recherches expérimentales sur les conditions physiques de la vie dans les eaux, Libr. Acad. Med., Paris (see Johnson, Eyring and Polissar, p. 835, 1954).

Rice, A. L., 1964, *J. mar. biol. Ass. U.K.* **44**, 163-175.

Sanders, H., 1963, *Proc. intern. Congr. Zool.* **4**, 311 only.

Sano, K., 1959, Intern. oceanogr. Congr. Preprints, edited by M. Sears, Amer. Ass. Advanc. Sci., Washington, 381-382.

Sars, M., 1858, Vidensk.-Selsk. Forhandl. f. 1868 (in Ekman, 1953).

Schjedrina, Z. K., 1959, *Trud. Inst. Okeanol. Akad. Nauk S.S.S.R.* **27**, 161-179.

Sokolova, M. N., 1959, Intern. oceanogr. Congr. Preprints, edited by M. Sears, Amer. Ass. Advanc. Sci., Washington, 382-384.

Squires, D. F., 1961, *Amer. Mus. Novit.,* No. 2046, 48 pp.

Sundnes, G., 1962, Repts. Norweg. Fish. Mar. Invest. **13**, 1-7.

Sverdrup, H. U., Johnson, M. W. and Fleming, R. H., 1942, *The Oceans,* Prentice Hall Inc., Englewood Cliffs, N. J., 1060 pp.

Thomson, C. W., 1873, *The Depths of the Sea,* Macmillan Co., London, 1874, 527 pp.

Uschakov, P. V., 1963, *Cahiers biol. mar.* **4**, 81-89.

Vinogradov, M. E., 1962, *Rapp. Cons. Explor. Mer* **153**, 114-120.

Vinogradova, N. G., 1959, *Deep-Sea Res.* **5**, 205-208.

Vinogradova, N. G., 1962a, *Deep-Sea Res.* **8**, 245-250.

Vinogradova, N. G., 1962b, *J. Oceanogr. Soc. Japan* **20**, 724-741.

Vinogradova, N. G., Birstein, J. A., and Vinogradova, M. E., 1959, *Itoginauki, Akad. Nauk U.S.S.R. Dostijema Okean.* **1**, 166-187.

Wolff, T., 1956, *Galathea Rept.* **2**, 187-214.

Wolff, T., 1960, *Deep-Sea Res.* **6**, 95-124.

Wolff, T., 1962, *Galathea Report* **6**, 1-320.

Zenkevitch, L. A., 1958, *Trud. Inst. Okeanol. Akad. Nauk S.S.S.R.* **27**, 192-203.

Zenkevitch, L. A. and Birstein, J. A., 1960, *Deep-Sea Res.* **7**, 10-23.

Zenkevitch, L. A., Barsanov, N. G. and Belyaev, G. M., 1960, *Dokl. Akad. Nauk* **130** (see Vinogradova, 1962b).

Zenkevitch, N. L., 1961, *Trud. Inst. Okeanol. Akad. Nauk S.S.S.R.* **45**, 5-21.

Zeuner, F. E., 1958, *Dating the Past,* Methuen & Co., London.

ZoBell, C. E., 1954, *Int. Union biol. Sci. Ser. B,* No. 16, 20-26.

ZoBell, C. E. and Morita, R. Y., 1959, *Galathea Rept.* **1**, 139-154.

ZoBell, C. E. and Johnson, F. H., 1949, *J. Bact.* **57**, 179-189.

Communication by Marine Animals

HOWARD E. WINN   1967

Studies of animal communication are concerned with signals produced by one individual that influence another individual. Signals may be passively produced, such as the red coloration of a male stickleback during the spawning season, or they may be controlled by the brain and produced in special or multipurpose organs, such as the air bladder of a fish.

We will concern ourselves here with sound communication in marine animals, particularly fishes and mammals.

At present there is little direct experimental evidence due to the scarcity of workers and the difficulty of establishing conditions for adequate experimental design. The suggestions in the popular press that porpoises and other cetaceans have a complex language, possibly human-like and far surpassing that of birds, primates, or other mammals, is without adequate foundation, despite some provocative data published recently.

Current fundamental studies of sound production by marine animals are attempt-ing to solve basic problems. We need to know the species of animal producing a biological sound, the distribution of the sound, and its contribution to noise levels. We need to understand the communication system of fishes and cetaceans, and this in turn requires knowledge of the controls of sound production and hearing and precisely the kind of information carried by the sound signal.

Can invertebrates "hear" water pressure or near-field waves or do they sense them entirely through the substrate? Is this a dominant signal system? We need to understand the relationship of acoustic signaling to other means of communication such as vision, olfaction, and touch, as well as its detailed relationship to the behavior of an animal.

The complete temporal, geographic and systematic distributions of sound production by fishes and cetaceans are not known as yet. When man tries to locate objects or to communicate underwater, his equipment is often jammed, his signals become garbled, and he cannot distinguish the signals of his

Dr. Howard E. Winn is a professor of oceanography and zoology with the Graduate School of Oceanography, University of Rhode Island. His research interests include electrophysiology, animal behavior, and acoustics. He has written numerous papers on sound production and associated behavior of fishes and crustaceans.

From *Oceanology International*, Vol. 2, No. 2, pp. 32-34, 1967. Reprinted with author's revisions and light editing and by permission of the author and Oceanology International.

transmitter from the tremendous array of other sounds. William Schevill, of Woods Hole Oceanographic Institution, has catalogued a large number of the natural sounds of seals, porpoises, and whales throughout the world. Their variety seems no greater than bird sounds and is generally restricted to several kinds of whistles and echolocation pulses.

Marie P. Fish, at the University of Rhode Island, has accumulated a library of sounds based on those produced by single species of fish in aquaria under a variety of stimulus conditions. Sounds recorded under other circumstances are included if the sound producer is identified. This library includes fishes from all over the world and is invaluable for the identification of underwater sound sources as well as for comparative biological studies.

An interesting fact is that most species in the sound-producing families such as toadfishes and squirrelfishes can be readily differentiated—they have "species signatures." The identification of reproductive choruses of croakers and many other bottom fishes still needs attention, because it is difficult to locate the specific producer and these sounds are frequently different from those produced in small tanks.

Much like frogs and birds, sea animals produce sounds that vary during the year. Dr. John Steinberg and his associates at the University of Miami's acoustic-video station at Bimini have published the only detailed quantitative studies on seasonal variation of sounds. Some fishes make sounds at night, some during the day, and others such as squirrelfishes have loud dawn and dusk choruses. The toadfish boatwhistle call is heard only for a month or so during reproduction.

These cyclic phenomena could be from either permanent or migrating populations. The roar-like sounds from conchs as they crawl on the bottom peak in July and finally there are some click-like sounds that last 6 to 10 days centered on the last quarter of the moon each month.

One of the important problems in communication by marine fishes is the separation of two basic parameters of sound, displacement and pressure. It has recently been shown by Dr. Wilhelm van Bergeijk, formerly of Bell Telephone, that the lateral line (a series of sensory receptors frequently seen as pores on the head and sides of fishes) responds to displacement, but not pressure.

Pressure decreases inversely as the first power of the distance from the source $(1/r)$, whereas displacement decreases inversely as the square of the distance from the source $(1/r^2)$ of a pulsating air bubble. If a response to decreasing amplitude follows one of these rates, then conclusions can be drawn. The basic transducer of the lateral line is the hair cell, which is found in all vertebrate ears.

In fishes, pressure is transduced by the gas bladder or possibly the otoliths (calcium carbonate stone-like structures in the ear) into displacement. There is current disagreement as to whether a fish can respond directionally to a distant pressure wave. Theoretically this is not possible because sound travels five times faster in the water than in air; the ears are separated by only a few centimeters; except for the gas bladder, a fish is transparent to sound, and fishes produce only low-frequency sounds.

There is theoretically no difference in amplitude or time of arrival of a pressure wave that can be compared by the two ears. On the other hand, at least within a few feet, hair cells can respond to displacement, a vector quantity.

The ear needs to be understood and behavioral experiments performed before the confusion can be resolved. It has been suspected that sharks and other predators home in on pressure waves from long distances.

If only a sound source within a few feet of a fish can be localized, then we must seek a more generalized function of mating calls, beyond their use by females locating males or territorial males locating one another. Many bottom fishes produce these types of sounds.

Moulton and one of my students, James Fish, independently mentioned that sound carried through the substrate might be heard. I suggest that fish may compare substrate and water-borne sounds in some unknown manner. The fact that most sound producers live on the bottom supports this idea as well as the concept that sounds are fundamental to territorial rights.

In general, fishes hear best at low frequencies. Man hears slightly higher frequencies, and porpoises much higher frequencies—in fact, far beyond human range. This correlates with a cetacean's use of a sonar system for locating food and other objects where good resolution is needed. The frequency range is greater and the threshold is lower in fishes that have a direct connection between air bladder and ear, as in squirrelfishes and catfishes.

It is tacitly assumed that sound signals emitted by fishes and whales transmit information from one individual to another. Sound production in specific situations and analogies with birds allow this assumption. Fishes produce sounds in aggressive encounters, spontaneously, during feeding, in aggregations, when confronted by predators, or escaping, while exploring new environments, and during courtship and spawning.

Tavolga studied a small tidepool goby. Brief bursts of sounds from the male attract females to the next site and stimulate reproductive activity. I have shown that the bright red squirrelfishes in tropical coral reefs produce short grunts when defending their cavities from other squirrelfishes and produce long chattering staccatos when approached by strange fishes such as moray eels.

When squirrelfishes hear sounds they first retreat into their cavities and later investigate the sound source. They may even mob a moray eel, making chattering calls reminiscent of small birds mobbing a hawk or cat. These sounds are produced at all seasons, and at dusk and dawn the fishes produce large choruses. At night they feed and make few sounds.

Several investigators have studied the Atlantic toadfish (Figure 4-9). In late spring

Figure 4-9. A hydrophone picks up the sounds of toadfish.

the male establishes a territory, which may be a crevice under a log or a tin can. From here he calls with the famous boatwhistle or foghorn sound until a female responds. He continues to call until he eggs are well developed. This sound is one of the loudest fish calls, comparable to nearby thunder. The male calls more rapidly when a female is near, and she is attracted to the sound source. When another male approaches, the response includes a series of aggressive grunts.

A theoretical organization of fish sounds prepared by me suggests that temporal coding is important in information transfer. Recorded toadfish boatwhistles played back at various rates evoked alternate responses from fishes in the natural environment. This confirmed temporal coding but was only part of the story. "Grunts" played back at the same rates drew no response. Discrimination involves other factors besides rate. Moulton found that searobins responded to playback of searobin sounds but were suppressed by other sounds.

The great choruses of croakers and drums along the Atlantic coast of the United States and in the tropics are poorly understood because of the depths involved, but they may be similar to those made by toadfishes and squirrelfishes. Pelagic fishes do not appear to be important sound producers but bottom fishes frequently produce sounds.

Strange sounds of apparent biological origin emanate from the depths of the ocean. Most of these are believed to come from fishes, but squids are possible contributors. N. B. Marshall, of the British Museum of Natural History, has shown that many benthopelagic fishes (those found swimming over the ocean floor) have swimbladder sound-producing mechanisms that are generally lacking in fishes found nearer the surface.

Moulton has described sounds made by schooling fishes and has presented evidence that predators may be attracted and prey repulsed by such sounds. Japanese investigators have suggested that such sounds can be used to drive and attract fishes over long distances. More evidence is needed before such results can be utilized in commercial fisheries.

As for invertebrates, there is no adequate study that strongly suggests water-borne sound communication. Fiddler crabs use sounds readily when out of water, and substrate vibrations may be utilized. Spiny lobsters produce sounds with their antennae, and the Maine lobster has a sound-producing muscle.

The snapping of the byssal threads of mussels, the rasping and roaring of conchs, the scraping of hermit crabs, and the snapping of shrimps, while not believed to be signals, do make significant contributions to ambient sound levels. In fact, they frequently interfere with scientific studies.

Porpoises produce a variety of vocalizations under specific circumstances, as many studies have shown. Such terms as barking, jaw snapping, jaw clapping, mewing, moaning, wailing, whistling, and clicking have been used for various calls. A distress call by a porpoise attracts others that push the head of the distressed animal to the surface. Echolocation calls are produced more rapidly as food and other objects are approached. These remarkable animals appear to call back and forth, and in captivity have been reported to mimic humans.

Although much has been said about the sounds of porpoises, little is understood about their communication. It is thought that they produce sounds to keep together, to summon others, to indicate distress or danger, and to make simple information transfers.

Although the press occasionally reports on the complex "language" of porpoises, the evidence indicates that their acoustic communication is not much more complex than that of some birds and primates, which use fewer than 30 types of calls. The number so far identified in porpoises is far less than that.

Investigators at Point Mugu, Cal., observed two porposes isolated from each other respond back and forth by a telephone system. This also has occurred with toadfishes, insects and frogs, however.

To know that animals will call back and forth is of little value in gauging the complexity of a communication system. The complex brain of cetaceans may represent enlargements necessary for processing echolocation data and may have little linguistic significance. Individual recognition by sounds, not even well documented for porpoises, is accomplished also by birds.

Apparent lack of a human-like language makes these animals no less interesting. As for mimicking, the data are completely lacking in controls, and its assessment will have to await corroborative results to see if it exists and, if so, how complex it is.

Dr. John Lilly of Communication Research Institute has quantified the close mimicking by porpoises of numbers of sounds produced by humans. If there were two sounds, the answer contained two bursts. Also, durations seemed to be closely matched. But anything new and startling must meet the requirements of scientific judgment. Other investigators must duplicate the results.

The language of bees was met with disbelief at first, so we should not turn a deaf ear to the possibility that other lower animals have complex languages. It is obvious that there is much to be learned about marine mammals, as well as fishes and invertebrates, and many startling things may be learned in the next 10 years.

The acoustic channel is not everything. We should not overlook the full range of communication by visual, olfactory, and other means. Certain male fishes secrete chemicals that attract females, anemones recognize their attachment site on clams by chemical means, and they move away from chemicals secreted by starfishes. Many such examples are known. Our appreciation of the olfactory world of aquatic animals is inadequate. Cetaceans are exceptional; they cannot smell.

Visual recognition of species and sex by brilliantly colored fishes is probably important. Investigators at Stanford Research Institute have demonstrated that sea lions have a keen sense of vision and that it is a more dominant sensory channel than their primitive echolocation system. The importance of vision has been underemphasized for some animals.

The evidence to date does not suggest that sound is a strong stimulus either for attraction or repulsion over distances that would be of value to a fishery. Its greatest value may lie in locating stocks.

We know little of the chemical senses and secretions of marine animals. Odorous substances may eventually be important for use in attracting fishes. After all, fishes are attracted to traps by the odor of bait.

Certainly an understanding of marine biological sounds is essential to training sonarmen, to the use of underwater sound communication gear in relation to changing ambient sound levels, and to the intrinsic value of understanding the strange noises of aquatic animals.

As man moves into the sea, porpoises might be trained to carry messages and to help divers keep their bearings. Lost divers may well be brought home by porpoises, which seem always to know their way. Dr. Kenneth Norris, of the University of California at Los Angeles, was able to train a porpoise, release it in the open sea, and call it back by means of a "porpoise whistle."

Much science fiction has been written about the possible use of marine animals to aid man. Sometimes the fiction of today is the fact of tomorrow. Possibly we will someday communicate with porpoises: perhaps we will learn to herd them like cows, or conceivably we will wage war with them for dominance of the oceans. As for the great whales, we are on the verge of watching them being exterminated for oil and dog food—and, sadly, we know so very little about them.

The Shark: Barbarian and Benefactor

PERRY W. GILBERT   1968

Ten years ago a group of scientists met in New Orleans to discuss basic research approaches for the development of better shark deterrents. Host was Tulane University and the conference was jointly sponsored by the American Institute of Biological Sciences and the Office of Naval Research. After 2 days of discussion, it was generally agreed that we knew all too little about "the enemy." A series of recommendations was drawn up that focused attention on the need for a better understanding of the biology and behavior of sharks. To activate and implement these recommendations, the AIBS Shark Research Panel* was formed and much of the impetus given to shark research in the past decade has stemmed directly from the activity of this panel.

* Members of the AIBS Shark Research Panel include: Perry W. Gilbert, Chairman; Capt. H. David Baldridge, USN; Sidney R. Galler; John R. Olive; Leonard P. Schultz; Stewart Springer; Albert L. Tester; and a representative from the Office of Naval Research — Deane E. Holt.

From its inception the Shark Research Panel has been concerned not only with the development of protective measures man might employ in shark infested waters but also with basic research programs that lead to a better understanding of the species. Under the able direction of Leonard P. Schultz, a Shark Attack File for the world was set up at the Smithsonian Institution and to date nearly 1500 case histories are on file. From the documentation in this file we hope to learn more about those environmental conditions under which attack takes place as well as the behavior patterns of swimmers that attract sharks. Information in the file is now being transferred to IBM cards for use in a data retrieval system and the file is presently located at the Mote Marine Laboratory in Sarasota in the charge of Captain H. David Baldridge, USN. Other activities of the Panel have included the testing of various chemical compounds and physical agents for their deterrent qualities on sharks and the coordination of an inter-

Dr. Perry W. Gilbert is director of the Mote Marine Laboratory in Sarasota, Florida and professor of neurobiology and behavior at Cornell University, New York. He is widely known for his studies on the physiology and behavior of the shark. He is chairman of the American Institute of Biological Sciences Shark Research Panel, which coordinates shark studies in all parts of the world.

From BioScience, Vol. 18, No. 10, pp. 946-950, 1968. Reprinted by permission of the author and the American Institute of Biological Sciences.

national shark tagging program headed by Stewart Springer. This latter study is expected to shed considerable light on the growth, distribution, and migrations of various species of sharks. Over 4000 sharks of several species have been tagged with AIBS tags on both coasts of the United States and 100 documented recoveries have been made. In one instance a blue shark, tagged at Los Angeles, was recovered off Central America, 1600 miles distant. An additional 4000 sharks have been tagged with Bureau of Sports Fisheries tags since 1963, on the coast of the United States by Panel consultant John Casey and his colleagues. During this period 130 recaptures of tagged specimens demonstrate that certain large sharks regularly move more than 800 miles between New Jersey and Florida, while other species exhibit a seasonal offshore-onshore migration between the Gulf Stream and the northeast coast.

In the past decade the AIBS Shark Research Panel has activated and coordinated a whole host of investigations that deal with various facets of the behavior and basic biology of sharks in many parts of the world. To acquaint those scientists who are studying sharks with the work of their colleagues, the Panel has sponsored a number of conferences. In 1966 a symposium on current investigations of elasmobranch biology was held at the Lerner Marine Laboratory, Bimini, Bahamas. Thirty-nine of the papers presented and discussed at this symposium were subsequently published in book form by The Johns Hopkins Press in a volume entitled *Sharks, Skates, and Rays*. In 1967 the California Academy of Sciences played host to a panel symposium that assembled more than 100 West Coast scientists working in the field of elasmobranch biology. A similar meeting took place in April 1968 in Washington, D.C., for 75 East Coast scientists at a Panel symposium with the Smithsonian Institution as host. Recently three Panel members participated in a meeting at West Palm Beach called by Congressman Paul G. Rogers to discuss shark control measures for a portion of the Florida coast where four shark attacks had taken place within a period of 8 months. Thus, the AIBS Shark Research Panel serves as both catalyst and coordinator for investigations of the basic biology of sharks and the development of more effective methods for their control.

Of the 250 species of living sharks about 35 are potentially dangerous to man. In a single year not more than 100 attacks by sharks on man occur in all parts of the world and approximately half of these are fatal. By comparison more than three times as many people die in the United States alone from bee stings or lightning. Why then is there such concern about the shark hazard? A single shark attack has implications that transcend the damage done to one victim. If the attack takes place on a popular bathing beach, the tourist industry for that area is seriously affected for months. The resultant monetary loss is calculated not in thousands but in millions of dollars. When an attack takes place on military personnel, morale problems of substantial proportions are created. During World War II probably far more servicemen in the water drowned from panic at the sight of a shark fin than from shark attack. The need for more effective protection from sharks is clear. Man, however, is not the only target for harassment by sharks. Many naval operations, such as gear recovery and underwater projects, have been curtailed or abandoned due to the aggressive activities of sharks. Expensive oceanographic instrumentation has been lost because sharks have severed the cables supporting it, and recent studies have shown that some sharks are capable of biting through ropes and cables of substantial size. The biting pressure of an 8 ft dusky shark may exceed 18 metric tons per square inch. The fishing industry in recent years has been seriously plagued by sharks. In 1966 mackerel fishermen in Florida had to give up their fishing operations in the Gulf of Mexico because of damage to nets and catch. For these reasons the search continues for a better shark deterrent than now exists—a deterrent which will protect not only a person who finds himself in shark infested waters but also expensive oceanographic equipment and fishing gear.

The most effective method for protecting swimmers at a bathing beach that has thus far been developed is "meshing." First introduced in Australia in 1937 and subsequently used at Durban, South Africa since 1952, this technique has virtually eliminated the danger of shark attack on beaches thus protected. Meshing, which consists of using gill nets for the capture of sharks, is an expensive operation, and the nets must be checked every other day to remove dead sharks. Moreover, the meshed beaches must have a certain configuration if the nets are to be effective.

Many other devices have been suggested and tested in recent years by members of the AIBS Shark Research Panel. The bubble curtain was at one time believed to provide an effective barrier to sharks. A perforated rubber hose lying on the bottom, off a bathing beach, with one end of the hose connected to an air compressor, produces a curtain of bubbles and thus screens off a bathing area. This device was tested at the Lerner Marine Laboratory on 12 adult tiger sharks, 11 of which promiscuously swam back and forth through the curtain. A twelfth tiger shark was repeatedly repelled and, had we worked with but that single shark, quite different conclusions concerning the effectiveness of a bubble barrier would have been reached. This pointed up the necessity of working with many individuals of a single species, as well as with many different species of dangerous sharks when testing any deterrent.

Variation in response pattern between species was dramatically brought out a few years ago when we tested a highly publicized electric repeller on large lemon and tiger sharks. The lemon sharks when shocked by the repeller were quickly turned away, but tiger sharks were attracted to the same electric pulse, and with repeated shocks could be held in close proximity to the repeller. Although electrical devices have considerable promise, none has been developed to date that works on all species of dangerous sharks. Moreover, electric barriers, such as the electric screen used at the mouth of the St. Lucia River, South Africa, while providing an effective barrier to most species of sharks, have been much too expensive to maintain. In due course electric barriers and other electronic devices may be developed that will be both practical and effective on most species of sharks.

More than 200 chemical compounds have been tested for their deterrent effects on sharks during the past 10 years. Dr. Albert Tester of the University of Hawaii and his colleagues have tested numerous chemical repellents in Pacific waters, and we have tested a wide variety of chemical compounds in the Caribbean and Bahamas. No chemical compound, including copper acetate (a component of Shark Chaser), has been found which will work on all species of sharks, and no chemical has been found which will deter sharks in a feeding frenzy. Recently, Captain H. David Baldridge, USN, has conclusively demonstrated that no drug, even one of very high toxicity, will provide satisfactory control for aggressive sharks in the open sea simply because rapid dilution requires an excessive quantity for action in the short time available. Attention has therefore been directed in recent years away from the development of chemical deterrents.

Very recently, Panel consultant Dr. C. Scott Johnson of the Naval Undersea Warfare Center, and currently working at the Mote Marine Laboratory on shark hydrodynamics, devised a Shark Screen (Figure 4-10) for the protection of victims of air and sea disasters who find themselves in shark-infested waters. The Shark Screen represents a new idea in shark deterrents and is designed to eliminate fear of sharks. It consists of a large bag made of thin, strong, very light material with three inflatable collars at the top. When not in use, the bag is folded into a small package and carried in a life vest or other survival gear. In the water the collars

Figure 4-10. An adult nurse shark turns away from dark colored shark screen occupied by its inventor, Dr. C. Scott Johnson. Sharks were attracted, however, to light colored bags having high reflectivity.

of the Shark Screen are inflated by mouth and the occupant, once inside the waterfilled bag, is completely concealed. Movement of limbs, or blood emanating from a wound, are confined within the bag, and reduce possible stimuli that would normally attract a shark to the occupant. Shark Screens of many colors have been tested by several members of the Shark Research Panel. They have been found effective under controlled conditions in pens and in the open sea when dark colored bags with low reflectivity are employed.

Spurred on in part by the quest for more effective shark deterrents, we have learned in the past decade more about the basic biology and behavior of sharks than was learned in the preceding two and a half centuries. A glance at *Sharks, Skates, and Rays* or at *Abstracts of Current Investigations in the United States Dealing With the Elasmo-* *branch Fishes* quickly reveals the wide range of investigations that have been completed or are in progress. Sharks possess both generalized and rather primitive features as well as several highly specialized structural and functional adaptations. This is not surprising in a group that first appeared 300 million years ago and has direct descendants that today reign virtually unchallenged in the sea. The skeletal, muscular, circulatory, respiratory, and central nervous systems are so generalized that they have long served as the classic prototype for students in comparative anatomy concerned with the evolution of vertebrate structure. The reproductive and digestive system, however, as well as a whole battery of sense receptors, exhibit specializations that have undoubtedly played as important a role in the success of the species as has the retention of many of its more generalized features. A wide range of studies is now in progress dealing with these more specialized systems. The role of sense organs in directing a shark to target is being studied at Cornell, Miami, and Hawaii Universities. The visual apparatus, long thought to be of secondary importance in guiding a shark to target, has been demonstrated to play an important role in finding prey in clear water at close range. Gruber and Hamasaki at the University of Miami Institute of Marine Sciences have recently described cones in the retina of several species of sharks. This discovery, coupled with electroretinogram studies, suggests that some sharks may have the ability to discriminate between colors. In all probability, however, the abundant rod cells of the retina, sensitive to varying degrees of brightness, are the principal photoreceptors employed for the detection of prey. The occlusible tapetum in the choroid layer of the eye, extensively studied by Nicol at the Plymouth laboratory, Kuchnow at Scripps, and by several investigators in Neurobiology and Behavior at Cornell, provides sharks with an unusual ability to discriminate between an object and its background in extremely dim light. Sharks have long been known to possess a keen sense of

Figure 4-11. Electroencephalogram is taken by Hodgson, Mathewson, and Gilbert of a young lemon shark while an odorous substance of known dilution perfuses its olfactory sacs. (Courtesy of Dade W. Thornton.)

smell, but there is great need for more precise information. Recent studies by Hodgson, Mathewson, and Gilbert at the Lerner Marine Laboratory on live lemon, bonnet, and nurse sharks suggest an approach to this problem. Electroencephalograms, taken when odorous compounds in various dilutions perfuse the olfactory sacs (Figure 4-11), compared with control perfusions of uncontaminated sea water, produce marked spikes in the EEG pattern, even when extremely dilute odorous substances are employed. Thus, electroencephalography provides a useful method for ascertaining thresholds of olfactory sensitivity and fatigue.

The sense organ studies are closely related to several investigations now in progress that deal with the behavior of sharks under partially controlled conditions in tanks and pens, and in the open sea. The work of the late David Davies of the Oceanographic Research Institute in Durham, South Africa, of Kenneth O'Gower and G. D. Satchell in Australia, and of Albert Tester and his graduate students at the University of Hawaii, has shed much light on the environmental stimuli that affect the response patterns of sharks.

Many species of sharks, especially young ones, are readily conditioned to come to a target for food as Eugenie Clark, Lester Aronson, and Dudley Klopfer have demonstrated recently at the Mote Marine Laboratory. The exploration and precise measurement of sonic signals that attract and repel sharks in the open sea, first described by Warren Wisby and his graduate students in 1964, are being continued by Arthur Myrberg and his graduate students at the University of Miami Institute of Marine Sciences. It is important that behavioral studies of sharks be continued and expanded if we are to better appreciate those factors which influence their control or survival.

Of paramount importance to any study of the anatomy, physiology, and behavior of a species is an accurate knowledge of its nomenclature and its taxonomic position in the evolutionary scheme. In the last 5 years a series of significant studies on the taxonomy of sharks has been carried out at the U.S. Bureau of Commercial Fisheries by Harvey Bullis, Stewart Springer, and Susumu Kato; and in the Smithsonian Institution's Department of Vertebrate Zoology by Jack Garrick, Robert Gibbs, Carter Gilbert, Leonard Schultz, and Victor Springer. Important papers on the phylogeny of sharks have been prepared in the last 5 years by Bobb Schaeffer of the American Museum of Natural History and Shelton Applegate of the Los Angeles County Museum of Natural History.

During the past decade sharks have been increasingly used as experimental animals for solving problems in human physiology, immunology, and virology, and have thus become important benefactors of man. A host of scientists at the Mt. Desert Island Biological Laboratory has continued to contribute significantly to our understanding of kidney physiology, osmoregulation, and metabolism through experiments on the spiny dogfish. A dozen investigators at the National Institutes of Health now regularly use sharks as experimental animals in their study of cerebrospinal fluid, cell metabolism, and cancer.

Figure 4-12. The commercial shark fisheries at Mazatlan, Mexico. Here the author saw fishermen bring to the beach over 5000 sharks of 17 species during a 5-day period in 1964.

Michael Sigel and L. W. Clem, working at the Lerner Marine Laboratory, and Edward Evans at the Mote Marine Laboratory have recently made important contributions to our understanding of immunoglobulins and antibody formation by using lemon and nurse sharks as experimental animals. The functional relationship of mother to developing young has been explored in a variety of viviparous sharks by Franklin Daiber and his graduate students at the University of Delaware and by several graduate students and myself at Cornell University. These are but a few of the scientists now making significant contributions to our understanding of basic life processes who employ sharks as experimental animals.

In addition to using sharks as experimental animals for the solution of human problems, the shark has long been recognized in many countries of the world as an economic asset. The school shark has a delicious flesh and when one eats "fish 'n chips" in Sydney, Australia, or London, England, one is eating shark meat. The Japanese make extensive use of shark flesh, and no fewer than 25 delicious shark dishes are prepared in Japanese homes. In many parts of Mexico shark meat is dried, salted, and sold extensively (Figure 4-12).

Recent statistics compiled by the Food and Agriculture Organization of the United Nations reveals that sharks comprise about 1% of the present world market for fish. Interestingly, the shark enjoys greatest popularity not only in those countries where food is in short supply but also in those countries that are noted for their fine cuisine. Success in promoting the use of shark flesh for food in the United States has been limited indeed, possibly because so many Americans are unadventurous in their eating habits, and also quite possibly because Americans have not experimented sufficiently with the preparation of shark flesh as a dinner entree. With the development of better methods for freezing and preserving shark meat, and more imaginative recipes for its preparation, we may anticipate greater use for this delicious product at the American dinner table.

Once the sharply pointed denticles are removed from a shark hide and it is tanned, an excellent quality leather may be prepared having several times the tensile strength of cow or pig leather. In recent years the Ocean Leather Company, Newark, New Jersey, exclusive tanners of shark leather, has had difficulty in procuring enough shark skins to meet the demand for their products. At the present time there are very few shark fisheries in the United States but, with an increased demand for shark flesh and hides, we might anticipate an increase in the number of profitable shark fisheries on our east and west coasts. These fisheries, if operated in popular resort centers, would perform the added function of reducing the shark hazard in those areas. The establishment of a commercial shark fishery in the Palm Beach area, for example, has been strongly recommended, and steps are now in progress to start one there.

Viewed dispassionately, the shark is both a hazard and a benefactor to man, just as is the automobile and the airplane. By exploiting fully its useful features and by taking adequate precautions when entering shark infested waters, there is reason to believe that in the next decade the shark will become less of a liability and more of an asset for man.

SUGGESTED READING

Gilbert, Perry W. 1963. *Sharks and Survival,* D. C. Heath & Company, Boston.

Gilbert, Perry W., Robert F. Mathewson, and David P. Rall. 1967. *Sharks, Skates, and Rays,* The Johns Hopkins Press, Baltimore, Md.

American Institute of Biological Sciences, Washington, D. C., 1967-1968. *Abstracts of Current Investigations in the United States Dealing With the Elasmobranch Fishes.*

McCormick, Harold W., Tom Allen, and Captain William Young. 1963. *Shadows in the Sea, The Sharks, Skates and Rays,* Chilton Books, Philadelphia, Pa.

Migration of the Spiny Lobster

WILLIAM F. HERRNKIND   1970

The migrations and mass movements of animals have long influenced man and stirred his curiosity. He has wondered about the cause and purpose of the periodic exodus by lemmings and the incredible navigational capabilities of salmon returning to their home stream or adult eels returning to spawn at their birthplace a thousand miles out in the Sargasso Sea. I, too, have been intrigued by the appearance overhead of hundreds of Canada geese in V-formations and by the swirling masses of wormlike black elvers in the estuaries near my childhood home. However, while I had occasionally reflected on these phenomena, it was not until 1963 that I began to direct my scientific research toward comprehending such events, particularly one that is, curiously, both spectacular and virtually unknown.

As a graduate student assistant working on a bioacoustic research program at the Institute of Marine and Atmospheric Science, University of Miami, it was my job one October day to trace a breakdown in the underwater cable linking our hydrophones to The American Museum's Lerner Marine Laboratory in Bimini. The installation was located on the edge of the Florida Strait at a depth of 65 feet. Sea conditions, which had been intolerable owing to an autumnal squall, had improved enough to permit my co-worker and me to maneuver out in a skiff to the vicinity of a main junction in the cable. We donned SCUBA gear and dropped into the murky water to begin a search for the splice box. Particulate matter, stirred up and held in suspension by the rough seas, limited underwater visibility to 15 feet in an area where it is usually about 100 feet. Peering through the haze below me as I sank, I saw what I first believed to be long, dark furrows or waterlogged timbers lying on the bottom, but as I dropped farther they resolved into lines of dozens of spiny lobsters, *Panulirus argus,* marching head to tail in single-file.

I was surprised because this delectable creature is typified by its sparse numbers in the open sand areas where we located our acoustic array. As I settled onto the bottom,

Dr. William F. Herrnkind is an assistant professor with the Department of Biological Sciences at Florida State University. He is interested in the behavioral ecology of marine animals, and he has studied marine crustaceans, and especially spiny lobsters, in Florida, Bahamas, and the Caribbean.

From *Natural History*, Vol. 79, No. 5, pp. 36-43, 1970. Reprinted with author's revisions and light editing and by permission of the author and *Natural History* magazine. Copyright by *Natural History* magazine, 1970.

still another column marched steadily by without missing a step. I realized then that I was witnessing a "crawl," or "crawfish walk," long known to professional fishermen and other old salts, but poorly known and, in some cases, disbelieved by marine scientists. I first heard of these mass movements of thousands of lobsters from biologist William C. Cummings, who witnessed a similar event off Bimini in 1961. I also realized, as another column of twenty lobsters went by, that we would be unable to locate the splice box in such murky water, so we spent as much time as possible studying the lobsters.

These mass, single-file marches by spiny lobsters are unique, the only known formation movements by bottom-living crustaceans. Furthermore, the marches markedly contradict the established view of this species' behavior pattern. Spiny lobsters are nocturnally active; they wander about at night to feed on mollusks and crustaceans but return before daybreak to shelter in crevices on the reef, under rock ledges, or among dense fronds of sea whips. Why, then, do all the lobsters in a region become active each fall, moving by day, in formation, over exposed areas where they are never seen at other times?

Perhaps the most striking feature of the mass movements, aside from the sheer numbers involved, is the single-file formation, which I call a queue (Figure 4-13). Some other crustaceans travel in more or less definable groups: fiddler crabs and soldier crabs form great droves, or herds, which scour the beach for food at low tide, while King crabs aggregate in clusters. However, none approach a stable, spatial configuration to match the long straight queues of *Panulirus argus*. All queuing lobsters maintain the precise course and speed of the leader and move through turns as though they were on rails. More amazing is that all the queues, no matter how far apart, travel in equivalent or parallel headings! Just how do the lobst rs organize themselves into queues, establish leadership, and maintain formation? What is the biological significance of the mass migra-

tions? Where do the migrants come from and where are they going? The questions seem endless.

The difficulty of answering them lies in the necessity of performing much of the research in the sea at a relatively unpredictable time, at a relatively unpredictable location, under conditions that severely restrict visual studies. In studying the Arctic tern or indigo bunting, at least we know where they come from, where they go, and what they do when they get there (although we still don't know precisely *how* they navigate). After observing that marching horde of spiny lobsters, I felt irresistibly challenged to discover their secrets.

At first my lobster research proceeded slowly since I was committed to a doctoral research problem on a distant relative of the spiny lobster, the fiddler crab. This doctoral research provided useful background when I later tackled the more formidable problem of lobster migration and orientation.

Other sources of help were the numerous professional lobstermen, conservation officers, and skin divers I spoke with, who related their observations of similar marches in different regions. In all cases the general descriptions were similar—the events took place in the fall after intense storms and involved large numbers of lobsters of approximately the same size, traveling in long queues. Each queue in a given march headed in the same compass bearing, and the compass bearing was specific to each location. The stories told to me also suggested striking behavior I had never witnessed myself. For example, a fisheries officer from Florida described a marching column that extended, with few breaks in rank, for nearly one-quarter mile; that would conservatively comprise one thousand spiny lobsters. Several Biminites independently told of an immense number of lobsters that wandered into the Bimini Lagoon and, upon reaching a cul de sac along the shore, swirled about in a great mass with many individuals walking out of the water onto the beach. During 1969 in Bimini a migration occurred in which about

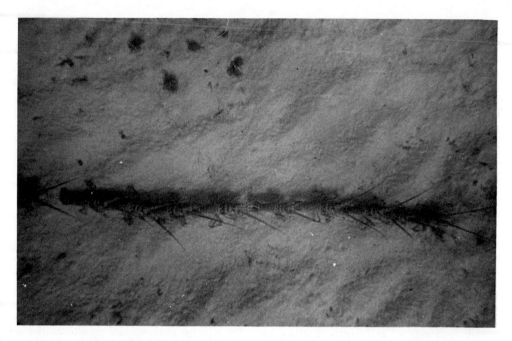

Figure 4-13. Single-file line, or queue, of spiny lobsters marching across open sand areas at a depth of nearly 35 feet near Bimini, Bahamas. Numbers of lobsters exceeding 100,000 take part in such autumnal mass movements. Each queue contains 2 to 50 individuals and travels with the same heading.

ten lobstermen captured an estimated 20,000 lobsters in five days. And they by no means caught them all, probably less than 10 per cent.

A colorful description of a march by one Bimini fisherman included the explanation that the lobsters migrate when they get "the spirit." I have subsequently found that this is a valid descriptive term for the internal state of the animals during these events. For my first opportunity to observe the persistence of this spirit I must thank a fellow student at Miami who called me at one o'clock one morning to invite me to witness a mass movement at Boca Raton. I gathered my diving paraphernalia and drove up in the wee hours to get overboard at daybreak. Sure enough, the columns of lobsters were marching alongshore and were literally piling up at a rock jetty, which looked from underwater like a pincushion of antennae. I mainly wanted some undamaged live specimens to bring back to the Marine Institute for

study, but had overlooked bringing my hand nets. A feverish chase ensued, during which I captured some undamaged live specimens and hauled them back to a vinyl-lined sea water pool, 15 feet in diameter. Upon release, the group formed a queue and marched clockwise around the pool almost continuously, day and night, for the next two weeks. All in all, it was almost five weeks, and an estimated 500 miles, before the marching activity halted along the endless migratory pathway presented by the perimeter of the circular pool.

During that time I fed the lobsters and attempted to induce them to enter a concrete block shelter. However, they would eat only for brief periods interspersed with marching and would not take up residence in the shelters. This matches the behavior exhibited by lobsters while marching in the sea. There, certain members of a file stop occasionally and grasp such objects as starfish and small sea cucumbers, then move on, eating as they march. Columns also cluster under rock ledges for some minutes, as many as 200 in a 10 cubic foot space, with groups continually forming and moving off as others arrive. All these actions are in strong contrast to the responses of both captive and wild lobsters at other times, when feeding lasts much longer and the shelters are inhabited through the daylight hours.

The extraordinary behavior of captive migrating lobsters under artificial conditions suggests that their "spirit" is a modified internal state, or drive, responsible for the maintenance of the migratory activity. It might be likened to the internal processes that cause birds to become restless at the time of migration, a condition termed Zugunruhe. In some birds Zugunruhe is brought on by a modification of the hormonal system, the result of changes in day length or, more simply, photoperiod. Thus, as fall days shorten and nights lengthen in the North Temperate Zone, changes in birds' internal processes are manifested as a general increase in activity and a tendency to fly southward. A similar Zugunruhe occurs in the spring increase of photoperiod, but brings on a tendency to fly northward. The seasonal nature of the mass movements by spiny lobsters, and the continuous hyperactivity of captive specimens, suggests control by some internal process brought on previously by environmental changes associated with autumn.

The autumnal storms always reported to precede the marches seemed a strong possibility at first as a cause of the internal changes. But present evidence suggests that this is not so. Violent storms also occur at other times of year—particularly in the winter, spring, and in association with summer hurricanes and tropical depressions—but marches have been reported only during the September through November period. And, in the area off Bimini where I had made my original observation, a small march of brief duration occurred in October, 1969, during a two-week period of almost uninterrupted calm.

It appears, then, that some other factor brings on the internal state preparatory to migration, and that storms at that time act to trigger and synchronize the movement of the population.

This past summer I investigated some nutritive factors as possible causes. This was suggested by marine biologist Robert Schroeder, who mentioned that captive lobsters would begin to march around their enclosure shortly after being switched from a mollusk diet to fish, as though fish lacked some necessary substance that inhibited the Zugunruhe or, perhaps, contained some inductive substance lacking in mollusks. To test this effect I placed groups of ten lobsters in three pools provided with running seawater, sand substrate, and terra-cotta pipes for refuge. Except for the dietary regime, each group had nearly identical conditions. We fed the mollusk group surf clams, the fish group chopped fish, and the third group nothing, and monitored the activity patterns daily for any changes. By the fourth day the fish group became hyperactive and exhibited queuing during the day, while the others retained the normal pattern of inactivity by day, thus confirming Schroeder's observations. The situation remained the same through the following week indicating a relatively long-lasting effect.

At this time we switched the diets of the fish group and mollusk group. Activity decreased daily in the new mollusk group (former fish group) and increased in the new fish group (former mollusk group). Thus, a change in diet is a factor that can control activity and is a possible cause of migration. However, the animals given no food behaved in accord with the normal nocturnal pattern, leaving us in a quandary. If the lack of some substance in fish causes the migratory state, why doesn't the absence of food have the same effect? It may be that fish flesh contains an induction substance, but we feel that this would be a remote possibility as a cause of mass movements in nature since spiny lobsters probably seldom eat fish; they simply are not equipped to catch them.

Fish-fed lobsters, however, may continually add body tissues lacking in some necessary substance(s) found in mollusks, which must be kept in balance with the added body material for normal growth. The imbalance causes modifications in the metabolic system that subsequently result in the migratory state. Starved lobsters gain none of this substance either, but they are losing, not adding body material and, therefore, are not affected in the same way. More studies of the type described, along with studies of internal processes, must be conducted to define the role of nutrition as a causal factor.

Other factors may also work independently of, or in conjunction with, dietary modifications to bring on the migrations. For example, photoperiod seems a likely possibility: light exerts a strong influence on the hormonal physiology and behavior of many crustaceans, including *Panulirus argus*. And temperature flux may be involved, since temperature drops of several degrees centigrade often result from autumn storm activities. Another possibility is increase in population density, which most of you may have already thought of in connection with lemming and locust emigrations. Being gregarious, lobsters tend to cluster by day in habitable crevices, so an increase in immigrants or an increase in the living biomass of lobsters in the population, as occurs during synchronous molting, might produce a density effect culminating in emigration from that area.

We do not yet have conclusive evidence about where the lobsters come from or go. At present, I believe that migrant lobsters off Bimini originate in shallow areas well to the north and east of the island group. Finding their ultimate destination is a problem in tracking. Using a sonic pinger tag, which pulses a signal detectable by directional hydrophone, I was able to follow a lobster at Bimini by boat for several miles. The paths suggest that migrants disperse into suitable habitats along the west edge of the Bahama Bank five to ten miles south of Bimini. An expanded tracking program should clarify this in the future.

When we turn to the striking feature of queuing, we find that it is a basic component of this species' behavioral repertoire, since even young lobsters two to three inches long sometimes form single-file lines. Queues also occur at times other than the mass movements—whenever a group of lobsters is deprived of shelter or is introduced to a novel habitat situation. This tendency to congregate—even in the large circular pools of the laboratory—has enabled me to observe the sequence of queue formation and the sensory mechanisms used in maintaining it.

I recorded data on queuing by means of an event recorder. This device has separate, manual pushbuttons for each of twenty pens, which trace paths on a moving chart. Thus a button coded for each separate action was depressed by the observer whenever that action took place and for as long as it lasted. Afterward, the chart was reviewed to determine the number of times an action occurred, its duration, and its sequential relationship.

We found that an isolated, stationary lobster visually perceives, and directs its antennae toward, a moving individual up to several yards away. It then queues up by approaching the moving lobster from behind until antennal contact is made. At this point its antennules are brought into contact with each side of the lead lobster's abdomen, completing the alignment. The queue is maintained by the almost constant contact of the antennules or by the hooking of the tips of the pereiopods (walking legs) around the telson (tail) of the lead lobster. This tactile locking into place enables all the queue members to walk at the same speed and in the same direction, resulting in strikingly straight columns of up to 50 lobsters. The contact also permits the queue to maintain its integrity when the leader changes course to detour an obstruction. The significance of constant contact is suggested by the effort that separated individuals immediately make to close up any gaps.

But could lobsters deprived of their antennae, antennules, or anteiormost pereiopods still queue? Losses of these appendages are common in nature. To test this, we recorded and compared the performance of individuals deprived of one of these receptor-appendages either by forcing autotomy (self-release of appendages at certain joints) or by taping them up so they could not receive stimulation. In all three cases the lobsters could still queue since they substituted usage of one of the other remaining appendages. The lobsters, like the Apollo moonships, possess redundant back-up systems to take the place of any one that should fail.

The strongest evidence for concluding that tactile cues are the most important for aligning and maintaining the queue formation came from studies on lobsters blinded by opaque tape. These would queue up only after some tactile contact was made with another lobster, at which time the taped lobster turned neatly into alignment and maintained position as effectively as untaped individuals. It seems likely, then, that spiny

lobsters can queue even in cloudy water and at night.

Since the queue involves a number of lobsters led by only one individual, it would seem that the leader should be outstanding for some noticeable attribute—perhaps size, indicative of age and experience, or peculiar behavior recognizable by the others. However, upon examination of over fifty "leaders" captured during one migration, we find them to fall by size and sex ratio right in the average for all the lobsters collected at that time. This, together with our laboratory studies, suggests that leadership is produced, not by exterior appearance, but by the behavioral manifestation of some inner drive.

Queues that I observed in the open sea formed in several ways, but the following was the most striking: Lobsters clustered together in a closely packed group or pod and turned about the center in a tight circle,

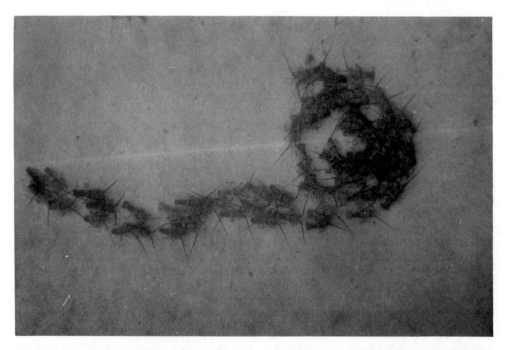

Figure 4-14. A queue may stop when the leader turns in a tight circle, the followers winding around one another in a spiral motion. Upon stopping, the individual lobsters rest facing the outside of the circle presenting a thicket of spiny antennae to an approaching diver or perhaps a predator. Locomotion resumes when one or more lobsters moves off and the rest fall into line.

giving the impression of rotation (Figure 4-14). At some point an individual moved off tangentially, pursued by the other lobsters. The formation then became a queue as individuals fell into single file. Moving queues form the pod in much the same way; i.e., the lead lobster turns in a tight circle and the followers wind about in an ever-widening spiral.

In the indoor pools, queues formed in back of those individuals that were most actively moving about the enclosure. Tests with "models" made from molted lobster carapaces mounted on glass rods proved that activity is the most crucial stimulus for leadership. I was able to induce queuing by merely moving the model about the test pool near the captive lobsters, in some cases even when they were inactive.

The question of the significance of queuing to survival arises. Answering this is far more difficult than determining the mechanism of queuing because it implies that the behavior has selective value in maintaining the species. As an educated guess, lobster queuing probably performs a defensive function. These lobsters are not offensive creatures of prey; they possess no claws or other weapons of attack. They rely on defense for their survival: hiding in a hard, horny armor, which deters all but the largest or hardest-mouth fishes; moving rapidly when in retreat, as the telson snaps forward propelling them backward; and keeping the abdomen, the least-protected portion of their anatomy, under a rock ledge. The pointed, hard front end is exposed to the predator, which usually gets either puncture wounds or a meal consisting of a spiny, almost hollow, autotomized antenna for its trouble. In open areas away from rock ledges, the abdomens of queuing lobsters are "protected" by the cephalothorax of the lobster behind. Of course, the last individual in line is at a decided disadvantage. Hiding under a rock doesn't always work either. My research assistant once came upon a lobster that had retreated from the open to the safety of a rock crevice only to find itself grasped tightly by a large *Octopus vulgaris* already in residence there.

The ability to move in a straight queue, whether or not the formation offers protection, seems of little value unless the movement is directed somewhere. This aspect of orientation brings up exciting problems because spiny lobsters are capable of feats that defy explanation. Witnesses of mass movements reported that all lobsters in a given area traveled in about the same direction although the direction varied from area to area: southerly at Bimini, northerly at Boca Raton, westerly at Grand Bahama. During the fall of 1969, at the Lerner Marine Laboratory in Bimini I observed a five-day mass movement and recorded the bearings of over 250 queues comprising some 2,000 spiny lobsters (Figure 4-15). The headings were strongly to the south over a distance of at least six miles. The lobsters maintained that bearing while moving over substrate of variable slope and at varying depths, in water visibility less than six feet, under completely overcast skies, and in areas of complex currents, all of which either occlude guidance cues or make them extremely variable. To appreciate this, consider yourself trying to walk on a direct course several miles through hill country, without a compass, in a dense London-style fog, while being buffeted by strong winds from different angles.

The most astounding performance by spiny lobsters occurred in experiments in Bermuda in 1949. Edwin Creaser and Dorothy Travis trapped and tagged lobsters, then released them at various distant locations. Afterward they regularly checked lobster traps located at the points of capture for any returns. Two lobsters released out at sea at a depth of 1,500 feet, two miles from the original point of capture, were retaken less than one week later! Since lobsters sink rapidly and are benthic creatures, it is unlikely they swam back near the surface. Returning two miles along the sea floor in virtually complete darkness suggests either a very effective guidance mechanism or a whole lot of luck. We have recently discovered that

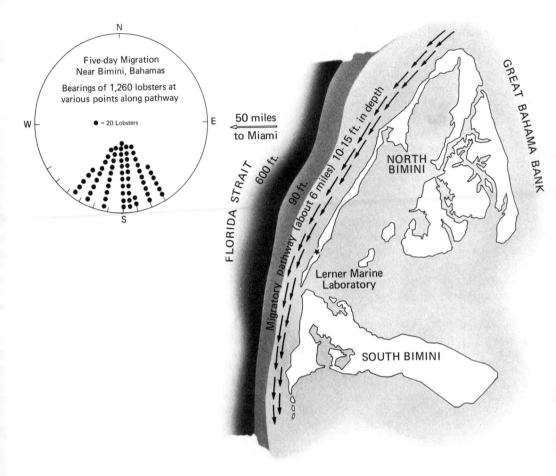

Figure 4-15. During November, 1969, migration, lobsters maintained strong southerly bearing, despite turbulent seas. Beginning and end of pathway are unknown.

reef lobsters typically live in a specific den for long periods, leaving by night to feed and returning before dawn. In addition, we corroborated the existence of homing ability by tracking the return of lobsters displaced up to 700 feet from their den.

To test for orientational mechanisms, I captured lobsters in nonmigratory condition from selected areas in the waters of Bimini. The animals were marked and released underwater at depths of 15 to 30 feet on level submarine sand plains devoid of vegetation. Each lobster's path was recorded on a plastic slate until it passed out of sight. The animal was then retrieved and released again to a

total of eight times, each time at a compass heading 45 degress from the preceding one to control for any bias introduced by the direction the lobster was facing at the time of release.

The typical behavior of 26 lobsters after 208 releases was forward locomotion for a few yards, followed by a turn to a new heading, then direct movement over the underwater horizon. Twenty of the 26 lobsters distributed their runs in a nonuniform manner, that is, all eight runs fell in a specific range of compass bearings (within the 90 degree sector from west to north). Several individuals followed parallel paths on each release, crossing within 10 degrees or a few yards of a given point on three or four runs. We interpret this behavior as an indication of their ability to orient themselves and maintain a bearing in a relatively featureless area.

Convinced that spiny lobsters were capable of establishing a course in the open sea, we next wondered whether the process was effected visually. In further orientation experiments using lobsters blinded by opaque tape, twelve of fourteen lobsters exhibited parallel headings on each of the eight runs. In fact, several showed stronger orientation than the unblinded lobsters. It was particularly startling to release to the south a lobster that had previously run north and have it immediately turn 180 degrees and walk across my plastic slate. Doubtless, the spiny lobster can establish bearings without vision, which is not too surprising since they are nocturnal animals. But we are left with the question of what sensory mechanism and what environmental guidance cues are used to accomplish orientation under nonvisual conditions.

At this point we are not presuming that a single cue or mechanism is the only possible guidance factor. Rather we are alert to having several physical stimuli operating to guide the migration under different conditions. We feel this way for two reasons. First, it is becoming more and more apparent that other orienting animals, such as salmon, bees, pigeons, and fiddler crabs, respond to several types of cues. For instance, fiddlers can and do orient their movements by the sun and polarized sky when underwater, in tall grass, and under other conditions when only the sky is visible. However, if they are on the open beach, they orient by local landmarks even when the sun is clearly visible. So, if landmarks aren't available they use the sun-compass, if the sun isn't visible they use landmarks, and if neither is visible, as during rain or fog, we have some evidence that they don't bother to move more than a few yards away from the sanctuary of their burrows. The environment of the spiny lob-

ster also has cues such as the sun, landmarks, and bottom slope available under some conditions, and these cannot be eliminated as possibilities.

A second reason is evident if one looks closely at the sensory capabilities of spiny lobsters. They have large eyes capable, at least, of recognizing other spiny lobsters; fine chemical and tactile sensors in each antennule, pereiopod tip, and also around the mouth; as well as tactile sensors in the telson and antennae. Internally there are proprioceptors, which sense not only movements of the appendages but also external forces such as gravity and uneven pressures. Additionally, there are indications of numerous receptors we have not yet characterized. If you look at the scanning electromicrographs taken of surface features, you will see numerous hairlike processes, clumps of setae, and pits in the chitin.

The functions of some of these are not yet known and one gets the strong impression that the spiny lobster is equipped as a walking undersea probe. Between the multitudinous physical cues of the ocean and the equally varied receptors, a self-contained guidance unit may be operating, exceeding any that man has yet devised.

SUGGESTED READINGS

Allen, J. A. 1966. The rhythms and population dynamics of decapod crustacea. *Oceanography and Marine Biology, Annual Review* 4, 247-265.

Herrnkind, William F. 1969. Queuing behavior of spiny lobsters. *Science* **164**, 1425-1427.

Herrnkind, William F., and R. McLean. 1971. Field studies of orientation homing and mass emigration in the spiny lobster, *Panulirus argus. New York Acad. Sci.* **188**, 359-377.

Waterman, T. H. (ed.) 1961. *The Physiology of Crustacea.* Academic Press, New York. Vol. 2.

Williams, A. B. 1965. Marine Decapod Crustaceans of the Carolinas. *U.S. Department of Interior Fishery Bulletin* **65**, 91-94.

5

MARINE GEOLOGY

Sea-Floor Spreading—New Evidence

F. J. VINE   1969

INTRODUCTION

The concept of sea-floor spreading was first formulated in some detail by Hess in 1962; however, only since 1966 with the advent of new and rather compelling evidence, has it attracted widespread attention. In light of these new data it seems appropriate to try to reformulate the hypothesis and starting with the current seismicity of the earth, develop it in terms of geologic time.

SEISMICITY OF THE EARTH

Earthquakes are not randomly distributed over the surface of the earth but are largely restricted to the young fold mountains and trench systems of the Alpine-Himalayan and circum-Pacific belts, and to the crests of the mid-ocean ridges (Figure 5-1). By far the greatest number of earthquakes occur in the circum-Pacific belt in association with the trench systems. Earthquakes are less frequent in the Alpine-Himalayan belt and less

frequent again on the mid-ocean ridge crests. The recent more accurate epicentral locations determined by the Environmental Sciences Services Administration—with data from the World Wide Standardized Seismograph Network—define these narrow belts in remarkable detail. Thus most of the current seismicity of the earth occurs in very restricted linear zones. This is where the action is.

The activity associated with ridge crests and strike-slip faults is confined to very shallow depths, probably not exceeding 10 or 20 km (Isacks et al., 1968). In the young fold mountains and trench systems, however, shallow focus earthquakes are present but in many of these settings intermediate and deep focus earthquakes, occurring to a maximum depth of 700 km, are also recorded (see Figure 5-1). According to the hypothesis of sea-floor spreading, these two seismic provinces—those in which only shallow earthquakes occur and those which are char-

Dr. Frederick J. Vine is a reader with the School of Environmental Sciences at the University of East Anglia, Norwich, Great Britain. His well-known hypothesis on magnetic reversals has provided marine geologists and geophysicists with a key for determining the rate of continental movement and the relative ages of different parts of the ocean floor. He has written over twenty publications and in 1970 received the Henry Bryant Bigelow Medal in Oceanography from the Woods Hole Oceanographic Institution, Massachusetts.

From *Journal of Geological Education*, Vol. 17, No. 1, pp. 6-16, 1969. Reprinted by permission of the author and the *Journal of Geological Education*.

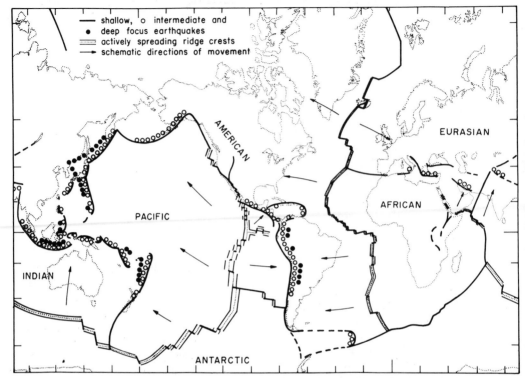

Figure 5-1. Summary of the seismicity of the earth (Gutenberg and Richter, 1954) and hence the extent of crustal plates bounded by active ridge crests, faults, trench systems and zones of compression. The six major crustal blocks assumed by Le Pichon (1968) are named. Spreading rates at ridge crests are indicated schematically and vary from 1 cm per year in the vicinity of Iceland to 6 cm per year in the equatorial Pacific Ocean (Heirtzler *et al.*, 1968).

acterized by deeper focus earthquakes—reflect very different but complementary processes at work in the upper mantle and being accommodated in the earth's crust. The mid-ocean ridge crests are extensional features along which new oceanic crust is created, and the trench systems are regions in which oceanic crust is partly resorbed. The oceanic crust is thus considered to be a surface expression of the mantle, derived from it by partial fusion and chemical modification beneath ridge crests, and in part returned to the mantle beneath the trench systems. Recent earthquake mechanism solutions have confirmed this picture of extension of the earth's crust at ridge crests in terms of normal faulting (Sykes, 1967), and compression and underthrusting of the crust

landward of the trench systems (Stauder, 1968).

MOVEMENT OF CRUSTAL PLATES

Within this basic framework of ridges and trenches—sources and sinks— several complications are produced by the distribution of continental crust. The simplest of these possibilities are illustrated in Figure 5-2. If an upwelling in the mantle is initiated beneath a continent, the continent will be rifted and passively drifted apart with the formation of a new ocean basin by lateral spreading from the original rift. The continent, coupled to a rigid conveyor belt of upper-mantle material, drifts either with a trench system at its leading edge (Figures 5-2A and 5-2B), an island arc system ahead of it (Figure 5-2B), or until it encounters a trench. In the last case, because of the continent's lower density, it is unable to sink and overrides the trench, forming a new trench system (Figure 5-2A). Such an ocean basin, with a median ridge and surrounded by the recently undeformed trailing edges of drifting continents, is reminiscent of large parts of the Atlantic and Indian Oceans. Alternatively, if rifting is initiated within a former oceanic area, new oceanic crust is formed and older crust resorbed in marginal trenches (Figure 5-2B). With time the trench systems will encroach on the oceanic area. There is no reason why a ridge of this type should be median within

the oceanic area. Such a picture is analogous to the situation in much of the Pacific at the present time. A third possibility, clearly, is that two continents may ultimately come together at a trench system, producing high mountain ranges such as the Alps and Himalayas which result from the collision of Africa and India with Eurasia (Figure 5-2C). Thus the distribution of the continents accounts for the marginal trench systems and island arcs of the Pacific and the median position of the Atlantic and Indian Ocean ridges.

The geometry of spreading on the real earth is a complex combination of these simple possibilities but it is, nonetheless, amenable to rigorous analysis in terms of the relative movement of rigid crustal plates bounded by the zones of current earthquake activity (Morgan, 1968). These plates or aseismic areas are outlined, therefore, by the seismic belts and spreading ridge crests shown in Figure 5-1.

TRANSFORM FAULTS

Turning now to specific new evidence for spreading, one might first consider the distribution of earthquake epicenters and the nature of faulting on mid-ocean ridge crests. Greatly improved epicentral determinations

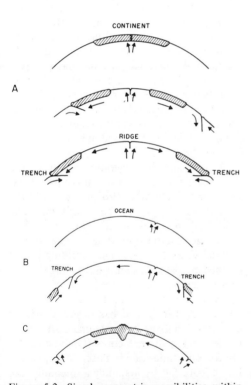

Figure 5-2. Simple geometric possibilities within the framework of sea-floor spreading, emphasizing the role of the continents. A and B illustrate the stages and possible configurations resulting from the initiation of rifting and spreading beneath continental and oceanic crust, respectively. C illustrates the possibility of two continental blocks coming together over a downcurrent.

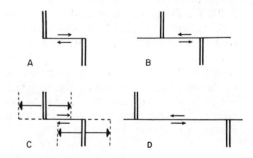

Figure 5-3. A right-lateral ridge-ridge type transform fault (A and C) contrasted with a left-lateral transcurrent fault (B and D). The ridge crest is indicated by the parallel bars, and the active trace of the faults by the thin solid line (after Wilson, 1965).

for these areas, available only in the last few years, reveal that the seismicity is concentrated on the transverse fracture zones between the offset points of the ridge crest. Mechanism solutions obtained for a number of these earthquakes indicate a strike-slip movement on a vertical plane paralleling the trend of the fracture. The sense of movement on this plane is the opposite to that suggested by a simple offsetting of the ridge crest (Sykes, 1967). These points are illustrated diagrammatically in Figure 5-3A. Classically the left-lateral offset of the ridge crest in Figure 5-3A would be interpreted as resulting from left-lateral movement along a transcurrent fault between two crustal plates as shown in Figure 5-3B. However, the implied sense of motion and distribution of earthquake activity along such a fault does not satisfy that which is observed. The sense of movement and distribution of epicenters summarized in Figure 5-3A is, however, compatible with the ridge-ridge type of transform fault formulated by Wilson (1965) within the framework of sea-floor spreading. This is illustrated in Figure 5-3C. An important corollary of this concept is that the offset of a ridge crest along a transform fault may well be an initial one in that the offset does not change with time as spreading occurs. In contrast, the offset of the ridge crest across a transcurrent fault would increase as renewed faulting occurs, as shown in Figure 5-3D. The faults and ridge crest in the equatorial Atlantic (Figure 5-1) parallel the continental margins of Africa and South America and may well represent the locus of initial rifting in this area. It was this point that led to the prediction of transform faults by Wilson (1965) prior to their verification by detailed epicentral determinations and focal mechanism solutions by Sykes (1967).

THE MAGNETIC RECORD

Further evidence for spreading of the sea floor at ridge crests comes from studies of disturbances in the earth's magnetic field recorded at or above sea level. These anoma-

lies are typically one or two percent of the total intensity in amplitude, and range from several kilometers to several tens of kilometers in wavelength. They are attributed to magnetization contrasts essentially within the earth's crust but potentially extending into the upper mantle beneath the ocean basins (i.e. to the depth of the Curie temperature isotherm—600-700°C—for the ferromagnetic rock forming minerals).

In 1963, Vine and Matthews suggested that if sea-floor spreading has occurred, it might be recorded by the remanent magnetization of the oceanic crust, especially within the basalt layer, because basalts have the highest intensities of remanent magnetization. It was postulated that if the earth's magnetic field reverses intermittently as spreading occurs, the resulting magnetization contrasts between normally and reversely magnetized material in the basalt layer should produce appreciable and predictable disturbances in the earth's magnetic field as measured at sea level. Such a crustal model for a ridge crest is shown schematically in Figure 5-4, and is discussed in some detail by Vine (1968).

At the time this suggestion was made the time scale for reversals was unknown (Cox, Doell and Dalrymple, 1963); it was still debatable as to whether the earth's magnetic field had reversed at all. However, since 1963, the reversal time-scale for the last 3.5 million years has become increasingly well defined by paleomagnetic and radiogenic age measurements on young lava flows from many parts of the world (Cox, Dalrymple and Doell, 1967), and has received striking confirmation from the paleomagnetic study of deep-sea sediment cores (Opdyke *et al.*, 1966). This time-scale has been incorporated into Figure 5-4 by assuming a spreading rate of 1 cm per year per ridge flank.

We can now, therefore, test the Vine and Matthews hypothesis. The central curve in Figure 5-5 shows the anomalies in the earth's magnetic field along a profile perpendicular to the crest of the East Pacific Rise at 51°S in the Pacific Ocean. The lower curve shows

RIDGE MODEL

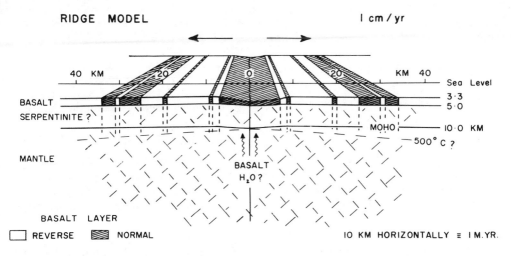

Figure 5-4. Diagrammatic representation of the oceanic crust at a mid-ocean ridge crest, assuming active spreading at a rate of 1 cm per year per ridge flank, and the geomagnetic reversal time scale of Cox *et al.* (1967).

the predicted anomalies which result from assuming the above model and a rate of spreading of 4.4 cm per year per flank. The upper curve in Figure 5-5 is simply the central profile plotted in reverse about its midpoint to emphasize the symmetry of the observed anomalies about the ridge crest. The crustal model is perfectly symmetrical about the ridge axis; the asymmetry of the computed profile is due to the dipolar nature of the earth's magnetic field.

RATES AND RELATIVE MOVEMENTS

This interpretation of the magnetic anomalies, coupled with the paleomagnetic and dating results obtained from subaerial lava flows, enables one to deduce recent rates of spreading at ridge crests for which magnetic data are available. Rates deduced in this way vary from 1 cm per year per flank near Iceland, to 6 cm per year per flank in the equatorial Pacific (Vine, 1966; Heirtzler *et al.*, 1968). These rates are summarized in Figure 5-1, and imply rates of separation of crustal blocks, i.e. rates of drift, varying from 2 to 12 cm per year.

It is now possible, therefore, to assign

rates of relative movement at certain of the boundaries between the crustal plates outlined in Figure 5-1. Having assumed these rates and a simple configuration of just six crustal plates, Le Pichon (1968) has deduced rates of compression and crustal shortening in the active mountain belts and trench systems by vectorially summing the relative movements between pairs of plates about their instantaneous centers of rotation. The rates of shortening deduced are highest in the Himalayas and in the trench systems. The rate for a trench system would appear to be directly proportional to the depth extent of the earthquake activity associated with it (Isacks *et al.*, 1968). This agrees well with the interpretation by Oliver and Isacks (1967) of the attenuation of seismic waves (notably shear waves) traversing the crust and upper mantle in the vicinity of the Tonga Trench. These authors envisage a cold rigid plate of crust and upper-mantle material, the lithosphere, being pulled or thrust down beneath the trench system and into the hotter, perhaps partly molten, asthenosphere (Figure 5-6). The degree of partial melting, if present, must be very small since the asthenosphere only slightly attenuates

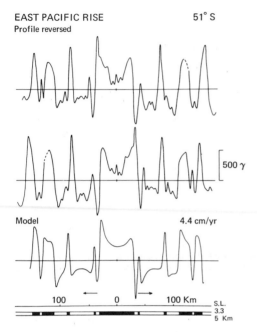

EAST PACIFIC RISE   51° S
Profile reversed

Model   4.4 cm/yr

100   0   100 Km

Figure 5-5. An observed magnetic anomaly profile across the East Pacific Rise (upper curves) compared with a computed profile for this area, assuming the model of Figure 5-4 and a spreading rate of 4.4 cm per year. Observed profile from Pitman and Hiertzler (1966). The computation assumes an intensity and dip for the earth's magnetic field of 48,700 gamma and –63°, and a magnetic bearing for the profile of 102° (1 gamma = 10^{-5} oersted). Normal or reverse magnetization is with respect to an axial dipole vector. Effective susceptibility assumed = ±0.01, except for central block (+0.02). S.L. = sea level.

seismic shear waves passing through it. The asthenosphere, however, correlates with the seismic low-velocity zone in the upper mantle, and convection within it may provide the necessary driving forces to move the lithospheric plates above. Presumably, stresses set up between and within the plates also determine their relative movements.

Current and recent poles of relative movement between plate pairs can be determined from slip vectors (deduced from large shallow focus earthquakes at plate boundaries), from the strike of transform faults, and from the variation in spreading rate along a ridge crest (Morgan, 1968; Le Pichon, 1968). Of these techniques clearly only the orientation of fossil transform faults and the variation in spreading rate at a given time (as revealed by the magnetic anomalies) can be used to determine past changes in the direction of spreading and relative movement (Vine, 1966; Menard and Atwater, 1968). The documentation of such changes is crucial to extending the plate theory back into the geologic past and gaining a clearer understanding of the detailed history of spreading and continental drift.

THE REVERSAL TIME SCALE

The geomagnetic polarity time scale has only been defined for the last 3.5 million years by independent techniques; it is only possible,

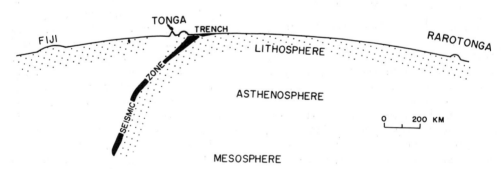

Figure 5-6. Postulated east-west section through the Tonga Trench, assuming that a relative lack of attenuation of seismic waves corresponds to rigidity. The lithosphere and mesosphere are believed to have appreciable strength, whereas the asthenosphere might flow more easily over geologic periods of time (Figure 14 of Oliver and Isacks, 1967).

therefore, to deduce rates of spreading at recently active ridge crests. Nevertheless, the symmetry of the observed magnetic anomalies persists out beyond this central region. It seems reasonable, therefore, to make a provisional extrapolation of the reversal time scale on the basis of the magnetic anomalies, assuming that the Vine-Matthews hypothesis continues to apply and that the spreading rate has remained constant. Such an extrapolation has been made using a profile obtained at 51°S in the Pacific Ocean, the central part of which is shown in Figure 5-5. This profile shows a remarkable degree of symmetry, and many of the details of the reversal time scale are apparent because of the high spreading rate (Pitman and Heirtzler, 1966; Vine, 1966; 1968).

The reversal time scale for the last 11 million years obtained in this way has been used to predict the anomaly profile one might observe over the Reykjanes Ridge south of Iceland (Figure 5-7), where the spreading rate is thought to be 1 cm per year per flank on the basis of the central anomalies. Above the simulated profile is an observed anomaly profile obtained across this ridge. Although the time scale used is derived from the South Pacific Ocean, 15,000 km away, the simulation shows many similarities with the observed profile—one of 58 traverses flown by the U. S. Naval Oceanographic Office in 1963 in making a detailed aeromagnetic survey of the ridge (Heirtzler, Le Pichon and Baron, 1966). The location of the Reykjanes Ridge and a summary map of the magnetic anomalies recorded over it are shown in Figure 5-8. The symmetrical pattern of anomalies provides striking evidence for sea-floor spreading and drift in this part of the North Atlantic Ocean (Vine, 1966).

EXTRAPOLATION ACROSS THE OCEAN BASINS

The magnetic anomalies and the extension of the reversal time scale enables us to map and date the ocean floor in the vicinity of active ridge crests. The question arises

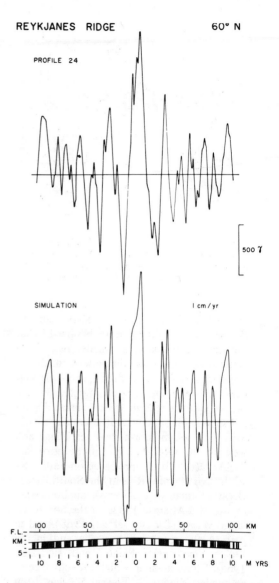

Figure 5-7. An observed aeromagnetic profile across the Reykjanes Ridge, southwest of Iceland, compared with a simulated profile, assuming a reversal time scale for the last 11 million years derived from a Pacific Ocean profile the central part of which is shown in Figure 5-5. Intensity and dip of the earth's magnetic field assumed to be 51,600 gamma and +74° respectively; magnetic bearing of profile 153°. (F.L. = flight level). Observed profile from Heirtzler, Le Pichon and Baron (1966).

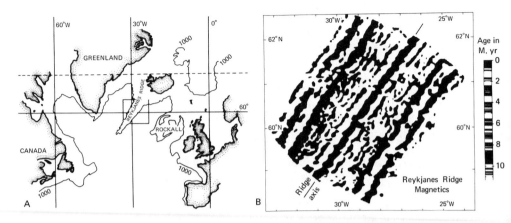

Figure 5-8. A. The location of the Reykjanes Ridge and the area of B. The 1,000-fathom submarine contour is shown together with the 500-fathom contours for the Rockall Bank. B. Summary diagram of magnetic anomalies recorded over the Reykjanes Ridge. Areas of positive anomaly are shown in black. The central positive anomaly of Figure 5-7 correlates with the ridge axis (after Heirtzler, Le Pichon and Baron, 1966).

whether this hypothesis is also applicable to the remainder of the ocean basins. In that the same sequence of anomalies away from active ridge crests is reproduced in all the ocean basins, it seems highly probable that sea-floor spreading accounts for the formation of most, if not all, oceanic areas. Figure 5-9 shows this sequence of anomalies as observed across the Juan de Fuca Ridge and west of the Gorda Ridge in the northeast Pacific Ocean. It is reproduced on both sides of the East Pacific Rise in the South Pacific Ocean (Pitman *et al.,* 1968), on both sides of the Mid-Atlantic Ridge in the South Atlantic Ocean (Dickson *et al.,* 1968), and in part to the south of Australia and in the Indian Ocean (Le Pichon and Heirtzler, 1968). Those areas in which the sequence of anomalies shown in Figure 5-9 has been recognized have been summarized by Heirtzler *et al.* (1968) and are shown in Figure 5-10. Provisional attempts to assign ages to the anomalies, and hence dates for the underlying oceanic crust and implied geomagnetic reversals, have been made by Vine (1966), and by Heirtzler *et al.* (1968) as indicated in Figure 5-9. There is little difference between the two time scales.

AGE OF THE OCEAN FLOOR

The shaded area in Figure 5-10 indicates a preliminary estimate of the area of oceanic crust created within Cenozoic time, i.e. during the last 65 million years; the lines drawn parallel to the ridge crests are the ten-million-year "growth lines" suggested by Heirtzler *et al.* (1968). (A 65 m. yr. growth line is also included in the Pacific.) Whereas the shaded area in the Pacific and South Atlantic Oceans is thought to have been formed throughout the last 65 million years, in the Indian Ocean and south of Australia it has been formed within the last 40 million years or so. The earlier part of the sequence is missing between Australia and Antarctica, implying that the separation of these two continents was the last stage in the fragmentation of the southern continents which formerly constituted Gondwanaland.

This evidence for spreading and the age of the ocean floors is clearly readily compatible with 'Continental Drift' as proposed by Wegener (1915), and Du Toit (1937). It is interesting and salutary to note that even such details as the post-Eocene separation of Australia and Antarctica, and some form of

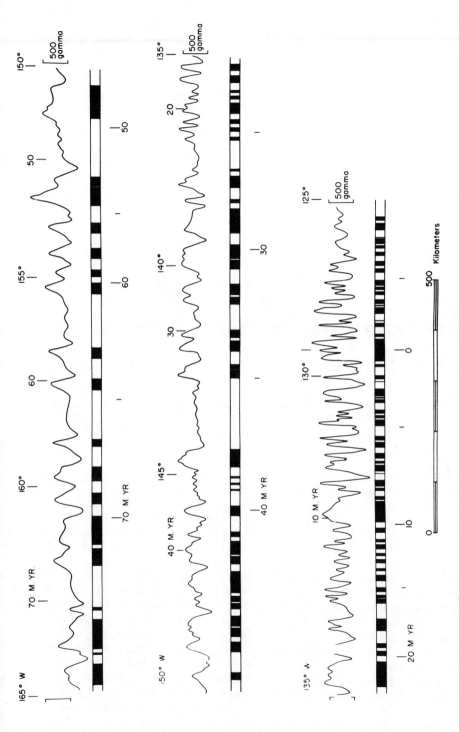

Figure 5-9. A composite magnetic anomaly profile across the Juan de Fuca Ridge, southwest of Vancouver Island, and to the west of the Gorda Ridge to the south, and immediately north of the Mendocino Fracture Zone (Vine, 1966). The time scales indicated are nonlinear. The upper one was suggested by Vine (1966) and the lower one by Heirtzler *et al.* (1968) and, for the last 11 million years, by Vine (1968).

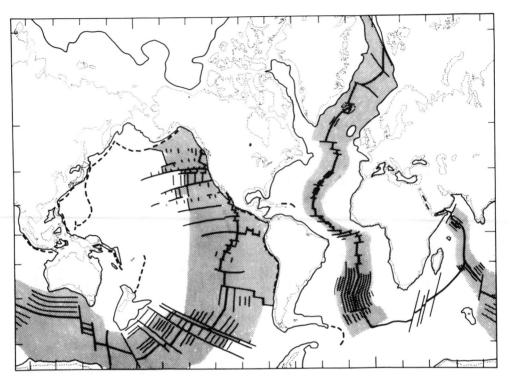

Figure 5-10. Provisional attempt to delineate areas of continental and oceanic crust. Within the ocean basins, trenches are indicated by thick dashed lines, ridge crests by thick solid lines, and fractures (transverse to the ridge crests) and correlatable linear magnetic anomalies (parallel to the ridge crests) by thin solid lines (Heirtzler *et al.,* 1968). Oceanic crust thought to have been formed within the Cenozoic (i.e. the last 65 million years) is shaded.

'sub-continental circulation,' producing new ocean basins and the separation of continental fragments, were proposed by Wegener (1915) and Holmes (1928) respectively more than forty years ago. The implications of these earlier ideas and the new data regarding the age and ephemeral nature of the ocean basins are clearly revolutionary and far-reaching. However, it has been known for some time that the relatively thin blanket of sediments on the ocean floors can be accounted for, in terms of recent sedimentation rates, in 100-200 million years; less than five percent of geologic time. Similarly, marine geologists have never dredged rock or cored sediment more than 150 million years old from the ocean floor. The detailed pattern of sediment distribution and thickness

now emerging from the results of seismic reflection profiling in no way conflicts with the age of the crust inferred from the magnetic anomalies. To some workers, however, it suggests pronounced discontinuities or irregularities in spreading in all areas (Ewing and Ewing, 1967; Le Pichon, 1968). The magnetic anomalies themselves suggest variations in spreading rates (Heirtzler *et al.,* 1968) and discontinuities in spreading in the North Atlantic Ocean and northwest Indian Ocean (Le Pichon and Heirtzler, 1968); but the way in which the complete sequence of anomalies is reproduced in the remaining oceanic areas necessitates that any other lengthy stoppages must be worldwide.

Thick wedges of apparently undeformed sediments in the troughs associated with certain transform faults (Van Andel *et al.,* 1967) and in parts of the trenches (Scholl *et al.,* 1968) have led their observers to question the validity of spreading in these areas. However, it must be acknowledged that neither sedimentation rates nor the precise near-surface expression of faulting and under-thrusting in these settings is known. Other troughs and trenches are essentially devoid of sediment; this is difficult to account for without spreading, because they form such pronounced topographic lows and ideal sediment traps.

The one place where the mid-ocean ridge crest is exposed subaerially is Iceland. Although atypical because of its anomalous elevation above the sea floor, Iceland fulfills many requirements of a spreading ocean floor: its bedrock is entirely igneous (mainly basalt), the age of these rocks probably nowhere exceeds 20 million years (Moorbath *et al.,* 1968), and there is evidence for crustal extension in the form of dike injection and normal faulting (Bodvarsson and Walker, 1964).

SUMMARY

Earthquake activity throughout the world is largely confined to the young fold mountains and trench systems of the Alpine-Himalayan and circum-Pacific belts and to the crests of the mid-ocean ridges. The depth extent of this activity in the mountain and trench systems is very much greater than beneath the mid-ocean ridge crests. The hypothesis of sea-floor spreading proposes that the belts of shallow focus seismicity only, i.e. the mid-ocean ridge crests, are loci along which new oceanic crust is derived from the mantle, and that the trench systems are regions in which oceanic crust is partly resorbed. Focal mechanism solutions for shallow-focus earthquakes in these areas confirm that the ridges are tensional features producing extension of the earth's crust, and that the trenches are compressional features characterized by underthrusting.

The creation of new crust and the resulting spreading from the median ridges of the Atlantic and Indian Oceans appear to have been initiated within a former supercontinent producing the present ocean basins and separation of continental fragments. The Pacific Rise also exhibits spreading, but was presumably largely initiated within former oceanic crust. The sinks in the system—the trenches—occur at the leading edges of continents and are generally marginal to the Pacific.

Recent more accurate determinations of earthquake epicenters on mid-ocean ridge crests have revealed that the activity is concentrated along the transverse fractures or faults which offset the ridge crest. The distribution and nature of the earthquakes on these fractures indicate a transform type of faulting, which is only explicable in terms of spreading of the sea floor. Disturbances in the earth's magnetic field recorded at and above sea level and due to the fossil magnetism of the oceanic crust also indicate that spreading occurs. The pattern of anomalies revealed in this way is strikingly symmetrical about ridge crests, and the anomalies correlate in every detail with reversals of the earth's magnetic field during the last few million years, as defined by paleomagnetic studies of terrestrial lava flows and deep-sea sediments. Such correlations with the reversal time scale, which has been dated by a refined radiogenic technique, enable one to determine recent rates of spreading at ridge crests. Rates deduced in this way vary from 1 to 6 cm per year per ridge flank.

The restricted belts of earthquake activity define essentially aseismic crustal plates bounded by active ridge crests, faults, trenches and mountain systems. If these plates are assumed to be perfectly rigid, then it is possible to calculate rates of shortening in the trench and mountain systems by assuming the rates of extension derived from the magnetic anomalies at the remaining boundaries. The predicted rates of crustal shortening correlate directly with the depth extent of the seismicity in these areas, particularly in the trench systems. Recent seismological studies of trench systems also suggest that a cold rigid plate of oceanic crust and uppermost mantle is being thrust down into the asthenosphere—the part of the mantle thought to be nearest its melting point.

The timetable for reversals of the earth's magnetic field has been accurately documented by independent techniques for only the last few million years. However, oceanic magnetic patterns appear to maintain their symmetry, parallelism and continuity beyond the ridge crests to the flanks and even across the deep ocean basins, implying that spreading has occurred earlier. The same sequence of anomalies away from ridge crests is reproduced in all the ocean basins, strongly supporting the concept that reversals of the earth's magnetic field are recorded in the oceanic crust as a result of sea-floor spreading. Recent spreading rates are consistent with the break-up of the continents and the formation of the present ocean basins within the last 200 million years, less than five per cent of geologic time. Despite this gross extrapolation, ages assigned in this way to the anomalies, and hence to the underlying oceanic crust, are entirely consistent with all that is known of the age of the ocean floors from sediment thicknesses and the age of sediment cores and dredged rocks.

Thus, sea-floor spreading is capable of explaining many of the first-order structural features of the earth's surface and is compatible with the results of geological and geophysical investigations in the ocean basins. It also provides, perhaps for the first time, a plausible mechanism for continental drift.

ACKNOWLEDGMENTS

I thank H. H. Hess, D. P. McKenzie and W. J. Morgan for valuable discussions, and Susan Vine for preparing the diagrams. This work was supported in part by the U. S. Office of Naval Research—contract no. NOO 014-67A-0151-0005AA.

REFERENCES

Bodvarsson, G., and Walker, G. P. L., 1964, Crustal drift in Iceland. *Geophys. Jour.* **8**, 285-300.

Cox, A., Dalrymple, G. B., and Doell, R. R., 1967, Reversals of the earth's magnetic field. *Sci. Amer.* **216** (2), 44-54.

Cox, A., Doell, R. R., and Dalrymple, G. B., 1963, Geomagnetic polarity epochs and Pleistocene geochronometry. *Nature* **198**, 1049-1051.

Dickson, G. O., Pitman, W. C., and Heirtzler, J. R., 1968, Magnetic anomalies in the South Atlantic and ocean floor spreading. *Jour. Geophys. Res.* **73**, 2087-2100.

Du Toit, A. L., 1937, *Our Wandering Continents.* Oliver and Boyd, Edinburg, 366 p.

Ewing, J., and Ewing, M., 1967, Sediment distribution on the mid-ocean ridges with respect to spreading of the sea floor. *Science* **156**, 1590-1592.

Gutenberg, B., and Richter, C. F., 1954, *Seismicity of the Earth:* Princeton University Press, Princeton, 310 p.

Heirtzler, J. R., Dickson, G. O., Herron, E. M., Pitman, W. C., and Le Pichon, X., 1968, Marine magnetic anomalies, geomagnetic field reversals and motions of the ocean floor and continents. *Jour. Geophys. Res.* **73**, 2119-2136.

Heirtzler, J. R., Le Pichon, X., and Baron, J. G., 1966, Magnetic anomalies over the Reykjanes Ridge. *Deep-Sea Res.* **13**, 427-443.

Hess, H. H., 1962, History of ocean basins: *Petrologic Studies: A Volume to Honor A. F. Buddington,* Geol. Soc. Amer., p. 599-620.

Holmes, A., 1928, Radioactivity and earth movements. *Trans. Geol. Soc. Glasgow* **18**, 559-606. (1965, p. 1001 in: *Principles of Physical Geology.* Nelson/Ronald Press, London/New York, 1288 p.).

Isacks, B., Oliver, J., and Sykes, L. R., 1968, Seismology and the new global tectonics. *Jour. Geophys. Res.* **73**, 5855-5899.

Le Pichon, X., 1968, Sea-floor spreading and continental drift. *Jour. Geophys. Res.* **73**, 3661-3697.

Le Pichon, X., and Heirtzler, J. R., 1968,

Magnetic anomalies in the Indian Ocean and sea-floor spreading. *Jour. Geophys. Res.* **73**, 2101-2117.

Menard, H. W., and Atwater, T., 1968, Changes in direction of sea-floor spreading. *Nature* **219**, 463-467.

Moorbath, S., Sigurdsson, H., and Goodwin, R., 1968, K-Ar ages of the oldest exposed rocks in Iceland. *Earth Planet. Sci. Letters* **4**, 197-205.

Morgan, W. J., 1968, Rises, trenches, great faults and crustal blocks. *Jour. Geophys. Res.* **73**, 1959-1982.

Oliver, J., and Isacks, B., 1967, Deep earthquake zones, anomalous structures in the upper mantle, and the lithosphere. *Jour. Geophys. Res.* **72**, 4259-5275.

Opdyke, N. D., Glass, B., Hays, J. D., and Foster, J., 1966, Paleomagnetic study of Antarctic deep-sea cores: *Science* **154**, 349-357.

Pitman, W. C., and Heirtzler, J. R., 1966, Magnetic anomalies over the Pacific-Antarctic Ridge. *Science* **154**, 1164-1171.

Pitman, W. C., Herron, E. M., and Heirtzler, J. R., 1968, Magnetic anomalies in the Pacific and sea-floor spreading. *Jour. Geophys. Res.* **73**, 2069-2085.

Scholl, D. W., Von Huene, R., and Ridlon, J. B., 1968, Spreading of the ocean floor: undeformed sediments in the Peru-Chile Trench. *Science* **159**, 869-871.

Stauder, W., 1968, Tensional Character of

Earthquake Foci beneath the Aleutian Trench with Relation to Sea-Floor Spreading. *Jour. Geophys. Res.* **73**, 7693-7701.

Sykes, L. R., 1967, Mechanism of earthquakes and nature of faulting on the mid-oceanic ridges. *Jour. Geophys. Res.* **72**, 2131-2153.

Van Andel, Tj. H., Corliss, J. B., and Bowen, V. T., 1967, The intersection between the Mid-Atlantic Ridge and the Vema Fracture Zone in the North Atlantic. *Jour. Marine Res.* **25**, 343-351.

Vine, F. J., 1966, Spreading of the ocean floor: new evidence. *Science* **154**, 1405-1415.

——— 1968, Magnetic anomalies associated with mid-ocean ridges: *History of the Earth's Crust,* Phinney, R. A. (ed.), Princeton University Press, Princeton, p. 73-89.

Vine, F. J., and Matthews, D. H., 1963, Magnetic anomalies over oceanic ridges. *Nature* **199**, 947-949.

Wegener, A., 1915, *Die Entstehung der Kontinente und Ozeane.* Vieweg und Sohn, Braunschweig. (1966, English translation of the fourth revised edition (1929), Dover, New York, 246 pp.)

Wilson, J. T., 1965, A new class of faults and their bearing upon continental drift. *Nature* **207**, 343-347.

Sea-Floor Spreading and Continental Drift

F. J. VINE   1970

Rather than attempt a further review of the vast subject matter now covered by the title of this paper, I shall endeavor to summarize developments and ideas not covered in my earlier short review in these pages (Vine, 1969).

As outlined in that article, the most specific and compelling evidence for sea-floor spreading in the past comes from the study of oceanic magnetic anomalies. The linear anomalies associated with ridge crests have been interpreted in terms of spreading of the sea floor accompanied by intermittent reversals of the earth's magnetic field. Since a reversal time-scale for the past few million years has been deduced by other independent techniques (Cox, Dayrymple, and Doell, 1967), spreading rates may be deduced at ridge crests and extrapolated across the flanks in an attempt to assign ages to the magnetic anomalies, underlying oceanic crust and reversals of the earth's magnetic field (Heirtzler, 1968). Clearly such an extrapolation is rather speculative and its validity was seriously questioned. However within the past year the results of the first phase of the JOIDES deep-sea drilling program have tended to confirm this magnetic time-scale.

On JOIDES leg III in the South Atlantic, eight sites were drilled across the Mid-Atlantic Ridge at approximately 30°S. These revealed a remarkably linear relationship between the age of the oldest sediment recovered, invariably immediately overlying or incorporated within basalt, and the distance of the site from the ridge axis (see Figure 5-11). This suggests continuous spreading throughout the Cenozoic at an essentially constant rate of 2 cm per year per ridge flank, precisely the rate assumed by Heirtzler et al. (1968) in calibrating the magnetic time-scale.

These points, together with similar results from leg V, on the flanks of the East Pacific Rise, and one fission track age from basalt

Dr. Frederick J. Vine is a reader with the School of Environmental Sciences at the University of East Anglia, Norwich, Great Britain. His well-known hypothesis on magnetic reversals has provided marine geologists and geophysicists with a key for determining the rate of continental movement and the relative ages of different parts of the ocean floor. He has written over twenty publications and in 1970 received the Henry Bryant Bigelow Medal in Oceanography from the Woods Hole Oceanographic Institution, Massachusetts.

From *Journal of Geological Education*, Vol. 18, No. 2, pp. 87-90, 1970. Reprinted by permission of the author and the *Journal of Geological Education*.

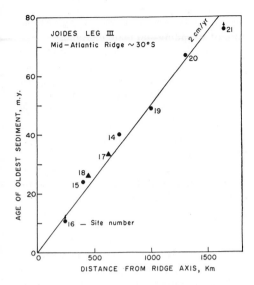

Figure 5-11. Age of oldest sediment substrate, "at drill site" plotted against distance of site from Mid-Atlantic Ridge axis, after Maxwell and Von Herzen (1969).

that this amounts to nearly 50 per cent of the area of the deep ocean basins. Alternatively one is maintaining that one-third of the present earth's crust has been created during the most recent one and one-half per cent of geologic time. It now seems very probable that all remaining oceanic areas are Mesozoic in age.

Recently, on the basis of paleomagnetic studies on land, it has been suggested that the Cretaceous is characterized by very few reversals and that throughout most of that period the earth's magnetic field was of normal polarity (Helsley and Steiner, 1969) (see Figure 5-13). This is in great contrast to the Tertiary, which is of essentially the same duration (i.e. 65-70 million years). Clearly if this is correct the magnetic signature of oceanic crust formed by spreading during the Cretaceous period should be very different to that formed during Cenozoic time. If we look beyond the oldest correlateable anomalies in the Pacific (Figure 5-14) or South Atlantic, where they are best documented, we do indeed find what have been termed magnetically "quiet zones" which appear to be devoid of the linear anomalies characteristic of Cenozoic crust. Actually

dredged beneath anomaly 10 in the Pacific (Luyendyk and Fisher, 1969), may be plotted on a graph of the age of the oldest sediment or hard rock against the age assigned on the basis of the magnetic anomalies (Figure 5-12). Clearly no points should lie significantly above the 45° line on such a plot if the magnetic time-scale is to be valid, but some points may be beneath it because the oldest material present has not been sampled. Thus although minor modification of the magnetic time-scale may ultimately be warranted, it now seems probable that it is essentially correct.

This geomagnetic time-scale is the best indication we have as yet of the timing and frequency of reversals during the Cenozoic. Ultimately, as a result of the JOIDES program, we should obtain an independent check on this time-scale from the magnetic stratigraphy of deep-sea sediments. Thus my earlier estimate of the extent of the present sea floor which was generated by spreading during Cenozoic time stands (Vine 1969, Figure 10), and it is rather sobering to note

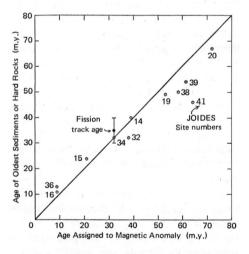

Figure 5-12. Fission track age from Luyendyk and Fisher (1969). JOIDES results from preliminary cruise reports. Magnetic time-scale assumed given by Hiertzler et al. (1968).

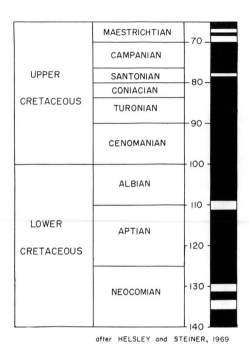

	MAESTRICHTIAN	70
	CAMPANIAN	
UPPER	SANTONIAN	80
	CONIACIAN	
CRETACEOUS	TURONIAN	
		90
	CENOMANIAN	
		100
	ALBIAN	
		110
LOWER	APTIAN	
		120
CRETACEOUS		
	NEOCOMIAN	130
		140

after HELSLEY and STEINER, 1969

Figure 5-13. Suggested geomagnetic reversal time-scale for the Cretaceous period. Shaded intervals—normal polarity.

this transition is not exactly at the Mesozoic-Cenozoic boundary, and both the continental and oceanic magnetics suggest that there may be several reversals in the uppermost Cretaceous. Thus a consistent picture is emerging and such magnetically quiet areas may well be indicative of Cretaceous ocean floor. In the North Atlantic and Western Pacific even older oceanic crust is present and may eventually enable us to extend the geomagnetic reversal time-scale back to the uppermost Triassic or at least the Lower Jurassic.

In the Atlantic and Indian Oceans the magnetic "growth lines" parallel the trailing edges of the separating continents. In the North Pacific, however, the guidelines of separating continents are absent and the geometry of spreading revealed by the magnetic anomalies is less readily understood. Although complex, I do not see that it is

necessarily incompatible with the global picture of spreading and plate tectonics.

Much of the Western Pacific is blanketed by the so-called "opaque layer" of reflection seismology. The upper surface of this layer is thought to approximate to the Mesozoic-Cenozoic boundary and its north-easterly extent is indicated by the pecked line on Figure 5-14 (Ewing *et al.,* 1968). Thus all crust to the southwest of this line must be at least Mesozoic in age, and again this is born out by the JOIDES results, although as yet the oldest material recovered is only Tithonian (uppermost Jurassic) in age. The Emperor Seamount Chain probably delineates a former transform fault, somewhat analogous to the Owen fracture zone in the northwest Indian Ocean at the present day. It must have accommodated a younger phase of spreading to the east. The Great Magnetic Bight at approximately 50°N 160°W was presumably formed at a triple junction of ridge crests, and again a present day analogy probably exists to the west of the Galapagos Islands (McKenzie and Morgan, 1969). It seems inconceivable that the Magnetic Bight could be formed in any other way, and it can be shown that the detailed geometry of the Bight is readily, and perhaps only, compatible with spreading from a triple junction (Vine and Hess, 1971). Finally one must remember that just as 50 per cent of the oceanic crust has been created at ridge crests during Cenozoic time, i.e. since the formation of this Bight, a similar area of oceanic crust has been destroyed in the trench systems, notably in the North and West Pacific. Paleomagnetic studies for the North Pacific (Vine, 1968) taken together with those for North America imply that as much as 5,000 km of oceanic crust may have been destroyed between the Alaskan peninsula and the Magnetic Bight during the Cenozoic. Clearly if resorption has gone on on this scale a great deal of the spreading geometry of the North Pacific has been lost, and it is perhaps no surprise that at present the Aleutian trench is consuming magnetic anomalies in the reverse order to what one might ex-

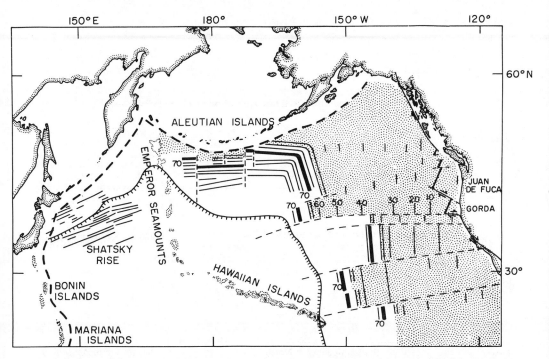

Figure 5-14. Summary of linear magnetic anomalies observed in the North Pacific (after Vine and Hess, in press). The extent of the "opaque layer" of reflection seismology, as given by Ewing *et al.* (1968), is indicated by the pecked line. The numbers refer to the age of the anomalies as suggested by Heirtzler *et al.* (1968). Crust formed by spreading during the Cenozoic (i.e., the past 65 million years) is shaded.

pect from the simplest formulation of spreading, i.e. younger anomalies first; although in some respects this is no different to the situation off California where presumably a trench system has overridden a ridge crest, and this explanation is fairly generally accepted.

To return to the gross picture of spreading and drift in the Atlantic and Indian Oceans, it is important to note that, as yet, we can trace the magnetic anomalies up to the continental margins, and hence date the initiation of drift, in only a few areas: in the extreme North Atlantic, south of Australia, and south of New Zealand. Thus in assigning an age to the initiation of drift in other areas one must turn to other criteria, particularly from the geologic record on the trailing margins of the continents. In drawing up the

suggested time-table for the opening of the Atlantic and the fragmentation of Gondwanaland shown in Table 5-1, the extrusion and intrusion of tholeiitic basalts and quartz diabases, marine transgressions and the formation of evaporite basins have been assumed to be precursors of the initiation of drift in any particular area (Vine and Hess, 1971).

I think that the significance of sea-floor spreading, continental drift, and plate tectonics in earth science education goes without saying. These concepts are fundamental to virtually all aspects of our science. Also, since they should lead to a better understanding of the cause and distribution of earthquakes and igneous activity, of mountain building and flow in the mantle, of continental margins and likely settings for

Table 5-1. A Time-Table for Continental Drift.

	Opening of the Atlantic	Fragmentation of Gondwana
— 0 —		
Tertiary	(Opening of the Red Sea, Gulf of Aden, Gulf of California, and initiation of Galapagos Rise)	
	Opening of the extreme North Atlantic and the Arctic Ocean.	Separation of Australia from Antarctica and India from the Seychelle Bank.
Upper Cretaceous	**North Atlantic extends Northwards to from the Labrador Sea and Bay of Biscay.**	**Initiation of Pacific-Antarctic Ridge between New Zealand and West Antarctica.**
— 100 m.y. —		
Lower Cretaceous	Opening of the South Atlantic	
Lower Jurassic	Partial opening of North Atlantic between North America and Africa.	Separation of Australia + Antarctica from Africa + South America.
— 200 m.y. —		

the concentration of oil and mineral resources, I think one might also claim that they have some "relevance."

REFERENCES

Cox, A., Dalrymple, G. B., and Doell, R. R., 1967, Reversals of the earth's magnetic field. *Sci. Amer.* **216**, 44-54.

Ewing, J., Ewing, M., Aitken, T., and Ludwig, W. J., 1968, North Pacific sediment layers measured by seismic profiling *in* The Crust and upper mantle of the Pacific area. *Geophys. Monogr.* **12**, *Amer. Geophys. Un.*, p. 147-173.

Heirtzler, J. R., 1968, Sea-floor spreading. *Sci. Amer.* **219**, 60-70.

Heirtzler, J. R., Dickson, G. O., Herron, E. M., Pitman, W. C., and Le Pichon, X., 1968, Marine magnetic anomalies, geomagnetic field reversals and motions of the ocean floor and continents. *Jour. Geophys. Res.* **73**, 2119-2136.

Helsley, C. E., and Steiner, M. B., 1969, Evidence for long intervals of normal polarity during the Cretaceous period. *Earth Planet. Sci. Letters* **5**, 325-332.

Luyendyk, B. P. and Fisher, D. E., 1969, Fission track age of magnetic anomaly 10: a new point on the sea-floor spreading curve. *Science* **164**, 1516-1517.

Maxwell, A. E., and Von Herzen, R., 1969, The Glomar Challenger completes Atlantic track—Highlights of Leg III. *Ocean Industry* **4**, 64-65.

McKenzie, D. P., and Morgan, W. J., 1969, Evolution of triple junctions. *Nature* **224**, 125-133.

Vine, F. J., 1968, Paleomagnetic evidence for the northward movement of the North Pacific basin during the past 100 m.y. [abstract]. *Trans. Amer. Geophys. Un.* **49**, 156.

Vine, F. J., 1969, Sea-floor spreading—new evidence. *Jour. Geol. Education* **17**, 6-16.

Vine, F. J., and Hess, H. H., 1971, Seafloor spreading, in The Sea, Vol. IV, pt. 2 Maxwell, A. E. (ed.), Wiley-Interscience, New York, pp. 587-622.

Reconstruction of Pangaea: Breakup and Dispersion of Continents, Permian to Present

ROBERT S. DIETZ   JOHN C. HOLDEN   1970

INTRODUCTION

Among advocates of continental drift, a definitive reconstruction of the continents into the original universal continent of Pangaea, or, alternately, into the two super-continents of Gondwana and Laurasia, has been a constant challenge but, as yet, an unsolved jigsaw puzzle. Most attempts made so far have been generalized sketch maps with distorted continental shapes, as was the original map of Pangaea by Wegener (1929). On the other hand, Carey (1958) and King (1962) produced rather precise reconstructions by using various geologic criteria, but without the benefit of the new global tectonics based on sea-floor spreading. Other reconstructions have also been realized with paleomagnetic data describing ancient pole positions (e.g., Van Hilten, 1964). Bullard *et al.* (1965) have effected a continental drift closing of the Atlantic Ocean, and Smith and

Hallam (1970) recently presented a relative closing of the Indian Ocean.

In this paper we not only attempt a reconstruction of Pangaea as of the Permian with considerable cartographic precision and in absolute geographic coordinates, but also present four charts showing the breakup and dispersion of continents at the end of subsequent geologic periods. The principal control utilized has been the morphologic outlines of the continents plus the concept of plate tectonics and sea-floor spreading. An additional constraint was provided by paleomagnetic pole positions, although we have utilized this as only a secondary constraint. To achieve our solutions, we have worked backwards in time from the modern crustal plate positions closing presumed zones of new ocean floor created by sea-floor spreading. Our principal technique consisted of using cutouts of continents and crustal plates superimposed on globes, the results being

Dr. Robert S. Dietz is a marine geologist with the Marine Geology and Geophysics Laboratory of the National Oceanic and Atmospheric Administration, Miami, Florida. He specializes in the geologic history and evolution of the ocean basins and is the co-originator of the sea-floor spreading concept. He has written and helped write nearly two-hundred publications.

John C. Holden is working on his doctoral dissertation in micropaleontology at the University of California, Berkeley.

From *Journal of Geophysical Research*, Vol. 75, No. 26, pp. 4939-4956, 1970. Reprinted with light editing and by permission of the authors and the American Geophysical Union.

plotted on Aitoff projections. Most of the earth's surface, but not all, is treated. The Pacific basin is excluded, since we have concentrated on those plates that have continents. We believe that if the concept of plate tectonics is right and our "rules" generally valid, the pattern of continent dispersion shown here must be roughly correct.

PLATE TECTONICS AND SEA-FLOOR SPREADING

As plate tectonics provides the primary rationale for our reconstruction, a brief summary of this hypothesis follows. The concept is an outgrowth of sea-floor spreading, the validity of which is implicitly assumed in this paper (Hess, 1962; Dietz, 1961). Plate tectonics includes also the concept of transform faulting as developed by Wilson (1965a). Evidence supporting this over-all concept has been assembled, as, for example, by Sykes (1967), Morgan (1968), Pitman and Hayes (1968), and Isacks *et al.* (1968). The suggestion of Vine and Matthews (1963) that the injection of basaltic magma combined with polarity changes in the earth's magnetic field, accounting for the bilaterally symmetrical anomaly patterns linking mid-ocean ridges, is now well established. These anomalies provide a pattern, somewhat analogous to tree rings, by which the growth of ocean basins can, in principle, be measured.

The concept of plate tectonics supposes that the earth's outer surface has a strong lithosphere about 100 km thick (probably thinner at the rifts and thickening toward the trenches), which is divided into a number of rigid plates that are margined by trenches, rifts, and great fractures or megashears. Plate tectonics may be envisioned as an extension of the sea-floor spreading concept in that it accepts the thesis of sea-floor spreading while adding that these "conveyor belts" may themselves also be moving. It assumes that the crustal plates are largely interlocked on the earth's carapace so that most motions are accommodated globally.

This presumably arises because there are few "ideal" plates, that is, plates that are rectilinear in shape and consist of an opposing rift and trench connected by opposing megashears. The Indian plate, however, may be an example. In general, the crustal plates appear to be interlocked into two plate systems, one of which is the Indian Ocean-Tethyan system and the other being the Atlantic-Pacific system, the first taking up northward motion and the second accommodating westward motion. Perhaps the lack of ideal plates is not too surprising, for it is impossible to subdivide the surface of a globe into spherical rectangles.

The concept of plate tectonics is in a state of flux with few agreed-upon definitions or assumptions. For our purposes here we define the term "trench" to mean a zone of crustal resorption, without necessarily implying a linear geomorphic depression. We also use the term "rift" in place of the more usual terms mid-ocean ridge (often a misnomer as to position) or simply ridge, since the extensional regime implied by the term rift accents its most fundamental geotectonic aspect. "Megashears" are fractures between plates where crust is neither added, as with rifts, nor destroyed, as with trenches; they are transform strike-slip faults on a planetary scale. Continental drift is a necessary consequence of plate tectonics in that the continents would be passively rafted on the backs of the conveyor-belt-like crustal plates. The drift of the continents may be conveniently thought of as a summation of sea-floor spreading plus any over-all motion of the entire conveyor belt itself.

Our assumptions or "rules" of plate tectonics are as follows:

1. Rifts ("mid-ocean ridges") can migrate (examples: the mid-Atlantic rift and the circum-Antarctica rift).

2. Trenches probably tend to remain fixed in position, although the leading edge of a drifting continent can displace them (example: the Andean trench). Also trenches appear to migrate slowly in the direction of

their convexity. The Japan Sea, for example, seems to be neo-oceanic, having opened by the migration of the Japanese trench away from Asia. Some other small ocean basins suggest a similar history.

3. Crustal plates can change in size and shape and can even be entirely destroyed by resorption, that is, the migration of a rift into a trench [example: the northeast Pacific plate has been almost entirely resorbed (Pitman and Hayes, 1968)].

4. Rifts, trenches, and megashears do not always appear in their pure form. Trenches may exhibit considerable strike-slip motion, as is postulated here for the European Tethys trench. A present-day example is the Tonga-Kermedee trench, which acts simultaneously as a megashear for the Australian plate and as a trench for the southern portion of the Pacific plate. Similarly, rifts may show strike-slip. For example, the southwest Indian Ocean rift acts as a megashear for the sinistral rotation of the Africa plate and also as a rift along which new sea floor has been emplaced by sea-floor spreading.

5. The upper mantle and the oceanic crust (sima) are capable of being largely resorbed into the mantle in trenches, but the continents (sial) are not, owing to their buoyancy and hyperfusible petrologic make-up.

6. The earth is essentially in steady-state both as to density and to surface area; hence, the area of new crust generated along a rift is offset by an equal amount of crustal uptake or resorption in a trench.

7. New crust is always sima, so that, for reasons of isostasy, new ocean basins or at least new ocean floor about 5 km below sea level is generated by sea-floor spreading.

8. The crustal plates are active, probably moving in response to body forces within the plates or in response to the vector sum of a multitude of small convection cells within the asthenosphere or rheosphere of the upper mantle. The injection of new crust along rifts, as zones of extension, appears to be a passive response such that the rifts readily migrate. In places, rift zones appear

to intersect trenches (off western Mexico) or have descended into trenches (in the Aleutian trench as indicated by the reverse order of magnetic anomalies (Pitman and Hayes, 1968)).

In summation, Figure 5-15 illustrates in time sequence some important aspects of plate tectonics following the rules set forth here:

1. Magma rises from the mantle, filling a rift zone, and lava-feeder dikes deposit a plateau basalt.

2. The resulting extension creates a new ocean basin. It also imposed a ridge with a central rift whose mid-ocean position arises, owing to the bilateral symmetrical generation of new ocean crust by sea-floor spreading. Plate b remains fixed for purposes of illustration. Both the mid-ocean rift and the continent on plate a move westward toward a trench, a crustal plate resorption zone.

3. As spreading continues, the drifting continent encounters the trench but is not resorbed into the mantle because of its buoyancy. With further drifting, the continent displaces the trench westward and probably reverses the sense of the Benioff zone within the trench. At this time plate a ceases to become a shrinking plate and becomes a growth plate at the expense of plate c, which now is reduced in size and may even disappear in the trench. The end result is the creation of three plates (a, b, and c) from a single plate. The plates are shown as ideal plates bounded by megashears lying in two planes parallel to the page.

Le Pichon (1968) regarded six rigid plates as sufficient to explain global geotectonics. We have found it necessary to subdivide two of these plates. We consider the India plate as distinct from the Australia plate with their common boundary being marked by Ninety-East ridge as a megashear. We also subdivide the Americas plate into a South America plate and a North America plate, the common boundary being a transverse

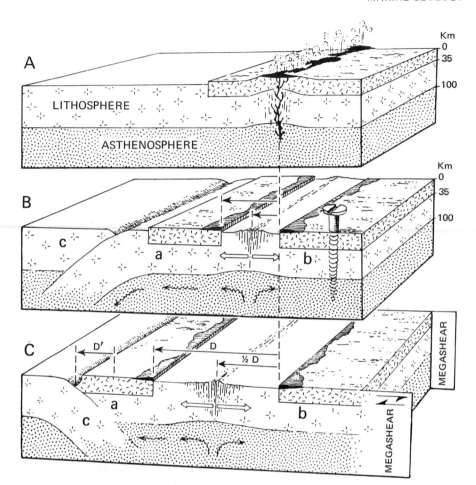

Figure 5-15. Aspects of plate tectonics showing the breakup of a continent by the formation of a rift, the creation of a new ocean basin, the migration of a mid-ocean rift, and continental drift. (A) a new rift forms beneath a continent in a single crustal plate. Plateau basalts are spilled out above the rift. (B) Three plates and a new ocean basin are created. Plate *b* is shown fixed for purposes of illustration. (C) The continent on plate *a* encounters and displaces the trench along its leading edge the distance D'. Because the continent on plate *b* is fixed, the mid-ocean rift has moved half the distance of any original segment of plate *a*.

zone (Funnel and Smith, 1968) or gore. The point of this is located at the bottom of the bulge of Africa, whereas the open end is subtended by the Lesser Antilles arc and then transforms into the Cayman megashear. The Pacific plate is probably subdivided into two plates, with the second plate including all the subplates east of the East Pacific rift together as one partially resorbed plate.

Thus, in total the earth would consist of nine major plates by our reconstruction.

The reconstruction is semiclosed in terms of plate tectonics, since the motion of the Pacific plate, or plates, is not considered. Because this region lacks a continent, there is little data with which to reconstruct the plate motion. Our reconstruction is, of course, but one of many possible ones, but we suppose that any other reconstruction would not be greatly different from the one proposed if our "rules" are followed and if the concept of plate tectonics is correct. The possibilities as to plate motions are necessarily constrained by their interaction. Sea floor can only be generated along a rift, and a plate with its superimposed continents must move away from a rift toward a trench. The continents must drift along free paths or be stopped by colliding with another craton.

In all our reconstructions, no continent or crustal plate has remained entirely fixed in position, which is a departure from previous solutions. In the construction of Pangaea by Wilson (1963), for example, Eurasia was kept stationary and the other continents were configured around its boundaries. By plate tectonics such a solution is at least intuitively unsatisfactory. Wilson's opinion is, however, reasonably satisfactory because Eurasia has moved the least (except for Antarctica). On the other hand, India has moved the most, and North America, South America, and Australia also have drifted for long distances across the globe.

PERMIAN RECONSTRUCTION OF PANGAEA

Figure 5-16 shows the continental reconstruction of Pangaea at the end of the Paleozoic (i.e., the Permian, -225 m.y.). Pangaea is shown on the Aitoff projection of the globe with the central meridian being 20°E. This same central meridian is used throughout this paper. We have attempted to depict the margins of continents accurately. The unusual outlines result from distortions inherent in the projection and from using the 1000-fm isobath instead of the shoreline. The tip of South America, for example, does

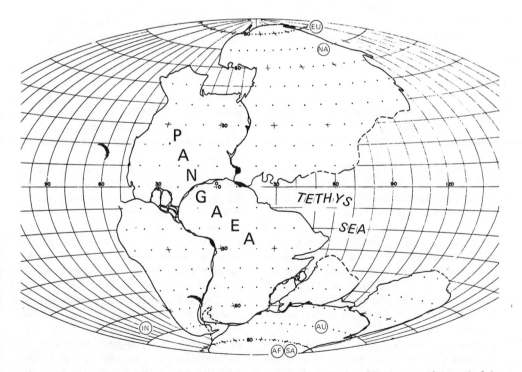

Figure 5-16. Reconstruction of the continents into the universal landmass of Pangaea as of the end of the Permian, 225 m.y. ago. See text for full explanation.

not taper to a point but has a broad base when this isobath is used. Certain boundaries that apparently have undergone great post-drift alteration, such as the former margin of Asia, are shown in dashed line. New Guinea, New Zealand, and southeast Asia have been omitted for cartographic convenience. Certain peripheral marginal plateaus have been deleted because they are considered to be post-breakup excrescences on the margins of the continents. This interpretation, for example, pertains to the Bahama platform (Dietz et al., 1970). The Gulf of Aden presumably did not open until the Cenozoic (Girdler, 1969), but we show it open as a visual geographic guideline. We recognize that limiting North America by the limits of Alaska is arbitrary; in fact, the geography of the polar basin in the Permian is quite uncertain. To provide well-known geographic reference points with respect to which the drift of the continents can be visualized, the Antilles and Scotia trenches are shown as hachured crescents.

The opaque areas on Figure 5-16 represent major overlaps of the fit while the underlaps are clear. The horn of Antarctica is shown in dashed line, since this is regarded as being nonexistent in the Permian (Hamilton, 1967). It appears to be a young foldbelt that later was accreted to Antarctica. Permian India is shown to be considerably larger than modern India, since it includes the Tibet plateau. We suppose this plateau is a double sialic plate in which the Indian craton has underridden the Asiatic craton, at least along its southern margin.

The relative position of the continents is obtained in the following manner. The North and South Atlantic oceans have been closed approximately according to the reconstructions of Bullard et al. (1965). In their fit, the position of Greenland is slipped considerably southward, with respect to the margin of Canada, and Spain is rotated so as to close the Bay of Biscay. We have chosen to close the northern portion of Greenland tighter against Canada, and we use the 1000-fm isobath rather than the 500-fm iso-

bath for delimiting all continents. Also, we close the American mediterranean by an arbitrary rotation of Yucatan and Honduras cratons into the Gulf of Mexico and by fitting the Greater Antilles into the Y junction between North America, South America, and Africa. The unfamiliar shape of the Gulf of Mexico and Florida is due to our choosing to close the cratonic elements against the Lower Cretaceous reef line (Meyerhoff, 1967) rather than the modern 1000-fm isobath.

The position of Antarctica within Pangaea is fixed morphologically with respect to Africa by nesting the shallow S-shape form of southeastern Africa into a similar configuration of East Antarctica from the Weddell Sea to Princess Astrid Coast on Queen Maud Land. Bathymetric data on the exact position of the 1000-fm line on this portion of Antarctica remains sparse, but the fit seems good based on an evaluation of new bathymetric data obtained from the Naval Oceanographic Office (Dietz and Sproll, 1970a). The position of Australia and Antarctica is based on computerized best fit study by Sproll and Dietz (1969), which provides a rather convincing unique position. This hooks the Tasmania toe end of Australia around Victoria Land and into the Ross Sea while the Great Bight of Australia is fitted along the eastern margin of East Antarctica.

This reconstruction creates a gap between India and Australia of unknown shape, since the northern limit of India is indeterminate today, having been thrust under the craton of Asia. It seems quite possible, however, that this gap was filled by the possible sialic microcontinents, Broken ridge and the Kerguelen plateau. The lavas of Kerguelen Island apparently contain no sialic xenoliths and give no evidence of sialic contamination; however, this island is but a small point on a large submerged plateau. Its blocky outline and abrupt margins suggest that it may be a microcontinent. A quasi-continental nature may also be inferred for Broken ridge from seismic refraction velocities (Francis and

Raitt, 1967). Similarly, the gap between India and Africa north of Madagascar was probably filled by the Seychelle microcontinent. Madagascar is fitted into the Tanzania rather than the Mozambique position of the African margin (as favored by some drifters) mainly because the latter is preoccupied by the location of Antarctica. The Tanzania position also appears preferable morphologically and for matching terrestrial geology.

It remains to explain the absolute positioning of Pangaea. For this we rely on paleomagnetic pole positions mainly as summarized by Irving (1964); these positions provide information on both paleolatitude and orientation. In the Permian reconstruction, however, we considered this as "soft" data, a secondary constraint after considerations of morphologic fit and plate tectonics have been satisfied. It was often not possible to satisfy both paleo-orientation and paleolatitude, but in an approximate manner the reconstruction does fit paleomagnetic data.

Of course, paleomagnetism provides no control in longitude. For this we have two constraints. First is a least-work consideration, inferring minimum drift of continents. Second, as will be explained later, we believe that the Walvis thermal center of "hot spot" provides an absolute fixed reference point at the end of the Jurassic at 135 m.y. ago. From this spot we have dead reckoned the position of West Gondwana back in time to the Permian by assuming only northward component to drift prior to the late Jurassic.

The Permian pole positions, shown as circles with letters, are not plotted in the conventional manner. The fossil pole positions shown are also reconstructions, their coordinates having been translated in the same sense as their respective continents. Ideally, in our Permian reconstruction, all these paleomagnetic poles should center at the present earth's rotational pole. They obviously do not; however, they are close, all falling within the Arctic and Antarctic circles. Jointly, if not individually, they support our reconstruction of Pangaea in a generalized manner.

We were unable to fix the Permian position of India morphologically, as we have done with the other large cratons of Pangaea. We place the east coast of India against the margin of central East Antarctica. This position follows largely by interpreting the topography of the Indian Ocean as megashears from the physiographic diagram of Heezen and Tharp (1964). Guided by seafloor structures, especially the fracture zones, we have inferred a migrating conveyor belt that lifted the east coast of India off of Antarctica and eventually drifted India to its modern position. This Permian reconstruction is also permissible from India's Permian paleomagnetic pole position. A more common solution is to fit India against western Australia, but this would infer the existence of a spreading rift zone between these two continents, a relationship not supported by known structures in the Indian Ocean. Our position is that suggested by Du Toit (1937) and King (1962).

Opinion among advocates of drift differs as to whether there was formerly one universal continent of Pangaea or two separate continents of Laurasia (North America plus Eurasia) and Gondwana (all the southern continents plus the subcontinent of India). We have accepted the view here that Pangaea existed at the end of the Paleozoic. Following Wilson (1966), we tentatively suppose that Laurasia and Gondwana were originally separate geologic entities, but that they collided in the late Paleozoic by moving toward a common trench in an ancient North Atlantic Ocean to form Pangaea and then subsequently parted once again along approximately the same geosuture with the formation of a new rift in the Triassic.

The choice of Pangaea versus Laurasia-Gondwana hinges in an important part on the goodness of fit between the bulge of Africa and North America. The congruency along their respective modern 100-fm isobaths is not very good. However, this fit becomes excellent once the misfit imposed by the Blake-Bahama plateau is removed by considering it as a post-drift growth (Dietz *et al.*, 1970). In fact, all three major underlaps and overlaps (the Bahama platform, the Ifni gap, and the Cape Hatteras-Cap Blanc overlap) probably are subject to reasonable geologic explanation (Dietz and Sproll, 1970b). The hypothesis that Gondwana and Laurasia were originally separate entities is supported by their remarkable equivalence of area, suggesting that they may have been born by differentiation from opposite hemispheres (Dietz and Sproll, 1968).

TRIASSIC

The stippled zones in Figure 5-17 show area of new sima or oceanic crust generated by sea-floor spreading throughout the Triassic. The open arrows are vectors revealing both the sense of direction and the distance of continental drift during the Triassic. Thus, the point at the tip of the arrow moved from

a point at the base of the arrow during the Triassic. This convention is followed on all the maps.

It suffices for convenience to speak of the drift of continents and of crustal plates in terms of a single compass direction, although any translation of a plate on a sphere is necessarily a rotation. By Euler's theorem, the translation of a plate on a globe can be described as a rotation about a fixed point or pole, but this is a geometric construction only. If the plate moves along a great circle (a "straight line"), the pole is $90°$ away; for motions along small circles of earth the pole is less than $90°$ away. The tighter the arc, the closer the pole position.

Although the time of the commencement of initial breakup of Pangaea remains tentative, the Triassic is most plausible, the midpoint of which would be -200 m.y. By our maps, Pangaea remains intact until the end of

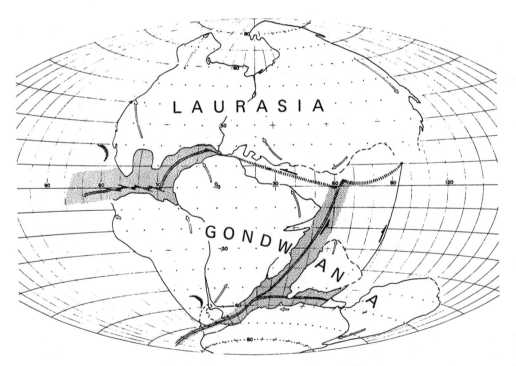

Figure 5-17. Initial rifting of Pangaea as of the end of the Triassic 180 m.y. ago. Open arrows are vectors showing the drift that occurred during the entire Triassic.

the Paleozoic, i.e., the end of the Permian, -225 m.y. Drake and Nafe (1969) suggested that the North Atlantic was never fully closed, with the limits of the quiet magnetic zone establishing the margin of a small Paleozoic North Atlantic Ocean. There is some paleomagnetic evidence to show that the extreme South Atlantic may have opened slightly in the Permian (Briden, 1967; McElhinny, 1969). Heirtzler and Hayes (1967) regard the magnetic quiet zone as marking the extent of Atlantic opening during the Kiaman interval of the Permian, when the earth's magnetic field remained uniformly reversed without switching dipoles for 50 m.y. Emery *et al.* (1970) in their study of the continental rise off eastern North America also propose initiation of opening to have occurred at the beginning of the Permian, 275 m.y. ago. However, Schneider *et al.* (1969) consider that opening by sea-floor spreading during a time when the North Atlantic was at a low magnetic latitude is a more likely explanation for the quiet zone; in this case a Triassic opening would be entirely satisfactory. Vine and Hess (1969) speculate that the first main phase of drift and breakup of Pangaea occurred as late as the Jurassic.

The probable synchroneity of the initial rifting with the effusion of Triassic basalts and the associated formation of taphrogenic basins lead us to begin the opening of the North Atlantic in the Triassic. To account for the position of the Bahama marginal plateau within the confines of a small ocean basin or mediterranean, we further suppose that this opening was initiated at the south end of the basin and proceeded northward to the Grand Banks (Dietz *et al.*, 1970). Based upon the paleomagnetic results, De Boer (1968) also proposed that the Triassic volcanism proceeded from south to north. Any choice as to the exact time of opening of the North Atlantic must be regarded as tentative, probably until the meaning of the magnetic quiet zone is resolved. Presumably the rift entered Pangaea from ancestral Pacif-ic, splitting North America away from South America and Africa.

The creation of the North Atlantic rift lifted North America off the bulge of Africa, but North America remained attached to Europe from the Grand Banks northward. Spain remained in point contact with West Gondwana. This rifting also defined the plate boundaries between North America and South America and established the primitive Caribbean region as a realm of extension. The boundaries of the Gulf of Mexico also were blocked out by the dextral (and retrograde) initial rotation of the Yucatan-Honduras subcratons. The continent of North America drifted northwest with respect to South America. Rotation of Laurasia, related to the opening of the rift, caused some closure of Tethys, indicating the presence of a Tethyan trench that also accommodated some left lateral shear between Europe and Africa.

The breakup of Gondwana also commenced in mid-Triassic with the creation of the pan-Antarctic rift system. We suppose that the rift system had the form of a Y junction, the arms of which are the Afro-Indian and the Indo-Antarctic rifts. This rift system split Gondwana into West Gondwana (Africa and South America) and East Gondwana (the remaining Gondwana continents) and also initiated the separation of India from Antarctica. The inception of the extentional pan-Antarctic rift system, together with the fixity of Antarctica in latitude, requires that West Gondwana and India migrate north, as does the rift itself. Because a part of Africa remains in point contact with Laurasia, this supercontinent must also move somewhat north. The fundamental scheme of rotation for the Gondwana plates that began in the Triassic appears to continue until today. This includes: (1) a large sinistral rotation of Africa (and, consequently, of South America, as long as it remains a part of West Gondwana); (2) a moderate sinistral rotation of India as it makes its long trek to the equator; and (3) a sinistral rotation of

Antarctica. We interpret the drift of Antarctica to always be one of tight rotation. This plate, in the Triassic as today, must have always been surrounded by the pan-Antarctic rift system. The pole of rotation of this plate must therefore be within the Antarctic continent and at no great distance from the geographic pole of rotation. There is no trench associated with the Antarctica plate to provide a locus of crustal resorption.

JURASSIC

The dispersion of continents as of the end of the Jurassic (-135 m.y.) is shown in Figure 5-18. In the northern hemisphere, additional generation of new oceanic crust by sea-floor spreading on opposing limbs of the North Atlantic-Caribbean rift caused a further widening of the North Atlantic basin. This rift grew northward, blocking out the Labra-

dor coastline and splitting off the western margin of Greenland from Canada. With this rift extension, the European coast as far as Scotland was also blocked out. This timing is, however, uncertain and these openings may have been deferred until the late Cretaceous.

Eurasia moved westward with respect to Africa and point contact was maintained between these continents by Spain. Interaction between these cratons caused a sinistral rotation of Spain, opening the Bay of Biscay (but perhaps only partially opening it at this time). North America continued its drift to the northwest, further widening the Atlantic Ocean and increasing the gap to both Africa and South America. The extent of new ocean in the North Atlantic and Caribbean is shown as a stippled zone, a new strip about 1000 km wide. Although there was no seaway to the Arctic and probably at most

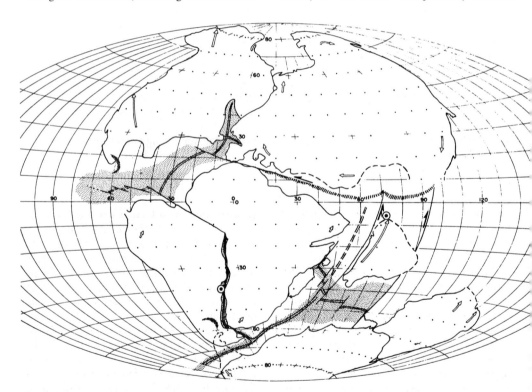

Figure 5-18. The continental drift dispersion of the continents as of late Jurassic, 135 m.y. ago.

only a limited connection to Tethys, the interconnection of the North Atlantic to the Pacific Ocean established a tropical and subtropical holo-oceanic regime. The eastern seaboard of the United States at this time had an almost east-west strike and was positioned between 20° and 30°N.

The interaction of Eurasia with Africa caused further closure of Tethys as crust was resorbed into the Tethyan trench. The compression associated with trenches is needed to raise deep water sediments to continental levels, so that one might expect deep water sediments of Jurassic age along Tethys. In his monograph on the Jurassic, Arkell (1956, p. 595) noted that: "All of the occurrences of the Jurassic formations chronicled in the preceding chapters amount to little more than relics of the marginal lappings of the sea around the edges of continents; the sole exception being Tethys. . . ." Jurassic deep water deposits are found along Tethys as far west as the Alpine region and even in southern Spain.

Splitting along the pan-Antarctic and the Indo-Antarctic rifts continued, translating South America and Africa northward. The rifts maintained a mid-ocean position by their own migration northward, whereas Antarctica remained relatively fixed in position, except for some sinistral rotation to the west. As with the present time, it appears likely that there are no trenches associated with the Antarctic plate toward which this plate would move.

In the late Jurassic or possibly in early Cretaceous, an incipient South Atlantic rift was created within West Gondwana (Africa plus South America) that commenced to drift these two continents apart. Most likely this rift was initiated as a spur off the pan-Antarctic rift commencing in the south and extending northward. Initially the North Atlantic and South Atlantic rifts were connected by a megashear, and, since megashears conserve crust, Africa and South America remained joined across the underbelly of Africa.

In latest Jurassic, with the initial rifting

of the South Atlantic, the Walvis hot spot formed. As explained later, this thermal center would provide a fixed point with which the later drift of the continents could be measured in terms of absolute geographic coordinates. The flight of the crust over the deep mantle would be subsequently marked by the Walvis and Rio Grande nemataths. This hot spot is shown by a circle on Figure 5-00 at latitude 30°S, longitude 14°W, this position being the modern point of intersection between the Walvis ridge and the mid-Atlantic rift, which is about 200 km southwest of Tristan da Cunha island. The incursion of sea water was restricted within this narrow rift so that the deposition of freshwater clastics followed by salt prevailed in the Jurassic and early Cretaceous even though the continental margins along the stems of these continents were now being blocked out by rifting. The situation resembled that of the Red Sea rift today.

Radiometric dating reveals that the Kaoko basalts of Southwest Africa fall in the interval from −136 m.y. to −114 m.y., with a peak effusion at −125 m.y. (Siedner and Miller, 1968). They are of the same age as Serra Geral lavas of Brazil (Amaral *et al.,* 1966), the most extensive and thickest plateau basalts on any of the continents with which they are brought into juxtaposition by continental drift reconstruction. Thus vast areas of South America and Southwest Africa were simultaneously inundated by tholeiitic basalts when a new "mid-Ocean ridge" was created under West Gondwana. With the subsequent spreading apart of Africa and South America, this new mantle hot spot poured out the lavas that have created the Rio Grande and Walvis ridges on the ocean floor.

These basalt flows are on opposing shores of the South Atlantic. Also, they are significantly younger than the plateau basalts of Rhodesia and South Africa (Stormberg, Karoo, and Swaziland flows), which are mid-Triassic to mid-Jurassic in age (−155 to −190 m.y.). This age difference is in agreement with our maps, which show the rifting of South America from Africa to be a much younger event (near the Jurassic-Cretaceous boundary) than the rifting of Africa from East Gondwana associated with the formation of the southwest Indian Ocean ridge.

India continued its remarkable drift northward, with the India crustal plate being bordered by the Ninety-East ridge and Maldive-Laccadive megashears. As with the Africa plate, the India plate was resorbed along its northern boundary by the Tethyan trench. Perhaps the India plate may be thought of as an ideal rectilinear plate bordered by well lubricated megashears, "pushed" from the south by a northward migrating rift and "pulled" from the north by a segment of the Tethyan trench (the future Himalayan uptake zone). The Indian conveyor belt, then, would be a low-friction system that could operate largely independently of the other, more interlocked terrestrial plates.

CRETACEOUS

Figure 5-19 shows the dispersion of the continents as of the end of the Cretaceous (~65 m.y.), during which the motions initiated in the Triassic continued. With the further opening of the Atlantic rift by sea-floor spreading, both North and South America moved westward, the North Atlantic opening about 1500 km and the South Atlantic 3000 km or more. The relative motion of these two continental plates was such that the Caribbean region closed slightly, making this a regime of compression rather than extension. Holo-oceanic conditions now fully prevailed throughout the North and South Atlantic oceans. The Antilles trench, the Scotia trench, and the horn of Africa are now shown in solid line rather than in ghost outline, indicating that they existed in much their modern form. At the end of the period, new extension of the North Atlantic rift commenced to block out the eastern margin of Greenland (Avery et al., 1968) but probably did not extend into the Arctic. The rotation of Spain was completed.

Although not shown on our maps, an extensive north-south Mesozoic trench system must have existed in the eastern Pacific Ocean to accommodate by crustal uptake the westward movement of both the North America and South America crustal plates. North America presumably encountered this trench in the Late Jurassic and was adjacent to it until the mid-Cretaceous, so that during this time the Franciscan foldbelt was accredited to North America as a collapsed continental rise prism. Presumably the trench was then overridden and destroyed by the continued westward movement of North America. Ocean crust, but not continental sial, can be consumed by trench resorption. Within the Cretaceous, South America encountered the Andes trench and commenced to displace this trench to the westward without stifling or overriding it. The Andean foldbelt resulted from this encounter. The Antilles trench may represent a sector of this extensive Pacific trench system, which did not encounter a continent and hence has remained in its primitive position.

During the Cretaceous, Africa drifted northward about 10° and continued its sinistral rotation, and this motion, together with a dextral rotation of the Eurasia plate to the north, caused an almost complete closure of the Tethys seaway.

The splitting away and dropping off of Madagascar from Africa is interpreted here as involving an early offshoot of the Carlsberg rift. Prior to the creation of this rift, the African plate was everywhere bounded by rifts except to the north, in which direction the plate moved and was resorbed within the Tethys trench. With the invasion of an offshoot of the Carlsberg rift into the African craton, the Madagascar subcontinent was arrested in its northern movement and rotated somewhat as a separate cratonic block.

The absolute motion of South America and Africa northward almost together over the deep mantle is traced out by the Rio Grande and Walvis nemataths. The westward rotation of Antarctica continued. By the end of the period the pan-Antarctic rift extended westward, initiating the rifting away of Aus-

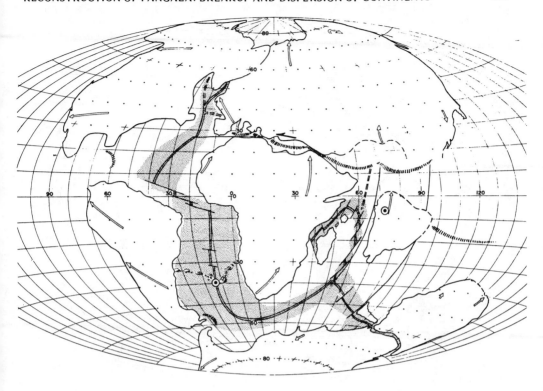

Figure 5-19. The continental drift dispersion of continents as of the end of the Cretaceous, 65 m.y. ago.

tralia from Antarctica. All the continents were now blocked out except for the join between Greenland and northern Europe. The universal continent of Pangaea was now highly fragmented. The supercontinent of Eurasia now almost ceased to exist except for a northern region of contact. Eurasia moved largely independently of North America and continued its dextral rotation, begun in the Triassic, with the consequent further closure of Tethys, especially in the Asian region. The European sector may have experienced little crustal uptake, since Eurasia was pivoting on a point farther to the east. The supercontinent of Gondwana was now fully dispersed.

CENOZOIC

In Figure 5-20 the modern positions of continents are shown with the shorelines also drawn in. The shaded areas represent new

ocean floor generated by sea-floor spreading since the beginning of the Cenozoic 65 m.y. ago. Unlike the previous map, we have included new oceanic crust generated in the Pacific basin (Vine and Hess, 1969). During the Cenozoic, about one-half of the ocean floor has been renewed, or about one-third of the entire surface area of the earth.

The subcontinent of New Zealand is included for the first time, presumably having separated from the east margin of Australia during the flight of this continent northward from Antarctica; however, the details are not shown because of map distortion in this portion of the drawing.

In the North Atlantic the ocean floor has continued to spread as North America has moved westward. The North Atlantic rift has now propagated into the polar basin separating Greenland from Europe, completing the breakup of Laurasia. Magnetic anomaly patterns suggest that this opening occurred in

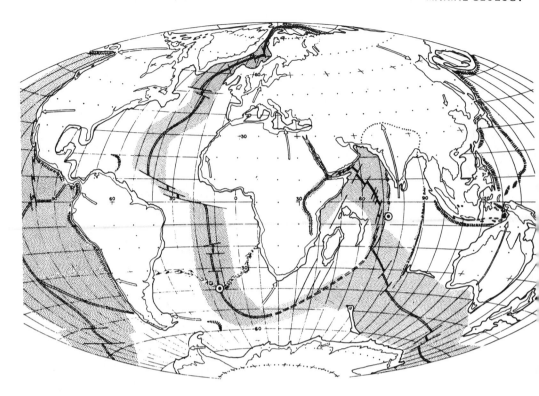

Figure 5-20. The position of continents today showing the amount of sea-floor spreading, etc., during the Cenozoic.

the early Cenozoic (Johnson and Heezen, 1967; Vine and Hess, 1969). It is tempting to speculate that this event happened in mid-Eocene and was marked by the deposition of Horizon A, a silicified zone as revealed by the JOIDES drilling program. The establishment of a connection with the polar sea would permit an interchange of cold water and enhance the circulation and hence the productivity of the North Atlantic, bringing in a flood of siliceous organisms—radiolarians and diatoms.

The Caribbean attained its modern configuration possibly owing to the rotation or translation of the Yucatan and the Honduras-Nicaragua blocks out of the Gulf of Mexico. The isthmus of Panama represents new ground caused by volcanism and uparching of the mantle that connected North and South America in Pliocene time (Lloyd, 1963).

Extensive spreading continued in the South Atlantic. South America moved westward, displacing the Andean trench as well. Although it continued to accommodate shear, the Afro-Indian rift became largely inactive during the Cenozoic, since the northward flight of Africa was blocked by Europe. This change of direction of movement in the Africa plate is reflected by the bent shape of the Walvis ridge, giving it a more east-west strike in the Cenozoic.

The opening up of the far north Atlantic produced a right lateral shear of Eurasia relative to Africa that is the sense of the "Tethyan twist" of Van Hilten (1964). This motion has left its imprint on the modern geotectonic style of the Mediterranean and the Middle East.

In the far east Indian Ocean, the Indo-Antarctic rift extended eastward, detaching Australia from Antarctica and producing the rapid northern flight of Australia and New Zealand. With this fast spreading, Antarctica has remained essentially still while both the rift and the Australia plate have moved to the north. In the eastern Indian Ocean, the region east of Ninety-East ridge is part of the Australia plate and apparently has moved along with Australia. It is noteworthy that the area between Broken ridge and the Kerguelen plateau shows the same degree of sea-floor spreading as the area between Australia and Antarctica. This supports our interpretation as to the tectonic autonomy of the Australian plate as distinct from the Indian plate.

The northward trek of India continued through the early Cenozoic until it encountered and underthrust the southern margin of Asia in the Neogene, throwing up the Himalaya foldbelt by the collision. About -65 m.y. (Cretaceous-Cenozoic boundary), the western margin of India crossed the equator and passed over a mantle thermal center at latitude 7°S, longitude 72°E. With this encounter tholeiitic magma welled up over the Indian subcontinent, laying down the Decca plateau basalts. After India had passed over the hot spot, magma continued to pour out, creating, by streaming out as the crust moved north, the Chagos-Laccadive nematath, a basalt ridge now covered by coral growth. The Chagos-Laccadive nematath has no mirror-image counterpart, as does the Rio Grande nematath with respect to the Walvis nematath. This would seem to be due to the Chagos-Laccadive nematath's occurring along a megashear plate boundary rather than along a rift with bilateral sea-floor spreading, resulting in a mononematath. The Hawaiian ridge appears to be another example of a mononematath, but one in which the hot spot punctured a crustal plate rather than being along any boundary.

The breadth of India appears slightly too broad to have passed between the fracture zone megashears within the Indian Ocean (see Heezen and Tharp, 1964), which apparently marked its drift north. Perhaps the Ninety-East ridge marks a compressional overlapping of oceanic crust such that the swath between the fracture zones was formerly broader than it is today.

The branch of the Carlsberg rift system that, in the Cretaceous, split off Madagascar from the mainland of Africa apparently now finds expression within the continent of Africa creating the rift zone of eastern high Africa. To occupy this new position, the rift must have taken up a new position within Africa, since normal sea-floor spreading would have kept it always east of the African margin and the Indian Ocean. The new Gulf of Aden-Red Sea spur of this same rift system has commenced to split off Arabia from Africa.

Antarctica and its associated plate underwent further westward rotation. The westward motion of Antarctica was slightly less than that of South America, so that the horn of Antarctica now lies somewhat to the east of the strike of the Andes.

ABSOLUTE GEOGRAPHIC GRID

Figures 5-16 to 5-20 show the continents and sea-floor features on an absolute geographic grid. This has not been attempted in any previous reconstructions; hence an explanation of our method is given here.

Paleomagnetic pole positions provide data only on latitude and orientation and supply no control in longitude. Also, it is quite impossible to dead reckon the path of continents back in time from the present by considerations of sea-floor spreading (magnetic anomaly patterns and strike of fracture zones), since these considerations provide only a relative solution of the motions between two plates. These plates may also move independently. It is necessary to find features that may have conserved their position over long periods of time. To this end, we have considered as likely possibilities both oceanic trenches unaffected by continental margins and thermal centers (hot spots). We have chosen the latter. Although we are aware that it is a somewhat arbitrary assumption, we find that the use of the Walvis hot spot as a fixed geographic point following the interpretation of Wilson

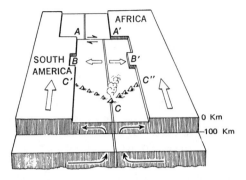

Figure 5-21. Block diagram to show how the two crustal plates on opposite sides of the South Atlantic may be moving relative to each other while both may be moving north with respect to a deep mantle "hot spot" [adapted from Wilson (1965b)].

(1965b) in fact appears to provide an acceptable solution.

As illustrated in Figure 5-21, the rationale is as follows. Information on the drift vectors of continents is provided by the fracture zones associated with ridge-ridge transform faults (A-A′) or by the matching of congruent indentations on opposing continental slopes across a rift ocean (B-B′). But this is only a relative solution as these plates in turn may both be moving, in this example to the north. However, the tracks of the Rio Grande and the Walvis ridges as nemataths may provide an absolute geographic solution to the motion of both plates. (We use the term nematath meaning "thread ridge" in the descriptive sense of Carey (1958), but not in his generic sense, for he regarded the Walvis ridge as sialic while we regard it as basaltic.) If the magma for the Walvis hot spot is derived from a deep-mantle source, and if this source is immobile or does not migrate appreciably, the nematath provides an absolute track for the crustal plates, a resultant direction combining both sea-floor spreading and the drift of the plates themselves over the deep mantle. This assumes that the lava dome above a point source for magma deep within the stagnant mantle is

streamed into a nematath by the over-all drift of the crust.

We have selected the position of the Walvis hot spot by extending the trend of the Walvis ridge until it intersects the mid-ocean rift, which places this point at latitude 39°S, longitude 14°W, or somewhat to the south of Tristan da Cunha Island. This hot spot then provides a fixed position on which to relocate West Gondwana subsequent to its breakup. As noted above, this thermal center appears to have been initiated at about -135 m.y., or at the Jurassic-Cretaceous boundary. Of course, we lose this position control prior to this date, so that we must dead reckon back in time to find the Permian position of Pangaea. For this purpose we have assumed that the motion of West Gondwana only has been northward. This drift direction is reasonable, as it is normal to the spreading of the Afro-Antarctic rift and toward the Tethyan trench.

SUMMARY OF PLATE EVOLUTION AND MOTIONS

Possibly one may visualize the entire world as consisting of but a single plate during the late Paleozoic when Pangaea existed. The continental drift dispersion of continents may be conceived in terms of breakup of the lithosphere into new crustal plates in the Mesozoic and Cenozoic. We cannot, of course, exclude the presence of one or more plates in Panthalassa (the ancestral Pacific), as rifts existing wholly within an ocean basin would not break up this universal continent. It appears quite likely that Pangaea drifted as a unit during the late Paleozoic.

Prior to the collision of the bulge of Africa with North America in the late Paleozoic (Wilson, 1966), Gondwana probably drifted separately and was moving north. Paleomagnetic evidence that the south pole migrated from a position near Dakar in the lower Paleozoic to a position near Johannesburg in the upper Paleozoic supports such a translation (Creer, 1968). Such a migration of the pole position would be consistent with the extensive continental glaciation in the Sahara in the Siluro-Ordovician (Rognon *et al.,* 1968) followed by the well known Permo-Carboniferous glaciation of South Africa and southern Brazil.

Initial rifting commenced in the Triassic, with the result that Pangaea was fragmented by the generation of the North Atlantic rift, pan-Atlantic rift system, and the Afro-Indian rift. The resulting continents were Laurasia, West Gondwana, and East Gondwana, with India being further split off from East Gondwana. Thus, four plates existed by the end of the Triassic (again excluding consideration of any holo-oceanic plates). By the end of the Jurassic a fifth plate was added by the incipient creation of a rift-megashear system that blocked out South America, thus splitting West Gondwana into two parts.

By the end of the Cretaceous, a new rift completed the fragmentation of Gondwana with the lift-off of Australia from Antarctica, creating a sixth plate. The dropoff of Malagasy and the Seychelles from Africa resulted in a new subplate. Although Greenland was still in contact with northern Europe, we suggest that a megashear already blocked out Greenland, generating a seventh plate such that North America and Eurasia moved differentially. The complete breakup of Laurasia then occurred in the Cenozoic as North America moved westward. Also, India collided into the southern margin of Asia. Although India is now attached to Asia, the India plate continues to exist as a separate entity. In the Neogene, the western portion of North America, westward of the San Andreas megashear, became stranded upon the northeast Pacific plate so that its future history, so far as drift dispersion is concerned, will be involved with that plate.

With the generation of the Indian and Atlantic oceans as rift basins, the ancestral Pacific or Panthalassa has grown smaller. Meservey (1969) has stated that continental drift is a topological impossibility unless the earth has expanded, since it requires that the continental fragments of Pangaea, the circumference of which would be a small circle

of earth, upon their dispersion toward the Pacific must pass over a great circle of earth according to Meservey. Actually, however, summing the sectors that make up the perimeter of the Pacific results in a length that is slightly greater than the 40,000-km meridianal circumference of the earth (K. Rudolfo, unpublished data). More importantly, this argument confuses the conservative size and shape of continents (although some accretion may occur along the leading edges) with the nonconservative crustal plates on which the continents are superimposed. These plates are capable of adding to or subtracting from their boundaries and so are non-topological. With the westward drift of the Americas, at least the Caribbean region appears to have first undergone extension and, subsequently, slight closure.

It appears that the ocean floor is essentially all Mesozoic or Cenozoic. Even the basins behind island arcs (the Japan Sea, etc.) appear to be neo-oceanic regions. Some possibility remains for Paleozoic ocean floor to exist in the Canadian basin of the Arctic Ocean, but this is unlikely if the Alpha ridge is an abandoned spreading rift associated with Mesozoic-Cenozoic drift event. There remains a possibility that Paleozoic ocean floor may be found off the Pacific margin of West Antarctica. By our reconstruction, the remainder of the Antarctica perimeter is a rift margin initially blocked out by the formation of a locus of spreading. It would be fortuitous if this were also true for the Pacific margin of West Antarctica, and it is more likely that the pan-Antarctic rift entered from the open Pacific. Since there is no trench around Antarctica and since the rift appears to have migrated outward, a sector of ancient crust may have been stranded along West Antarctica. The most likely large sector of pre-Mesozoic sea floor is the Wharton Sea basin, as outlined by Ninety-East ridge, Java trench, western Australia, and the Diamantina fracture zone. Our reconstruction shows this as an embayment of ancient Tethys that, owing to the late detachment of Australia from Antarc-

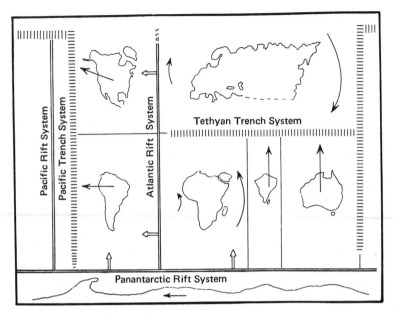

Figure 5-22. Plate movement schematic of the crustal plates with reference to present-day rift and trench positions. The trenches are represented in hachured lines; the rifts are shown as double lines and the megashears as a simple straight line. Migrational directions of the rifts are indicated by open arrows as all the rifts are approaching trenches. Single-line arrows indicate directions that individual plates have moved.

tica, has not yet been subducted into the Java trench.

An over-all summary of the plate motions is illustrated in the schematic shown as Figure 5-22. The trench systems, where crust is resorbed, may be visualized as an inverted U (the Pacific trenches) with a stem attached (the east-west Tethys trench). The complimentary system of rifts forms a small circle on the earth within the Antarctic Ocean with two north-south spurs, the Pacific rift system, and the Atlantic rift system. In response to these rift and trench systems, the crustal plates with their included continents have a history of mostly northward and westward displacement. The trenches have tended to remain quite fixed in position, whereas the rifts have migrated so that the plates have changed size. Having no trench or zone of resorption, the Antarctic plate has grown considerably larger; other plates have grown differentially in size and shape, depending on the amount of new crust

added with respect to that resorbed. The pan-Antarctic rift has migrated northward while the Atlantic rift system has migrated westward. Carried like a raft stranded on a moving ice floe, North and South America have moved large distances westward toward the Pacific trench system. Africa, India, and Australia have moved northward as if drawn by the Tethys trench system. Antarctica and Eurasia have remained relatively fixed, although undergoing some rotation.

It is interesting that Wegener recognized a westward drift of the continents (*Westwanderung*) and a flight away from both poles (*Polarfluchtkraft*). We agree with the former but would replace the latter with a flight only away from the south pole—a *Sudpolarfluchtkraft*.

REFERENCES

Amaral, G., U. Cordani, K. Kawashita, and J. Reynolds, K-Ar dates of basaltic rocks from southern Brazil, *Geochim. Cosmochim. Acta* **30**, 159 (1966).

Arkell, W. J., *Jurassic Geology of the World*, 806 pp., Oliver and Boyd, Edinburgh, 1956.

Avery, O. E., G. D. Burton, and J. R. Heirtzler, An aeromagnetic survey of the Norwegian Sea, *J. Geophys. Res.* **73** (14), 4583-4600 (1968).

Briden, J., Recurrent continental drift of Gondwanaland, *Nature* **215**, 1334-1339 (1967).

Bullard, E. C., J. E. Everett, and A. Gilbert Smith, The fit of the continents around the Atlantic. *Phil. Trans. Roy. Soc. London* **A258**, 41-51 (1965).

Carey, S. W., A tectonic approach to continental drift, in Continental Drift, A Symposium, edited by S. W. Carey, pp. 177-355, University of Tasmania Press, Hobart, Tasmania, Australia, 1958.

Creer, K. M., Arrangement of the continents during the Paleozoic era, *Nature* **219**, 41-44 (1968).

De Boer, J., Paleomagnetic differentiation and correlation of the Late Triassic volcanic rocks in the central Appalachians, *Bull. Geol. Soc. Amer.* **79**, 609-626 (1968).

Dietz, R. S. Continent and ocean basin evolution by spreading of the sea floor, *Nature* **190**, 854-857 (1961).

Dietz, R. S., and W. P. Sproll, Equal areas of Gondwana and Laurasia, *Nature* **212**, 1196-1198 (1968).

Dietz, R. S., and W. P. Sproll, Fit between Africa and Antarctica: A continental drift reconstruction, *Science* **167**, 1612-1614 (1970a).

Dietz, R. S. and W. P. Sproll, Overlaps and underlaps in the North America to Africa continental drift fit, in Proceedings SCOR symposium, Geology of the East Atlantic Continental Margin, Her Majesty's Stationery Office, 1970b.

Dietz, R., J. C. Holden, and W. Sproll, Geotectonic evolution and subsidence of the Bahama platform, *Bull. Geol. Soc. Amer.* **8**(7), 1915-1928 (1970).

Drake, C. L., and J. E. Nafe, Geophysics of the North Atlantic region, in Symposium on Continental Drift, edited by J. T. Wilson, UNESCO, Paris, in press, 1969.

Du Toit, A. L., *Our Wandering Continents; an Hypothesis of Continental Drifting*, 366 pp., Oliver and Boyd, London, 1937.

Emery, K. O., E. Uchupi, J. D. Phillips, C. O. Bowin, E. T. Bunce, and S. T. Knott, Continental rise off eastern North America, *Amer. Ass. Petrol. Geol. Bull.* **54** (1), 44-108 (1970).

Francis, T., and R. Raitt, Seismic refraction measurements in the southern Indian Ocean, *J. Geophys. Res.* **72**(12), 3015-3041 (1967).

Funnell, B., and A. Smith, Opening of the Atlantic Ocean, *Nature* **219**, 1328-1333 (1968).

Girdler, R. W., The Red Sea—a geophysical background, in *Hot Brines and Recent Heavy Metal Deposits in the Red Sea*, E. Degens and D. Ross ed. pp. 38-58, Springer, New York, 1969.

Hamilton, W., Tectonics of Antarctica, *Tectonophysics* **4**, 555-568 (1967).

Heezen, B., and M. Tharp, *Physiographic Diagram of the Indian Ocean*, Geological Society of America, New York, 1964.

Heirtzler, J. R., and D. E. Hayes, Magnetic boundaries in the North Atlantic Ocean, *Science* **157**, 185-187 (1967).

Heirtzler, J. R., G. O. Dickson, E. M. Herron, W. C. Pitman, and X. Le Pichon, Marine magnetic anomalies' geomagnetic field reversals, and motions of the ocean floor and continents, *J. Geophys. Res.* **73**(6), 2119-2136 (1968).

Hess, H. H. History of Ocean basins, in *Petrologic Studies: A Volume to Honor A. F. Buddington*, A. E. J. Engel, H. L. James, and B. F. Leonard, eds. pp. 599-620, Geological Society of America, New York, 1962.

Irving, E., *Paleomagnetism*, 399 pp., John Wiley, New York, 1964.

Isacks, B., J. Oliver, and L. Sykes, Seismology and the new global tectonics, *J. Geophys. Res.* **73**(18), 5855-5899 (1968).

Johnson, G. L., and B. C. Heezen, Morphology and evolution of the Norwegian-Greenland Sea, *Deep-Sea Res.* **14**, 755-771 (1967).

King, L. C., *The Morphology of the Earth*, Hafner, New York, 699 pp., 1962.

Le Pichon, X., Sea-floor spreading and continental drift, *J. Geophys. Res.* **73**(12), 3661-3697 (1968).

Lloyd, J., Tectonic history of the south Central American orogen, in *Backbone of Americas,* Amer. Assoc. Petrol. Geol. Mem. 2, 88-100, 1963.

McElhinny, M., Paleomagnetism of the southern continents, in Symposium on Continental Drift, edited by J. T. Wilson, UNESCO, Paris, in press, 1969.

Meservey, R., Topological inconsistency of continental drift on the present-sized earth, *Science* 166, 609-611 (1969).

Meyerhoff, A. A., Future hydrocarbona provinces of Gulf of Mexico-Caribbean region, *Amer. Ass. Petrol. Geol.* 17, 217-260 (1967).

Morgan, W. J., Rises, trenches, great faults, and crustal blocks, *J. Geophys. Res.* 73, 1959-1982 (1968).

Pitman, W., and D. Hayes, Sea-floor spreading in the Gulf of Alaska, *J. Geophys. Res.* 73(20), 6571-6580 (1968).

Rognon, P., O. de Charpal, B. Biju-Duval, and O. Gariel, Les glaciations "silureinnes" dans l'Ahnet et le Mouydir (Sahara Central), in *Geol. Serv. Algiers Bull.* 38, 53-81 (1968).

Schneider, E. D., P. R. Vogt, and A. Lowrie, Diapiric structures and magnetic anomalies of the pre-Cenozoic Atlantic Ocean (abstract), EOS, Trans. AGU, 50, 212, 1969.

Siedner, G., and J. Miller, K-Ar determinations on basaltic rocks from southwest Africa and their bearing on continental drift, *Earth Planet, Sci. Lett.* 4, 451-458 (1968).

Smith, A., and A. Hallam, The fit of the southern continents, *Nature* 223 (139), 139-144 (1970).

Sproll, W. P., and R. S. Dietz, Morphological continental drift fit of Australia and Antarctica, *Nature* 222, 345-348 (1969).

Sykes, L., Mechanisms of earthquakes and nature of faulting in the mid-oceanic ridges, *J. Geophys. Res.* 72, 2131-2153 (1967).

Van Hilten, D., Evaluation of some geotectonic hypotheses by paleomagnetism, *Tectonophysics* 1, 3-71 (1964).

Vine, F. J., and H. H. Hess, Sea floor spreading, in *The Sea,* vol. 4, Interscience, New York, in press, 1969.

Vine, F. J., and D. H. Matthews, Magnetic anomalies over ocean ridges, *Nature* 199, 947-949 (1963).

Wegener, A., *The Origin of Continents and Oceans* (transl. by J. Biram), 4th ed., 246 pp., Dover, New York, 1929.

Wilson, J. T., Continental drift, *Sci. Amer.* 208, 86-100 (1963).

Wilson, J. T., A new class of faults and their bearing upon continental drift, *Nature* 207, 343-347 (1965a).

Wilson, J. T., Submarine fracture zones, aseismic ridges, and the ICSU line: Proposed western margin of the east Pacific ridge, *Nature* 207(5000), 907-911 (1965b).

Wilson, J. T., Did the Atlantic close and then re-open? *Nature* 211, 676-681 (1966).

Pleistocene Climates and Chronology
in Deep-Sea Sediments

DAVID B. ERICSON   GOESTA WOLLIN   1968

INTRODUCTION

The Pleistocene epoch was a time of emergent continents, deep ocean basins, active volcanoes, and high mountains; a time of drastic climatic changes marked by repeated spreading of great ice sheets over wide regions; and a time of exceptionally rapid evolution of living things. The Pleistocene has been a battleground for scientists since it was named more than 100 years ago. A reason for great interest and heated controversy regarding the Pleistocene is probably its close connection with the evolution of man. If the climatic and topographical changes of the Pleistocene had not occurred, it is doubtful that *Homo sapiens* could have developed within its short span.

The Pleistocene is known to most people as the great ice age, the epoch of cavemen and woolly mammoths and huge continental ice sheets covering much of northern Europe and North America. Most early glaciologists refused to believe in more than a single glaciation. Explorations around the turn of the century, however, led to the recognition that the Pleistocene was divided into four successive major glaciations separated by long intervals of temperate climate. The view that the base of the Pleistocene coincides approximately with the earliest glacial deposits is generally accepted by glaciologists.

The concept that the Pleistocene is synonymous with glaciation has been widely discussed. Recently Selli[1] observed that direct correlation of the Pleistocene with the glacial epochs can no longer be maintained. He suggests that the preglacial Pleistocene appears to have lasted 1 to 1.5 times as long as the glacial Pleistocene. Berggren *et al.*[2,2a] and Hays and Berggren[3] have come to conclusions very similar to Selli's.

1. R. Selli, *Progr. Oceanogr.* 4, 67 (1967).
2. W. A. Berggren, J. D. Phillips, A. Bertels, D. Wall, *Nature* 216, 253 (1967).
2a. W. A. Berggren, J. D. Phillips, A. Bertels, D. Wall, *Deep-Sea Res.* 14, No. 2 (1968).
3. J. D. Hays and W. A. Berggren, *Micropaleontology of Marine Bottom Sediments* (Cambridge Univ. Press, New York, in press).

David B. Ericson and Goesta Wollin are, respectively, senior research associate and research consultant with the Lamont-Doherty Geological Observatory of Columbia University, New York. Both are well known for their cooperative study and documentation of the most complete record of the Pleistocene Epoch preserved in deep-sea sediment cores collected from the world's oceans.

From *Science*, Vol. 162, No. 3859, pp. 1227-1234, 1968. Reprinted with light editing and by permission of the authors and the American Association for the Advancement of Science. Copyright 1968 by the American Association for the Advancement of Science.

Our estimate for the duration of the Pleistocene is about the same as that of these investigators, but instead of including a pre-glacial interval and subsequent series of glaciations, our results indicate that the entire Pleistocene was divided into four major glaciations separated by three interglacial stages.

Because of the nature of the depositional process—that is, the alternating expansion and melting of the ice sheets—the deposits readily accessible to study that were left on the continents by the ice provide a discontinuous record at best. Then the difficulty of interpreting the record is compounded by the fact that, as each succeeding ice sheet spread over the land, it tended to destroy the evidence left by earlier glaciations. Furthermore, the long interglacial stages are often represented by nothing more than a weathered, or chemically altered, zone on the surface of glacial detritus left by a preceding ice sheet. As if this were not enough, in many regions the alternative to weathering has been total destruction of the record by erosion—the ceaseless washing away of unconsolidated sediment by rainfall, streams, and high winds which, little by little, transfer material from the continents to the ocean basins. Because the record of the Pleistocene on the continents is so nearly illegible, some investigators turned to the sediments of the deep ocean basins.

Deep-sea sediments offer an alternative approach. The extremely slow and continuous rain of fine mineral particles and hard parts of temperature-sensitive microorganisms on the ocean floor provides an ideal recording mechanism.

The pioneer work of Schott[4] on sediment cores collected in the equatorial Atlantic by the *Meteor* expedition opened up a new approach to the problem of the Pleistocene climates. Schott concluded that vertical variations in the abundance of the planktonic foraminiferans *Globorotalia menardii* in the cores probably corresponded with the moderating of climate at the end of the last ice age. This suggested the probability that a legible record of all the major climatic events of the Pleistocene lay within the deep-sea sediments of the Atlantic Ocean.

Lamont-Doherty Geological Observatory has a unique collection of cores of deep-sea sediment from all the oceans. The collection contains more than 5000 cores, obtained in more than 50 expeditions. In order to cope with this quantity of material, it was necessary to make use of a rapid method of foraminiferal analysis of the climatic record, based on relative numbers of warm-water and cold-water species. We published[5] results of an extensive study based on this method and gave a composite climatic record correlated with the glacial and interglacial stages of the Pleistocene. We defined the Pliocene-Pleistocene boundary from changes in species of planktonic foraminiferans and the extinction of Discoasteridae, the organisms responsible for secreting the minute star-shaped objects called discoasters. Dates determined by the radiocarbon, the protactinium-ionium, and the protactinium methods provided an absolute time scale from the present back to about 175,000 years ago. We established a time scale of about 1.5 million years for the entire Pleistocene epoch by extrapolation beyond 175,000 years ago.

The discovery of paleomagnetic reversals in deep-sea sediments and their correlation with the reversal record of continental rocks is one of the most important recent contributions to marine geology. Significant results of studies made to determine the dates of sediment cores by this new method have been published by Harrison and Funnel,[6] Fuller *et al.*,[7] Opdyke *et al.*,[8] Berggren et

4. W. Schott, Wiss. Ergeb. Deut. Atlantischen Expedition *Vermess. Forschungsschiff 'Meteor' 1925-27* 3, 43 (1935).

5. D. B. Ericson, M. Ewing, and G. Wollin, *Science* 146, 723 (1964).
6. C. G. A. Harrison and B. M. Funnell, *Nature* 204, 566 (1964).
7. M. D. Fuller, C. G. A. Harrison, and Y. R. Nayudu, *Amer. Ass. Petrol. Geol. Bull.* 50, 566 (1966).
8. N. D. Opdyke, B. Glass, J. D. Hays, and J. Foster, *Science* 154, 349 (1966).

al.,[2] Glass *et al.,*[9] and Hays and Opdyke.[10] The sequence of reversals provides a long-sought-for means of dating stratigraphical levels as far back as the beginning of the Pleistocene.

The chronology of the glacial Pleistocene is probably the most important because it provides a time scale against which to study the rate of biological and geological development, and it supplies anthropologists with a dated context in which to place the emerging races and cultures of man and his primitive predecessors. With the exception of our earlier estimate of about 1.5 million years for the duration of the glacial Pleistocene,[5] estimates of its duration range from 300,000 to about 1 million years. Now, with paleomagnetic dates of the deep-sea cores available which give an absolute chronology for the entire Pleistocene and with a different method for interpreting the climatic record,

9. B. Glass, D. B. Ericson, B. C. Heezen, N. D. Opdyke, and J. A. Glass, *Nature* 216, 437 (1967).
10. J. D. Hays and N. D. Opdyke, *Science* 158, 1001 (1967).

the results indicate that the duration of the glacial Pleistocene was about 2 million years.

THE CORES

Figure 5-23 shows the geographical locations of the coring stations, and Table 5-2 gives geographical locations, depths of water, and lengths of the cores, chosen because they contained correlatable faunal zones for which dates were available.

Our studies of thousands of cores have shown that two main classes of deep-sea sediments are (i) sediments that accumulate slowly and continuously and (ii) sediments that are laid down almost instantaneously by intermittent turbidity currents. We have found that there are three main processes that may cause confusion in the interpretation of the sediment record. Turbidity currents may deposit, almost instantaneously, layers of sediment several meters deep; slumping and submarine erosion due to deep currents may remove parts of the section; and deep-current scour may transport and

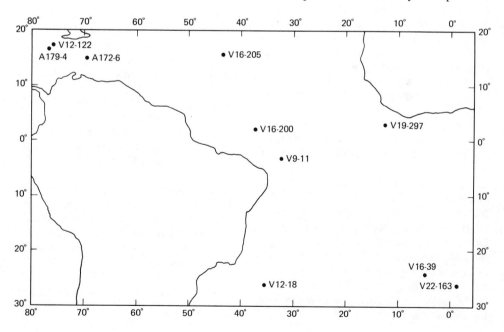

Figure 5-23. Coring stations where the cores of Table 5-2 were obtained.

Table 5-2. Geographical Location, Water Depth, and Core Length for the Ten Cores of our Study.

Core[a]	Latitude	Longitude	Water Depth (m)	Core Length (cm)
A172-C6	14°59′N	68°51′W	4160	935
A179-C4	16°36′N	74°48′W	2965	690
V9-C11	3°13′S	32°12′W	4120	1207
V12-C18	28°42′S	34°30′W	2935	1080
V12-C122	17°00′N	74°24′W	2730	1090
V16-C39	24°43′S	4°45′W	4510	802
V16-C200	1°58′N	37°04′W	4095	989
V16-C205	15°24′N	43°24′W	4045	1257
V19-C297	2°37′N	12°00′W	4120	1300
V22-C163	26°22′S	00°56′E	4440	1080

[a]The letter that precedes a core number indicates the research vessel by which the core was taken: (A) Atlantis, (V) Vema. The number directly following the letter is the number of the expedition; the number that follows the hyphen and is preceded by C is the serial number of the core for the expedition in question.

redeposit both organic and inorganic components of older sediments. More often than not, cores from the deep ocean basins contain evidence of the effect of one or more of these disturbing processes. In view of the prevalence of disturbing processes, we have felt it necessary to check carefully the continuity of our Pleistocene sections by cross correlation.

The correlation of the variation in the abundance of the *Globorotalia menardii* complex, of changes in coiling direction of *G. truncatulinoides,* of paleontological boundaries, and of the paleomagnetic stratigraphy of the cores used in this study indicates that they contain a continuous sediment record.

All the cores consist of foraminiferal lutite, a mixture of fine mineral particles from the continents and particles of calcium carbonate secreted by planktonic organisms. Except for minor variations in color in some of the cores, the sediment is uniform from top to bottom.

FREQUENCY-TO-WEIGHT-RATIO METHOD

The method generally employed in foraminiferal studies is a determination of percentages of species in the total number of tests in the sample. This is time-consuming and, in the case of climatic studies, to some extent wasteful, in that it necessarily involves counting all species, even though some of the most abundant are of little or no significance from the standpoint of climate.

In deep-sea sediments of normal particle-by-particle deposition, the material coarser than 74 micrometers in particle diameter consists almost entirely of the tests of planktonic foraminifera. Therefore, the ratio of the number of tests of a particular species to the weight of the material coarser than 74 micrometers may be substituted for percentages (see reference 10a). Experience shows that plots of variation in this ratio from level to level in sediment cores closely parallel curves for variation in percentages of species calculated in the conventional way. The new method has several advantages: (i) it results in a great saving of time; (ii) it makes it possible to plot variations in the relative abundance of particular climatically sensitive species; and (iii) this, in turn, makes it possible to count larger samples, thereby increasing the statistical validity of the data.

The first step is to weigh to the nearest milligram the washed sample to be analyzed. Then the tests of the particular species to be studied are counted, and the ratio of the count to the weight (in milligrams) of the washed sample is calculated. This ratio, which may be called the "frequency" of the species, is, then, an index of its productivity with respect to the total productivity of all planktonic species at the particular core level being studied.

10a. For our first report on this method, see D. B. Ericson and G. Wollin, *Micropaleontology* 2, No. 3 (1956).

Our counts of the *Globorotalia menardii* complex and of *G. truncatulinoides, G. scitula, Globigerina inflata, Pulleniatina obliquiloculata, Sphaeroidinella dehiscens,* and *Globigerinella aequilateralis* indicate that the *Globorotalia menardii* complex is the most sensitive climatic indicator in the cores. This complex, which includes the three subspecies *Globorotalia menardii, G. m. tumida,* and *G. m. flexuosa,* was chosen because it characterizes certain well-defined faunal zones that are widespread in the equatorial Atlantic and in the Caribbean and the Gulf of Mexico. In addition, the frequency is highly variable from sample to sample, and this suggests that the complex is especially sensitive to changes in the environment. In

the cores discussed here, the number of specimens counted in each sample varies from zero to over 2000. Samples were taken at 10-centimeter intervals in the cores.

Counts of left- and right-coiling shells of *Globorotalia truncatulinoides* have been made, and recorded as percentages of shells coiling in the dominant direction in the total count of the species. This method, described in detail by Ericson *et al.,*[11] is especially useful as a check on other methods, for it permits identification and cross correlation of layers no more than a few centimeters thick. Such precise correlations provide a valuable test of continuity of accumulation.

11. D. B. Ericson, G. Wollin, and J. Wollin, *Deep-Sea Res.* 2, 152 (1954).

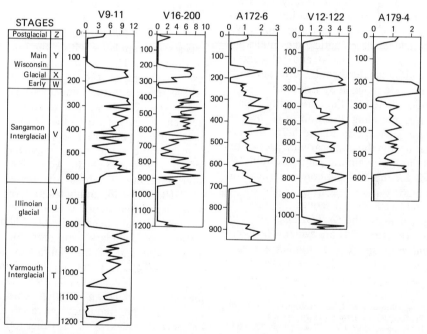

Figure 5-24. Frequency curves for the *Globorotalia menardii* complex in five deep-sea sediment cores. Samples were taken at 10-centimeter intervals from top to bottom of the cores. The scales at the tops of the columns are ratios of the number of shells of the *G. menardii* complex to the total population of foraminifera in the samples. Since the magnitudes of these ratios differ consistently from core to core, we have plotted the ratios for each core on a different scale in order to show the climatic zones more clearly. We have correlated the climatic zones indicated by the curves with glacial and interglacial stages. The numbers to the left of the columns are depths in cores, in centimeters.

CLIMATIC INTERPRETATION

Studies by Schott,[4] Cushman and Henbest,[12] Phleger,[13] Ovey,[14] Parker[15] Wiseman,[16] Ericson et al.,[17] and others, of the planktonic foraminiferans in deep-sea sediment cores have shown that the relative abundances of the species that are sensitive to temperatures vary from level to level. These investigators agree that the variations record shifts in the geographical ranges of the species and that these shifts were a consequence of the climatic changes of the late Pleistocene. Like Schott, we believe that the *Globorotalia menardii* complex is the most sensitive and reliable of the climatic indicators.

Figure 5-24 shows the ratios of the number of shells of the *Globorotalia menardii* complex to the total population of foraminifera in the sample, for five cores. We have correlated the climatic zones indicated by the curves with glacial and interglacial stages, and designate them by letters according to the system begun by Ericson.[18]

The correlation of the cores with respect to changes in coiling direction of *G. truncatulinoides* is shown in Figure 5-25. Age determination by the radiocarbon, protactiniumionium, protactinium, and thorium-230 methods for three of the cores are shown in Figure 5-26.

Several well-defined zones in the frequency curves of the *Globorotalia menardii* complex (Figure 5-24) are evident. *Globorotalia menardii* and *G. m. tumida* are very abundant, but *G. m. flexuosa* is absent, in the Z zone, which we believe represents the

12. J. A. Cushman and L. G. Henbest, U.S. Geol. Survey Profess. Paper 196-A (1940), p. 49.
13. F. B. Phleger, *Bull. Geol. Soc. Amer.* 53, 1073 (1942).
14. C. D. Ovey, Centenary Proc. Roy. Meteorol. Soc. 1950, 211 (1950).
15. F. L. Parker, *Rep. Swedish Deep-Sea Expedition* 8, 219 (1958).
16. J. D. H. Wiseman, *Proc. Roy. Soc. London Ser. A* 222, 296 (1954).
17. D. B. Ericson, M. Ewing, G. Wollin, and B. C. Heezen, *Bull. Geol. Soc. Amer.* 72, 193 (1961).
18. D. B. Ericson, *Ann. N.Y. Acad. Sci.* 95, 537 (1961).

postglacial section. In the Y zone the *G. menardii* complex is abundant, particularly the subspecies *G. m. flexuosa*. This zone characterized by *G. m. flexuosa* is present in cores from the Gulf of Mexico and the Caribbean, and in many cores from widely scattered stations in the Atlantic. Thus the evidence that the Y zones are equivalent in the five cores is good. In the W zone the *G. menardii* complex is absent or rare. In these five cores from low-latitude stations, this layer is relatively thin. Further north we have found it to be thicker, probably because of more rapid accumulation of the fine terrigenous fraction. The Y, X, and W zones we correlate with the Wisconsin glacial stage.

The *Globorotalia menardii* complex is consistently abundant (although the frequency varies markedly) in the V zone, which we correlate with the Sangamon interglacial stage. *Globorotalia menardii flexuosa* appears in abundance within the lower part of this zone. The *G. menardii* complex is rare or absent in the U zone, which we believe represents the Illinoian glacial stage.

Figure 5-27 shows the variation in the frequency curves for the *Globorotalia menardii* complex in five cores which penetrate more climatic zones than the cores shown in Figure 5-24 do. According to our interpretation, in three of these cores (V12-18, V22-136, and V16-205) the complete Pleistocene section is represented. Core V16-39 almost penetrates the Pliocene-Pleistocene boundary, and core V19-297 reaches the end of the Nebraskan glacial, about 1.8 million years ago. The Jaramillo event was not detectable in V19-297, but the good correlation of that core, with respect to the frequency curve of the *G. menardii* complex, with the other four cores convinces us that absence of evidence of the Jaramillo event in V19-297 must be due to the difficulty of detecting magnetic reversals in unoriented cores from stations near the equator.

The coiling of *Globorotalia truncatulinoides* is also a valuable indicator of climate in the earlier zones of the Pleistocene, left coiling being dominant during times of cold climate, and vice versa.

The correlation of the three cores of this study from the South Atlantic (Figure 5-23) with respect to changes in coiling direction of *Globorotalia truncatulinoides* is shown in Figure 5-28. The species was too infrequent in one of the cores, V19-297, to yield meaningful counts. The coiling curve of core V16-205 from the North Atlantic is not shown in Figure 5-28 because correlation is poor between this core and the three cores from the South Atlantic. Since, as has been

shown by Ericson *et al.*,[11] coiling of the living populations of *G. truncatulinoides* varies with latitude, similar variations probably occurred, at least at some periods, in the past. Therefore, close correlation of coiling curves in cores from stations widely separated in latitude is not to be expected.

As shown in Figure 5-27, the T zone, which we correlate with the Yarmouth section, is somewhat different from the V zone, and from the R zone, which, we believe, represents the Aftonian section. The population of the *Globorotalia menardii* complex

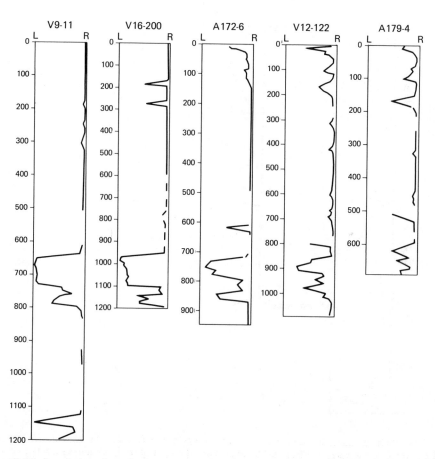

Figure 5-25. Correlation of the five cores of Figure 5-24. The correlating levels are defined by changes in the direction of coiling of *Globorotalia truncatulinoides* as observed in core samples taken at 10-centimeter intervals. The scale runs from 100 per cent left-coiling at the left-hand margins of the columns to 100 per cent right-coiling at the right-hand margins. Numbers to the left of the columns are depths in cores, in centimeters.

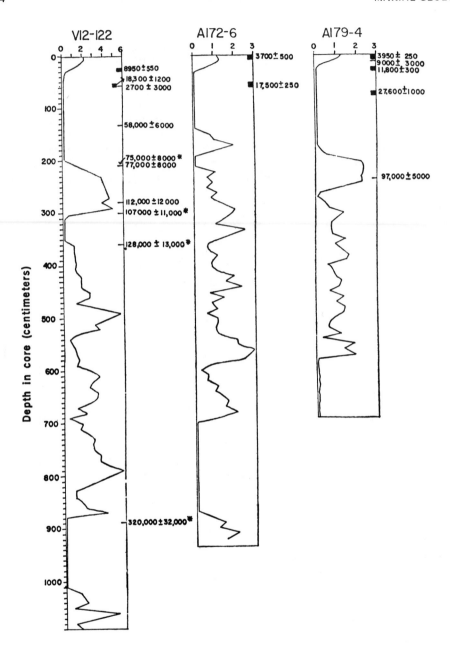

Figure 5-26. Radiochemical age determinations and frequency curves for the *Globorotalia menardii* complex in three of the cores of Figure 5-23. Numbers to the right of the black squares (sections of cores used for dating) are radiocarbon ages, in years. Numbers to the right of the horizontal lines of core A179-4 are dates determined by Rosholt *et al.*[32] by the protactinium-ionium method; those to the right of core V12-122 are dates determined by Sackett[33] by the protactinium method, and by Ku and Broecker[34] by the thorium-230 method (those obtained by Ku and Broecker are marked with an asterisk). The horizontal lines opposite the dates indicate the mid-depths of the samples used for dating by the three methods.

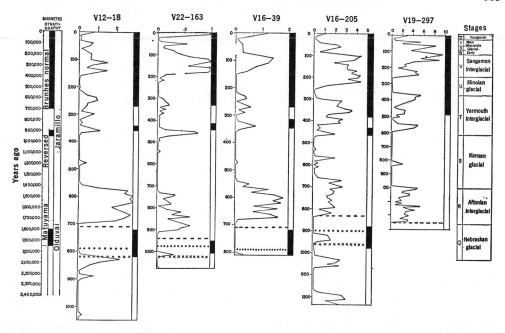

Figure 5-27. Frequency curves for the *Globorotalia menardii* complex and magnetic and paleontological stratigraphy in five cores. Samples were taken at 10-centimeter intervals from top to bottom of the cores. The scales at the tops of the columns are ratios of the number of shells of the *G. menardii* complex to the total population of foraminifera in the samples. Since the magnitudes of these ratios differ consistently from core to core, we have plotted the data for each core on a different scale in order to show the climatic zones more clearly. We have correlated the climatic zones indicated by the curves with glacial and interglacial stages. The time scale for the magnetic stratigraphy in the column at left is from Pitman and Heirtzler.[35] The magnetic stratigraphy for each core is designated by black (normal) and white (reversed) areas to the right of the columns (dashed lines). The level at which *Globorotalia* sp. 1 became extinct; (dotted lines) the level at which *Discoasteridae* became extinct; (X-lines) the first appearance of abundant *G. truncatulinoides*, the criterion used for defining the Pliocene-Pleistocene boundary. Numbers to the left of the columns are depths in cores, in centimeters.

in the T zone makes up, on the average, a smaller proportion of the total assemblage of planktonic foraminiferans. Especially in the three cores from the South Atlantic (V12-18, V22-163, and V16-39), the Yarmouth sections, as indicated by the frequency curves of the *G. menardii* complex, are not so prominent as they are in cores from the North Atlantic and the equatorial Atlantic (cores V16-205, V19-297, and V9-11) (Figure 5-24). The evidence suggests that the climate of the Yarmouth differed from that of the other two interglacials; we infer that the temperature of the water of the Atlantic was lower in Yarmouth time

than it was during the other two interglacials. Another difference between the three interglacials is the fact that *G. m. flexuosa* occurs in the major part of the Sangamon section but is absent in the Yarmouth and Aftonian sections.

The *Globorotalia menardii* complex is absent or rare in the S and Q zones, which we correlate, respectively, with the Kansan and Nebraskan glacial stages. These two glacial sections differ in that the Discoasteridae and *Globorotalia* sp. 1[15,19] become extinct in

19. F. B. Phleger, F. L. Parker, and J. F. Peirson, *Rep. Swedish Deep-Sea Expedition* 7, No. 1, 1 (1953).

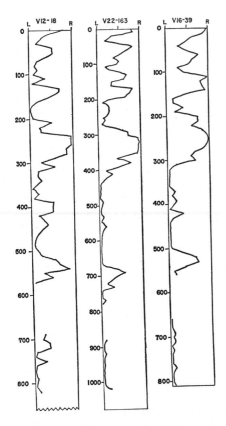

Figure 5-28. Correlation of three of the cores of Figure 5-27. The correlating levels are defined by changes in the direction of coiling of *Globorotalia truncatulinoides* as observed in core samples taken at 10-centimeter intervals. The scale runs from 100 per cent left-coiling at the left-hand margins of the columns to 100 per cent right-coiling at the right-hand margins. Numbers to the left of the columns are depths in cores, in centimeters.

the Nebraskan, whereas they are, necessarily, quite absent from the Kansan section. Only the lower part of one of the cores, V16-205, contains *Globigerinoides sacculifera fistulosa,* and this species disappears at about the level where the discoasters do. The change from right to left in the coiling direction of the members of the *Globorotalia menardii* complex occurs in two of the cores, at depth of 1030 centimeters in V12-18 and at 1170 centimeters in V16-205. The Pliocene-Pleistocene boundary is indicated by the first appearance of abundant *Globorotalia truncatulinoides.*

The *Globorotalia menardii*-complex curves obtained from cores V12-18 and V16-205 show cold zones extending beyond the Pliocene-Pleistocene boundary, as determined by the first appearance of abundant *G. truncatulinoides.* Thus, it seems that the change toward a colder climate began before the beginning of the Pleistocene.

The fact that the geographical spread of the cores used in this study is from 74°W to 1°E and from 17°N to 26°S is significant. Evidently variations in the frequency curve of the *Globorotalia menardii* complex occurred on a wide scale. Also, zone-by-zone correlation of the major features of the curves shown in Figures 5-24 and 5-27 are continuous over a time interval of more than 1.8 million years. Furthermore, there is nothing in the curves to suggest that evolutionary variation in the *G. menardii* complex has caused an increase in sensitivity to change in ecological conditions; the change from "absent" to "frequent" that occurred about 1.7 million years ago is as clearly defined as later changes in the frequency curve of the complex.

THE PLIOCENE-PLEISTOCENE BOUNDARY

The crux of the problem of finding a complete record of the Pleistocene has been to find tangible evidence of the beginning of the epoch. The Pliocene-Pleistocene boundary in deep-sea sediments was first defined by Arrhenius,[20] primarily on the basis of a sharp upward increase in calcium carbonate content in some cores from the eastern equatorial Pacific. On the basis of evidence from the cores, Arrhenius estimated the duration of the Pleistocene to be about 1 million years. Riedel[21] suggested that the level of

20. G. Arrhenius, *Rep. Swedish Deep-Sea Expedition* 5, pt. 3, 194 (1952).
21. W. R. Riedel, *Rep. Swedish Deep-Sea Expedition* 6, No. 3, 65 (1957).

extinction of two radiolarian species (*Pterocanium prismatium* and *Eucyrtidium elongatum peregrinum*) in tropical Pacific cores may serve to mark the "top of Pliocene." Riedel *et al.*[22] subsequently found that *P. prismatium* ranged higher than *E. e. peregrinum.*

We established[5,23] the following faunal criteria for defining a Pliocene-Pleistocene boundary in deep-sea sediments: (i) extinction of Discoasteridae; (ii) first appearance of abundant *Globorotalia truncatulinoides;* (iii) change of *G. menardii* complex from diverse below the boundary to a more uniform lineage above it, increase in average test size and reduction in number with respect to total population above the boundary, and change from 95 per cent sinistrally coiled tests above it; and (iv) extinction of *Globigerinoides sacculifera fistulosa* above the boundary. The thickness of the zone within which these changes occurred differed somewhat in the various cores.

Having examined the coccoliths in seven of the cores we had studied, McIntyre *et al.*[24] concluded that the boundary we had defined was the boundary between Aftonian interglacial and Nebraskan glacial. Bandy[25] expressed the opinion that the sections in the cores which we believed to be of Pliocene age were actually of latest Miocene age, and that the marked extinction of Discoasteridae in the cores was due to an unconformity, separating Miocene and Pleistocene sediments and representing a gap of some 10 million years of Pliocene time. However, Glass *et al.*[9] made paleomagnetic measurements on some of the cores and reported evidence that sediment accumulation across the boundary was continuous in two of the five cores in which meaningful magnetic measurements could be made. Glass *et al.*

concluded that the Pliocene-Pleistocene boundary, as indicated by first evolutionary appearance of *Globorotalia truncatulinoides,* occurred at the base of the Olduvai event, about 2 million years ago, instead of about 1.5 million years ago as Ericson *et al.* had estimated.

Emiliani,[26] on the basis of oxygen isotope analysis of cores, has estimated the base of the glacial Pleistocene variously at 300,000 to 425,000 years ago, and the base of the Pleistocene epoch at 600,000 to 800,000 years ago.

Recently Bolli and his associates[27] have identified the faunal changes of the Pliocene-Pleistocene boundary in a sediment section partially cored on the Nicaragua rise in the Caribbean. Because the core recovered included only 40.5 per cent of the sediment section, it is perhaps not surprising that these investigators misidentified the magnetic reversal that coincides with the faunal changes as that marking the beginning of the Brunhes Normal Epoch. Accordingly they mistakenly assign an age of 700,000 years to the boundary.

In a study of radiolarians in deep-sea cores from Antarctic seas, Hays[28] recognized four faunal zones, the boundary between the lower two being marked by a striking faunal change and by a change from red clay to diatom ooze. He suggested that this boundary coincided approximately with the Pliocene-Pleistocene boundary of Ericson and his associates. Opdyke *et al.*[8] mined the paleomagnetic stratigraphy of several Antarctic cores containing Hays's boundary and found that the boundary occurs near the base of the Olduvai event, about 2 million years ago.

Banner and Blow,[29] working in the Cala-

22. W. R. Riedel, M. N. Bramlette, and F. L. Parker, *Science* 140, 1238 (1963).
23. D. B. Ericson, M. Ewing, and G. Wollin, *Science* 139, 727 (1963).
24. A. McIntyre, A. W. H. Be, and R. Preikstas, *Progr. Oceanogr.* 4, 3 (1967).
25. O. L. Bandy, *Science* 142, 1290 (1963).

26. C. Emiliani, *Ann. N.Y. Acad. Sci.* 95, 521 (1961); J. Geol. 74, 109 (1966).
27. H. M. Bolli, J. E. Boudreaux, C. Emiliani, W. W. Hay, R. J. Hurley, and J. J. Jones, *Bull. Geol. Soc. Amer.* 79, 459 (1968).
28. J. D. Hays, *Amer. Geophys. Union Antarctic Res. Ser.* 5, 125 (1965).
29. F. T. Banner and W. H. Blow, *Micropaleontology* 13, 133 (1967).

brian type section in Italy, have shown that *Globorotalia truncatulinoides* evolved from *G. tosaensis* and have suggested that the first appearance of *G. truncatulinoides* may be a suitable criterion for defining the Pliocene-Pleistocene boundary. Their boundary, therefore, is approximately the same as ours,[5,23] since we placed the Pliocene-Pleistocene boundary in deep-sea sediment cores at the point where *G. truncatulinoides* first appears in abundance. Thus, deep-sea sediments can be correlated with the continental type Pliocene-Pleistocene section in Italy. Selli,[1] by extrapolation of sedimentation rates, arrived at a date of 1.8 million years ago for the Pliocene-Pleistocene boundary, on the basis of the first evolutionary appearance of *G. truncatulinoides,* in the Calabrian type section.

In a recent study of deep-sea sediment cores by Berggren *et al.,*[2] the Pliocene-Pleistocene boundary, as determined by the first evolutionary apperance of *Globorotalia truncatulinoides,* was found to occur within the Olduvai event, about 1.85 million years ago.

As shown in Figure 5-27, the Pliocene-Pleistocene boundary in our cores, as indicated by the first appearance of abundant *Globorotalia truncatulinoides,* occurs at the base of the Olduvai event, about 2 million years ago.

CHRONOLOGY

Paleomagnetic studies of volcanic rocks on land have established the occurrence of a series of reversals of polarity of the earth's magnetic field and have provided dates for these events, obtained by means of the potassium-argon method.[30] It is known that the earth's field has had its present, or normal, polarity for the last 700,000 years. This period has been named the Brunhes normal epoch. For approximately 1.7 million years

before the Brunhes normal epoch the earth's field had an opposite or reversed polarity; this period has been named the Matuyama reversed epoch. Within the Matuyama reversed epoch, two short periods of normal polarity occurred, at about 900,000 and 1.9 million years ago; these have been termed the Jaramillo and Olduvai events, respectively. It has recently been shown that the remanent magnetism of deep-sea sediments is sufficiently strong and stable that these polarity reversals can be used to date and correlate geological events recorded in these sediments

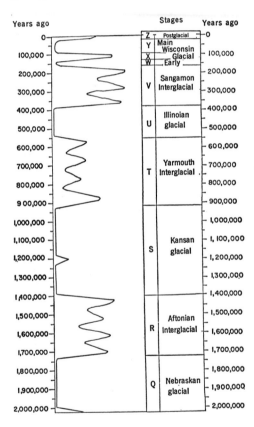

Figure 5-29. Pleistocene time scale, based on magnetic reversals in the five cores of Figure 5-27, a generalized curve based on variation in the frequency of the *Globorotalia menardii* complex in the ten cores used in this study, and a proposed correlation with the stages of the Pleistocene. The beginning of the Pleistocene is defined by the first appearance of *G. truncatulinoides* in abundance.

30. A. Cox, R. R. Doell, and G. B. Dalrymple, *Science* 144, 1537 (1964); A. Cox and G. B. Dalrymple, *J. Geophys. Res.* 72, 2063 (1967).

throughout the world during the past 5 million years.[2,8-10,31]

Paleomagnetic studies of four of the cores shown in Figure 5-27 have been made by Glass, and the fifth core, V220163, has been studied by N. D. Opdyke of the Lamont-Doherty Geological Observatory.

As shown in Figure 5-27, the Pliocene-Pleistocene boundary, as defined by the first appearance of *Globorotalia truncatulinoides* in abundance, occurs as or near the base of the Olduvai event, about 2 million years ago. Discoasteridae became extinct within the Olduvai event about 1.9 million years ago, and *Globorotalia* sp. 1 became extinct near the top of the Olduvai event, about 1.8 million years ago.

31. D. Ninkovich, N. D. Opdyke, B. C. Heezen, and J. Foster, *Earth Planet. Sci. Letters* 1, 476 (1966).
32. J. N. Rosholt, C. Emiliani, J. Geiss. F. F. Koczy, and P. J. Wangersky, *J. Geol.* 69, 162 (1961).
33. W. M. Sackett, in "Symposium on Marine Geochemistry," Graduate School of Oceanog. Univ. Rhode Island Occasional Pub. No. 3 (1965).
34. T. L. Ku and W. S. Broecker, *Science* 151, 48 (1966).
35. W. C. Pitman III and J. R. Heirtzler, *Science* 154, 1164 (1966).
36. We are grateful to Maurice Ewing for his generous support of our research work and his active participation in it. We thank Janet Wollin, Julie McGowan, and Anne Riley for able assistance in the laboratory investigations. We also thank Arnold Finck and other members of the staff of Lamont-Doherty Geological Observatory who, directly or indirectly, contributed to this study. We are grateful to N. D. Opdyke, B. Glass, and J. Foster for magnetic stratigraphy determinations and to W. S. Broecker, D. L. Thurber, and Hans Suess for radiocarbon determinations. The collection, preservation, and analysis of the cores were supported in large part by grants and contracts with the National Science Foundation (GP-4376, GA-982, and 1193) and the Office of Naval Research (TO-4). This is Lamont-Doherty ·Geological Observatory contribution No. 1278.

Figure 5-29 shows a time scale based on magnetic reversals in the five cores of Figure 5-27; our generalized climate curve, based on variation in the frequency of the *Globorotalia menardii* complex in the ten cores used in our study; and our proposed correlation with the stages of the Pleistocene. In comparison with the time scale of 1.5 million years that we had proposed earlier,[5] this revised Pleistocene time scale of 2 million years increases the durations of the three earliest climatic events, Q, R, and S. The S zone, considered by us to be equivalent to the Kansan glacial stage, now seems to have been of longer duration than the T zone, which we equate with the Yarmouth interglacial.

CONCLUSION

Our Pleistocene climatic record of four major glaciations and three interglacial stages is the result of our study of variations in the frequency of the *Globorotalia menardii* complex in ten cores from the Atlantic and the Caribbean. Evidence for continuity of the Pleistocene record is also provided by changes in the coiling direction of *G. truncatulinoides.*

The most important criteria which distinguish Pleistocene pelagic sediments from sediments of earlier epochs of the Cenozoic Period are the general occurence of *Globorotalia truncatulinoides* in abundance and the absence of discoasters.

Our time scale, based on magnetic reversals, dates the beginning of the Pleistocene, as defined by the first apperance of *Globorotalia truncatulinoides* in abundance, at about 2 million years ago.

6

OCEAN RESOURCES

Aquaculture

JOHN E. BARDACH   1968

INTRODUCTION

Optimistic forecasters of the world food supply in the year 2000 predict that serious regional food shortages will exist; pessimists warn of widespread famine as populations in many countries increase more rapidly than agricultural production. The one bright spot in this otherwise gloomy picture was the rate of growth of 8 per cent per year of the world fish harvest during the last decade.[1] The annual harvest in the mid 1960's amounted to 53 million metric tons[2] and is expected to increase for some time to come, though estimates of the possible total sustained yield differ widely, ranging from 120 to 2000 million metric tons.[3] All are based on extrapolations from scanty data on primary productivity throughout the world's

seas and on incomplete knowledge of the food relations of many harvested organisms.

Aquatic harvests supply almost nothing but proteins, since the bulk of aquatic plants are plankton algae that are uneconomical to harvest[4] and hold no promise for furnishing carbohydrate staples. A breakdown of the total yield from aquatic ecosystems into marine, fresh, and brackish water moieties shows that the yields from fresh and brackish water make up about 14 per cent of the meat fare.[2] Much of man's fishing has been in estuaries and bays of the sea, and in ponds, lakes, reservoirs, and streams, where he has long practiced aquaculture.[5]

Aquaculture resembles agriculture rather than fisheries in that it does not rely on a common property resource but presumes ownership or leased rights to such bases of production as ponds or portions of, or sites in, bays or other large bodies of water. Prod-

1. W. M. Chapman, *Food Technol.* 20, 895 (1966).
2. Editorial, *Yearb. Fish. Stat.* 20 (1965).
3. President's Science Advisory Council, The World Food Problem (U.S. Government Printing Office, Washington, D.C., 1967), vol. 2, pp. 345-361.

4. L. A. Walford, *Living Resources of the Sea* (Ronald, New York, 1958), pp. 121-132.
5. I prefer the spelling aquaculture to aquiculture because the former is etymologically more correct.

Dr. John E. Bardach is director of the Hawaii Institute of Marine Biology and professor of zoology on the Manoa campus of the University of Hawaii. He studies the ecology of aquatic natural resources with particular reference to fisheries development.

From Science, Vol. 161, No. 3846, pp. 1098-1106. Reprinted with author's revisions and light editing and by permission of the author and the American Association for the Advancement of Science. Copyright 1968 by the American Association for the Advancement of Science.

ucts of aquaculture must compete successfully with those of fisheries and of animal husbandry; in Western food economy, aquaculture products such as trout, oysters, and shrimp bring good returns because they fall in the luxury class, whereas in developing countries various kinds of raised fish command a high price,[6] since animal protein, including that derived from marine catches, is generally scarce. Although subsidized small home or village ponds may be justified in certain underdeveloped areas to help alleviate malnutrition, aquaculture, wherever it is practiced, should be examined primarily as a commercial enterprise that must compete with other protein supplies to be successful.

The organisms now being raised in aquaculture comprise several bivalve mollusks (mainly oysters of the genera *Ostrea* and *Crassostrea*), a few crustaceans (predominantly shrimp, in particular *Peneus japonicus*), and a limited number of fish species.[6] Among the fish species, the carp, *Cyprinus carpio,* and selected other members of the same family, Cyprinidae (minnows), are the most important. Trout and salmon are also important in aquaculture, especially the rainbow trout, *Salmo gairdneri,* as are the Southeast Asian milkfish (*Chanos chanos*), and mullets, especially *Mugil cephalus,* and yellowtail (*Seriola quinqueradiata*), in Japan. Also noteworthy are the channel catfish industry (*Ictalurus punctatus*) in the southern United States and the use of *Tilapia* as pondfish, mainly in Africa.[6] Most of these species are adjusted to life in fresh or brackish

water, but the culturing of some marine fishes is being attempted, notably in Great Britain with plaice (*Pleuronectes platessa*) and sole (*Solea solea*)[7] and with pelagic (high seas) schooling species in Japan and the United States, among them the Pacific sardine and mackerel (*Sardinops caerula, Pneumatophorus diego*[8] and the pompano (*Trachinotus carolinus*).[9] Some attached algae are also produced under semi-cultivation, both in temperate and tropical waters and certain phytoplankton species are cultured as food for oyster and shrimp larvae. These represent a special crop and are of minor or indirect nutritive value; they are omitted from this article, in which the potential of aquaculture for supplying high-quality protein is assessed.

In 1970, aquaculture furnished the world with over 3 million metric tons, mainly fish. Mainland China alone reports annual production of 1.5 million metric tons of carp and carplike fishes.[6] Three million metric tons represent about 5 per cent of the total world catch and far exceed the United States food fish harvest, although they are produced from a fraction of 1 per cent of the world's waters.

Aquaculture ranges in intensity from simple weeding of natural stands of algae to complete husbandry of domesticated fish like trout or carp. It is sometimes difficult to distinguish intensive management from culture. The term, as used here, comprises practices that subject organisms to at least one, and usually more than one, manipulation before harvest. In addition, as in agriculture, the harvest in aquaculture takes most, if not all, the organisms tended. Most often only one species is raised, although a few to several compatible species may be cultured simultaneously.

To be productive for husbandry, aquatic animals should have the following characteristics. (i) They should reproduce in captivity

---

6. In 1970, 23,000 metric tons of channel catfish are expected to be harvested in the lower Mississippi River states, and this production will double again in 1972 [editorial, Comm. Fish. Rev. 30 (5), 18 (1968)]; J. E. Bardach and J. H. Ryther, The Status and Potential of Aquaculture, Particularly Fish Culture, prepared for National Council on Marine Resources and Engineering Development 1967, PB 177768 (Clearinghouse Fed. Sci. Tech. Info., Springfield, Va., 1968); J. H. Ryther and J. E. Bardach, The Status and Potential of Aquaculture, Particularly Invertebrate and Algae Culture, prepared for National Council of Marine Resources and Engineering Development, PB 177767 (Clearinghouse Fed. Sci. Tech. Info., Springfield, Va., 1968).

7. J. E. Shelbourne, *Advan. Mar. Biol.* 2, 1 (1964).
8. G. O. Schumann, personal communication.
9. C. P. Idyll, personal communication.

or semi-confinement (for example, trout) to make selective breeding possible or yield easily to manipulations that result in the production of their offspring (for example, carp). Failing ease of breeding, their larvae or young should be easily available for gathering (for example, oysters). (ii) Their eggs or larvae or both should be fairly hardy and capable of being hatched or reared under controlled conditions. (iii) The larvae or young should have food habits that can be satisfied by operations to increase their natural foods, or they should be able to take extraneous feeds from their early stages (iv) They should gain weight fast and nourish themselves entirely or in part from abundantly available food that can be supplied cheaply, or that can be readily produced or increased in the area where the cultured species lives.

Few aquatic organisms have all these attributes, and substantial expansion of the aquacultural crop depends in part on how biological and engineering skills can make the missing characteristics less crucial; other constraints are economic. I discuss here several operations and problems common to the raising of aquatic organisms,[10] and I attempt to appraise realistically the potential of aquaculture on a world scale.

SELECTIVE BREEDING OF AQUATIC STOCK

Even before Jacob tended Laban's flocks, livestock had been subjected by man to selection for one or another desirable attribute, and breeding of domestic birds and animals has produced spectacular results. The first treatise on fish culture was reportedly written in China by Fan Li in 475 B.C.,[11] but there are still only two aquatic animals over which genetic control has been

exercised. These are carp and several species of trout; trout has a shorter and less varied history of breeding than carp. No true breeding programs exist with invertebrates, though oyster culture is advancing so rapidly that experiments in oyster genetics are likely to begin soon.

The breeding of aquatic animals, compared with terrestrial animals, has peculiar problems. Spawning habits often make the isolation of pairs difficult; isolation of numerous offspring requires many replicate ponds or other instruments of confinement; and there is rarely more than one mating a year. Moreover, the environment has an overriding influence on the growth of poikilothermous animals; consequently, many different-sized animals of the same age are found together. Many aquatic animals require special environmental or social conditions for mating and reproduction, which are not easily duplicated under human control. Manipulations of water temperature or flow have triggered spawning; however, the development during the last two decades of the practice of hypophyzation, or treatment with pituitary hormone,[12] to make some fishes spawn helps alleviate constraints on breeding for some species. This practice has influenced fish culture all over the world, from catfish growers in the southern United States and sturgeon breeders in the Ukraine, to the fishpond cooperatives of mainland China, where it is of paramount importance in making common carp produce eggs three times a year and in facilitating the propagation of its cyprinid pond mates, whose eggs were difficult to collect in rivers before the process of hypophyzation was developed. Use of pituitary material may also produce advances with the breeding of two species of fish important in brackish water culture—the milkfish and the gray mullet. Aquatic animals have one advantage over terrestrial animals from the breeder's point of view—a pair

10. The study of the status and potential of aquaculture was financed by a contract with the National Council on Marine Research and Engineering Development, Executive Office of the President.
11. W. A. Dill, Proc. World Symp. Warm Water Pond Fish Culture, *Fish. Rep.* 44 (1), i (1967).

12. H. P. Clemens and K. E. Sneed, Bioassay and Use of Pituitary Materials to Spawn Warm-Water Fishes, Res. Rept. 61 (U.S. Fish and Wildlife Service, Washington, D.C., 1962), 30 pp.

have large numbers of offspring, which permit mass selection.

Carp are readily adaptable to selective breeding because their eggs are large for fish eggs; they are not too delicate; and they are easily secured. Carp have been bred for fast growth, a body shape with more flesh than is found on the wild type, reduction of scale cover for greater ease in preparing the fish for the table, resistance to disease, and resistance to crowding and to low temperature.[13] With such breeding practices as progeny testing (selection of parents according to the performance of their offspring) and diallele analysis (a system of mating that determines separately the genotypes of each parent),[14] further improvements on already well-domesticated strains may be expected. It is necessary to prevent reversions to the wild type. That these can occur rapidly is illustrated by the fate of carp introduced to America. After being brought to the New World in 1877,[15] carp was allowed to escape into lakes and rivers where indiscriminate mixtures of its prolific stock resulted in bony and scaly fish which soon became a nuisance in waters used for game fishing. There was no incentive for carp culture in the United States, where protein was abundant from land livestock. However, since carp has become a prized angling trophy in Western Europe, and because of the rapid eutrophication of American lakes and rivers (a process which favors carp) and the predicted narrowing of the gap, even in America, between the supply of terrestrial protein and the demand, it is not farfetched to think that carp may be cultured in the United States.

Trout, at least in America, were until recently raised mostly for stream stocking; consequently, disease resistance was the main concern in hatcheries not equipped for experiments in fish genetics. Demonstration of what may reside in the trout's gene pool has come mainly from two sources: the Danish table trout industry[16] and the experimental trout and salmon breeding program of the University of Washington in Seattle,[17] where specially fed rainbow trout stock, continuously graded by selection during 30 years, grows to as much as 3 kilograms in 18 months while a wild rainbow trout in a lake at that age rarely weighs 200 grams (Figure 6-1);[18] these fish tolerate higher temperatures than their wild congeners.

Trout and salmon eggs are larger than those of the carp; they develop slowly and are hardy, combining several advantageous properties. Salmon permit the establishment of hatcheries on suitable streams because they return to spawn to the stream with the odor of which they were imprinted as fry. In such hatcheries inadvertent selection from the spawning run of the largest—fastest growing—brood fish has produced strains that returned to the hatchery 1 year earlier than the offspring of their wild congeners. Salmonid fishes can be selected for higher fecundity, larger egg size, and better survival and faster growth of fry, and for exact timing of their return to the parent stream.[19] These breeding potentials should be used to increase the abundance of salmon especially since improved techniques now feasible in United States salmon hatcheries could produce about ten times as many young fish as are now released.[20]

Salmon-fishing regulations are still based on propagation potentials in natural streams and require that 50 per cent of the run be

13. W. Steffens, *Verh. int. Ver. Limmol.* 16 (3), 1441 (1967).
14. R. Moav and G. Wohlfarth, *Bamidgeh* 12, 5 (1960).
15. Departments of Commerce and Labor, Fisheries of the U.S. 1908; Special Report (U.S. Government Printing Office, Washington, D.C., 1911), p. 49.

16. F. Bregnballe, *Progr. Fish. Culturist* 25 (3), 115 (1963).
17. L. R. Donaldson and D. Manasveta, *Trans. Amer. Fish. Soc.* 90, 160 (1961).
18. K. D. Carlander, *Handbook of Freshwater Fishery Biology* (Brown, Dubuque, Iowa, 1950), pp. 30-36.
19. L. R. Donaldson, Proc. Pac. Sci. Congr. Tokyo Sci. Counc. 11th 7, 4 (1966).
20. N. Fredin, personal communication.

Figure 6-1. Wild and mass-selected, hatchery-fed rainbow trout, at the University of Washington School of Fisheries, Seattle. Fish are 2 years old, and the large one is the result of over 30 years of selective breeding. Their respective lengths can be estimated from the diameter of the bucket base, which is 22 centimeters.

allowed to escape the fishery. Salmon runs will increasingly depend on hatcheries that program their fish to return for stripping and the raising of a well-protected progeny whose rate of survival at the time of release is many times greater than that attained in nature, where maximum fish mortality takes place during the first few months of life. Thus, the salmon harvest of certain river mouths may almost be doubled in view of the fact that hatchery-dependent runs need only a few fish to supply the next generation. Salmon are highly valuable fish ($65 million for the United States catch in 1965),[21] and it may be worthwhile to press for regional revision of escapement regulations and to examine the economic requirements and consequences of hatchery improvements.

Another advantage of breeding fishes is the ease with which many of them hybridize.[22] At the University of Washington at Seattle, male steelhead (that is, seagoing rainbow) trout were crossed with fast-growing freshwater rainbow females. The growth rate of the offspring was intermediate between that of the parents, their shape was more fusiform than that of the female, and they migrated to sea. They had a voracious appetite and adopted parent streams to which they returned as 2-year-olds, weighing 2 to 3 kilograms on the average and occasionally as much as 5 kilograms.[23] They would probably not breed true in the second generation, and they should therefore be hatchery-produced, but they represent an interesting use of the sea's unused fish food.

Difficult as it may be to raise the progeny of one pair of parents of carp and trout, to do so with oysters is still more complicated. Mass spawning is usually done on oyster beds, and although the female of the genus *Crassostrea* to which the American oyster belongs, retains the eggs inside her shells until after fertilization, paternity on the oyster bed is impossible to ascertain. Al-

21. Fishery Statistics of the United States 1965, Statistical Digest No. 59 (U.S. Department of Interior, Washington, D.C., 1967), pp. 541-547.
22. C. L. Hubbs, in *Vertebrate Speciation*, W. F. Blair, Ed. (Univ. of Texas Press, Austin, 1961), pp. 5-23.

23. L. R. Donaldson, personal communication.

though there are thousands of eggs for each carp or trout, there can be up to 100 million for each female oyster.[24] This fact, however, has aided in mass selection. Progressive growers of Long Island oysters raise the larvae in warmed water and use cultured algal food. They also give proper attention to stirring and other manipulations simulating planktonic conditions. The many eggs and improved survival of free-floating larvae permit a filter screen to be used to select only 20 per cent of the largest, most rapidly growing early larvae. These larvae exhibit good growth throughout life.[6] But to achieve true selective breeding, growers of Long Island oysters now plan to rear single oyster progeny; since oysters reverse their sex from male to female halfway through their adult lives, the possibility of freezing sperm from a functional male is being tested, and it may be possible to use it to fertilize the same oyster later when it becomes a female.[25]

THE RAISING OF AQUATIC LARVAE

Many aquatic animals go through larval stages which do not resemble their adult phase; some larvae, including those of shrimp or oysters, and of many fishes, are planktonic and minute and feed on the smallest organisms. More than with domestic birds or mammals, nursing them through their early lives poses difficult technical and nutritional problems to growers. In British experiments with raising plaice and sole larvae in captivity, as much as 66 per cent survival through the stage of metamorphosis has been accomplished. Ultraviolet treatment of the water decreased the danger of bacterial infection; tanks without corners minimized encounters with solid obstacles; and salinity, temperature, and pH were controlled. The size of the first food offered was geared to the tiny mouthparts of the larvae,

but was increased with their capacity to take larger live food. Nauplii of the barnacle (*Balanus balanoides*) were used at first; they were replaced by nauplii of the brine shrimp (*Artemia*) with subsequent admixtures of small oligochaetes (*Enchytrea*) when the small fishes had metamorphosed and were resting on the bottom. Finally, chopped mussels (*Mytilus*) were used. Since plaice larvae, just before settling, consume 200 brine shrimp nauplii per day, the production of several hundred thousand young plaice posed serious technical problems in continuous food culture.[7]

Obtaining and correctly supplying food was a significant part of the experiments at the U.S. Bureau of Fisheries at La Jolla, California, with larvae of high seas schooling species such as Pacific sardines and mackerels. In these experiments very small food organisms had to be supplied at the precise time of complete yolk absorption, and in sufficient quantities to allow larvae of limited mobility to find food in all parts of the aquarium. Because sardine larvae search in only about 1 cubic centimeter of water per hour, at the onset of feeding, but require a minimum of four food organisms per hour to replace energy lost in swimming and body functions, the rearing of 2000 larvae in 1,800,000 cubic centimeters of water (500 gallons) meant replacing 7,200,000 food organisms removed by larval predation each hour or approximately 86,400,000 food organisms during a 12-hour day.

The large quantities of food organisms in varying sizes needed for these experiments were collected mostly at night. A 1000-watt underwater lamp connected to a submersible pump was suspended several feet below the surface of the sea. Copepods were attracted from a wide distance and concentrated near the pump where they were sucked up with water and transported to the surface. Plankton-enriched water was then passed through a series of filters, which further concentrated food organisms, and the highly enriched filtrate was piped to a 760-liter storage tank. Organisms with a cross-sectional diameter of

24. P. S. Galtsoff, "The American Oyster," U.S. Dept. Interior Bull. 64, 297-323 (1964).
25. J. H. Ryther, personal communication.

0.028 millimeter and larger were thereby collected. Before being fed to fish larvae, concentrated plankton was graded by filters to remove organisms larger than 0.1 millimeter. The portion containing large copepods, crab larvae, chaetognaths, and the like was fed to advanced fish fry and juveniles.[6,8]

Comparable techniques may help to achieve survival, after forced spawning in captivity, of milkfish and mullet. Inasmuch as these two species of economic importance in Asia are now raised from fry collected on the shores and as the fry are becoming scarce regionally, domestication of the two species, including manipulations ensuring high survival of fry, will be an important advance for fish culture.

Although fish larvae are recognizable as fish even though they are not like the adults, invertebrates undergo more profound transformations from egg to adult. Oysters spend their first 2 weeks before they "set" as ciliated trochophores and veliger larvae needing flagellate algae for food. About 2000 cells of two or more species, for instance *Isochrysis galbana, Monochrysis lutheri,* and *Rhodomonas* and *Nannochloris* species, have to be available for each larva per day, and larger species are required to replace the smaller species as the larvae grow. Algae must be cultured en masse when oyster larvae are raised indoors, an innovation largely developed at the U.S. Bureau of Commercial Fisheries Biological Laboratory at Milford, Connecticut,[26] and now expanded by progressive growers of Long Island oysters.

In shrimp raising, which is successful on a commercial scale only in Japan, there are problems with the larval stages before the animals can be fed chopped trashfish and shellfish. The operation, initiated by M. Fujinaga in 1934, begins in the spring with the collection of "berried" (egg carrying) females ready to release the stored spermatophores from their seminal receptacles; raising

the water temperature speeds this and subsequent processes. After three distinctly different planktonic stages and 12 molts in about as many days, the postlarvae begin to crawl on the bottom; they still have to undergo some transformations and another 20 molts before they become adults. The early part of the life cycle of the cultured shrimp takes place indoors in ceramic tile-lined wooden tanks and in water heated to between 26° and 30°C. Diatom—mainly *Skeletonema costatum*—and flagellate cultures are maintained for feeding the early larvae, which are later given finely chopped mussel or clam flesh. When they have reached a length of between 15 and 20 millimeters or a weight of about 10 milligrams, they are stocked in outside ponds with arrangements for aeration and circulation (Figure 6-2).

In October or November, the shrimp, though not fully grown, are ready for market. They are about 10 centimeters long and weigh 20 grams having been fed once daily, converting 10 to 12 kilograms of food into 1 kilogram of shrimp. When the water later cools down to below 15°C, the animals no longer feed, but many of them may be retained without feeding for a later more favorable market.[6]

The oldest shrimp-farming enterprise is now located near Takamatsu on Shikoku. It covers almost 10 hectares and has a staff of 30 men, including some in management research. Ten million shrimp were produced there in 1967, a quarter of which were raised to adult size; the rest were sold for stocking. The cost of production of cultured shrimp is certainly higher than that in any other aquacultural enterprise, but the wholesale price in Japan for tempura-sized shrimp of 6 to 10 centimeters can seasonally exceed $50 per kilogram, and the supply does not meet the demand. Shrimp farming of this type in a country whose material or labor costs are less favorable than those of Japan would not be possible.[6] There are, however, opportunities for greater mechanization and for feeding innovations that will simplify the most laborious parts of culture operations for lar-

26. V. L. Loosanoff and H. C. Davis, Adv. Mar. Biol. 1, 1 (1963).

Figure 6-2. The 28 running-water ponds (91 by 9.1 meters) for culturing adult shrimp at the Shrimp Farming Co., Takamatsu, Japan.

val as well as postlarval shrimp. The use of the most improved shrimp-culturing methods with fast-growing species may hold some promise for a number of regions in the world.[27]

MAKING FULL USE OF THE WATER

Aside from selecting the best suited strains, a practice not yet widely followed in aquatic husbandry, aquaculture should make use of the entire water column where possible and be three dimensional, as it is in China and other Asian countries, where common carp is stocked with other species of the minnow

family (Cyprinidae) such as the grass carp (*Ctenopharyngodon idellus*), the silver carp (*Hypothalmichthys molitrix*), and the big-head carp (*Aristichthys nobilis*).[28] The success of this method is based on the different food habits of the respective species; the carp is a bottom feeder; the grass carp and the silver carp feed on plants (banana leaves, even) and beanmeal or rice bran supplied to them from outside the ponds; and the big-head carp uses the plankton surplus in the well-fertilized water. Thus, the various water layers and all potential food sources are used.[29]

The culture of oysters in Japan's best oyster-growing district, Hiroshima Bay, also

27. Research on shrimp rearing in the United States is carried on at the Laboratory of the Bureau of Commercial Fisheries in Galveston, at the Bears Bluff Laboratory of the South Carolina Wildlife Resources Commission, and at the Institute of Marine Sciences of the University of Miami in Florida.

28. S. L. Hora and T. V. R. Pillay, Handbook on Fish Culture in the Indo-Pacific Region, FAO Fish. Biol. Tech. Paper No. 14 (Foreign Agriculture Office, Rome, 1962), pp. 124-132.
29. Yun-An Tang, personal communication.

Figure 6-3. Seed-oyster production near Hiroshima, Japan; top, general view of area; bottom, detail of above; bottom right, close-up of oysters on scallop shell, several weeks after setting (J. H. Ryther and author, photo).

are added as the oysters grow, and before harvest require several times the support they needed at the beginning of the growing season. Long lines instead of rafts are an innovation in the method of suspension, but they are still only a variant on the hanging-culture technique, which uses the water column efficiently and which protects the oyster from its bottom-living predators, such as starfish and oyster drills.

A typical raft, about 20 by 25 meters carries 600 rens and produces more than 4 metric tons of shucked oyster meat per year (Figure 6-4). On a per-hectare basis, this harvest amounts to about 25 metric tons, if it is assumed that only one-fourth of a certain area of intensive cultivation is covered by rafts, as is the current practice. Such yields result from intensive care and high primary productivity in the water that is

is an illustration of the use of the entire water column.[6] Seed oysters are collected on scallop shells suspended on wires from a bamboo framework driven into the bottom (Figure 6-3). Biologists from the prefectural and municipal laboratories monitor the plankton during the spawning period and advise the growers on the best time for spat collection. It is not uncommon to collect several thousand spat per scallop shell, although the average is about 200. The shells are removed from the collecting frames after 1 month when the surviving oysters have reached a size of about 12 millimeters. They are then cleansed, culled, and restrung on heavier wires separated by bamboo (and more recently by plastic) spacers. These wire rens are suspended from bamboo rafts, buoyed by floats of various kinds, and extend to a depth of 10 to 15 meters. Floats

Figure 6-4. Harvesting oysters in Hiroshima Bay, Japan. The rens of scallop shells with their attached oysters are strung from the boom.

dependent on tidal exchange and fertile terrestrial runoff. By comparison, the average is 5 metric tons per hectare of well-managed, leased oyster ground in the United States and the peak harvest of 300 metric tons per hectare of mussels (*Mytilus edulis*) also grown with hanging culture in the bays of Galicia in Spain. On public oyster grounds in the United States, where the mollusks are a minimally managed common property resource, the average per hectare is only 10 kilograms (0.001 metric ton) or less.[6]

FERTILIZATION AND FEEDING

Fertilization of bays, fjords, or enclosures has led to increases in phyto and zooplanktons, but favorable cost-benefit ratios for use with fish have not been proved.[30] In ponds (including brackish ones) organic and inorganic fertilization has been efficacious. In Israel, fertilized carp ponds, some with admixtures of *Tilapia* and mullet, produce twice the tonnage per hectare of unfertilized ponds, and fertilization and additional feeding doubles the yield again.[31] Fertilized *Tilapia* ponds in which the fish were also fed have yielded as much as ten times the crop of unenriched ones.[32]

Many kinds of inorganic or organic fertilizers can be used, but sewage which produces dense invertebrate populations certainly works well. Munich sewage ponds with a slow exchange of water produce 500 kilograms of carp per hectare per year and a profit for their operator, the Bavarian Hydropower Company; the method requires large tracts of land, however; under contemporary conditions rising land values threaten to make it obsolete.[6]

In a much warmer, rapidly flowing stream in West Java, with a high sewage content, carp, confined in bamboo cages to graze on the dense carpet of worms and insect larvae in the sandy substrate, grow rapidly to yield 50 or more kilograms of fish per square meter of cage surface, or 500 metric tons per hectare.[33] Even with allowances made for only partial use of the stream surfaces, this practice clearly represents an extremely efficient and ecologically sound use of sewage, especially in warm waters. The main sanitary drawbacks to this practice arise because the fish are not always well cooked before they are eaten.

In addition to the fertility of the water, its temperature, especially in a colder climate, is also very important. The most spectacular use of naturally warmed water for fish culture is in Idaho's Snake River valley trout-farming district, where springs of an even 16°C (optimum temperature for trout) gush forth from the canyon wall year in and year out. A thousand tons of trout can be raised in a year on every 2830 liters per second (100 cubic feet per second) that flow from these springs. Such unprecedented results in fish husbandry depend on high-density stocking, fast growth, mechanization, and cheap feed—the latter being locally procurable since the Snake River valley is also a stock feed-growing area.[34] Most of all, however, the high yield depends on the flushing of growth-inhibiting wastes from the trout raceways. Hence, it is more appropriate to relate weight gain to water flow rather than to water surface or volume. By such a measurement, production would be around 170 kilograms per liter per second.

Naturally warmed water is not prevalent, but man-made heated effluents occur with increasing frequency. In fact, thermal pollution may become a threat to some natural waters because it hastens eutrophication. Heated power plant effluents, however, can also be used to the advantage of the aqua-

30. F. Gross, S. M. Marshall, A. P. Orr, J. E. G. Raymont, *Proc. Roy. Soc., Edinburg Ser. B* 63, 1 (1947); F. Gross, S. R. Nutman, D. T. Gauld, J. E. G. Raymont, *Proc. Roy. Soc. Edinburgh Ser. B* 64, 1 (1950).
31. A. Yashouv, *Bamidgeh* 17(3), 55 (1965); A Yashouv, personal communication.
32. M. Huet, Rech. Eaux Forêts Groenendaal-Hoeilaart Belgique, *Trans. Ser. D* 22, 1-109 (1957).
33. K. F. Vaas and M. Sachlan, Proc. Indopacif. Fish. Counc. 6th (1956), pp. 187-196.
34. Th. Rangen, personal communication.

culturist. At the atomic energy plant at Hunterston, Scotland, cooling water, ascertained to be nontoxic to fish, was fed into cement troughs for sole and plaice raising. Both species were grown to marketable sizes in 6 to 8 months at between 15° and 20°C, as compared with the 3 or 4 years needed for the same growth under natural conditions.[35]

A progressive grower of Long Island shellfish used about 57,000 liters per minute of cooling water discharge of the Long Island Lighting Company. The cooling water is taken from a deep section of the bay and has a high nutrient content, which favors oyster growth as does its warmth. Year-round production in a near 3-hectare lagoon of both oysters and hard clams (*Mercenaria mercenaria*) has been achieved, and seed oyster production in the heated lagoon promises to be highly successful. Summer water temperatures above 30°C, first feared to arrest growth or to be lethal, in fact, promoted exceptionally rapid growth.[6,25] At the atomic plant at Turkey Point (Florida), replicate feeding trials by the University of Miami with shrimp (*Peneus duorarum*) and pompano (*Trachinotus carolinus*) are in progress to compare the effects of different levels of water temperature and consequently of different levels of heated water admixture.[9] Heated waste water is also used for freshwater fish culture in the Soviet Union, Poland, and East Germany.[36]

STATUS AND POTENTIAL OF AQUACULTURE

Aquaculture, practiced with a far wider range of species than mentioned here, is found in most of the world. In many areas it occurs at a subsistence level, and its potential contribution to the food supply has not been assessed.[37] Village ponds, once a hopeful development in Africa, for instance, are now in disrepair and their potential is not being realized.[32] Local fish ponds can be important, however, as has been demonstrated in Taiwan, mainland China, and Indonesia.[6]

Husbandry of aquatic animals brings increasing financial returns as it is practiced on a larger scale. Culture intensities vary, as do the fixed and variable costs of the operations and the yields (Table 6-1). From a commercial point of view, the return on the investment is of most interest; in milkfish culture the annual return ranges from 10 to 20 per cent or more, and increases with the intensity of cultivation. Malayan mixed pig and fish farms yield 30 per cent, and similar returns are noted in the oyster business.[6]

Aquaculture can be not only a lucrative business but it may even produce yields high compared with the harvest of comparably sized land surfaces. The relative scarcity of such peaks in aquacultural production, especially in the tropics, is caused by a lack of biotechnical engineering and managerial skills, the absence of suitable credit or seed capital for even low-cost installations, and the absence of transport and marketing facilities that might encourage the development of a product for a certain market, and so forth.

This is well illustrated by a comparison of Indonesian and Taiwanese milkfish culture in brackish water. Milkfish feed predominantly on bluegreen algae and are raised in pond complexes on land cleared of mangroves. Canals permit the control of water level and salinity by means of sluices, which regulate tidal or freshwater flow[38] (Figure 6-5). Average Indonesian and Philippine annual harvests are 300 to 400 kilograms per

35. British Whitefish Authority, Annual Report and Accounts 1967 (Her Majesty's Stationery Office, London, 1967).
36. L. V. Gribanov, Use of Thermal Waters for Commercial Production of Carp in Floats in the U.S.S.R., Working MS 44060, World Symposium on Warm Water Pondfish Culture (Foreign Agriculture Office, Rome, 1966).

37. T. V. R. Pillay, personal communication.
38. W. H. Schuster, Fish Culture in Salt-Water Ponds on Java (Dept. of Agriculture and Fisheries, Div. of Inland Fisheries, publ. 2, Bandung, 1949), 277 pp.

Table 6-1. Selected Ranges of Aquacultural Yields[6] per Year.

Type of Cultivation		Location	Yield (kg/hectare)	Approximate Wholesale Value of Annual Crop ($/hectare)
	Oysters			
Common property resource (public grounds)		U.S.	9	38
Intensive cultivation, heated hatchery, larval feeding		U.S.	5,000	21,000
Intensive care, hanging culture		Japan[a]	25,000	30,000
	Mussels			
Intensive care, hanging culture		Spain[a]	300,000	49,000
	Shrimp			
Extensive, no fertilization, no feeding		S.E. Asia	1,000	1,200
Very intensive, complete feeding		Japan	6,000	43,000
	Carp			
Fertilized ponds, sewage ponds		Israel	500	600
		S. Germany	500	
Fertilized ponds, accessory feeding		Israel	2,100	
Sewage streams, fast running		Indonesia[a]	125,000	
Recirculating water, intensive feeding		Japan	100[b]	114[b]
	Catfish			
Ponds, no fertilization or feeding		Southern U.S.	200	70
With fertilization and feeding in slowly flowing water			3,400	2,400 (net profit 300)
	Milkfish			
Brackish ponds, extensive management		Indonesia	400	
With fertilization and intensive care			2,000	600
	Trout			
Cement raceways, intensive feeding, rapid flow		U.S.	170[b]	168[b]

[a]Values for raft culture and comparable intensive practices based on 25 per cent of the area being occupied.
[b]Per liter per second.

hectare, whereas Taiwanese milkfish raisers attain nearly 2000 kilograms on the average, in spite of a cooler climate.[39] Cooperatives, rural reconstruction agencies, a good layout of the farms, control of predators of the fry, some fertilization, and prevention of siltation of ponds and connecting water bodies are some of the secrets of successful milkfish farming in Taiwan. For similar reasons there occur in mollusk culture the aforementioned wide range of yields, from nearly 10 kilograms per hectare on public oyster grounds

39. Yun-An Tang, Philippines Fish. Yearb. (1966), p. 82.

in the United States to the 25,000 kilograms per hectare in Hiroshima Bay.

Filter feeding mollusks and milkfish are brackish water plankton- or algae-feeders, respectively. These hold more promise for protein-deficient regions than do the carnivores of the same environment because it is more sound to increase the fertility of the water than to produce extraneous feed, let alone to raise one aquatic animal with scrap from another, which is perhaps already being used, or could be used, directly for human consumption.

Most products of aquaculture could be called luxury foods, whether they are sold as

Figure 6-5. Milkfish pond complexes near the coast of East Java (W. H. Schuster, photo).

high-priced items in a food economy with wide consumer choice (for example, shrimp in Japan, trout in the United States) or boost the scant animal protein supply of developing nations (for example, milkfish in Southeast Asia), where they also bring a good return to the producer. It might seem unrealistic, therefore, to expect aquaculture to help alleviate the world protein deficiency, but such is not necessarily the case. Luxury foods stop being a luxury when they can be mass produced, a case well exemplified by the broiler chicken industry in the United States and Western Europe.

Differences in biology between chickens and aquatic animals notwithstanding, some of the latter could well become mass-produced cheap and abundant foods at conversion rates of two parts of dry feed to one part of fish flesh. Among fresh and brackish water fish, especially trout, carp, and catfish can be raised with pellets. Chinese carp and certain tilapias eat leaves and stems of leafy plants; other fish feed on algae. In Southeast Asia well over 200,000 hectares of ponds now lie in former mangrove areas; there are in the tropics vast mangrove regions, some of which could be turned into pond complexes for the culture of fish. Mollusk production, though limited eventually by the suitability of grounds, could be expanded, and above all intensified in the areas where it is now prevalent. Aquaculture is only beginning to develop such practices as manipulation of the temperature regime to achieve best growth, devising simple automated feeders that fish can learn to activate themselves, and building machines that simplify harvesting. Several disciplines are expected to contribute to the development of aquaculture.

Since intensive husbandry alters the conditions of nature, a knowledge of the ecology of the cultured organisms in both natural and artificial states is essential. Engineering can also make increasingly important contributions to aquaculture development as it has in the successful pilot-scale raising of plaice and sole by the British Whitefish Authority.[40] It was the basis for the as yet theoretical calculation that "the annual British catch of plaice could be housed in shallow ponds covering 1¼ square miles in extent."[7]

Japanese yellowtail fish are now raised at high density, and with sequential cropping have already achieved yields of 28 kilograms per square meter (280 metric tons per hectare) and shown that it is economical to use small portions of the sea under very intensive management. The success with this oceanic schooling species and the fact that other species of similar habits had become adapted to confinement led to the speculation that still others, such as tuna, might behave similarly and that their mass culture under controlled conditions might become possible. In fact, Inoue of the Fisheries Research Laboratory, Tokay University, Japan, urged that Japan take the initiative in launching a tuna-rearing project in the equatorial Pacific, where atolls and lagoons could be used as sea farms.[8]

Such projections say nothing of the problems of translating small to modest enterprises into much vaster ones—the main one likely to be the procurement of many millions of tons of suitable food. Trash fish, in part now used for fish meal, krill and other marine organisms lower in the food chain than the highly prized fish to be cultured have been thought suitable, provided that they can be produced at a low enough cost.

There are also potential use conflicts. Fish culture lagoons would be fertile, perhaps almost eutrophic; they would be valueless for recreation. The theoretical potential of marine fish culture also rests on the assumption that marine fish can be induced to function sexually under artificial conditions, as have many freshwater fish. Hormone stimulation is expected to be one of the solutions to this problem along with rearing an initial breeding stock born and adjusted to life in artificial environments.

But even without further advances through research, a considerable increase of aquaculture yields appears attainable soon by consistent application of already known techniques on inefficiently managed fresh and brackish water bodies. It has been advocated[4] that millions of hectares of ponds be constructed in Asia, Latin America, and Africa to help satisfy the protein needs of these areas. If local economic and socio-political constraints were removed, these new waters and the upgrading of presently existing ones could yield by the year 2000 a harvest of 30 to 40 million metric tons[3, 41] produced near areas of need, which are still likely to be deficient in refrigeration.

Long-term and large-scale projections of yields attainable through practicing aquaculture with marine animals, outside the brackish water zone, can hardly be attempted; true mariculture is in its infancy. However, experiments in several locations have established that it is technically feasible, and no doubt the intensive development and success of brackish water aquaculture will lead to further efforts to develop mariculture on a large scale. It is too early, however, to tell

40. The U.S. Atomic Energy Commission Laboratory at Oak Ridge studies the feasibility of agro-nuclear complexes as shore installations in arid regions to produce cheap power, fresh water, and fertilizer; see New York Times, 10 Mar. 1968, p. 74. The agronuclear complexes will furnish ideal conditions for advanced aquaculture on a large scale.

41. S. J. Holt, in The Biological Basis of Freshwater Fish Production, S. D. Gerking, ed. (Wiley, New York, 1967), pp. 455-467.

42. I thank for assistance and information J. H. Ryther, G. O. Schumann, L. R. Donaldson, S. J. Holt, T. V. R. Pillay, W. Beckman, Th. Rangen, M. Fujiya, A. Yashouv, F. Bregnballe, E. Bertelsen, C. Mozzi, K. Kuronuma, S. Y. Lin, S. W. Ling, Y. A. Tang, M. Ovchynnyk, C. F. Hickling, I. Richardson, S. H. Swingle, R. V. Pantulu, F. P. Meyer, J. Donahue, M. Bohl, M. Delmendo, H. H. Reichenback-Klinke, and D. E. Thackrey.

where or under what conditions such efforts could become economically sound.

SUMMARY

The role of aquaculture in producing high-grade animal proteins for human nutrition is discussed. Raising and tending aquatic animals is mainly practiced in fresh and brackish waters although there are promising pilot experiments and a few commercial applications of true mariculture. Yields vary with the organisms under culture and the intensity of the husbanding care bestowed on them. The products are now mainly luxury foods, but there are some indications that upgrading of the frequently primitive culture methods now in use could lead to increasing yields per unit of effort and to reduced production costs per unit of weight. Under favorable conditions, production of animal flesh from a unit volume of water far exceeds that attained from a unit surface of ground. With high-density stocking of aquatic animals flushing is important, and flowing water or tidal exchange is essential. Combinations of biological and engineering skills are necessary for full exploitation of aquacultural potentials; these are only partially realized because economic incentives may be lacking to tend aquatic organisms rather than to secure them from wild stocks, because of social, cultural, and political constraints. Nevertheless, a substantial development of aquaculture should occur in the next three decades and with a several-fold increase in total yield.

Fish Meal and Fish Protein Concentrate

C. P. IDYLL   1970

In recent years world fish landings have increased twice as fast as human population (in contrast to the trends for other kinds of food), and from 1958 to 1964 fully 60 per cent of this increase consisted of varieties made into fish meal—species such as sardines, anchovies, and hakes.

For many decades fish meal was used largely as fertilizer and was referred to as "fish guano." Useful as it is for fertilizer, fish meal is now far too valuable to be used for this purpose. Instead, its enormously increased popularity is a consequence of the remarkable things it does in increasing the growth rates, the vigor, and the general health of farm animals at a cost much below that of most other supplements.

Farmers have known for centuries that their livestock relish—and thrive on—fish. In 325 B.C. Nearchus, one of Alexander the Great's generals, attacked a town on the Persian Gulf to restock his larder. According to the account given by the Greek historian Arrian, "the natives showed freely their flour, ground down from dried fish. . . . Even their flocks are fed on dried fish so that the mutton has a fishy taste like the flesh of sea birds." In the fourteenth century Marco Polo reported that some Asian peoples "accustom their cattle, cows, sheep, camels and horses to feed upon dried fish . . . of a small kind which they take in vast quantities during the months of March, April and May; and when dried they lay up in their houses for food for their cattle." Farmers in such far-flung places as Malaysia and the Shetland Islands of Scotland learned long ago to feed fish to pigs and sheep. The first bulletin issued by the United States fish commission in 1881 contained a letter from Isaac Hinkley describing the fish-eating cows of Provincetown, Massachusetts, which crowded around fishermen cleaning their catch to browse on the offal. Ling or blen-

Dr. C. P. Idyll is chairman and professor of the Division of Fishery Sciences and Applied Estuarine Ecology with the Rosenstiel School of Marine and Atmospheric Sciences, University of Miami, Florida. He is currently on sabbatical leave as senior consultant with the Food and Agricultural Organization of the United Nations in Rome, Italy. His research interests include ecology and fishery and estuarine biology. He is a member of the National Academy of Sciences' International Marine Science Affairs Panel. He has published nearly fifty books, articles, and reports.

From *The Sea Against Hunger*, pp. 128-144, 1970. Reprinted by permission of the author and the Thomas Y. Crowell Company, New York. Copyright 1970 by C. P. Idyll.

nies of three pounds and more were "freely eaten." Farmers taught cows to accept fish by mincing it and including it in their rations.

The use of fish meal for farm animals picked up momentum when its high content of nitrogen and phosphorus compounds was noticed in the early decades of this century. Feeding experiments met with remarkable success. Fish meal began to have general use as animal food in Europe, but this trend lagged in the United States until after World War I.

The American poultry industry has grown enormously with the aid of fish meal. More than a third of the United States broiler production can be attributed to it. More than half the fish landed in the United States goes to feed farm animals, and in 1968 this enormous bulk (more than 200,000 tons of fish meal) was supplemented by imports of 855,000 tons for the same purpose. In America turkeys and pigs also benefit from fish-supplemented diets. In Europe more meal is fed to swine than to chickens; in Germany, for example, 70 per cent goes to pigs.

For all farm animals—and for humans too—growth, vigor, and general health are dependent to a very important extent on the amount and quality of proteins available. These are the substances making up the muscle and other parts of the body mass. In addition, the minerals that are essential components of the skeleton and the vitamins that control the chemistry of the body are other important food constituents. Fish, whether fresh or reduced to meal, is one of the world's best sources of all three of these food groups.

Proteins occur in many kinds of foods, including plants: cereal grains, fruits, soybeans, and many more. They occur also in foods of animal origin: meat, milk, eggs, fish. The ruminants—cows and sheep—flourish with proteins derived from any of these sources. Other animals, including chickens, pigs, and man, may be undernourished on a diet high in protein if the sole source is cereals or other plants. The difference is related to the composition of the proteins from the two food sources.

Proteins are among the most complex of chemical substances. Their large molecules are made up of networks of smaller molecules of various amino acids. There are about eighteen amino acids, occurring in varying proportions in different kinds of protein. Nine or ten of these, called the essential amino acids, must be supplied ready manufactured, since most animals cannot synthesize them. Foods containing these essential amino acides are more nutritious than others in which some or all of them are deficient or out of balance.

Of all the essential amino acids three are likely to be missing in the grain rations of poultry. These are lysine and two sulfur-containing amino acids, methionine and cystine. For swine rations, lysine and tryptophan are those likely to be deficient. Cereal grains are low in all of these, so that chickens fed only with corn have to be fed larger amounts of grain to achieve optimum growth rate, maximum size, survival, and general health than if a better balance of amino acids can be supplied. Soybean meal does a better job than the grains, having ample quantities of lysine and tryptophan, but it is low in methionine and cystine. Sesame and sunflower meals, used increasingly in feeds in recent years, are rich in the sulfur amino acids but are low in lysine.

The rapidly rising use of fish meal for animal feeds has taken place because it is rich in all of these essential amino acids. One of the great strengths of fish meal is that only small amounts are required to supplement the essential nutrients available in cereal grains. Fish meal has by no means replaced the grains or soybean meal in animal feeds but instead has been used as a feed supplement. Soviet scientists have discovered that a metric ton (2,204 pounds) of fish meal added to the ration fed to pigs increased the yield of pork by 700 to 800 kilograms (1,540 to 1,760 pounds). When the same amount of fish meal was added to poultry feed, production of eggs increased by 25,000. In addition the meal replaced three tons of vegetable feeds. In Norwegian experiments, 7 per cent of fish meal added to the diet of chicks during their first six weeks increased their growth 11 per cent over that of animals fed only vegetable food. Pigs given a fish meal supplement to only their diet of grains increased in weight an extra 5 to 12 per cent by the time they reached market size. In Denmark the number of eggs per hen in ten months of production was 153 for birds fed 15 per cent fish meal compared to 126 for those on vegetable protein only. Soybean meal produced a 45 per cent hatch of eggs, but when the ration was supplemented by condensed fish solubles (the dissolved and suspended materials from fish reduction) the percentage rose to 74.

In the United States small portions of fish meal—ordinarily about 2 to 3 per cent but sometimes as much as 10 per cent—are used in the diet of chickens. More fish meal gives better results, but the cost rises when the fish replaces grain.

Meat, eggs, or milk fed to farm animals would give the same desirable effects as fish, and chemists have learned to synthesize methionine and other amino acids, as well as vitamins, in the laboratory. Hence it is possible to formulate fully balanced rations without fish meal. The modern poultry farmer lets a computer tell him what to use. The computer juggles the amino acid content of various components and comes up with a formulation that may include fish meal one day but exclude it the next because a cheaper protein source may be available. But fish meal is usually the cheapest source of high-grade animal protein.

Fish meals are rich in minerals, especially calcium and phosphorus, which are essential to the formation of bones. All such meals include iodine, copper, manganese, zinc, iron, and cobalt; they contain large quantities of the B vitamins, including B_{12}, riboflavin, niacin, and choline, all necessary for proper nutrition.

Finally, there is an air of pleasant mystery to the nutritional qualities of fish meal. After all the beneficial effects—the high protein content, the good amino acid balance, the high mineral and B vitamin content—have been accounted for, an additional value has been reported by some investigators. This has been called the "unidentified growth factor." It makes fish meal unique in nutritional value and has helped push it to its present heights of popularity.

There are a number of ways of making fish meal, but the largest quantities by far—some 95 per cent worldwide—are manufactured by the "wet reduction process." This is used when large volumes of oily fish are available. The fish are cooked with steam to denature the protein and break the cell walls to release the oil. The cooked mass is squeezed in a continuous screw-type press to remove water (up to 80 per cent of the raw fish) and oil (up to 20 per cent). The resulting "press cake" is dried and ground, an antioxidant is added, and the meal is bagged. Fish meal has a moisture content of 6 to 10 per cent, a fat content of 5 to 12 per cent, a protein content of 60 to 75 per cent, and an ash (mostly mineral) content of 10 to 20 per cent. It takes five pounds of raw fish to make a pound of fish meal; the difference is the water and oil removed.

This removal has two important results: it reduces the weight (and hence the shipping costs) to a fifth of the original fish weight, and it reduces spoilage, since bacteria must have moisture to operate. Hence fish meal can be stored at room temperature in simple containers such as bags, and can be shipped without expensive refrigeration.

Two other products, oil and solubles, result when fish meal is made this way. The oil-water mixture, called press liquor, is centrifuged, and oil and stickwater result. The stickwater is centrifuged again, to produce solubles. The market for these products is poor at present.

Fish oils are relatively unsaturated, meaning that their molecules are capable of picking up hydrogen atoms. This is of importance in human nutrition—and presumably for farm animals as well—and consumption of unsaturated instead of saturated fats may be a factor in the prevention of heart disease. Fish oils also have ready markets as ingredients for soap, paint, linoleum, lipstick, table and cooking fats, ink, and a surprising variety and number of other products.

Stickwater (or gluewater in Europe, both names describing its most obvious physical characteristic) is the residual liquid from the pressing of the cooked fish. It contains soluble and suspended materials—protein, minerals, and in particular, large amounts of the water-soluble B vitamins. Stickwater, once discarded, is now evaporated to about half its original volume to produce "condensed fish solubles." It is a valuable supplement to animal feeds, sometimes being added back to fish meal to produce "whole meal."

There has been a phenomenal rise in the use of fish meal throughout the world in recent years (Table 6-2). From a production of 571,000 metric tons in 1948, production increased eight and a half times by 1968 to about 4.8 million tons. The proportion of the world fish catch used for meal in 1948 was less than 8 per cent, rising to 14 per cent eight years later and to a level of about a

third of the total catch in 1966. The price of fish meal before World War II was $30 to $40 per ton. By 1948 this had risen to $125, and in 1969 it ranged from about $150 to $195 per ton, depending on season and market.

This worldwide surge in demand for fish meal has been made against strong, sometimes bitter opposition. That this opposition should exist in the face of the obvious benefits in making better use of the desperately needed resources of the sea and in the creation of new food and new wealth, is only one more illustration that man is the world's most baffling and inconsistent animal.

Opposition to the use of fish for reduction to meal often takes this tack: By transforming fish into meal, costs are raised. More important, fish meal is fed to chickens and pigs, and only a fraction of the food value is passed on to human beings. It is cheaper and more efficient to feed fresh fish to people. Hence it is immoral and it should be illegal to manufacture fish meal.

This argument is a complex mixture of the truth and of blind sentimentality. It is true that costs are raised when fish is processed, and that there is loss of energy, protein, and other nutrients when it is cycled through chickens and pigs on its way to the

Table 6-2. World Production of Fish Meal in Metric Tons.[a]

1938	627,000
1948	571,000
1952, 1953, 1954 average	995,500
1955, 1956, 1957 average	1,198,400
1958	1,396,000
1960	1,955,800
1961	2,496,000
1962	2,885,000
1963	2,890,000
1964	3,660,000
1965	3,549,000
1966	4,196,500
1967	4,500,000
1968	4,802,000[b]

[a]Source: Food and Agriculture Organization of the United Nations.
[b]Estimate.

human belly. But most of life's goals are reached by zigzag paths. The day will never come when mankind can use all fishes just as they come from the sea, wasting nothing by processing or by transforming the fishes to other kinds of flesh. The manufacture of fish meal permits the utilization of fishes that cannot be used in any other way at the present state of our skill.

The efficiency of conversion of food value by farm animals is already good and it is increasing as agricultural research progresses. When fish meal is added to a hen's ration, it lays eggs or grows to a plump broiler more rapidly, and with a recovery of food value higher than the rule-of-thumb 10 per cent usually used in calculations of transfer from one trophic level to another. A hen's recovery of energy from fish meal ranges from about 20 to 25 per cent of the calories; for proteins it is a high 40 to 50 per cent; and for the all-important essential amino acids it is even greater. Turkeys, pigs, and other farm animals also do good jobs of conversion.

It was pointed out earlier that more than a third of the broilers produced in the United States can be attributed to fish meal. In 1964, for example, fish were responsible for some 720,425 broilers on American tables; to this should be added substantial numbers of turkeys, pigs, and other food animals. Similar results are produced in Germany, in Japan, and in many other countries. In the face of such figures the argument that fish should be fed to humans instead of animals, implying that beasts are being fed at the expense of people, misses the point.

Humans probably will continue indefinitely to benefit from feeding fish to farm animals. Yet, in the face of a desperate and worsening shortage of food throughout the world, men should learn to use directly as human food as much as possible of the fish now fed to livestock. We may be on the threshold of such a development. It is possible now to produce high-quality fish meal and tasteless, odorless (or suitably flavored) fish protein concentrate (FPC) acceptable to many palates.

There are two principal barriers to rapid and widespread use of edible fish meal and FPC. The first of these is technological and economic: methods available are still relatively costly for most markets. The second barrier is psychological: they are unfamiliar foods, and therefore, in the minds of many people, unacceptable. Both of these situations can be changed—the first relatively easily and quickly, in direct proportion to the amount of effort put into research and development; the second only slowly and painfully, since human prejudice is far less amenable to manipulation than machines or chemicals.

Fish meal can be made for human food by the same methods as those used to make the meal now sold by the millions of tons for animal feed. But so much more care must be taken in its manufacture, and the cost is therefore so much higher, that this approach is probably not feasible. The raw material must be delivered fresher to the plants than is now usually done; the work must be conducted more hygienically and at a lower temperature than is common. Such edible fish meal is highly nutritious, but it has very little appeal to most palates, since its odor and flavor are strong. In the future it may be possible to improve the manufacture of fish meal sufficiently and to conduct energetic campaigns to persuade people to eat it. This would be a cheap and satisfactory method of improving the nutrition of mankind, but it will be long in coming, if it ever does.

A much more likely possibility is that the use of fish protein concentrate, or "fish flour," will be greatly increased. This is a whitish powder with high animal protein and mineral contents, a nutritionally well balanced amino acid composition, and a low fat content. It can be made tasteless and odorless if this is required by its consumers; it can have a fish flavor if this is preferred. It must be made with a process different from and more expensive than that used for fish meal.

FPC has some exceptionally weighty credentials as food for humans. It is nutritious and wholesome; two ounces contain as much animal protein as a twelve-ounce steak. It can be shipped and stored in cheap containers without refrigeration; in Canada herring flour has been stored in polyethylene bags for three years without noticeable change in flavor. It is acceptable in a variety of foods in many parts of the world. It is already moderately cheap and it will become cheaper.

At least 500 million people throughout the world are short of proteins, but 5 million tons of FPC could supply them with enough animal protein for a year. Made into fish protein concentrate, the unharvested fish of the United States coastal waters alone could raise significantly the nutritional standard of 1 billion people for 300 days.

Despite all this, there is no substantial commercial production of FPC anywhere in the world. This is largely because it is a new and unfamiliar material, and men are afraid of strange things.

FPC has had a difficult youth in the United States. Perhaps this is because ours is a country, unlike a considerable portion of the rest of the world, with very little need for new, cheap sources of animal protein.

The opposition of greatest consequence has been to the sale for human food of fish protein concentrate manufactured from whole fish. This opposition has been important because it came from the United States Food and Drug Administration, whose word on such matters is law and whose edicts produce repercussions in the rest of the world. The Food and Drug Administration has served the people of the United States well by protecting them against unscrupulous or careless manufacturers of drugs and processors of food; it has undoubtedly saved countless lives and dollars. But in the case of FPC made from whole fish the FDA has been wrong. Until early in 1967 it argued that such a product, containing as it does scales, fins, and entrails of fish, as well as the muscles and other parts of the body, would be "aesthetically unacceptable" and could not be sold in this country for human consumption. But the American public eats canned sardines and other fishes, entrails and all; it greedily swallows oysters and clams without cutting out any parts of them; it consumes a great many other foods that are truly "unaesthetic" if they are viewed in a detached and dispassionate manner. FPC, on the other hand, is wholesome and no more objectionable to the unprejudiced palate than a chicken leg or a spoonful of boiled spinach. The FDA came under heavy fire from nutritionists, biologists, members of Congress, and government groups for its stand, and it eventually retreated.

Research by scientists of the Bureau of Commercial Fisheries confirmed FPC's low level of toxicity (content of fluoride) and high nutritive value. These results were important in persuading the Food and Drug Administration to change its position, as was a strong stand taken by the Advisory Committee on Marine Protein Resources of the National Academy of Sciences.

One reason that FPC is more expensive to make than fish meal is that a great proportion of the fat must be removed. Fat and its decomposition products are the principal causes of objectionable odors and flavors, and the length of time the product can be stored before use depends heavily on low fat content to prevent it from becoming rancid. Thus, in the manufacture of FPC, processes must be employed to remove nearly all the fat. For example, the pioneer VioBin process uses ethylene dichloride, and the method developed by the Bureau of Commercial Fisheries (BCF) in its laboratory at College Park, Maryland, employs isopropyl alcohol. The fish used by the BCF is red hake, *Urophycis chuss,* a cheap and abundant species caught on the Atlantic coast of the United States. It is sold only in small amounts in this country, since it is too small and too soft to fillet, freeze, or can.

To make FPC by the Bureau process, whole fish are minced and treated three times with separate batches of isopropyl alcohol, once cold and twice hot. The alcohol extracts water as well as fat, and the residue of fish is dried, ground, and packaged. One hundred thousand pounds of raw hake yield 15,000 pounds of fish protein concentrate—a ratio of about 6 to 1. The concentrate is a white powder with a yellowish cast. It contains 80 per cent protein, 13.5 per cent ash (mostly calcium, phosphorus, and other minerals), and has virtually no odor or flavor. If the fish were bought for 1 cent a pound, a plant with a daily capacity of fifty tons of raw fish could probably produce this flour at 13.9 cents a pound and sell it at a profit for 20 cents a pound. If two extractions instead of three were sufficient (that is, if the market would accept a small residue of fat in the product and thus a faint fish taste) the selling price per pound would be about 13.5 cents. Canadian technologists estimate that FPC from herring would cost 15 cents a pound to manufacture in their country. In other parts of the world the selling price would be about the same as for dried skim milk, and in most countries fish protein concentrate could probably be produced more cheaply per unit of protein than any other animal material.

Fish protein concentrate is already highly acceptable to many people, especially children. Adults, with the usual stubborn adherence to the ways and tastes developed in youth, take to it much less readily, as they do to any other new product. But the Food and Agriculture Organization and the Children's Fund of the United Nations have carried out tests in thirty or more countries—often with encouraging success.

Fish protein concentrate can be used in a great many ways: in breads, pastas (that is, macaroni, spaghetti, and similar products), cakes, cookies, sauces, cereals, pastries, candy, soups, baby foods, and beverages. If a neutral bland product is required, without taste or odor (for markets in Europe, the Americas, and India), it can be produced; various strengths and kinds of fish flavors can be added for markets in central and southern African countries and those in Southeast Asia. With additives, fish flour can even be made to taste like cheese or beef.

Chile started testing FPC as early as 1958. Children in Santiago schools like bread made with an FPC content of 7 to 10 per cent. At the upper limit the color of the bread was affected, and above that the flavor was noticeable, but it will be remembered from the experiments with poultry that amounts of fish meal considerably less than 10 per cent were remarkably effective in improving nutrition, and with children similarly small proportions had beneficial effects.

In Kuala Lumpur, Malaya, children fed a standard diet enriched with skim-milk solids showed twice the rate of gain in growth as those without the milk; better still, those fed a standard diet supplemented with cookies made of FPC, cereal, sugar, and flavoring showed a *triple* gain. Moreover, the children (but not their parents) liked the cookies. In Senegal foods containing fish protein concentrate were successfully fed to children. In Burma it has been incorporated into soups, sauces, and vegetable dishes with high acceptability. In the Belgian Congo and in Ghana FPC was in good demand when the price was kept low by subsidy; later it was commercially successful at competitive prices.

In Mexico Dr. Federico Gomez carried out several years of experiments in the Hospital Infantil in Mexico City, and in Tlaltizipan, with impressive results. In 1960 he declared that "10 to 15 years after supplementation with 30-40 grams [about 1 to 1.5 ounces] of animal protein in the form of fish flour to the daily Mexican diet of corn, beans, and chili, the characteristics of Mexican people will change physically, mentally, and emotionally."

Sweden and the Union of South Africa have FPC plants in operation. The United States established a pilot-scale plant in Grays Harbor, Washington, in 1969.

The U. S. Agency for International Development (AID) has launched a vigorous campaign to persuade people in developing countries to eat FPC despite their reluctance to try strange foods. The first step in this program is to get an insight into consumer psychology in several countries (including Brazil, the Philippines, Korea, Thailand, and India), then to launch a campaign that will encourage voluntary use of FPC on a scale sufficient to support a profitable industry.

When the world is ready to accept it, immense amounts of FPC can be manufactured. A great proportion of the fish now landed and consumed in various fresh or processed forms is suitable for this purpose. Obviously no one is going to close the salmon canneries of British Columbia and Alaska, or the fillet and fish stick freezers of New England, and convert them into FPC factories, and highly regarded species such as halibut, sole, red snapper, shad, and others will continue to be marketed in their present forms. But more than a third of the world fish catch is now used to make fish meal, and all the same fish could theoretically be made into fish flour.

It is unlikely that all fish meal plants will be converted for fish protein concentrate, but once a demand is created and handling methods are improved, such huge stocks as the Peruvian anchovetta (now harvested at the rate of 22.5 billion pounds a year) would be suitable for making fish protein concentrate. Next door, Chile has produced 22 million pounds of anchovettas. At its peak the California sardine fishery produced 1.5 billion pounds. Whether it will ever do so again apparently depends on whether the ocean off California and adjacent areas warms up again to the sardines' liking, and whether they can shoulder their way back into a living space lost to the anchovies when their numbers dwindled. But if the sardines do come back, they would make millions of tons of good FPC. The menhaden industry of the United States Atlantic and Gulf of Mexico waters has produced as much as 2.25 billion pounds of fish. The Alaska herring populations produced 261 million pounds at their peak, and even this may have been less than the stocks could sustain. The British Columbia herring fishery peaked at 96 million pounds. In South Africa 880 million pounds of pilchard are landed in some years for the fish meal plants; in South-West Africa the peak amount has been nearly 1.5 billion pounds; in Angola, 238 million pounds. And so the roster grows.

These figures are taken from maximum catches of fisheries exploited now at varying levels: some fully exploited like that of the Peruvian anchovy, some overexploited like that of the California sardine, some underexploited to various unknown degrees. Of course, it is misleading to quote maximum catches as though these were the amounts available every year from their respective areas, and greatest catches may represent unusually favorable years instead of average years, or years when overfishing occurred. But some inkling of the immense total potential is gained in this way, and by no means all the fishable stocks have been included in the list above.

There are also stocks of fish whose existence is known but whose size can only be guessed, since fishing has not tested them. Estimates here may be very inaccurate, but judging from previous experience, assessments are more likely to err on the conservative side than otherwise. This is to be expected, since what is not seen is not counted.

In California alone there may be 30 billion pounds of anchovies, hake, lanternfishes, deep-sea smelt, and other species now unused. From one-quart to one-half—probably closer to the latter—of this quantity of fish is available on a sustained-yield basis. There are millions of pounds of hake and smaller but substantial quantities of other species to be taken off the coasts of Argentina and other southern South American coasts. West African nations are just beginning to exploit the fish off their coasts, and there are millions of pounds to be had there on a sustained basis. The total for these and dozens of other fish stocks over the world is impressive.

Another enormous resource, the squids, might also contribute great supplies of fish meal or FPC. Less is known about squids than about many other sea creatures; because relatively few people eat them, they have not been studied sufficiently. But it is clear from a few isolated fisheries and from limited scientific investigations that the sea contains enormous quantities of them.

Squids are popular food in some parts of the world, notably Japan and southern

Europe. Fisheries exist in waters off Hokkaido (the northern island of Japan), off Newfoundland, in the Mediterranean, and in a few other places. Japan landed 14.4 billion pounds in 1963, and its stocks of squids may be no larger than some others in various parts of the world. It is certain that vastly greater quantities could be caught worldwide. This is largely a matter of developing markets. One large market would be created by the manufacture of meal and protein concentrate from squids.

The protein of squids is of high quality, nearly equal to that of fish. They have a strong advantage over many other animals in the large proportion of edible parts of the whole body: 80 per cent compared to 40 to 70 per cent for fish. The water content of squid flesh is about the same as that of white fish meat, ranging from 70 to 80 per cent. The oil content of the flesh is low, ranging from 1 to 1.5 per cent.

On my desk I have meal made from squids by the American VioBin process, containing 77.6 per cent protein, 10.2 per cent ash, 9.0 per cent moisture and 1.9 per cent fat. It is a tan powder, like very fine beach sand, with a faint fishy odor. There is no reason to doubt that it would be an excellent supplement to either animal or human diets deficient in animal protein.

Of course, there is an eventual limit to the amount of fish, and even squids, that can be brought to shore. At that point the only place to go, if mankind is to get more food from the sea, is down one or more steps in the oceanic food pyramid, to exploit zooplankton. The quantities available are enormously larger than those of the nearly untouched squids, or even of the swarming little fishes that have formed the basis for the greatly expanded landings of recent years. In harvesting plankton the conclusion is reached that although it looks unlikely that man can soon make use of most of these resources, some of the larger animal-plankton organisms, notably the antarctic krill and the red crabs of the eastern Pacific, seems promising. Fantastic totals of millions of tons of krill may be available in far-southern waters, and the Soviets have shown that nutritious meal can be made from this raw material. Enormous quantities of red crabs are likewise available for capture off California and northern Mexico.

Thus we are not short of raw material for fish meal and fish flour. If we have the skill to produce acceptable products and to catch the animals cheaply, we have the opportunity to supply immense quantities of human food.

The Biologist's Place in the Fishing Industry

J. L. McHUGH   1968

A generation ago the biologist was dominant in fishery science. The problems of the fisheries seemed to be—and indeed were—much simpler in those days. The principal concern was with fluctuations in the supply of living resources. Biologists were asked to explain these fluctuations and to recommend measures to eliminate them as far as possible. The theory of fishing stated that each species or stock of fish has an innate capacity to grow in numbers, and that, provided certain precautions are taken, each fishery resource can sustain certain rates of fishing forever. The biological objective of fishery management was to find out what level of fishing effort would produce the maximum sustainable yield for each resource and to develop scientific management measures which would maintain the maximum sustainable yield.

Today, fishery science is recognized to be much more complicated. The biologist still plays an important part in fishery management, as he always must. The interactions between living resources and their environment are now known to be exceedingly complex, especially since man has developed such awesome powers to change the physical and biological environment. But the biologist's original concept of objectives has been challenged by the economist, who sees no logic in a commercial fishery goal based on biological yields alone. In a country where commercial fishermen fish for profit rather than for protein, the economist's view is that the objective must be maximum economic return to the fisherman and to the national economy. Implicit in this definition is the conclusion that if protein or other products available in the sea can be produced in some other way with greater return to the economy, there is no point in supporting a fishing industry. This ignores some social values in the fishing industry, probably because they are poorly understood, and it leaves the question of biological research on sport fisheries to be decided by other criteria.

Pragmatic economic values notwithstand-

Dr. J. L. McHugh is a professor of marine resources at the Marine Resources Research Center, State University of New York, Stony Brook. He is the former deputy director of the Bureau of Commercial Fisheries, Fish and Wildlife Service, U.S. Department of Interior, Washington and currently holds the post of U.S. Commissioner of the International Whaling Commission. He has written more than ninety technical articles.

From BioScience, Vol. 18, No. 10, pp. 935-939, 1968. Reprinted with the author's note and by permission from the author and the American Institute of Biological Sciences.

ing, it must be assumed that there are important reasons why we should be concerned about the condition of fishery resources. Growing experience with the Federal program planning and budgeting system (PPBS) should allow the fishery establishment to improve its justification of biological research. It is now clear that unregulated fishing can have profound effects upon ecological communities, as demonstrated by the reversal in abundance of Pacific sardine and anchovy and by the dramatic changes in some fishery stocks in the Great Lakes. The role of the biologist is to provide scientific information necessary to maintain the living resources in healthy condition, to ensure an end product of high quality with maximum economic return to the industry and to the Nation.

PRESENT CONTRIBUTIONS BY BIOLOGISTS TO INDUSTRY PROBLEMS

The principal roles of the biologist in fisheries are to describe the distribution and abundance of the resources and their variations in time and space; to estimate the harvest that each resource can sustain; to understand the effects of man's activities, including fishing, upon the supply and to explain the effects of natural environmental variations; to forecast abundance, and place and time of arrival upon the fishing grounds; and to develop other information that can be used to increase the supply, reduce fluctuations in abundance, or otherwise reduce the costs of locating and catching fish.

Fishery biologists have made many valuable contributions to our understanding of oceanic and freshwater ecology. They have pioneered in the study of population dynamics in the natural environment. They have developed many ingenious scientific techniques for the artificial culture of fish and shellfish and for control of predators. The substantial fund of available biological knowledge, if it could be put to full use, would restore to historic levels of productivity some of our most important, but

dwindling American fisheries, such as the oyster industry. But practical application of the work of biologists has been limited seriously by political and social conflicts based on prejudice and tradition. The United States, in common with many other countries, lacks effective institutional arrangements for scientific fishery management.

For these reasons, the biologist alone, although he can have an important influence in fishery management, cannot solve all problems of the fisheries. Recognizing this long before other scientific and social disciplines were aware of the breadth of fishery problems, biologists have pioneered in applying to fishery management the principles of engineering, economics, law, and the social-political sciences. In today's complex world the fishery biologist, once a jack-of-all-trades, no longer can cope with the diverse subjects that must be brought together for systematic development of domestic and international fishery management. It is encouraging that the immensity of the problem is being recognized by universities and government. The Food and Agriculture Organization of the United Nations (FAO) has been convening expert conferences on the diverse problems of world fisheries for many years. Recently some American universities have been organizing broadly-based conferences on domestic fishery problems. The University of Rhode Island and University of Washinton have sponsored particularly effective recent meetings.

Fishery biologists, or fishery laboratories employing scientists of other disciplines to work with biologists, have made many important contributions to general knowledge of aquatic ecology. They have discovered or described major ocean currents such as the subsurface Cromwell Current in the tropical Pacific Ocean and a similar current below the surface in the tropical Atlantic. They have developed most of the knowledge now available about ocean circulation in the Central Pacific and the sub-Arctic North Pacific and Bering Sea. They have studied oceanic fronts in many parts of the world ocean and have observed their effects on animal distribution and aggregation. They have mapped seasonal and annual variations in ocean temperature, salinity, and sea levels, noted the ways in which these properties are affected by changes in atmospheric circulation, and have observed the effects of these changes upon the distribution and migrations of life in the sea. They have provided charts of bottom topography containing details of seamounts and other features around which fish congregate, or bottom types favored by certain demersal animals.

Fishery biologists have been equally productive in biological research. They have made important contributions to the literature on plankton, especially in taxonomy and distribution of fish eggs and larvae, and also of many important invertebrates. They have developed ingenious techniques for detecting races or populations of fish species, thus demonstrating lack of intermingling between stocks of many widely distributed species. These techniques have been applied successfully to the problem of distinguishing continental origin of Pacific salmon caught on the high seas, and this has brought handsome benefits in preventing major exploitation of North American salmon by foreign fishermen. They have provided information on distribution, general biology, and sustainable yields of major fishery resources harvested jointly by American and foreign fishermen off U.S. coasts. Such information has served to protect American interests in international negotiation of fishery agreements. Fishery scientists also have established the great importance of the estuarine environment as a nursery ground for many of our most important fishery resources. This information is now being used with effect in developing methods to control man-made changes in the estuaries. Scientific knowledge of molluscan shellfish has been advanced greatly by fishery biologists, who have developed successful techniques for predator control and artificial culture and have made major contributions to knowledge of the pathways followed by radio-nuclides and pesticides in the aquatic environment and their effects upon aquatic animals and plants.

In fresh waters, fishery biologists also have many important achievements to their credit. They have studied the behavior of anadromous fishes, especially salmon, near dams, in fishways, and in reservoirs. Their findings have improved the design of fishways, reduced their construction costs considerably, and have made it possible to guide adult salmon successfully over any dam. Scientific hatchery techniques developed by biologists have renewed our interest in the hatchery as a management tool and have led recently to a phenomenal revival of silver salmon runs in the Columbia River. Fishery biologists, active on the Great Lakes for 75 years or more, have documented the remarkable changes in environment and aquatic life hat have affected all five lakes with increasing tempo, and which led to the disastrous disturbance of ecological balance exemplified recently by massive deaths of alewives in Lake Michigan. Fishery biologists also have accomplished in the Great Lakes one of the most striking examples of predator control ever achieved, by developing effective methods for selective destruction of the exotic sea lamprey and restoration of the endemic lake trout resource.

These are but a few of the many solid accomplishments of fishery biologists. Many achievements equally important have necessarily been omitted for lack of space. The practical benefits of many biological findings are virtually unrecognized by most laymen, for various reasons. Since most applications of biological knowledge to industry problems cannot be conducted as controlled experiments, the onlooker has no way of knowing how much worse off he might have been if the biological research had not been done. In addition, the economic and social-political structure of the fisheries and fishery management tends to nullify many of the potential gains from biological research. Methods for improving the situation are known, but politically many of them are extremely difficult to achieve. The biologist cannot be blamed for our extremely poor record in commercial fishery management, because application of the results of his research is usually beyond his control. But he *can* be blamed for his poor record in communicating the results vigorously in language that a layman can understand.

TRAINING OF FISHERY BIOLOGISTS

As the problems of the fishing industry become more complicated, concepts of fishery management are changing rapidly. As a consequence, opinion on the ideal curriculum for training fishery biologists is far from unanimous. Some believe that specialized fishery training should come early in the undergraduate period. Others, including myself, believe that the best preparation is a broad education including social sciences, humanities, and the arts, with stress on the skills of communication. Probably neither extreme is practical in the complex world of today. The best compromise must be to balance the limitations of time against the conflicting needs for specialized knowledge and for broad understanding of man and his environment.

Fishery science is ecology in the broadest sense. The immediate concern of the fishery biologist is with the living resource and its interactions with the aquatic environment. The fisheries, interacting with many other human effects on the resource and its environment, introduce man as a very potent force which can alter the biological and physical features of the ecosystem profoundly. The fishery biologist will not be well prepared for his job unless he understands how man affects the resources and recognizes the institutional barriers that make it difficult to control these effects. Furthermore, he must be sensitive to the social-political peculiarities of fishery management. In the last 30 to 50 years fishery science has evolved from relatively simple consideration of the effects of fishing on single species or stocks of fish to a realization that man and

his environment are no longer in dynamic balance and that fishery management requires broad understanding of, and ability to control, the biosphere, of which man is an important part. Training of fishery biologists must include exposure to all the professional skills that bear on the solution of fishery problems. This includes economics, engineering, law, sociology, communication, and language, to mention only a few of the more important disciplines other than the natural sciences. Conferences organized by FAO and by American universities, already cited, are beginning to produce a comprehensive fishery literature to supply a growing need. It is indicative of the relative immaturity and rapid evolution of fishery science that there is yet no adequate textbook or truly comprehensive reference work on the subject. Training of fishery biologists also must include practical experience with fishermen, with the fishing fleet, and in processing plants.

SPECIFIC GOALS OF INDUSTRY

The American fishing industry is so fragmented that it can scarcely be said to have well defined goals for which biological research would be helpful. In 1956, when the Fish and Wildlife Service was reorganized and the Bureau of Commercial Fisheries was created by an Act of Congress, the national policy was stated thus:

The Congress hereby declares that the fish, shellfish, and wildlife resources of the Nation make a material contribution to our national economy and food supply, as well as a material contribution to the health, recreation, and well-being of our citizens; that such resources are a living, renewable form of national wealth that is capable of being maintained and greatly increased with proper management, but equally capable of destruction if neglected or unwisely exploited; that such resources afford outdoor recreation throughout the Nation and provide employment, directly or indirectly, to a substantial number of citizens; that the fishing industries strengthen the defense of the

United States through the provision of a trained seafaring citizenry and action-ready fleets of seaworthy vessels; that the training and sport afforded by fish and wildlife resources strengthen the national defense by contributing to the general health and physical fitness of millions of citizens; and that properly developed, such fish and wildlife resources are capable of steadily increasing these valuable contributions to the life of the Nation.

The Marine Resources and Engineering Development Act of 1966 reiterated national fishery policy:

It is hereby declared to be the policy of the United States to develop, encourage, and maintain a coordinated, comprehensive, and long-range national program in marine science for the benefit of mankind to assist in . . . rehabilitation of our commercial fisheries, and increased utilization of these . . . resources.

These objectives have been reiterated and elaborated in the two annual reports of the National Council on Marine Resources and Engineering Development.

The growing problems of the domestic fishing industry have led several fishery associations or labor unions recently to develop national programs or to sponsor broad legislation to aid the industry. The well-organized processor, although he has problems, usually is able to diversify his operations and sources of supply to avoid the worst difficulties. The primary producer is in much worse condition. Most fishermen in the United States are independent operators with limited capital. They face serious competition from highly organized foreign fleets, often fishing the same grounds. They must compete with an excess of domestic fishermen, usually more than are necessary to take the allowable catch. Their foreign competitors work for lower wages and are subsidized liberally by their governments, thus the fish they catch often end up in U.S. markets at prices that domestic fishermen cannot meet. American fishermen must pay very high insurance rates, if they take out insurance at all. They are required by law to build their vessels in U.S. shipyards where costs are the highest in the world. Many of the laws regulating American fishing prevent adoption of efficient catching methods, to spread the catch among the maximum number of fishermen or to protect traditional methods of fishing or other special interests. Their costs also are increased because the supply of fish fluctuates from causes that are only partially predictable. No one has yet devised a solution for all these problems. Indeed, some of the most popular and obvious solutions, such as import tariffs or quotas, are contrary to administration policy. Economists point out that even if all of the obvious problems were solved in one way or another the effort would be nullified in an economic sense unless some way is found to limit fishing effort to the minimum amount necessary to make the desired catch.

Various segments of the fishing industry recognize various possible benefits from biological research. Some of these benefits are impossible to attain, others are feasible, and some have been at least partially realized. Others among industry, particularly many of the fishermen themselves, see no present or possible gain from biological research and consider it a waste of money. This attitude is not surprising in view of the many economic and institutional constraints which make it impossible for the average fisherman to realize a return on his investment, even if the resources were to be maintained in healthy condition as a result of biological research.

The goal of many industry members is to have assurance that the abundance and availability of fish can be held at maximum historic levels. This is an impossible objective for several good biological reasons which are very difficult to communicate to industry. Attainment of this objective is further complicated by economic and social interactions. A fishery which depends upon a single widely fluctuating species tends to overinvest capital in periods of high biological productivity, then to fish harder when biological productivity drops, to protect its capital investment. The almost inevitable result is overfishing, with biological consequences which intensify the economic and social-political problems. What is needed by industry is awareness of the universal phenomenon of fluctuations in abundance of species, of the dynamic interactions between species, and the economic and biological advantages of a broad resource base containing several alternative species. What is needed by biologists is a new science of fishery population dynamics based on ecological reality, leading to theoretical development and practical establishment of multiple species management techniques.

FISHERY PROBLEMS AMENABLE TO BIOLOGICAL SOLUTIONS

It should be clear from the foregoing discussion that although fishery biologists have a distinguished record of scientific achievement, most fisheries are not under scientific management and many segments of industry see little or no benefit from biological research. The next great advances in domestic fishery management necessarily must come in the fields of economics and the other social sciences. This will be necessary even to put the present fund of biological knowledge to effective use. The question might then be asked: "Why continue to invest substantial sums in biological research while the social system remains hostile to application of the results?"

Man is essentially optimistic about his ability to achieve miracles. Some of the social changes necessary as a prelude to an effective fishery management system will come about only through virtual miracles. In many respects fishery problems are a microcosm of the total range of human problems, all of which must be solved if mankind is to survive. It must be assumed that man will be wise enough to get his world in order before it is too late, otherwise there would be little point in any human endeavor. Although there is much we know about the biology of fishery resources, there is much more we do

not know. Anticipating the day when scientific fishery management is possible nationally as well as internationally we should be gathering the necessary biological data according to a well-developed plan so that biological science will not be found unprepared when the social sciences have opened the way.

The Utopia of fishery management will require a much more sophisticated management model than any available today. This model can be described in general terms now, but its full development will require much quantitative information not now available. It is unlikely that the present highly selective fishery in the United States, in which a dozen kinds of fish and shellfish contribute about 80% of the catch, can or should persist. It may well be that many of the preferred kinds, for which the consumer is apparently willing to pay a high price, should be produced under private control. Indeed, fish farming in privately controlled ponds or impoundments is attractive today to the fishery manager because it is the only practical way of avoiding the economic and social-political barriers that hamper fisheries on wild stocks. There will still be a need for adequate management information for commercial or recreational fisheries on wild resources, however, and the ocean for a long time to come may be the major source of cheap fish protein to feed developing nations.

Since it is now known that fishing and other human forces can cause major ecological imbalances even in the ocean, we must learn how to control the harvest in such a way that desirable balances can be maintained. Development of products like fish protein concentrate (FPC) has important ecological implications, for it improves the possibility of balanced harvesting, which can be achieved only if economic uses can be found for all pertinent aquatic resources. Incidentally, the most promising methods of producing FPC would be based on biological fermentation because this would produce a water-soluble product. There is important work ahead for biologists in developing biological methods for FPC and in improving the quality and shelf life of fishery products generally.

We need to know a great deal more about the relationships between fishery resources and their environment and their interactions with each other, to discover the causes of fluctuations, to improve our ability to forecast, and if possible to control fluctuations. One method of control might be to return to hatchery methods, using biological research to develop economical techniques for holding the young in a controlled environment through the early stages critical for survival, to release them in the sea when their chances to reach maturity are high. Another method would be to develop truly balanced fisheries, controlling fishing effort on each, and switching from one resource to another according to plan so that desirable yields can be attained and ecological balances preserved. In developing the knowledge needed for these purposes the fishery biologist will become the informed ecologist that he has always realized he should be. The present world fishery catch from the ocean is such an infinitesimal part of potential oceanic productivity that it is almost certain that the harvest could be increased severalfold without fear of overfishing, provided that the necessary precautions were in effect for individual fish stocks. Estimates by competent scientists suggest that the present ocean harvest of about 50×10^6 metric tons could be increased by a factor of 4, using present methods of fishing, or a factor of 40 if we learn how to catch and utilize economically the smaller, more abundant fishes and invertebrates. The accuracy of these estimates must be tested by biological research and exploration. Theoretically, a much larger catch than even the larger of these estimates could be sustained.

New problems may arise if the level of exploitation of marine resources is increased by an order of magnitude or more. Generally speaking, the maximum sustainable yield of a single species is reached when the standing crop of catchable fish is reduced to about 50% of its virgin magnitude. If the smaller organisms on which primary, secondary, and climax predators feed are reduced by such amounts, this may affect the abundance of predators. Since many of the higher-level predators are especially desirable in commerce, and certainly of primary interest to sport fishermen, new conflicts of interest may arise. Understanding the biology of these interactions will require ecological analyses of the highest quality.

WHY IS FISHERY BIOLOGY ATTRACTIVE TO THE BIOLOGIST?

This question cannot be answered with any assurance of accuracy because no one has ever polled the profession. It might be more useful to attempt to answer the question: "Why should fishery biology be attractive?"

It was noted recently at a Congressional hearing by Dr. Ripley, Secretary of the Smithsonian Institution, that ecology is an extremely complex, formless, and untidy branch of science. The implication was that biologists are avoiding ecology because it is so difficult and frustrating. Fishery biology, as a special branch of ecology, is equally challenging yet frustrating. It demands the best available scientific talent, and it does have some highly competent practitioners, but it is doubtful that fishery science is getting its share of top level professional men.

The attraction of fishery biology lies in the challenges it offers to unravel the complicated relationships between man and living resources and between the resources themselves. The opportunity to contribute to human welfare generally and to contribute to a desirable union between biological and social phenomena should add another challenging dimension to the already exciting science of ecology. If fishery scientists and managers could resolve the major fishery problems successfully, their achievement might well serve as a model for solution of broader problems of man and his environment. Some of the best scientific minds believe that man's present problems with his changing environment must receive the highest priority in scientific research and development. If the importance and excitement of these challenges can be communicated to promising students, a continuing supply of talent should be available. Perhaps one of the most important needs is to develop effective ways of recruiting excellence.

SUGGESTED READINGS

Cushing, D. H. 1968. *Fisheries Biology, A Study in Population Dynamics.* University of Wisconsin Press, Madison, Milwaukee & London, 200 pp.

Gulland, J. A. 1968. The concept of the maximum sustainable yield and fishery management. Food and Agriculture Organization of the United Nations, Fisheries Technical Paper No. 70, FRs/T70: i+13 pp.

Gulland, J. A., and J. E. Carroz. 1968. The state of world fisheries. *Advances in Marine Biology,* **6**, 1-71.

Murphy, Garth I. 1966. Population biology of the Pacific sardine (*Sardinops caerulea*). *Proc. Calif. Acad. Sci.* **34**,(1), 1-84.

Schaefer, Milner B. 1965. The potential harvest of the sea. *Trans. Am. Fish. Soc.* **94**(2), 123-128.

Sette, O. E., and J. D. Isaacs (Eds.). 1960. Symposium on "The changing Pacific Ocean in 1957 and 1958." *Calif. Coop. Oceanic Fish. Invest. Reports* **7**, Pt. 2, 13-217.

Smith, Stanford H. 1968. Species succession and fishery exploitation in the Great Lakes. *J. Fish. Res. Bd. Canada* **25**(4), 667-693.

AUTHOR'S NOTE IN 1972

Although this paper was written about four years ago the views and conclusions expressed are still valid. If anything, the complexity of the problems and the importance of finding solutions have come into sharper focus. The world catch of fish is still growing, and with a few notable exceptions the marine fisheries are still relatively uncontrolled. Concern about the environment has reached a fever pitch in the United States, as exemplified by the current excitement over the killing of marine mammals. It is proper to advocate effective controls on the harvest of all living marine resources. This is the objective of research. But to argue that because they have a special emotional appeal to an, certain resources should not be taken at all, is to carry the question far beyond the bounds of resource management.

Many people believe, for example, that world whale resources are near extinction, that whaling is proceeding unchecked, and that the only solution is a worldwide moratorium on whaling. Many scientists, as well as laymen, believe that the International Whaling Commission has been a complete failure. Yet the record shows, if it is examined carefully and objectively, that the whaling commission has made considerable progress in the last six or seven years. This encouraging progress has been made possible by careful scientific analysis of the condition of the whale resources, and general, if not total, acceptance of the scientific conclusions by the member nations. The whaling commission still has some difficult problems to solve, and it cannot be denied that it moved too slowly to save some species from substantial overexploitation, but it is not sitting idly by, as many people believe. Indeed, the achievements of the whaling commission must be counted among the pitifully few examples of reasonably successful management of living marine resources around the world.

The current controversy over whaling, especially in the United States, is a prime example of how opinions can override facts. The alarming aspect of this development is that even many scientists have taken a stand on the issue without bothering to be adequately informed. If they had sought the facts they might have learned that many of our domestic fishery resources are in much worse condition than the whale resources, even in territorial waters over which the United States have complete control (McHugh, 1972). Our performance under international fishery agreements, by and large, has been far better than our sorry performance at home, despite the many shortcomings of international management of living marine resources. We are doing far better with whales than we have done with sardines, menhaden, oysters, and most other strictly coastal fishery resources. If that fact were to be recognized, and national attention were to be directed at the problem with the same powerful force presently directed at harvesting of marine mammals, the accomplishments of fishery scientists might begin to pay off handsomely, and the image of the United States as a leader in the conservation movement might be established firmly.

To protect the coastal fisheries of the United States many people today believe that unilateral extension of national fishery jurisdiction to 200 miles would solve all the problems. It is easier to blame the ills of the domestic fisheries on certain nations which have been successful in developing distant water fisheries than to face the uncomfortable fact that we alone have been responsible for most of the troubles of our fishing industry. Almost all of our successes in management of coastal fisheries in the United States have been under international agreements, in which the United States has committed itself to research and management in return for concessions by the other nations party to the agreements. A good example is the growing success of Pacific salmon management by the United States and Canada under the terms of the International Pacific Salmon Fisheries Convention and the International North Pacific Fisheries Convention. Unless it were accompanied by a drastic revision of domestic fishery management policy and institutional arrangements, unilateral extension of national fishery jurisidiction to any distance beyond the present 12 miles would be in effect a hunting license to destroy resources over a wider area off our coast. Without the incentives and obligations provided by solemn international agreement, it appears that we have been unable to apply existing scientific knowledge to successful management of living marine resources.

The need for increased attention to the social sciences as an adjunct to fishery research in the natural sciences becomes more acute. One of the principal reasons is that improved knowledge of economics, sociology, and law as they relate to fishery ecology is essential to create an informed constituency, which in turn is mandatory if the institutional problems are to be corrected. Such knowledge is needed also as background for the public educational and extension programs that must be a part of any attempt to bring scientific findings to the attention of those who will use them. Economic research, for example, is needed to evaluate the practical implications of assertions like the following, quoted from my article above:

The substantial fund of available biological knowledge, if it could be put to full use, would restore to historic levels of productivity some of our most important, but dwindling American fisheries, such as the oyster industry.

This statement is true as far as it goes, but unless there is a market for the increased production, application of this scientific knowledge will add nothing to the national economy.

The need for a multidisciplinary approach to environmental science is now gaining ground in the universities. Programs like the Marine Environmental Sciences Program in the State University of New York at Stony Brook are producing students with a broad understanding of the problems of resource and environmental management, including fisheries, that today's students need.

It is no longer correct to say that "There is yet no adequate textbook" on fishery science (see Royce 1972).

REFERENCES

McHugh, J. L. 1972. The fisheries of New York State. *U. S. Dept. Commerce, Natl. Marine Fish. Serv., Fish. Bull.* **70**(3), 579-604.

Royce, William F. 1972. *Introduction to the Fishery Sciences.* New York: Academic Press, x +351 pp.

Power from the Tides

EDWARD P. CLANCY   1968

ORIGIN OF TIDAL POWER
AND HOW THE POWER IS DISSIPATED

To a bather stretched idly on the warm sand, the coming of the tide hardly suggests vast power. The sea is calm; as the hours pass the water's edge inches its way up the beach, then down again.

Is not tidal energy a negligible factor in the scheme of things? All along, we have emphasized how very small are the tide-generating forces comparing with other forces which operate in our planetary system.

Very small they are, and yet such is the scale of the cosmos that tidal effects result in an energy dissipation—in the oceans, the atmosphere, the "solid" earth—which is enormous. Enormous, that is, in terms of man's measure of energy. Energy of the tides is continuously being dissipated at a rate whose order of magnitude is a billion horsepower!

This vast power output comes at the expense of the kinetic energy of the earth-moon system. Part of the kinetic energy exists in the rotation of the earth on its own axis; loss of this energy means slower rotation, and a consequent lengthening of the day. There is abundant evidence that the earth is slowing. We shall examine this evidence later.

But what becomes of this awesome expenditure of power? The law of the conservation of energy says that the lost kinetic energy must appear in some other form. The loss takes place in frictional effects and therefore appears ultimately in the form of heat. Many opportunities exist for such effects: viscous forces in the fluid interior of the earth, nonelastic properties of the solid parts of the earth which result in the phenomenon called hysteresis, and frictional loss experienced by tidal currents in the seas and the atmosphere.

At present, no one can assign relative weights to these various processes. Only the ocean tidal currents lend themselves to direct observation. One might hope to make an intelligent guess about them. But they occur throughout the world in most complicated forms. They flow in depths which vary

Dr. Edward P. Clancy is chairman and professor of the physics department at Mount Holyoke College, Massachusetts. He is currently on sabbatical leave with the physics department at the University of Aukland, New Zealand. He is the author of numerous articles in leading journals of physics.

From *The Tides*, pp. 133-147, 1968. Reprinted with light editing and by permission of the author.
Copyright 1968 Doubleday and Company, Inc., New York.

greatly; their velocities change hourly and monthly; they move over bottoms which may be rough or smooth; internal turbulence, even without contact with land, is always present. To assess energy loss due to tidal currents is, as you see, almost impossible. Only the roughest sorts of estimates have been attempted.

HARNESSING THE TIDES

Whatever the energy in the ocean tides, it is enormous. Can we somehow get power from this energy? Can we, in other words, construct a system whose end result is to change into some other form of energy, power which would otherwise have been dissipated in tidal friction?

Here is a challenge to you, before you read further. Imagine yourself an engineer, asked to devise ways of harnessing tidal power for the generation of electric current. How many methods can you think of? Write them down, and for each plan some suitable machinery.

Probably the most unsophisticated approach—almost childish, as it turns out—would be to say, "All right, the tide goes up and down. Build some large floating object—an enormous, heavy, rectangular barge. Let the tide lift it. Have a supporting structure overhead with steel cables wound around drums; let the ends of the cables drop vertically to eyes in the deck of the barge. As the barge descends on a falling tide, the cables will turn the drums. With suitable step-up gearing, the turning shafts of the drums could rotate an electric generator."

For his own amusement, the author figured out how much power could reasonably be expected from such an arrangement. Take one of the largest floating objects in existence—say a modern oil tanker weighing 100,000 tons, or 2×10^8 pounds. Place it where the tidal range is extreme. In descending through a distance of, say, 40 feet, work of 2×10^8 lb $\times$ 40 ft = 8×10^9 foot-pounds will be done. Since power is the rate of doing work, we must divide this figure by the time involved, which is roughly 6¼ hours, or 2.25×10^4 seconds. Then

$$\frac{8 \times 10^9 \text{ ft-lb}}{2.25 \times 10^4 \text{ sec}} = 3.55 \times 10^5 \text{ ft-lb/sec}$$

To convert this number into horsepower, we recall that one horsepower is 550 footpounds per second.

$$\frac{3.55 \times 10^5 \text{ ft-lb/sec}}{550 \text{ ft-lb/sec per hp}} = 647 \text{ hp}$$

This is the *average* power output to be expected during the 6¼ hours. It assumes no frictional loss in the machinery.

The mountain has labored mightily and brought forth a mouse. Four ordinary automobile engines could do the same job! Furthermore, they could to it around the clock, whereas our mechanical monstrosity would work only during a falling tide.

Our approach to the problem of power from the tides has been too naive. We failed to recognize that since tidal rise and fall are so small, the secret of significant power generation is to process large quantities of water. We must, in other words, allow a great volume of water to descend through whatever tidal difference in levels is available or can be made available. The water, as it descends, must do work—say by turning a waterwheel.

ANCIENT TIDAL POWER INSTALLATION

Waterwheels, at least in their simpler forms, have been known almost from the dawn of history. Ancient Egyptian drawings show paddlewheels dipping into the current of the Nile. Buckets attached to their rims lifted water above the level of the bank, to dump into sluices and thereby irrigate fields. It is not surprising that where swift tidal currents exist they have sometimes been used to turn wheels—though in Europe and America the power has ordinarily been used for other purposes, such as grinding corn.

Paddlewheels are rather inefficient devices; a much better machine is the typical overshot waterwheel, which can be installed wherever a head of water naturally exists or can be created by a dam. Roughly speaking, the term *head of water* means simply the available difference in levels of water between which the waterwheel works. It is the amount by which the water behind the dam stands higher than that at the foot of the dam.

In a few places a natural head of water exists along the ocean shore, where tidal currents must enter and leave a bay through a narrow opening in the rocks. More often, a dam is built across the mouth of a tidal basin. A typical example of such a power installation—used for grinding spices—was "Slade's Mill" in Chelsea, Massachusetts. Here four waterwheels generated about fifty horsepower when the full head of tide existed.

ASSESSMENT OF TIDES AS AN EFFECTIVE SOURCE OF POWER

Many drawbacks are inherent in the use of tidal power. One is the lack of a large head of water. Another—and worse—is the inconstancy of what head there is. Power nowadays is demanded almost solely for the generation of electricity, a demand which varies somewhat with the hour of the day and the day of the week, but which is never below a minimum value for a given region. Tidal power—unless special arrangements can be made, as we shall see later—is obviously available for only a fraction of each day.

So perhaps it is not surprising that the development of tidal power plants has languished in favor of ordinary hydroelectric installations on rivers, where dams are usually easier to build, and where a constant head of water can be maintained. Rivers themselves, of course, have inconstant flow. But if there is a sufficiently large reservoir behind the dam, the operators can "average out" the flow. If the reservoir is too small, one can (a) install only enough generating capacity for the minimum seasonal flow of the river or (b) draw water from the reservoir only for short periods of the day, when a demand for *peaking power* exists. Peaking power is that extra power necessary for times of high electrical consumption—during times of day when both industrial and domestic demands are large, or during hot days in summers when use of air conditioners represents an enormous load.

If tidal power is not constant, why not use it for peaking purposes? Alas, the tides are on a lunar cycle, whereas man lives on a solar cycle. Only by coincidence would tidal peaking power become available at the right time.

But one can look at the thing from another point of view. For a conventional thermal power plant—burning coal, or oil, or natural gas—the capital cost is ordinarily a small fraction of that for a hydroelectric installation of the same capacity. Fuel costs are high, however, whereas water comes free. For certain regions where fuel is particularly expensive, a system in which a tidal plant complements a thermal plant now turns out to be economically attractive. Output of the thermal plant can be reduced, and fuel saved, while the tidal plant is now operating.

The word *now* is significant. Only in recent years has imaginative hydraulic engineering created machines and structures which lend themselves to the special problems of the tidal power plant.

Perhaps you are puzzled by this statement. After all, has not a tremendous technology for hydroelectric plants existed for some time? Yes, but there are subtleties in tidal power generation which do not appear in ordinary waterpower installations. One we have already mentioned: the head of water is always small. Typical vertical-shaft watermotors are not feasible since their use means loss of a few precious feet in the effective head of water.

Furthermore, the head of water is constantly changing, and in fact reverses itself every six hours. For the simplest cycle—that of the ancient tidal mills—this presented no problem. These mills functioned merely on the outgoing tide, and could therefore operate only twice in twenty-four hours, for a few hours each time. In a modern tidal plant, however, power must be generated both on the rising and on the falling tide since the output per day is thus doubled. (This statement applies to the ordinary situation, in which the generating equipment is located in a dam across the mouth of a single estuary.)

DESIGN REQUIREMENTS FOR AN EFFICIENT TIDAL POWER INSTALLATION

How then shall we design our waterwheel, or in modern terms our *turbine?* Since the head of water is small, we must allow large quantities to flow through *horizontal* channels in the dam. As the water flows, it must impinge on the blades of a turbine whose shaft is also horizontal, and on which is mounted the rotor of an electric generator. But water flows both ways as the tide goes through its cycle. How shall we manage this? Simple enough! Make the blades of the turbine reversible!

Another difficulty arises. Electric generators must run at constant speed; they are electrically "locked in" with the generators in other stations of the power network. The effective tidal head is constantly changing, however, and the speed of the water impinging on the turbine blades varies correspondingly. How do we keep the turbines from slowing as the head decreases? We have our answer already. By designing the turbine blades to be reversible, we have also made it possible to adjust their pitch, or angle with respect to the flow of water. As water speed decreases, the blades are set flatter. The power output is reduced, but the shaft maintains its necessarily constant speed. Automatic machinery controlled by a computer can take care of the whole business; no human factor need be involved.

One difficulty is inescapable. The maximum head is small, so the dam is not high. Channels through the dam cannot be of large diameter. The size of the turbine and therefore its power output are limited. The only solution: install many channels and many turbines. This is too bad, for it means violation of a basic engineering maxim that the bigger a machine the more efficient it is.

THE POWER PLANT AT LA RANCE

At the time of this writing only one large-scale tidal power plant exists. So far as the author is aware, no other is even under construction, though some proposals are being vigorously pushed. The existing plant, newly completed, is in a barrier built across the mouth of La Rance estuary in Brittany. The design follows the basic principles we have been discussing.

The estuary, shown in Figure 6-6, lies between the towns of St. Malo and Dinard. It has an area of about nine square miles, and is located in a region of exceptional tidal range. The basin extends inland some thirteen miles.

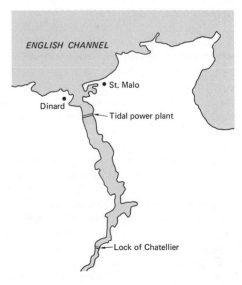

Figure 6-6. Tidal power plant across the mouth of La Rance estuary in Brittany.

Calculations show that usable tidal energy is proportional to the area of the basin involved and to the square of the amplitude of the tide. The maximum tidal range at La Rance is 44 feet. For these equinoctial tides, water flows in or out of the estuary at the maximum rate of 630,000 cubic feet per second. (This is three times the flow of the Rhone River in flood.) A useful volume of over six billion cubic feet is created.

The barrier was built at a point where the channel is 2500 feet wide. Rocks two miles downstream at the mouth of the estuary protect the works from ocean storm waves.

The installation consists of four main parts: the dam, the power plant, locks for navigation, and a barrage of sluice gates to accelerate the filling and emptying of the basin. Much geological reconnaissance was done beforehand. Borings revealed a substratum chiefly of gneiss, covered in places by thin layers of sand and gravel. The deepest water encountered was 39 feet at lowest tide, and 83 feet at maximum tide. The engineering difficulties faced were substantial but not formidable. Design of the dam involved an unusual factor, of course—the structure must be able to withstand pressure from both directions!

Before construction was started, hydraulic engineers built a working model of the whole estuary. Everything—the basin with all its contours, the proposed dam and associated works, and the tides themselves—was reproduced with the greatest fidelity. The model showed the intensity and direction of currents to be expected, and helped enormously with various aspects of the planning.

To get a "dry field" so that foundations of the different structures could be placed on bedrock, the working areas were surrounded by a string of caissons. The caissons were linked with watertight joints, and the enclosure they formed was then pumped dry. An engineering operation of this magnitude is not done quickly; some six years were needed for completion of the whole enterprise.

Vanes 33 feet high and 49 feet wide control water flow through the sluices. They, too, must resist pressure from both directions. Unlike gates in ordinary dams, which are moved only a few times a year, the vanes at La Rance are in almost constant motion. The reason: for the most efficient generation of power there must be a continuous and very close control over the water level within the basin.

The power plant, of reinforced concrete, looks inside like a vast tunnel, 1200 feet long. It contains all the control equipment. The arched roof gives the structure rigidity to resist water pressure from either side. Traveling cranes within the tunnel service the generating units. Each unit lies in its individual conduit beneath the floor; the bottom of the conduit is some 33 feet below the lowest water level. Figure 6-7 shows the structure in cross section. It is interesting that at high water an observer outside and at some distance sees only a thin white line projecting above the surface of the bay. This is the roof of the power plant; everything else is submerged! To give some idea of scale: each conduit is 174 feet long, and its cross-sectional area at either end is some 1000 square feet. There are twenty-four of these conduits.

Construction of a project such as this had to wait upon design and development of a most unusual and ingenious device—the "bulb-type" turbine-generator unit which can also act as a pump. The device resembles, on an enormous scale, a short, fat torpedo. It is entirely surrounded by water. The

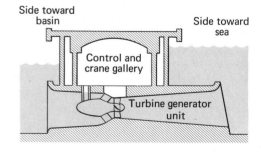

Figure 6-7. Cross section of the tidal power plant.

body of the torpedo contains the electric generator, and the propeller exactly fits the constricted throat of the conduit (see Figure 6-7). Struts projecting radially inward from the walls of the conduit support the unit. One of the struts is hollow, and large enough for a man to climb down a ladder from the control room into the interior of the bulb to perform servicing.

Design studies on the device were started in 1952. Smaller models of differing characteristics were installed in existing dams, and hydraulic circuits were constructed so that the units could be tested with water going both ways. In 1955 sizable units were set up in an abandoned lock system at St. Malo in order to test under actual tidal conditions.

The machines which evolved and which are now installed at La Rance each have a generating capacity of 10,000 kilowatts. The propellers, with a diameter of 17.5 feet, turn at 94 revolutions per minute. Inside the bulb, air pressure is kept double that of the normal atmosphere. The result: better cooling of the generator. At atmospheric pressure the output is limited to 7500 kilowatts, showing the great importance of proper temperature control.

The units generate power at 3.5 kilovolts. Step-up transformers increase this voltage to 225 kilovolts; the current then enters three transmission lines which leave in different directions to link with the French national power grid.

So much for the technical details of construction. A most fascinating question remains: How can the whole complex be managed for most efficient power output? Figure 6-8 shows the sequence of various phases of operation when no pumping is done. Sea level is denoted by the solid line; it is essentially a sine curve. Level within the basin is shown by the dashed line. A crucial aspect is the vertical difference in height of these two curves, i.e., the head of water. When it is appreciable, power generation is possible. The cross-hatched sections represent the times of turbine operation.

The sequence is as follows: as the ocean starts to rise from its minimum level and the head decreases to the point where efficient generation is no longer possible, the turbines are shut down, but the sluice gates open to allow continuing outflow from the basin. When basin and sea are at the same level, all gates are closed and there is a waiting period until sufficient head builds up. Then power generation begins, and continues until the combination of decreasing sea level and increasing level in the basin results in too little head. The turbines are stopped, but the sluice gates allow further filling of the basin.

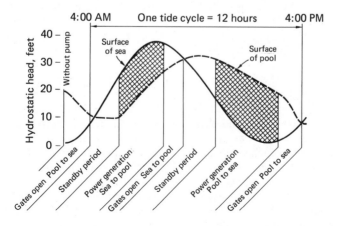

Figure 6-8. Phases of operation of tidal power plant without pumping.

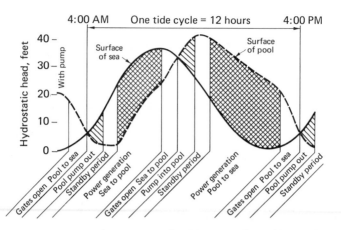

Figure 6-9. Phases of operation of tidal power plant with pumping.

When levels within and without are equal the gates are closed and there is a waiting period until sea level falls sufficiently to create a new head—this time in the opposite direction. Meanwhile the blades of the turbines have been reversed and a new cycle of power generation begins, continuing until the head is again no longer adequate.

We have referred to the fact that the bulb units can also act as pumps. If power from outside the plant is fed into the generator, it is no longer a generator but a motor! It can then drive the turbine blades to act as a pump forcing water through the conduit, just as the motor of an electric fan turns the blades to create a stream of air.

But under the circumstances, what is the advantage of pumping? The answer is a little subtle but not too hard to see. Figure 6-9 is like Figure 6-8, but with the superposition of a dotted line which shows the basin level as it can be managed if pumping is done. We see that the cross-hatched area, and therefore the amount of power generated, is appreciably increased. Part of each waiting period is used for pumping—in one part of the cycle for building up the basin level, in the other part for lowering it. In both cases the available head is increased.

You can't get something for nothing, though. Since the efficiency of any machine

is less than 100 per cent, don't you end up with a net *loss* of energy? No, for the following reason. As you can see from the graph, one always pumps against a small head. The water pumped will later be used at a much greater head, however. Here we have the key to the worth of pumping. Another, and not altogether minor, factor is that electric energy has different monetary values at different times of day. During hours of low usage, extra power is available at reduced rates. (You already know this if your home has a clock-operated electric water heater on a separate meter.) When a waiting period at a tidal plant occurs in these hours, pumping becomes particularly rewarding.

There are several variables, as you see: basin level, sea level, value from hour to hour of energy generated and of energy bought for pumping. Proper operation of a tidal plant for maximum efficiency requires a constant and rather complicated calculation of all factors involved. The result of this running calculation must be continuously translated into control of the various components: the vanes of the sluice gates, the pitch of the turbine propeller blades, the starting and stopping of generating or pumping cycles. The whole situation calls for a computer-controlled operation, and in the plant there is indeed a computer programmed not only to perform the necessary calculations but also to control all components according to the results of the calculations.

In each of the 24 conduits at La Rance is a turbogenerator of 10,000 kilowatts capacity. Maximum power output is therefore 240,000 kilowatts. An ordinary hydroelectric plant running continuously at this capacity would produce 2100 million kilowatt-hours per year. The annual energy output at La Rance is about 540 million kilowatt-hours without pumping, and an additional 130 million with pumping. The figures starkly underline the nonconstant output of a tidal power plant. Note well, however! This is an *assured production*. There are no dry years for a tidal installation! And no worries about damage from disastrous floods.

Recovery of Chemicals from the Sea and Desalinization

RALPH A. HORNE   1969

INTRODUCTION

The seas and their immediate environment, both existing and ancient, represent an enormous, but not inexhaustible, reservoir of chemical substances. Here we will be briefly concerned only with those materials which come from the waters of existing seas, and we will make no further mention of the vast deposits of salts, petroleum, sulfur, limestone, and other mineral substances laid down by ancient seas, nor will we discuss the problems of recovery of the great treasure of manganese nodules at the bottom of the sea—a subject treated with thoroughness and enthusiasm in Mero's book, *The Mineral Resources of the Sea* (1965).

SALT

The most obvious chemical recovered in large quantities from the sea is NaCl. About 6,000,000 tons of salt are produced annually from sea water by evaporation in shallow basins. About 5% of the salt consumed in the United States is produced in this manner, largely in the San Francisco Bay area. As the sea water evaporates, various dissolved salts begin to sequentially crystallize out. First the $CaSO_4$, then the brine is concentrated until NaCl begins to precipitate; finally, when the specific gravity exceeds 1.28, magnesium salts begin to precipitate. The remaining concentrated brine solution is called "bitterns" and may be further processed to recover magnesium compounds, halides, and other salts.

HALIDES

The element iodine was first discovered by Courtois in 1811 (isolated from the ash of seaweed), and in 1825 another Frenchman, Balard, discovered the sister element bromine in bitterns prepared from salt marsh waters (Weeks, 1945). The production of iodine from seaweed ash has long since been replaced by other sources and processes, but bromine is still obtained from sea water both

Dr. Ralph A. Horne most recently was principal scientist of the JBF Scientific Corporation, Massachusetts. He is the author of more than eighty technical papers in physical chemistry, molecular biology, and environmental sciences.

From *Marine Chemistry: The Structure of Water and the Chemistry of the Hydrosphere*, pp. 444-454, 1969. Reprinted with light editing and by permission of the author and the publisher. Copyright 1969 John C. Wiley & Sons, Inc., New York.

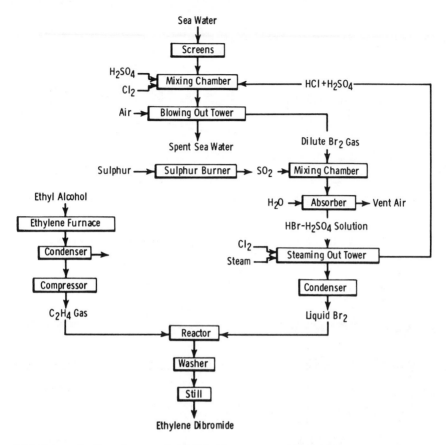

Figure 6-10. Schematic flow diagram of the Ethyl-Dow process for recovering bromine from sea water. From Shigley (1951), with permission of Metallurgical Society of AIME.

as a by-product of salt production and by direct precipitation of the insoluble tribromoaniline formed on treating unconcentrated but acidified sea water with aniline and chlorine. The latter process has been refined and modified, notably by the use of SO_2 and air in the stripping steps (Figure 6-10), and it now accounts for the majority of the bromine production in the United States. Chlorine is prepared by the electrolysis of fused NaCl (Downs process)

$$2 Na^+ + 2 e^- \rightarrow 2 Na \text{ (cathode)}$$
$$2 Cl^- \rightarrow Cl_2 \uparrow + 2 e^- \text{ (anode)}$$

or of concentrated brine (Vorce process)

$$2 H^+ + 2 e^- \rightarrow H_2 \uparrow \text{ (cathode)}$$
$$2 Cl^- \rightarrow Cl_2 \uparrow + 2 e^- \text{ (anode)}$$

but mined salt from ancient seas rather than solar-evaporated sea salt is largely used. This chlorine production forms the nucleus of a cluster of heavy industrial chemicals.

MAGNESIUM

Next to Na, Mg is the most abundant metallic cation in sea water. Since World War II the Mg metal consumed in the United States

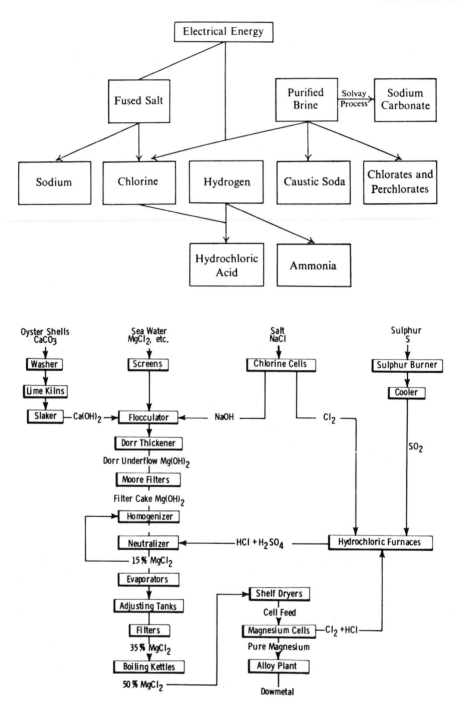

Figure 6-11. Schematic flow diagram of the Dow Process for recovering magnesium from sea water by the electrolysis of $MgCl_2$ solutions. From Shigley (1951), with permission of Metallurgical Society of AIME.

has come largely from the sea. In 1964 the annual world production of this strong, lightweight metal was about 150,000 tons. The Dow Chemical Company Mg production plant at Freeport, Texas, processes 1,000,000 gal/hr of sea water from the Gulf of Mexico (Figure 6-11). In addition to the metal, MgO, $Mg(OH)_2$, and $MgCl_2$ are also produced from sea water and brines.

GOLD

Human nature is tempted by the lure of something for nothing and, ever since its presence was detected there, some men have schemed to devise ways to recover gold from sea water. The oceans of the world are said to contain enough gold to make everyone on Earth a millionaire. The recovery of gold from sea water, curiously enough, appears particularly to appeal to the Germans who, we recall, were also avid alchemists. No less a German chemist than Fritz Haber dreamed of paying his nation's World War I debt with gold extracted from the sea. But after the expenditure of ten years of effort he was forced reluctantly to conclude that the gold concentration is far too small and the quantity of sea water that would have to be processed far too large to make the extraction profitable. Some years ago a colleague of mine at M.I.T. (incidentally also a German) dragged a sack containing a very strong anion exchange resin through the waters off Plumb Island in order to recover $AuCl_4^-$, but all he ever got for his pains was a bad chill. After processing 15 tons of sea water in their North Carolina bromine extraction facility, the Dow Company succeeded in recovering 0.09 mg of gold worth about $0.0001.

BIOMATERIAL

The marine biosphere yields many materials useful to man, the most obvious being, of course, the food materials, but also seaweed as a source of sodium alginate and fertilizer as well as food. Not so well known but attracting increasing scientific attention is the marine biosphere as a source of physiologically active substances ranging from toxins (see the magnificent volume on toxic marine animals by Halstead and Courville, 1965) to antibiotics (Olesen et al., Maretzki, 1964; Aubert, Aubert et al., 1966; and Nigrelli, Stempien et al., 1967). In August of 1967 the Marine Technology Society sponsored a conference on "Drugs from the Sea" at the University of Rhode Island, readers interested in this fascinating topic are referred to the detailed bibliography prepared by A. D. Marderosian for that conference.

DESALINATION

In the not too distant future by far the most important chemical recovered from the sea may very well be water. The unchecked spread of the human infestation of this smallish planet not only is making an ever-increasing demand on water resources but is at the same time diminishing the usability and debasing the quality of those resources by pollution. Just as he has turned to the oceans as a panacea for his food problem, man is turning to the seas as a panacea for the water shortages he has created.

We know the powerful solvent properties of water, the ease with which substances dissolve, the large energies involved, and the great strength of their solvation envelopes. Clearly, to reverse this process, to separate these affianced substances, the salts from the water, is not an easy task. But, in spite of all this difficulty, so frightening is the urgency that processes designed to accomplish this end have already been pushed from research and development into engineering and production phases (for a good review see Spiegler, 1966).

Glueckauf (1966) classifies processes for desalting water into three main categories:

1. Process involving a phase change
   Evaporation
   Distillation
   Freezing
2. Processes involving semipermeable membranes

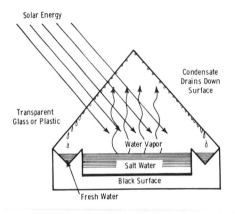

Figure 6-12. A simple solar still.

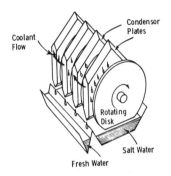

Figure 6-13. A diffusion humidification unit.

Electrodialysis
Reverse osmosis (hyperfiltration)
3. Processes involving chemical equilibria
Ion exchange
Hydration

The method of obtaining fresh water from salt water with a minimum of equipment that first comes to mind is probably solar distillation. In a simple solar still such as shown in Figure 6-12 solar energy is utilized to evaporate the water which is then condensed on a cooler surface and collected. More complicated but closely related in principle is the diffusion humidification process (Figure 6-13) in which the water evaporates from a rotating disk, diffuses across a narrow airspace, and collects on a stationary, cooled

condenser plate. The equipment requires an energy source both to warm the salt water and to rotate the apparatus, but nevertheless it appears to represent a feasible type of installation for small-scale applications.

The classical method of purifying water is, of course, distillation. Distillation is one of the most important large-scale desalination techniques, and the method admits of many engineering ramifications such as multistage flash distillation (Figure 6-14), long-tube vertical distillation, multistage-multi-effect distillation, and vapor compression distillation. But water molecules are very disinclined to be separated and, compared to other substances, enormous quantities of energy are required to convert water from the liquid to the gaseous state. The conversion of the liquid to the solid, however, requires less than one-sixteenth as much energy as evaporation. As an additional dividend, the lower temperatures tend to minimize scale and corrosion problems; hence freezing appears to offer an attractive phase transition method of desalinating sea water. If an aqueous electrolytic solution is not frozen too rapidly, crystals of pure ice are formed and the remaining liquid becomes more concentrated in salt. A major problem in the freezing process which, like distillation, also admits of several engineering variations, is the mechanical separation of the solid and liquid phases.

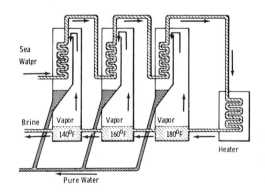

Figure 6-14. Multistage flash distillation process.

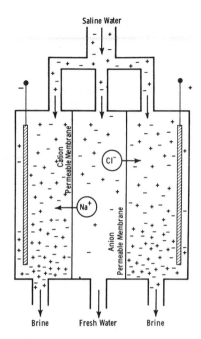

Figure 6-15. Electrodialysis.

In electrodialysis, under the influence of an imposed electric field, ions are transported out of the saline solution through ion-permeable membranes (Figure 6-15). Inasmuch as the current requirements depend on the electrolyte concentration, at the present time the process appears to be better suited for the desalination of less concentrated brackish waters than of sea water. The anion-permeable membrane tends to deteriorate under operating conditions, and in a variant on the process—transport depletion—it is replaced by a nonselective membrane.

If pure water and a saline solution are separated by a membrane permeable only to water molecules and not electrolytes, water molecules diffuse through the membrane into the saline solution, diluting it and building up an osmotic pressure (Figure 6-16). Conversely, if pressure in excess of the osmotic pressure is applied to the solution, water molecules can be forced out of the solution through the membrane; and this forms the basis of desalination by reverse osmosis. This process also is presently most applicable to brackish waters, although it is hoped to be able to extend its use to seawater.

Ion exchange can also be used to purify waters with relatively low salt contents. A cation exchange resin replaces the Na^+ with H^+

$$R_cH + Na^+ + Cl^- \rightleftharpoons R_cNa + H^+ + Cl^-$$

and an anion exchange replaces Cl^- with OH^-

$$R_aOH + H^+ + Cl^- \rightleftharpoons R_aCl + H^+ + OH^-$$
$$\updownarrow$$
$$H_2O$$

but regeneration of the exchangers tends to make the process costly.

Finally, desalination processes based on the phenomenon of hydration have been suggested. Crystalline hydrate forms of substances, such as propane (C_3H_8) or carbon dioxide, are formed, separated from the concentrated brine, washed, and then decomposed by a temperature and/or pressure change to give separated hydrating agent, which is then reused, and pure water.

The cost of pure water from conventional sources is slowly rising, whereas that from desalination, thanks to research, development, and engineering, is rapidly falling, es-

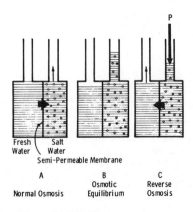

Figure 6-16. Reverse osmosis.

Table 6-3. World's Major Saline Water Conversion Plants as of 1965[a] (From *Business Week*, No. 1883, 120, Oct. 16, 1965, with permission of McGraw-Hill Pub. Co.)

In operation or under construction
  Middle East

Quatar, Doha	360,000 gal/day	Multistage flash distillation
Quatar, Doha	1,800,000	
Saudi Arabia, Dhahran	105,000	Electrodialysis
Saudi Arabia, Dhahran	200,000	Vapor compression
Neutral Zone, Arabia	500,000	
Kuwait	5,520,000	
Kuwait	2,400,000	Multistage flash distillation
Ahmadi	360,000	
Ahmadi	720,000	
Al Shaiba	6,000,000	
Kuwait	240,000	Electrodialysis

  Atlantic Ocean–Mediterranean

United Arab Republic, Sinai	600,000	
Italy, Taranto	1,200,000	Multistage flash distillation
Malta, Valletta	1,200,000	
Israel, Eilat	1,000,000	
Israel, Eilat	240,000	Freezing
Morocco, Centa	1,060,000	
Guernsey	600,000	Multistage flash distillation
Canary Islands	650,000	
Libya	200,000	

  Caribbean–South America
    Dutch West Indies

Curacao	3,400,000	
Curacao	1,600,000	
Aruba	2,650,000	
Aruba	795,000	
Virgin Islands		Multistage flash distillation
St. Thomas	1,000,000	
St. Thomas	275,000	
St. Croix	1,500,000	
Venezuela		
Port Cardon	1,270,000	
Bermuda Air Base	200,000	Vapor compression
Cuba, Quantanamo	2,250,000	
Chile, Charnaral	240,000	Multistage flash distillation
Bahama, Nassau	1,440,000	

  United States

California, Catalina	150,000	Multistage flash distillation
Arizona, Buckeye	650,000	Electrodialysis
S. Dakota, Webster	250,000	
N. Carolina, Wrightsville	200,000	Freezing
New Mexico, Roswell	1,000,000	Vapor compression
Texas, Freeport	1,000,000	Long-tube vertical
Texas, Chocolate Bayou	900,000	
Virginia, Possam Pt.	188,000	
California, Moss Landing	376,000	Multistage flash distillation
Tennessee, Paradise	418,000	
Florida and Key West	2,620,000	
U.S. Mexican Border	100,000,000	Undecided

Table 6-3, continued

Soviet Union		
20 plants (location unknown)	1,000,000	Unknown
Kazakhstan, Shevchenko	1,300,000	Long-tube vertical
Being planned		
Egypt	50,000,000	
Hong Kong	3,000,000	
New York, Riverhead	1,000,000	
Southern California	150,000,000	
Israel	100,000,000	Unknown
Saudi Arabia	2,000,000	
Spain, Southeast Coast	2,000,000	
Soviet Union, Don Basin	150,000,000	

[a]A more recent and detailed listing can be found at the end of *Saline Water Conversion Report for 1966* Office of Saline Water, U.S. Dept. of Interior, Washington, D.C., 1966.

pecially for large-capacity installations. At present the costs of both sources are falling in the 50¢/1000 gal range. At this price the water in a cubic mile of seawater is worth some 10^{10}, whereas the gold is worth only a paltry 10^5. Gold is a luxury, but without water we shall perish. Table 6-3 summarizes the location, capacity, and method of the world's major desalination installations. I apologize for the table for, although it is only four years old, so rapidly are developments occurring in the field that it is already out of date.

REFERENCES

Aubert, M., J. Aubert, M. Gauthier, and S. Daniel, *Rev. Intern. Oceanogr. Med.,* **1,** 10 (1966).

Glueckauf, E., *Nature,* **211,** 1227 (1966).

Halstead, B. W., and O. A. Courville, *Poisonous and Venomous Marine Animals of the World,* U.S. Government Printing Office, Washington, D.C., 1965.

Mero, J. L., *The Mineral Resources of the Sea,* Elsevier Pub. Co., Amsterdam, 1965.

Nigrelli, R. F., M. F. Stempien, Jr., G. D. Ruggieri, V. R. Ligouri, and J. T. Cecil, *Fed. Proc.,* **26,** 1197 (1967).

Olesen, P. E., A. Maretzki, and L. A. Almodovar, *Botan. Marina,* **6,** 224 (1964).

Shigley, C. M., *J. Metals,* **3,** 25 (1951).

Spiegler, K. S., ed., *Principles of Desalinization,* Academic Press, London, 1966.

Weeks, M. E., *Discovery of the Elements,* J. Chem. Educ., Easton, Pa., 5th ed., 1945, Chap. 24.

World Subsea Mineral Resources

V. E. McKELVEY   FRANK F. H. WANG   1969

INTRODUCTION

Maps showing the world distribution of potential subsea mineral resources are based on sparse information. Samples and photographs that show the character of the bottom sediment and seismic surveys that provide information on the thickness of sediments are widely spaced. Drill holes that reveal the composition and structure of the rocks at depth are largely confined to nearshore areas where petroleum exploration has been undertaken. Information on the bathymetry of the sea bottom, which tells something of its geologic character and its suitability for dredging and other operations related to mining or drilling, is scant in many parts of the ocean basins, and large areas have not been surveyed in any fashion. According to an assessment by the International Hydrographic Bureau in Monaco, even at a reconnaissance chart scale of 1:1,000,000 only 15 to 20 per cent of the oceans and continental margins are adequately covered by bathymetric data, and the data are almost entirely lacking for nearly 50 per cent of the areas (United Nations, Economic and Social Council, 1969). Detailed bathymetric charts at scales from 1:250,000 to 1:50,000 now only cover a few shelf areas. Because of this paucity of information, and particularly that directly related to subsea minerals, the distribution of potential subsea minerals is highly conjectural. Further exploration doubtless will substantially alter the projected and inferred distribution and in addition may reveal kinds of subsea mineral occurrences not now known or anticipated.

In spite of its inadequacy, the mass of information on the seabed is large and is growing rapidly. Academic and government institutions have collected hundreds of thousands of bottom samples, and thousands of shallow cores and bottom photographs (the data filed at the National Oceanographic Data

Dr. Vincent E. McKelvey is an economic geologist and director of the U.S. Geological Survey, Washington. He has published numerous articles on the geology of phosphate, uranium, and mineral resources.

Dr. Frank F. H. Wang is a marine geologist with the Office of Marine Geology, U.S. Geological Survey, Menlo Park, California. He has studied the technology of offshore petroleum and marine mineral resources and oceanographic data processing. He has written over a dozen technical reports.

From a discussion to accompany U. S. Geological Survey's Miscellaneous Geologic Investigations Map I-632, 17 pp., 1969. Reprinted with light editing and by permission of the authors and the U. S. Geological Survey.

Center alone now exceeds 76,000 bottom samples, including grab, shallow cores, and dredge samples. See National Oceanographic Data Center, 1968). Beginning in recent years with the Mohole and JOIDES projects, 62 holes have been drilled to depths of several hundred meters in the deep ocean floor (Figure 6-17). In addition, millions of miles of geophysical traverses and a few tens of thousands of drill holes and wells in the shallower parts of the continental shelves have been completed by oil companies, the results of which, although generally not released in raw form, are published in syntheses on regional geology and stratigraphy and add greatly to total knowledge. This fund of data has helped develop an advanced understanding of the geologic character of the sea bottom and many of the processes that operate within it. Coupled with knowledge of coastal geology and of the geologic occurrence of minerals on land, the information available makes it possible to ascertain

most of the kinds of minerals that the seabed may be expected to contain, to identify many of the geologic environments in which they may occur, and to make meaningful generalizations about their potential magnitude. The projections and generalizations in the following pages about the distribution and magnitude of subsea resources are not accurate or detailed enough to serve as a guide to exploration or to make quantitative appraisals of the extent and value of resources in specific areas, but they may be useful to those concerned with the broad aspects of the character and distribution of seabed resources and the prospects for their development.

We briefly describe the main subsea geologic and physiographic provinces and their bearing on potential seabed resources, discuss the classification of mineral resources estimates, and give a rough indication of the magnitude and potential production of seabed resources.

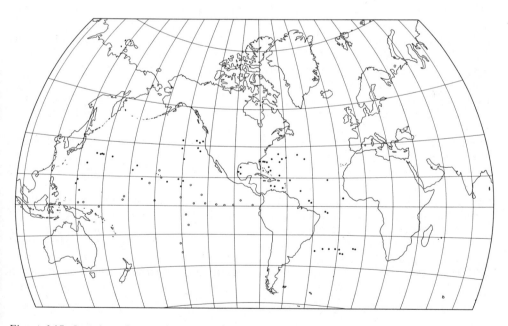

Figure 6-17. Location of core holes being drilled under the Deep Sea Drilling Project of the JOIDES program sponsored by the National Science Foundation. Those completed by mid-July 1969 shown by solid dot; remainder are scheduled for completion by early 1970.

GEOLOGIC AND PHYSIOGRAPHIC PROVINCES AND THEIR BEARING ON POTENTIAL SEABED MINERAL RESOURCES*

The solid earth's surface consists of two great physiographic divisions, the ocean basins, and the continents that rise to mean heights of 4300 to 5800 meters above the ocean floor. The ocean basins, of course, are filled with sea water—more than filled, in fact, for the ocean extends over the margins of the continental masses for distances ranging from a few to more than 1300 km. The boundary between the continental masses and the ocean basins thus lies beneath the sea, generally at depths ranging from 2000 to 4000 meters.

The physiographic contrast between the continents and the ocean basins reflect fundamental geologic differences between them. The continental crust is richer in silica and the alkalis and poorer in iron and magnesia than the oceanic crust. Although the continental crust averages about 35 km in thickness compared with about 5 km for the oceanic crust, its density is less (Figure 6-18). The continental and oceanic masses are in flotational equilibrium with the underlying mantle, and the lighter continents rise above the ocean basins, much as does an iceberg in the sea.

The igneous rocks of the continental crust consist mainly of granite and related rocks relatively rich in silica and the alkalis, although some basalt and other rocks rich in iron and magnesia are also present. In many areas these granitic rocks intrude, or are overlain by, thick accumulations of sediments deposited in ancient seas that spread over the continents, and in marginal oceanic basins—sediments derived in large part from the weathering and erosion of adjacent land. Oil, gas, sulfur, saline minerals, coal, and other deposits occur in these sedimentary

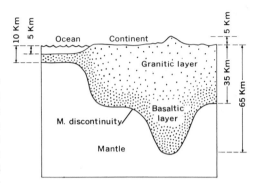

Figure 6-18. Idealized cross section showing the flotational equilibrium between oceanic and continental crust (Takeuchi and others, 1967).

basins. Oceanic crust, in contrast, is largely composed of basalt and related rock; and, except near the continental margin where erosional debris from the land may be present, it is generally overlain by no more than a few hundred meters of sediments.

Both the continents and the ocean basins display marked variations in their physiography and geology. In broad terms, the continental masses consist of several kinds of geologic provinces: (1) Mountain chains composed of highly folded and faulted sedimentary rocks many thousands of meters in thickness, metamorphic rocks, and intrusive and extrusive igneous rocks. (2) Shield areas, where ancient and generally highly deformed and metamorphosed sedimentary and igneous rocks are exposed in large areas of relatively low relief. (3) Ancient basins and embayments, where hundreds to tens of thousands of meters of sediment accumulated from seas that spread over ancient shields and platforms—now generally areas of low relief underlain by flat-lying or gently dipping sedimentary rock. (4) Coastal plains and continental shelves, the relatively flat or gently dipping surfaces along the continental margin[†] that have been planed off by wave

* This section is repeated in part from McKelvey, Stoertz, and Vedder (1969).

† Continental margin is used here as a geologic term referring to the submerged part of continental crust. It includes the continental shelf and slope, and in many places part of the continental rise where its apron of debris spreads over and conceals continental crust. The same term is sometimes used as a physiographic term to include the continental shelf, the continental slope, and the continental rise.

erosion or prograded by deposition of marine sediment. (5) Continental slopes, the surfaces of the continental margin that slope more steeply from the edge of the shelves to the ocean floor. A demarcation between the shelves and slopes does not everywhere exist, and the two are often described together as the continental terrace, or, in areas of irregular topography, as the continental borderland. (See Emery, 1968, for a description of the structural varieties of shelves and slopes.)

The outer limits of the continental margin in many places are concealed beneath the continental rises—gently dipping surfaces underlain by an apron of erosional debris derived from the continent and extending from the slopes onto the adjacent abyssal plains of the ocean basins (Figure 6-19); see also the excellent physiographic diagrams of the Atlantic and Indian Ocean floors published by the National Geographic Society, based on the work of B.C. Heezen and Marie Tharp.)

These provinces are the products of various geologic processes—processes which have also resulted in the differentiation of the continental crust into a diverse assemblage of rocks and minerals. Petroleum, coal, sulfur, salt, potash, phosphate rock, limestone, and many other minerals have been concentrated by biologic and sedimentary processes in sedimentary rocks. Copper, lead, zinc, nickel, gold, silver, mercury, fluorspar, beryl-

lium, tungsten, tin, and many other minerals have been concentrated by the igneous and hydrothermal processes that operate within the continental crust. Many minable concentrations of iron, alumina, manganese, gold, tin, and other minerals have been formed as a result of weathering processes.

The physiography of the large ocean basins beyond the continental margin and rise is also varied but is dominated by the following features: (1) Oceanic ridges and rises, often called "midoceanic" although they do not everywhere occur in midocean. These form a nearly continuous but branching worldwide mountain chain with a total length of about 75,000 miles. A rift valley along the crest is a prominent feature of the ridges in many places, as are volcanoes and volcanic fields, many of which are islands. (2) Abyssal plains and hills, lying on both sides of the oceanic rises and underlain by a thin veneer of pelagic sediments. (3) Individual volcanoes and composite volcanic ridges formed by overlapping volcanoes, and scattered over the ocean basins but often clustered to form groups of islands or seamounts and linear chains along oceanic margins. (4) Trenches, commonly present along volcanic island arcs or young mountain chains at the periphery of the large ocean basins.

In some places small ocean basins lie be-

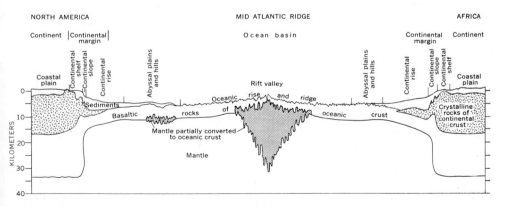

Figure 6-19. The major physiographic provinces of the ocean basins, illustrated by a trans-Atlantic profile from New England to the Spanish Sahara (modified from Heezen, 1962).

tween two continents or between continents and offshore island arcs. Characteristically, they have an abyssal plain below a depth of 2000 meters, and a few also have trenches along the concave side of the bordering island arcs. Those that border land areas with large surface runoff have trapped erosional debris and hence contain thick accumulations of sediments analogous to those beneath the continental rises.

Many physiographic features of the ocean basins are related to volcanism, crustal subsidence, and possibly to a process of ocean-floor spreading that brings basaltic igneous rock to the surface along the oceanic ridges, and carries new crust away from the mid-oceanic ridge at the rate of a few to 10-15 cm a year. Much is still to be learned about this process, but many now believe that the continents have split apart, along what are now these oceanic ridges, and drifted away from each other, and that the continental mass is still in the process of separation in the Atlantic Ocean, Gulf of California, Red Sea, and African rift-valley system (for a review see Vine, 1969).

In contrast to continental crust, oceanic crust is thin and relatively uniform in composition. Basaltic bedrock lies at or near the surface over much of the oceanic rise and ridge; and, although sediments increase in thickness toward the edge of the oceanic

basin (reflecting the longer time available for their accumulation on the older parts of the ocean floor), they probably do not attain a thickness of more than a few hundred meters in most places. The basalt originates by partial melting of rocks in the underlying mantle; and while other rocks and minerals are produced in this process, they are not as varied in composition as those that result from the action and interaction of the processes operating on and within the continental crust. Nevertheless, the oceanic basins contain large concentrations of manganese, nickel, copper, cobalt, and other metals in manganese oxide nodules, crusts, and pavements on the surface of the sea bottom; and the ocean basins probably also contain these and other metals, such as zinc, mercury, chromite, and platinum, in other kinds of deposits formed from the differentiation of the mantle material from which the basalt itself was derived.

The areas of the subsea physiographic provinces are summarized in Table 6-4.

Many uncertainties are involved in the identification of these provinces from the sketchy knowledge presently available on the character of the ocean floor. For example, K. O. Emery (1969) independently estimates that the total area of continental rises of the world is about 30 million km^2 compared with the 19 million km^2 estimated by

Table 6-4. Areas of Subsea Physiographic Provinces (after Menard and Smith, 1966).

Province	Area (millions of km^2)	Per cent of Total Area
Continental shelf and slope[a]	55.4	15.3
Continental rise	19.2	5.3
Abyssal plains and hills	151.5	41.8
Trenches and associated ridges	6.1	1.7
Oceanic ridges and rises	118.6	32.7
Volcanic ridges and cones and other features	11.2	3.2
	362.0	100.0
Small ocean basins (included above in continental rise and abyssal plains)	7.5	2.1

[a]The extent of the physiographic shelf has never been estimated, but is often taken to approximate the area landward of the 200-meter isobath. Menard and Smith estimate this area to be 27.1 million km^2, about half that shown above for the shelf and slope combined.

Menard and Smith (1966). Nevertheless, it is clear that most of the seabed is geologically a part of the large ocean basins rather than the continental masses.

In spite of their smaller size compared to the floor of the large ocean basins, the submerged parts of the continents, the continental rises, and the small ocean basins contain the greater part of the potential subsea mineral resources, both in terms of the variety of minerals that exist and the value of those that are likely to be recovered within the next few decades. Most important among these minerals is petroleum, the bulk of which likely occurs in the continental terraces and rises and the small ocean basins. Only a relatively small number of other minerals are presently recoverable from the continental shelf, but potentially it contains virtually the same large array of minerals now mined from the land. In contrast, the seabed of the large ocean basins contains a much smaller number of minerals—mainly the metals in the manganese oxide nodules and such related metals as zinc, mercury, gold, platinum, and chromium that may occur in other kinds of deposits. The total amount of these metals in the ocean basins and crust is very likely far larger than in the continents, but much technologic advance will be required to find them and make them economically available.

CLASSIFICATION OF MINERAL RESOURCE ESTIMATES

Although subsea resources of petroleum and several other minerals are potentially large and widely distributed, only a small part are likely to be economically recoverable within the next few decades, and an unpredictable part may never be recoverable. In order to give economic and geologic perspective to estimates of resources, it is desirable to view them in a framework that takes account of the degree of certainty of knowledge about their existence and character, and the feasibility of their recovery and sale. In the classification below (McKelvey, 1968), the degree

of certainty of knowledge of the dimensions and quality of mineral deposits is shown on the abscissa, and the feasibility of recovery and marketing is shown on the ordinate. The classification of individual deposits shifts with progress in exploration, advance in technology, or changes in economic conditions. Recoverable reserves are marketable materials that are producible under locally prevailing economic and technologic conditions. Paramarginal resources are prospectively marketable materials that are recoverable at prices as much as 1.5 times those prevailing now or with a comparable advance in technology. Submarginal resources are materials recoverable at prices higher than 1.5 times those prevailing now but that have some forseeable use and prospective value.

Classification of reserves and resources

Classification of Reserves and Resources

Seen in this framework, the presently recoverable proved reserves of most minerals are relatively small compared to the resources that may eventually be found by exploration or become recoverable as a result of technologic advances or changes in economic conditions. This is particularly true for subsea resources, because only a small part of the seabed has been explored and most of the resources it contains are not yet economically recoverable.

DISTRIBUTION, MAGNITUDE, AND FUTURE PRODUCTION OF SUBSEA MINERAL RESOURCES

Petroleum

Subsea petroleum (oil and gas), produced off shore 25 countries, presently contributes 17 per cent of the world's output and makes up nearly 90 per cent of the total value of current subsea mineral production. Through the remainder of the present century and probably longer, petroleum will continue to be the principal mineral produced from the seabed. Offshore sources may come to supply 30-35 per cent of the world's petroleum production by 1980, and the annual value of subsea petroleum production probably will soon exceed that of all other marine resources combined, including sea-water chemicals and fish.

Petroleum resources are largely confined to the continental shelves, continental slopes, continental rises, and the small ocean basins. Because these areas in general contain a greater thickness of marine Tertiary sediments, from which most of the world's petroleum production comes, than do the lands, taken as a whole the shelf and slope areas are more favorable for petroleum than the exposed parts of the continents. Environments favorable for petroleum are highly localized; and as mentioned earlier, only a small part of the broadly favorable areas actually contain producible petroleum accumuations. Among the geologic provinces considered broadly favorable, the incidence of petroleum accumulations in the shelves, slopes, and the small ocean basins may be greater than in the continental rises bordering the large ocean basins (L. G. Weeks, personal commun., 1969). The rises contain great thicknesses of sediments (Emery, 1969), probably including organic-rich source rocks deposited when the proto-ocean basins were narrow and had restricted circulation (Schneider, 1969, and Rona, 1969), but in many places they may not contain suitable reservoir rocks.

Although the geologic provinces named are the only ones that can be identified now as broadly favorable for petroleum, the possibility that it occurs in other parts of the ocean basins cannot be ruled out (Hedberg, 1969). The abyssal plains for example, are believed to contain insufficient thicknesses of sediments to yield petroleum accumulations, but the basement now identified by seismic reflection probes is, of course, an acoustic basement, and its composition is uncertain. In most places it is probably basaltic crust, but in some places it may prove to be merely a hard layer that conceals thicker sediment below. Parts of the deep trenches may also be favorable for petroleum, as may some of the subsea ridges or plateaus of unknown composition and presently unidentified foundered remnants of continental blocks broken off during the process of continental drifting. The potential in such areas, however, seems likely to be small compared to that of the seabed bordering the continents.

No complete estimates of potential world subsea petroleum resources have been made, but enough is known to be certain that they are large, perhaps even larger than those of the continents. World subsea proved recoverable reserves are 90 billion barrels, and Weeks (1969) estimates that world "offshore petroleum resources" (including proved reserves) beneath a water depth of as much as 1000 feet (300 m) amount to 700 billion barrels of petroleum liquids, plus 350 billion barrels recoverable by secondary methods, and the equivalent of 350 billion barrels in natural gas. In the classification shown above, all of this would be presently in the known and undiscovered recoverable and paramarginal categories. The area covered by Week's estimate is 28 million km^2, of which he believes 16.1 million km^2 is the favorable area. The area beyond the 300-meter depth to the toe of the continental rise is 46.6 million km^2. A larger proportion of this area is probably underlain by a thicker accumulation of sediments than in the area covered by Week's estimate, and the total volume of sediments beneath the continental rises and the small ocean basins may be far larger than under the shelves and slopes. (See K. O. Emery, 1969.) Until more is known about the composition and structure of these sediments, it is impossible to judge their potential. Whatever may be their magnitude, potential resources in such areas must be classed as undiscovered submarginal, and they are likely to remain so for a few decades or longer. Known and undiscovered recoverable and paramarginal resources in the continental shelves and slopes, however, are large and probably will supply most of the offshore production during the next few decades.

Areas favorable for the local occurrence of subsea petroleum resources lie adjacent to nearly every coastal nation, and in fact, geologic or geophysical exploration is already underway off the coast of more than 75 countries and drilling is in progress off 42 of them. Wide shelves, where petroleum in large accumulations, if they are present, can be recovered economically now, occur off the coasts of Greenland, Norway, the United Kingdom, Canada, Mexico, Trinidad-Tobago, Venezuela, Guyana, Surinam, French Guiana, Brazil, Uruguay, Argentina, Australia, New Zealand, mainland China, Korea, Taiwan China, and the Soviet Union as well as along the Atlantic, Gulf of Mexico, and Alaskan coasts of the United States. The continental rise is especially wide* in the Arabian Sea, the Bay of Bengal, off eastern Africa, off most of western Africa, and off much of the eastern coasts of North America and South America. Small ocean basins that have a large petroleum potential include the Gulf of Mexico, the Caribbean Sea, the Mediterranean Sea, the Black Sea, the Caspian Sea, the Bering Sea, the Sea of Okhotsk, the Sea of Japan, the South China Sea, and the seas within the Indonesian Archipelago. Several of these favorable areas

reach depths of as much as 5500 meters and extend 1500 km or more from shore. Petroleum occurrence at such depths was shown in August 1968 by the *Glomar Challenger* drilling in the Gulf of Mexico, when one hole encountered a show of oil and gas (and sulfur also) in the Sigsbee Deep beneath a water depth of 3582 meters (Burk and others, in press).

Not only does the seabed contain a large part of the world's petroleum potential (seabed petroleum resources may exceed those of the lands), but its development in regions where little or none is now produced may change significantly the outlook for petroleum production and supply for individual countries and regions.

Current offshore petroleum production comes from water depths of less than 105 meters and from areas within 120 km of the coast. The technologic limit of offshore petroleum production, however, may be extended to water depths of as much as 6000 feet (1830 m) by 1980, although at much higher costs (National Petroleum Council, 1969). Because of the higher cost of deepwater production and the wide availability of petroleum in shallower parts of the shelves, production from areas beyond the 200-meter isobath is likely to be largely restricted during the next decade to giant fields in the most favorable locations. It probably will not amount to more than 0.5-1.0 billion barrels a year by 1980, but might increase to a few billion barrels a year by the end of the century when deep-water exploitation technology is further advanced.

Potash and Other Saline Minerals

Most of the world's deposits of anhydrite and gypsum (calcium sulfates), common salt, and potash-bearing minerals are formed by evaporation of sea water and other natural brines in basins of restricted circulation. Im-

* Although the thickness of sediments in continental rises may increase with increasing width, the same is not true for the shelves. Width in itself is not a measure of favorability for either (L. G. Weeks, personal commun.).

portant deposits of magnesium-bearing salts are also deposited in such basins, and elemental sulfur forms in some of them by biogenic processes involving the alteration of anhydrite. Because rock salt tends to flow at relatively low temperature and pressure, salt in thick beds squeezed by the weight of a few thousand feet or more of younger sediment often pierces or intrudes the younger sediments, forming salt domes, plugs, and other structures. Such masses, which may be a few miles in diameter, may bring salt to or near the surface. They form structures in the intruded sedimentary layers that may be favorable for the accumulation of petroleum, and the limestone cap rock associated with some of these masses may be the site of sulfur decomposition. Elemental sulfur in these deposits can be recovered by the Frasch process, in which the sulfur is melted by the injection of hot water into drill holes. Salt, and some potash and magnesium minerals, can also be recovered by solution-mining methods.

Saline deposits formed in ancient marine basins are extensive on the land. Many deposits extend beneath the sea, not only under the continental shelves but also under some of the small ocean basins. For example, in the Gulf of Mexico the 1968 drilling by the *Glomar Challenger* in the Sigsbee Deep confirmed the previously held belief that the Sigsbee Knolls are salt domes, and showed that these deep structures may contain petroleum and sulfur. Because of the widespread anhydrite, gypsum, and salt on the land, and the ease of obtaining salt by evaporation from sea water in many coastal regions, these minerals are widely available at low cost. Consequently, there is little need to seek them from subsea sources, except perhaps in local areas far removed from other sources. A new and potentially important use of salt deposits in offshore areas, however, may be as underground storage chambers (presumably opened subsea by solution mining through drill holes) for petroleum and radioactive waste (Halbouty, 1967; Pendery, 1969).

No attempt has been made to estimate potential subsea resources of salt and anhydrite, but assuredly they amount to at least tens of trillions of tons.

Potash deposits in the salt basins are not as widespread as salt and gypsum, but individual deposits are large—generally in the range of hundreds of millions or billions of tons. World supplies from land sources are presently abundant, but because potash is a relatively valuable mineral, there are opportunities for the development of strategically located subsea deposits, particularly those amenable to solution mining or to underground mining from a land entry. The feasibility of producing potash from deposits beneath the North Sea is now being investigated, and deposits suitable for mining may exist in salt basins in other coastal regions. Potential world resources in subsea deposits are probably in the range of tens of billions of tons of K_2O, some of which may be economically recoverable.

Thick beds of a magnesium salt—tachhydrite ($CaCl_2 \cdot MgCl_2 \cdot 12H_2O$)—previously known only in trace amounts, have recently been found associated with potash in the Sergipe salt basin along the eastern coast of Brazil and in the Congo basin along the southwestern coast of Africa. Tachhydrite is highly soluble, forms a concentrated brine, and probably can be mined by solution methods (R. J. Hite, personal commun.). Because magnesium is now recovered economically from sea water and other natural brines, presently recoverable reserves are enormous. Even so, if it can be produced more cheaply from tachhydrite, these and other favorably situated deposits may prove to be valuable.

Sulfur

Nearly 60 per cent of the world's production of sulfur comes from Frasch-type deposits associated with anhydrite, either in bedded deposits or salt domes. Subsea production, however, is presently limited to two salt-dome deposits offshore Louisiana, which yield about 20 per cent of United States production. Growth in demand for sulfur has exceeded supply in recent years, and while new discoveries are helping to ease world shortages, new sources are needed. In part, these may be met by development of new processes—already in an advanced state—for the recovery of sulfur from gypsum on land, and for the distant future these and other land sources—including by-product sulfur from sour natural gas, from asphalt-base petroleum, and from stack gases of various kinds—are likely to supply the bulk of world needs. During the next decade or so, however, subsea sulfur may become an important factor in world production.

Individual sulfur deposits tend to be much smaller than those of the saline minerals, generally in the range of a few millions to a few tens of millions of tons. Known recoverable reserves offshore the United States amount to about 37 million tons, and a similar magnitude may exist in undiscovered but recoverable deposits. Potential world resources of subsea sulfur have not been estimated, but are likely to amount to scores of millions of tons or more, much of which might be recoverable under present economic conditions. Not enough is known about the origin of sulfur to focus exploration on the most favorable environments within the salt-dome and anhydrite basins, but the numerous offshore occurrences, some of which may be more extensive, are certain to contain many recoverable deposits that are as yet undiscovered.

Heavy Mineral Concentrates, Coal, and Other Subsea Minerals Currently Mined by Dredging or Underground Methods

The production potential for heavy-mineral concentrates (placers), sand, gravel, shell, and lime mud, currently mined by dredging and for coal, iron ore, copper, limestone, and other minerals currently mined underground from a land or artificial-island entry is limited to the shallow nearshore parts of the continental shelves. Favorable areas for

the subsea occurrence of these minerals are difficult to outline in a meaningful way. Often only the location of known or producing coastal or offshore deposits is known. For the placers, however, these occurrences are in themselves among the best clues to the general location of favorable areas offshore (Emery and Noakes, 1968; McKelvey and Chase, 1966; both papers also discuss geologic guides that may be helpful in selecting favorable areas). For some of the deposits mined underground, the location of existing subsea mines is also one of the best indications of the general areas in which other deposits are likely to be found and mined within the next decade or so. One reason for this is that subsea prospecting methods for most bedrock minerals are as yet so inefficient that coastal occurrences, together with coastal geology, are the best clues to offshore prospects. Another is that existing subsea mines of some minerals—coal, for example—identify regions in which subsea underground mining may be economically viable.

Although the value of the annual world production of the minerals now mined by dredging or underground methods totals more than $500 million, it represents less than 2 per cent of the onshore production of these minerals. World subsea reserves and resources of these minerals have not been estimated, but as with petroleum, known recoverable reserves are probably small compared to undiscovered paramarginal and submarginal resources. Even though their potential may be large, however, it is not likely to be nearly as large as that of onshore deposits. During the next few decades the economic development of these minerals probably will be almost entirely limited to shallow water (less than 100 m deep) for the deposits in sea-floor sediments minable by dredging, and to nearshore areas (less than 50 km from land entry) for large bedrock deposits mined by underground methods. Eventually, technologic advance may make possible seafloor entry from a vertical shaft (Austin, 1967) or may permit solution mining of certain metals through drill holes. For both economic and technologic reasons, however, it appears probable that during the next few decades land sources will be preferred for most of these minerals in many areas. Exceptions may be offshore deposits that (1) are extremely large, (2) are of high grade and easy access (such as tin deposits off the coasts of Thailand, Indonesia, and Malaysia), (3) contain minerals in short supply (such as gold and platinum), (4) are desired by individual countries to reduce balance-of-payments deficits or to provide for security of mineral supply, or (5) are in local demand because of the high cost of transport from distant sources or the need to conserve land resources and preserve land environment. Sand and gravel, shell, and coal in some places are examples of minerals in the last category. The growing demand in coastal cities for sand and gravel especially is likely to stimulate offshore production in many areas.

Phosphorite

Phosphorite is widely distributed on the continental shelves and upper slopes in areas of upwelling currents at low latitudes. Other than the record of their occurrence and perhaps their composition, published information about them is scant. It is sufficient, however, to say that they are abundant in some areas such as off the coast of Baja California, southern California, and east of New Zealand, but that in many places consist of scattered nodules too sparsely distributed to be recoverable. Their phosphate content varies considerably also, but is seldom more than 29 per cent P_2O_5 —a few per cent lower than the present commercial cutoff grade. No offshore deposits are being mined now because of the availability of lower cost and higher quality deposits on land. Land deposits are large enough to meet world demands for many decades; even so, subsea production may prove to be economic in local areas far removed from land deposits, particularly for developing countries having difficulty in making foreign payments.

World subsea resources are probably at least of the order of hundreds of billions of tons. Possibly a few billion tons may be classed as paramarginal now, with some prospects of becoming recoverable within a decade, but the bulk of subsea phosphorite resources must be classed as submarginal.

Manganese Oxide Nodules

Surficial deposits of manganese oxide nodules, crusts, and pavements, which are currently of more interest for their content of nickel, copper, and cobalt than for manganese, are largely confined to the deep ocean floor, generally at depths of 3500 to 4500 meters, and to the seamounts within it. In several areas, however, they occur near land—notably on the Blake Plateau off eastern United States, off the west coast of Baja California, near some of the islands in large ocean basins as well as scattered occurrences in several inland seas such as the Baltic. In most of these nearshore areas, however, their metal content is much lower than that of the nodules far from land. Although bottom photographs and closely spaced samples show that nodules are extensive in many areas, the same kinds of data show that both their abundance and composition vary. Available information is not sufficient to infer their continuity between stations from which they have been reported or their absence in areas where they have not yet been found. Cores show that they are present in the sediments a few meters beneath the surface at some stations where they are not present at the surface. Inspection of the results of available bottom photographs in the Pacific, however, suggests that nodules are absent from the surface of the bottom in some large areas and that they are particularly abundant in others. In a broad way this distribution tends to confirm the pattern shown in Figure 6-20 compiled by Skornyakova and Andrushchenko. Even if subsequent exploration shows that the nodules and related deposits are continuous over large areas, their minability may be adversely affected locally by irregularity of the bot-

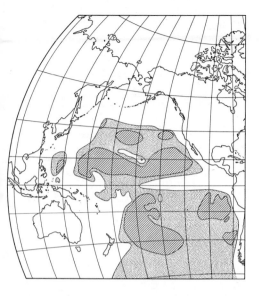

Figure 6-20. Generalized regional variations in abundance of manganese oxide nodules on the surface of the sea floor in the Pacific (Skormyakova and Andrushchenko, in Strakov and others, 1968, p. 128). Nodules are absent or sparse in blank areas and where present may be small or consist of films or coatings of oxides on other materials. They may cover as much as 20 per cent of the bottom in stippled areas, and 20 to 50 per cent of the bottom or more in ruled areas. Although the outlines of these provinces are generally consistent with the available data, bottom photographs and samples are not sufficient to infer continuity or absence of nodules in any given area.

tom surface, presence of extensive crusts of pavements troublesome to break and lift, and deleterious impurities.

The composition of the nodules also varies greatly. Again there appear to be regional variations, but published attempts to define the pattern of variation have not been satisfactory. The averages below are indicative of the range in metal content over large areas, but should not be taken as characteristic of the composition of the nodules in the entire region from which the component samples were collected (Figure 6-21). The content of each of the metals may be much higher in individual samples. Manganese is nearly 50 per cent in some samples, which are generally low in the other metals. Copper

and nickel tend to vary in rough proportion to each other and may be as much as 2 per cent; cobalt also may be as much as 2 per cent. Whether or not minable quantities contain such concentrations remains to be demonstrated.

The nodules constitute a huge resource. Mero (1967) estimates that they aggregate 1.7 trillion tons and contain 400 billion tons of manganese, 16.4 billion tons of nickel, 8.8 billion tons of copper, and 9.8 billion tons of cobalt. Zenkevich and Skornyakova (1961, quoted by Mero, 1965, p. 175) place a much lower figure on the total tonnage of the nodules—90 billion tons. Whatever the aggregate tonnage proves to be, the amount in deposits of suitable quality, abundance, and environmental setting to warrant dredging is likely to be much smaller.

The production of manganese oxide nodules is not economically feasible now, partly because of the high cost of dredging from such great depths and partly because suitable refining methods have yet to be demonstrated. The availability of large and relatively low-cost sources of the component metals on land—adequate at least for several decades—is another obstacle to their development. Added to these constraints are uncertainties about which of the metals could be recovered and sold, and hence bear some part of the cost of the overall operation. The metals are not present in the nodules in the same ratio in which they are used. Thus, the ratio of copper, nickel, and cobalt in present use is 266:27.5:1, whereas in nodules of average composition it is about 3:4:1. Hence, in an operation designed to produce large quantities of copper, for example, it might not be possible to sell all the nickel and cobalt contained in the nodules mined. Similarly, if the costs of refining one or more of the metals prove too high—and manganese itself appears to pose such a problem—the value of the recoverable coproducts would be reduced proportionately. The significance of these uncertainties in evaluating the prospects for exploration of the nodules may be seen in the range of value of potentially saleable products from nodules of aver-

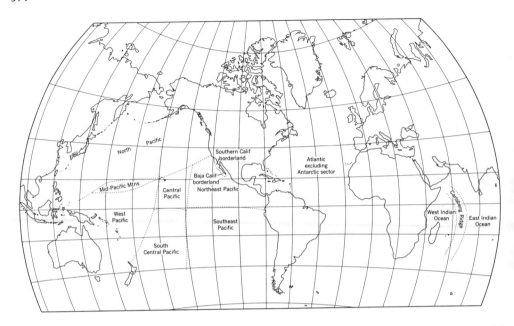

Figure 6-21. Regions from which analyses of manganese oxide nodules shown by the averages in Table 6-5 were collected. Analyzed samples are insufficient to define the pattern of regional variation of the metal content.

age composition at present prices—$6.75 per ton of nodules if only copper is recoverable and saleable to $137 per ton if all four metals are viable coproducts.* The variation in the composition of the nodules from place to place may make it possible to select deposits that give optimum yield under a given method of treatment—if only copper is recoverable nodules with 1.5 per cent copper will be the target, and so on. Although the nodules must be classed as submarginal for the present, Deepsea Ventures, Inc. (Flipse, 1969)—a company which is actively pursuing both exploration and research on mining and process development and plans to begin a pilot mining operation in 1970—believes that the nodules in some environments will be paramarginal by 1970 and economically recoverable within a few years.

If sufficiently low-cost production and refining methods are developed to permit nodule mining, subsea production could conceivably satisfy a large part of the world's needs for each of the component metals. Production of nodules would be a gathering operation with no permanent installation at the mining site. Although the operation would sweep a large area compared to mining operations on land, the area involved would be only a tiny fraction of the ocean floor. For example, an operation designed to supply the world's nickel requirements in the year 2000—say 1.7 million tons a year—from nodules of average nickel content (1.0 per cent) and with an average concentration on the sea floor (31,000 tons per mi^2, according to Mero), would need to sweep about 11,000 mi^2 (28,500 km^2) a year if only 50 per cent of the nodules were recovered. Taking an output of 20,000 tons per

---

* This does not represent the range in value to the producer, of course, for refining and other costs attach to each metal produced. But each additional coproduct that can be recovered profitably helps to defray basic costs such as exploration, dredging, transportation, and crushing.

Table 6-5. Average Metal Content of Surface Manganese Oxide Nodules from Different Regions (after Cronan, 1967, in the Pacific and Indian Oceans, and Manheim, 1965, in the Atlantic).

Metal	1	2	3	4	5	6	7	8	9	10	11	12	13
Mn	34.0	14.9	22.4	18.4	15.9	16.4	16.1	14.0	12.6	13.7	15.8	15.1	13.5
Fe	1.6	12.7	8.9	10.9	9.6	14.1	13.5	13.1	12.1	15.9	11.3	17.7	15.5
Ni	.1	.4	1.1	.9	.9	.4	.5	.4	.5	.3	.5	.3	.3
Co	<.1	.5	.2	.2	.2	.6	.4	1.1	.2	.3	.1	.5	.2
Cu	.1	.1	.7	.3	.6	.2	.4	.1	.3	.1	.3	.1	.2
Depth (m)	3535	1131	4553	4141	5025	3551	5024	1756	5142	3722	5046	3240	

(1) Southern California borderland (3 samples).
(2) Baja California borderland-seamount (6 samples).
(3) Northeast Pacific (10 samples).
(4) Southeast Pacific (11 samples).
(5) Central Pacific (12 samples).
(6) South Central Pacific (12 samples).
(7) West Pacific (30 samples).

(8) Mid-Pacific Mountains (5 samples).
(9) North Pacific (5 samples).
(10) West Indian Ocean (13 samples).
(11) East Indian Ocean (14 samples).
(12) Carlsberg Ridge (10 samples).
(13) Atlantic Ocean (excluding Antarctic sector).
(sector).

day as the kind of unit operation that would be involved (Ensign, 1968), 23 mining units would be required, each sweeping 470 mi^2 (1200 km^2) per year. Of the 281 million km^2 that makes up the ocean floor beyond the continental rise, the area being mined would be trivial, and the producers might be widely separated and relatively few in number. If technology and other factors impose special requirements on the composition of the nodules and other characteristics of the deposit or mine site, however, the few areas selected for mining initially may have to be chosen from a relatively small range of possibilities.

Other Metalliferous Deposits

Mud rich in copper and zinc was first reported by a Woods Hole expedition in 1965, in the deeps of the Red Sea (Miller and others, 1966); iron and manganese precipitates have been reported from the submarine Banu Wuhu Volcano, Indonesia (Zelenov, 1964); and sediments containing as much as 5 per cent Mn and 0.1 per cent Cu have been found in the rift zone of the East Pacific Rise (Bostrom and Peterson, 1966). There is reason to believe that deposits of these and other metals may occur in similar surficial

deposits or in relatively shallow bedrock deposits associated with rift or fracture zones in other parts of the deep ocean basins. Thus, it is becoming evident that oceanic basalt has formed from the magmatic differentiation of rocks of the underlying mantle. Such differentiation processes may be expected to yield other rocks (some of which have already been found in fracture zones) and metals. Abetting this process of concentrating metals by magmatic differentiation or acting essentially independently of it, very likely is another—namely, the leaching of metals from the bedrock of the ocean floor by heated circulating sea-water brine (D. E. White, personal commun., 1969). The known occurrences of metal-bearing mud are all in regions of high heat flow—which appears to be characteristic of the rift and fracture zones in general. Where fractures in the bedrock permit sea water to enter the crust, such high heat flow is likely to set up a convection system, drawing cold sea water into the crust and heating it. The higher temperature of the water increases its capacity to dissolve metals, as shown by the high metal content of the hot brines of the Red Sea and Salton Sea. As this enriched brine is returned to the bottom surface, metals in solution might be precipitated along fractures in the bedrock or at the surface of the sea bottom. The process of forming localized and concentrated metalliferous deposits would be more effective where the fracture systems intersect salt deposits. This apparently occurs in the Red Sea where the circulating brines become concentrated and heavy enough to pond in depressions on the bottom surface. Without the formation and ponding of such heavy brine, metals brought to the surface would more likely be dispersed and perhaps be deposited elsewhere in the form of manganese oxide nodules or much lower grade deposits.

Although a substantial part of the metals released from the oceanic crust by these processes may now be in the surficial deposits, minable concentrations of some of these metals are also known to be associated with oceanic crust on land where it has been uplifted by tectonic processes, such as in Cyprus, the Philippine Islands, and New Guinea (see, for example, H. L. Davis, 1968). Similar deposits very likely exist beneath the deep ocean floor.

The assemblage of metals in surficial muds or bedrock deposits probably is qualitatively similar to that in the nodules, but individual deposits might contain only one or two metals, and their concentration in many deposits might be higher. Their recovery from muds or other surficial deposits might prove to be easier than from the nodules, but the technology for mining bedrock deposits at abyssal depths is not in sight and may not be available for many decades or longer. The availability now of these metals in low-cost deposits on land and the prospects for their availability later in nodules and muds gives little incentive to develop technology for mining bedrock deposits beneath the deep ocean floor.

On the basis of sparse data, Bischoff and Manheim (1969) have estimated that the upper 10 meters of mud in the Atlantis II deep contains 2.9 million tons of zinc, 1.1 million tons of copper, and much smaller amounts of other metals. In the aggregate, metals in undiscovered surficial deposits elsewhere might amount to billions of tons, and those in bedrock deposits might be far larger. Some of the metal-bearing muds may prove to be minable within a decade or so, but the bedrock deposits are submarginal and may remain so for many decades or longer.

Fresh Ground Water and Geothermal Energy

The potential distribution of subsea ground water and geothermal energy is not well known, but a few observations are appropriate here.

Fresh ground water occurs in some aquifers beneath the continental shelf—off the southeast coast of United States, for example (Manheim and Horn, 1968)—and might be produced for local use in some areas. Most such aquifers probably lie landward of the 200-meter isobath.

Large potential resources of geothermal energy in the form of hot water and steam may be present in areas such as the Gulf of Mexico (Jones, 1967) where there are thick geosynclinal accumulations of geologically young sediments, in the zones of high-heat flow associated with the rift and fracture zones mentioned above, and in areas of present or recent volcanic activity. Geothermal energy associated with rift zones and volcanic activity has little prospective value in areas far from land because of the difficulty of utilizing it. In coastal regions, however, geothermal energy has a potential use in the generation of electric power, in the desalination of water (the superheated brines beneath the Gulf Coast are self-flashing), in solution mining of sulfur and potash, and in the secondary recovery of petroleum. Of these potential uses, the last may be the most significant in offshore production. The potential magnitude and value of geothermal energy in such use cannot be appraised now, but whatever resources are present are likely to lie within (but be smaller than) the areas known as favorable locally for petroleum.

SUMMARY

Subsea petroleum (oil and gas), produced from the seabed offshore 25 countries, makes up nearly 90 per cent of the total value of current subsea mineral production; and through the remainder of the present century and probably longer, it will continue to be the principal mineral produced from the seabed. Subsea sources already contribute 17 per cent of the world's output of petroleum and may supply 30-35 per cent of it by 1980. The areas favorable for petroleum are mainly the continental shelves and slopes, the small ocean basins, and the continental rises. Potential subsea resources of petroleum in these geologic provinces are as large and perhaps larger than those of the continents. Areas favorable for its occurrence lie adjacent to nearly every coastal nation. Development of petroleum in areas where it is not now produced could change significantly the outlook for petroleum production and supply for many countries and regions. All the petroleum now produced offshore comes from water depths of less than 105 meters, but the technologic limits for subsea production may be extended, at much higher cost, to about 1800 meters by 1980. Because of the higher cost of deepwater production and the wide availability of petroleum in the shallower parts of the shelves, the bulk of future production probably will continue to come from shallow-water areas, but production from beyond the 200-meter isobath might reach 0.5-1.0 billion barrels a year by 1980 and perhaps a few billion barrels by the end of the century.

Subsea production of other minerals from the continental shelves now includes heavy-mineral concentrates (placers), sand, gravel, shell, and lime mud mined by dredging in shallow water near shore; coal, iron, copper, limestone, and a few other minerals mined underground from a land or artificial-island entry; and sulfur and salt mined through drill holes. Phosphorite (minable by dredging) and potash, magnesium, fresh ground water, and geothermal energy (recoverable through drill holes) are other minerals that may be brought into production from the continental shelves in the future. Potential resources of these minerals are large, but because of the availability of lower cost land sources for most of them, subsea production is likely to furnish only a small part of their world output. Nevertheless, offshore production of them may have an important impact on local and regional economies, and for a few (such as tin) it may supply a substantial part of world requirements.

The manganese oxide nodules and other metalliferous deposits that occur on and beneath the deep ocean floor are an enormous potential source of manganese, copper, nickel, cobalt, zinc, and other metals. Production of the nodules is not economically feasible now because of the high cost of mining and refining the nodules and the availability of large lower cost land sources. One company actively engaged in exploration and research, however, believes that the nodules in some environments may become economically recoverable within a few years. If they can be brought into profitable production in competition with the large and low-cost land sources, they could conceivably come to satisfy a large part of the world's needs for the component metals. If production does take place it will be a gathering operation with no permanent installation at the mine site. The area being mined at a given time would only be a tiny fraction of the ocean floor; and inasmuch as the demand for these metals is limited, the operators might be few in number and widely separated.

REFERENCES

Alverson, D. C., Cox, D. P., Woloshin, A. J., Terman, M. J., and Woo, C. C., 1967, Atlas of Asia and eastern Europe to support detection of underground nuclear testing vol. II, Tectonics: Clearinghouse for Federal Scientific and Technical Information, Springfield, Va. 22151.

Austin, Carl F., 1967, In the rock—a logical approach for undersea mining of resources. *Eng. and Min. Jour.* **168** (August), 82-88.

Australia Bureau of Mineral Resources, 1965, Geological Map of the World: Australia and Oceania, scale 1:5,000,000, 13 sheets, Canberra.

Aymé, J. M., 1965, The Senegal salt basin, in *Salt Basins around Africa.* Inst. Petroleum, London, pp. 83-90.

Belmonte, Y., Hirtz, P., and Wenger, R., 1965, The salt basins of the Gabon and the Congo (Brazzaville)—a tentative palaeogeographic interpretation, in *Salt Basins around Africa.* Ins. Petroleum, London, pp. 55-74.

Benavides, Victor, 1968, Saline deposits of South America, in Mattox, R. B., ed., Saline deposits: Geol. Soc. America Spec. Paper 88, pp. 249-290.

Bischoff, J. L., and Manheim, F. T., 1969, Economic potential of the Red Sea heavy metal deposits, in E. T. Degens and D. A. Ross, ed., *Hot brines and Recent Heavy Metal Deposits in the Red Sea.* Springer-Verlag, New York, pp. 535-541.

Bostrom, K., and Peterson, M. N. A., 1966, Precipitates from hydrothermal exhalations on the East Pacific Rise: *Econ. Geology* **61**, 1258-1265.

Burk, C. A., Ewing, M., Worzel, J. L., Beall, A. O., Jr., Berggren, W. A., Bukry, D., Fischer, A. G., and Pessagno, E. A., Jr., 1969, Deep-sea drilling into the Challenger Knoll, Central Gulf of Mexico: *Am. Assoc. Petroleum Geologists Bull.* **53**(7), 1338-1347.

Commission for the Geological Map of the World, Subcommission for the Tectonic Map of the World, 1962, Mezhdunarodnaya Tektonicheskaya Karta Yevropy (International Tectonic map of Europe), scale 1:2,000,000, 16 sheets and legend sheet, Moscow.

Cronan, D. S., 1967, The geochemistry of some manganese nodules and associated pelagic deposits: Univ. London, England, Ph. D. thesis, 342 pp.

Cruickshank, M. J., 1962, The exploration and exploitation of offshore mineral deposits: Colorado School Mines, Golden, M. S. thesis, 185 pp.

Dalgarno, C. R., and Johnson, J. E., 1968, Diapiric structures and late Precambrian-early Cambrian sedimentation in Flinders Ranges, South Australia, in Braunstein, Jules, and O'Brien, G. D., eds., Diapirism and diapirs: Am. Assoc. Petroleum Geologist Mem. 8, pp. 301-314.

Davis, H. L., 1968, Papuan ultramafic belt. *Internat. Geol. Cong., 23d, Prague 1968* **1**, 209-220.

Demaison, G. J., 1965, The Triassic salt in the Algerian Sahara, in *Salt basins around Africa.* Inst. Petroleum, London, pp. 91-100.

Donovan, D. T., ed., 1968, *Geology of Shelf Seas.* Oliver & Boyd, London, 160 pp.

Douglas, R. J. W., Norris, D. K., Thorsteinsson, R., and Tozer, E. T., 1963, Geology and petroleum potentialities of nothern Canada. World Petroleum Cong., 6th, Proc. sec. 1, pp. 519-572.

Emery, K. O., 1968, Shallow structure of continental shelves and slopes. *Southeastern Geology.* **9**, 173-194.

Emery, K. O., 1969, Continental rises and their oil potential: *Oil and Gas Jour.* **67** (19), 231-243.

Emery, K. O., and Noakes, L. C., 1968, Economic placer deposits of the continental shelf. *Technical Bull. ECAFE* **1**, 95-111.

Emery, K. O., Uchupi, Elazar, Phillips, J. D., Bowin, C. O., Bunce, E. T., and Knott, S. T., 1969, The continental rise off eastern North America. Woods Hole Oceanographic Inst. Contr. 2324. (In press)

Ensign, C. O., Jr., 1968, Address to the joint meeting of the National Security Industrial Association and State of Oregon, Dec. 11, 1968.

Flipse, J. E., 1969, An engineering approach to ocean mining: Offshore Technology Conf. preprint, paper OTC 1035, pp. I-317-I-328, 10 figs.

Ford, A. B., 1964, Review of Antarctic geology: *Am. Geophysical Union Trans.* **45** (2), 1-19.

Geological Institute of the Academy of Sciences of the USSR and Ministry of Geology of the USSR, 1966, Tectonic map of Eurasia, scale 1:5,000,000, 12 sheets, Moscow.

Gill, W. D., 1965, *The Mediterranean Basin,*

in Salt Basins around Africa. Inst. Petroleum, London, pp. 101-111.

Goddard, E. N., chm., and others, 1965, Geologic map of North America: U.S. Geol. Survey, North Am. Geol. Map Comm., scale 1:5,000,000.

Gorrell, H. A., and Alderman, G. R., 1968, Elk Point Group saline basins of Alberta, Saskatchewan, and Manitoba, Canada, in Mattox, R. B., ed., Saline deposits: Geol. Soc. America Spec. Paper 88, pp. 291-318.

Gould, D. B., and Mille, George de, 1968, Piercement structures in Canadian Arctic Island, in Braunstein, Jules, and O'Brien, G. D., eds., Diapirism and diapirs: Am. Assoc. Petroleum Geologists Mem. 8, pp. 183-214.

Halbouty, M. T., 1967, *Salt Domes—Gulf Region, United States and Mexico.* Houston, Tex., Gulf Publishing Co., 425 pp.

Hedberg, Hollis D., 1969, Jurisdiction over offshore mineral resources: Address before the American Association of Petroleum Geologists, Dallas, Texas, April 15.

Heezen, B. C., 1962, The deep sea floor, in S. K. Runcorn, ed., *Continental Drift,* vol. 3, International Geophysical Series: Academic Press, New York and London. pp. 235-288.

Heezen, B. C., and Tharp, M., 1965, Tectonic fabric of the Atlantic and Indian Oceans and continental drift: *Royal Soc. London Philos. Trans.,* **258,** 90-100; also, *Saturday Review,* Jan. 10, 1969.

Heezen, B. C., and Thorp, M., 1967, Physiographic diagram of the Indian Ocean floor: Washington, D. C. Natl. Geo. Soc.

Heezen, B. C., and Thorp, M., 1968, Physiographic diagram of the Atlantic Ocean floor: Washington, D. C., Natl. Geo. Soc.

Hutchinson, G. E., 1950, Biogeochemistry of vertebrate excretion, no 3 of Survey of contemporary knowledge of biogeochemistry: *Am. Mus. Nat. History Bull.* **96,** 554 pp.

International *Petroleum Encyclopedia.* 1968, Tulsa, Okla., Petroleum Publishing Co.

Ivanov, A. A., and Levitzky, Y. F., 1960, Geologiya galogennykh otlozheny (formatzy) SSR [Geology of the halogen deposits of the USSR]: *Vseesaiozni Geol. Inst. Trudy, n. s.* **35,** 424 pp.

Jones, Paul H., 1967, Hydrology of Neogene deposits in the northern Gulf of Mexico Basin: 1st Symposium on Abnormal Subsurface Pressure, Proc., pp. 81-205.

Kent, P. E., 1965, An evaporite basin in southern Tanzania, in *Salt Basins around Africa.* Inst. Petroleum, London, pp. 41-54.

King, P. B., 1969, Tectonic map of North America, scale 1:5,000,000: U.S. Geol. Survey.

Kozary, M. T., Dunlap, J. C., and Humphrey, W. E., 1968, Incidence of saline deposits in geologic time, in Mattox, R. B., ed., Saline deposits: Geol. Soc. America Spec. Paper 88, pp. 43-58.

Lefond, S. J., 1968, *Handbook of World Salt Resources.* New York, Plenum Press.

Liechti, P., 1968, Salt features of France, in Mattox, R. B., ed., Saline deposits: Geol. Soc. America Spec. Paper 88, pp. 83-106.

Manheim, F. T., 1965, Manganese-iron accumuations in the shallow marine environment: Univ. Rhode Island Narragansett Marine Lab., Occasional Pub. 3-1965.

Manheim, F. T., and Horn, M. K., 1968, Composition of deeper subsurface waters along the Atlantic Continental margin: *Southeastern Geology* **9,** 215-236.

McKelvey, V. E., 1963, Successful new techniques in prospecting for phosphate deposits: U.S. papers for U.N. Conf. Application Sci. and Technology for the Benefit of the Less Developed Areas, v. 2, Natural Resources, Minerals and Mining, Mapping and Geod. Control, Washington, U.S. Govt. Printing Office, pp. 163-172.

McKelvey, V. E., 1968, Mineral potential of the submerged parts of the continents, in Mineral Resources of the World Ocean: U.S. Geol. Survey, Univ. Rhode Island, U.S. Navy, Occasional Pub. 4, pp. 31-38.

McKelvey, V. E., and Chase, Livington, 1966, Selecting areas favorable for subsea prospecting: Marine Tech. Soc. Trans. 2nd Ann. Conf., pp. 44-60.

McKelvey, V. E., Stoertz, G. E., and Veder, J. G., 1969, Subsea physiographic provinces and their mineral potential, in U.S. Geol. Survey Circular 619.

McNaughton, D. A., Quinlan, T., Hopkins, R. M., and Wells, A. T., 1968, Evolution of salt anticlines and salt domes in the Amadeus basin, central Australia, in Mattox, R. B., ed., Saline deposits: Geol. Soc. America Spec. Paper 88, pp. 229-248.

Menard, H. W., 1964, *Marine Geology of the Pacific.* New York, McGraw-Hill Book Co., Inc., 271 pp.

Menard, H. W., and Smith, S. M., 1966, Hypsometry of ocean basin provinces: *Jour. Geophys. Research,* **71**, 4305-4325.

Menard, H. W., Smith, S. M., and Pratt, R. M., 1965, The Rhone deep-sea fan: Colston Research Soc. Symposium, 17th, Proc., Butterworths London, pp. 271-286.

Mero, J. L., 1965, *The Mineral Resources of the Sea.* New York, Elsevier Publishing Co., 312 pp.

Mero, J. L., 1967, Ocean-mining—a potential major new industry: World Dredging Conf., 1st, New York 1967, Proc. pp. 625-641.

Meyerhoff, A. A., 1967, Future hydrocarbon provinces of Gulf of Mexico—Caribbean Region. *Trans. Gulf Coast Assoc. of Geol. Soc.* **XVII**, 217-260.

Meyerhoff, A. A., and Hatten, C. W., 1968, Diapiric structures in central Cuba, in Braunstein, Jules, and O'Briend, G. D., eds., Diapirism and diapirs: Am. Assoc. Petroleum Geologists Mem. 8, pp. 315-357.

Meyerhoff, A. A., 1970, Continental drift: implications of paleomagnetic studies, meteorology, physical oceanography, and climatology. *Jour. Geol.* **78**, (1). (In press)

Meylan, M. A., 1968, The mineralogy and geochemistry of manganese nodules from the southern ocean: Florida State Univ., Dept. Geology, Tallahassee (M.S. thesis) Contr. 22, 172 pp.

Miller, A. R., Densmore, C. D., Degens, E. T., Hathaway, J. C., Manheim, F. T., McFarland, P. F., Pocklington, R., and Jokela, A., 1966, Hot brines and recent iron deposits in deeps of the Red Sea. *Geochim. et Cosmochim. Acta* **30**, pp. 341-359.

Murray, G. E., 1968, Salt structures of Gulf of Mexico Basin—a review, in Braunstein, Jules, and O'Brien, G. D., eds., Diapirism and diapirs: Am. Assoc. Petroleum Geologists Mem. 8, pp. 99-121.

National Geographic Society, 1968, Map of the world: Washington, D.C., Natl. Geog. Soc.

National Oceanographic Data Center, 1968, National Oceanographic Data Center highlights: Wash. D. C., 19 pp.

National Petroleum Council, 1969, Petroleum resources under the ocean floor: Washington, D.C., 107 pp.

Pearson, W. J., 1963, Salt deposits of Canada, in Bersticker, A. C., ed., Symposium on salt: Northern Ohio Geol. Soc., pp. 197-239.

Pendery, E. C., 1969, Distribution of salt and potash deposits—present and potential effect on potash economics and exploration: Symposium on Salt, 3d, Cleveland, Ohio, Northern Ohio Geol. Soc. (In press.)

Physical-geographic atlas of the world, 1964: Moscow.

Rios, J. M., 1968, Saline deposits of Spain, in Mattox, R. B., ed., Saline deposits: Geol. Soc. America Spec. Paper 88, pp. 59-74.

Rona, P. A., 1969, Possible salt domes in the deep Atlantic off north-west Africa. *Nature* **224** (5215), 141-143.

Sannemann, D., 1968, Salt-stock families in northwestern Germany, in Braunstein, Jules, and O'Brien, G. D., eds., Diapirism and diapirs: Am. Assoc. Petroleum Geologists Mem. 8, pp. 261-270.

Sarocchi, C., 1965, Exploration et exploitation pétrolière en mer: état actuel et perspectives: Annales des Mines, no. 4, pp. 13-16.

Schneider, E. D., 1969, The deep sea—a habitat for petroleum? *Undersea Technology* Oct. 1969, pp. 32-34 and pp. 54-57.

Stöklin, Jovan, 1968, Salt deposits of the Middle East, in Mattox, R. B., ed., Saline deposits: Geol. Soc. America Spec. Paper 88, pp. 157-182.

Strakhov, N. M., Shterenberg, L. E., Kalinenko, V. V., and Tikhomirova, E. S., 1968, Geokhimiia osadochnogo margantsovorudnogo Protessa [Geochemistry of a sedimentary manganese ore-forming process]. *Akad. Nauk SSSR Geol. Inst., Trans.* **185**, 459 pp.

Takeuchi, Hitoshi, Uyeda, S., and Kanamori, H., 1967. *Debate about the Earth; Approach to Geophysics through Analysis of Continental Drift.* San Francisco, Freeman, Cooper, and Co., 253 pp.

Tortochaux, Francois, 1968, Occurrence and

structure of evaporites in North Africa, in Mattox, R. B., ed., Saline deposits: Geol. Soc. America Spec. Paper 88, pp. 107-138.

UNESCO, 1968, Carte tectonique international de L'Afrique, scale 1:5,000,000, New York.

United Nations ECAFE, 1959, Geological map of Asia and the Far East, scale 1:5,000,000, 6 sheets, Bangkok, Thailand.

United Nations Economic and Social Council, 1969, Mineral resources of the sea: Report of the Secretary-General, no. E/4680, 124 pp.

Vine, F. J., 1969, Sea-floor spreading—new evidence: *Jour. Geol. Education* **17**, 6-16.

Weeks, L. G., 1963, Worldwide review of petroleum exploration: Address before the 38th annual meeting of the Society of Petroleum Engineers of AIME, New Orleans, La., Oct. 6-9.

Weeks, L. G., 1966, Assessment of the world's offshore petroleum resources and exploration review, in *Exploration and Economics of the Petroleum Industry*, vol. 4, Gulf Publishing Co., Houston, Texas, pp. 115-148.

Weeks, L. G., 1969, Offshore petroleum developments and resources. *Jour. Petroleum Technology*, pp. 337-385.

Withington, C. F., 1962, Gypsum and anhydrite in the United States, exclusive of Alaska and Hawaii: U.S. Geol. Survey Mineral Inv. Resource Map MR-33, scale 1:3,168,000.

World Petroleum Congress, 1967, Origin of oil, geology, and geophysics: World Petroleum Cong., 7th, Mexico City 1967, Proc., v. 2.

Zelenov, K. K., 1964, Iron and manganese in exhalations of submarine Banu Wuhu volcano (Indonesia). *Akad. Nauk SSSR, Doklady* **155**, 1317-1320 (AGI Transl.).

Deep-Water Archeology

WILLARD BASCOM   1971

INTRODUCTION

The greatest remaining treasure-house of information about the ancient world—the bottom of the Mediterranean—is about to become accessible. In this article I explain why I believe that there may be many ancient wooden ships in reasonably good condition on the deep-sea floor, how these ships can be located and recovered, why they sank, and why their cargoes are clearly worth a substantial search and recovery effort. In addition, I present the rationale for deciding where to search in order to optimize the chances of finding long-lost ships.

A new form of underwater archeology will begin when a new kind of scientific ship and new techniques are used to explore the deep-sea floor for sunken ancient ships. The *Alcoa Seaprobe*[1] is such a ship—capable of

1. *Alcoa Seaprobe* is owned by the Aluminum Corporation of America and operated by Ocean Science and Engineering, Inc., for Ocean Search, Inc., a wholly owned subsidiary of Alcoa. It was designed by Ocean Science and Engineering, Inc., and built by Peterson Builders, Inc., in Sturgeon Bay, Wisconsin. It has just begun operations.

reaching down with its sensors, which are in a pod at the tip of a pipe, and making a detailed examination of the bottom in water several thousand meters deep (Figure 6-22). It is equipped with sonar to systematically search the sea floor at a rate of about 1 square nautical mile every 6 hours (1 square nautical mile = 3.4 square kilometers). Men at the surface will be able to inspect objects on the bottom with television, dusting away sediment by means of jets and propellers. Photographs can be taken and objects of interest identified (and perhaps recovered) by means of grasping devices. Because *Alcoa Seaprobe* will be capable of lifting from deep water loads weighing 200 metric tons, it may be possible, under some circumstances, to recover entire small ships in one piece. The overall capability of this new ship will be substantially greater than that of any previous device for search and recovery in deep water.

I deal here only with ships that sailed the Mediterranean Sea during the pre-Christian era, but it is evident that there are many

Willard Bascom is president and chairman of Ocean Science Engineering, Inc., California. He is also director of Ocean Mining, Inc., Deep Oil Technology, and Oil Search, Inc. He is former director of the National Academy of Sciences' Mohole Project. He has written numerous scientific and popular books and articles.

From *Science*, Vol. 174, No. 4006, pp. 261-269, 1971. Reprinted with light editing and by permission of the author and the American Association for the Advancement of Science. Copyright 1971 by the American Association for the Advancement of Science.

Figure 6-22. The *Alcoa Seaprobe.* (Photograph courtesy of Ocean Science and Engineering, Inc.)

other, more recent (but still very old) ships in that sea, and elsewhere in the world, to which these search and salvage methods can be applied.

Every ship is a small sample of the life and times during which it sails. On ancient ships, as on today's ships, the people aboard had all the utensils required for living, the weapons for fighting, and the tools for working. Much of this hardware and part of the cargo is virtually imperishable. Some of these objects can be dated and associated with peoples and ports to give information about the culture and commerce of the period. The hulls of the ships reveal the state of the art of marine technology in naval architecture and shipbuilding. Since each ship represents the surrounding civilization, and since most sink quickly, carrying down

everything except minor flotsam, a complete ancient ship in good condition is the marine archeologist's dream—a sort of undersea Pompeii, in which dark, cold water instead of hot ashes has preserved a moment in time.

Until now, most of the old ships excavated have been represented mainly by piles of amphorae and the part of the ship's bottom structure that was protected by being submerged in mud. Most of these ships were wrecked on reefs in shallow water, and only the most resistant materials, such as ceramics, bronze, and glass, have survived.[2] Ex-

2. The following objects can survive under the sea for 2000 years. General: metal ingots (bronze, tin, copper, and silver), tools, coins, amphorae, pottery, glass, and inscribed stone tablets. Ship's gear: anchors, bronze tools, tiles, utensils, metal fasteners, ballast rock, firepots, leg-irons, navigational

posed wood rarely survives the attacks of boring animals and waves for more than a dozen years. Only the wood buried safely beneath mud and sand can be studied and recovered.

Our picture of ancient shipping is, therefore, somewhat skewed by the materials that survive and by the corresponding blank spots in recorded history. A great deal is known about some aspects of the old ships, but virtually nothing is known about others;[3] for example, knowledge of the upper structure is derived primarily from the generally inadequate renderings on bas-reliefs and vases. The scientific argument about how the rowers and oars were arranged on warships is still active after 100 years of debate. Although we are not certain what it meant to say a ship was an "eight" or a "thirteen," details such as the cost and length of oars are known. No example of the trireme, the most common warship for some 400 years, has yet been found and excavated, although the sinking of thousands of them has been recorded.[4] Much remains to be learned about how ancient ships fought, were constructed, sailed, loaded, and crewed.

SURVIVAL OF WOODEN HULLS IN DEEP WATER

In deep water (more than 1000 meters below the surface), the chance of survival of both the wooden and the fragile parts of a ship should be enormously improved; the temperature on the bottom is near freezing, and chemical reactions proceed slowly. It is dark, and the currents are usually minimal, thus allowing protective silt to accumulate on the upper surface. A ship lying there will

be well beneath wave action, trawl nets, and divers. Since the cargo will not have moved after its original landing, it should not be scattered about nearly as much as it is in ships wrecked in shallow water.

The most uncertain aspect of wreck preservation in deep water has to do with the possible damaging effects of marine borers. Teredos are not known to go deeper than 200 meters, but in some places *Xylophaga* a borer clam, lives at 2000 meters.[5] Old wood untouched by borers has been found on the Pacific continental slope south of Panama, and an old ship was seen in deep water off Gibraltar by a small submarine looking for the lost H-bomb. In one Atlantic location off the coast of Florida, exposed wood from a Spanish wreck, 400 meters deep and about 300 years old, has not been attacked.[6] Probably the reason that some hulks have borers and others do not is because of local variations in oxygen, nutrients, metallic ions, or currents. The very action of a ship landing on the sea floor might cause a cloud of mud to swirl up and around, falling back on the ship as a protective cover. Some ships had resistant wood, lead sheathing, paint, copper parts, and tarred decks or seams, all of which may have helped protect against borers, especially where currents are low. A wooden ship landing in an anaerobic area, such as the Black Sea, should endure indefinitely.

The amount of mud that accumulates on a ship because of normal sedimentation will vary considerably, since the rate of sedimentation depends on distance from a land mass, amount of runoff, and local currents. In the central basins of the Mediterranean where the water is 4000 meters deep, the rate of sedimentation is believed to be about 10

equipment, and lead sheeting. Military equipment (if bronze): armor, swords, spears, shields, helmets, chariots, axes, boarding grapples, Greek fire tubes, and ramming beaks. Art: sarcophagi; columns; obelisks; statues of bronze, stone, or clay; bracelets, rings, and other jewelry; and musical instruments.
3. L. Casson, *The Ancient Mariners* (Minerva New York, 1959).
4. H. Frost is excavating a hull of Sicily, believed to be that of a trireme.

5. J. S. Muroka, *Deep Ocean Biodeterioration of Materials—Six Months at 6,000 feet* (Technical Note N-1081, U.S. Naval Civil Engineering Laboratory, Port Hueneme, Calif., 1970); R. D. Turner, *Some Results of Deep-Water Testing* (Annual reports, 1965, American Malacological Union, Allen, Lawrence, Kans., 1965).
6. R. Marx, personal communication. In deep water off Florida, Marx found a Spanish wreck with the wood in good condition.

centimeters per 1000 years.[7] In embayments close to shore, the rate may be more than ten times that amount;[8] but intermediate depths, say 1000 meters, where muds, clays, and calcareous oozes are deposited, 20 centimeters per 1000 years is a reasonable estimate. Forty centimeters of sea dust since the start of the Christian era should give an ancient ship, if otherwise whole, the blurred appearance of a wagon after a snowstorm.

The degree of preservation of a ship in shallow water apparently becomes stabilized after about a dozen years, and then seems to remain unchanged for centuries.[8,9] One has good reason to hope for much less deterioration in deep water, both in those first years and afterward.

Almost certainly, special circumstances exist in the many places beneath the sea where very old ships have not decayed substantially where they still sit upright on the bottom and have barely changed from the way they looked when they first sank.

SALVAGING WRECKS IN SHALLOW WATER

The salvage of ancient ships and cargoes for their archeological information and art treasures is not new. In 1832, the 5th century bronze Apollo which is now in the Louvre was brought up by a trawler from the waters off the island of Elba. In 1901 a wreck at Antikythera was accidentally discovered by sponge fishermen who had made an exploratory dive while waiting for the weather to moderate and found a spectacular collection of bronze and marble statues. They reported their find to the government, and an expedition was launched. Many statues were salvaged, along with an astronomical computer which set the date of the voyage at 80 B.C., suggesting that this special cargo was

7. Y. Herman, *Late Quaternary Mediterranean Sediments,* in press.
8. P. Throckmorton, *Shipwrecks and Archaeology* (Little, Brown, Boston, 1970), pp. 152-155.
9. H. Frost, *Under the Mediterranean* (Routledge, London, 1963), pp. 122-129.

probably loot from Greek cities en route to Rome. This wreck was on top of an undersea cliff, and some of the cargo had slid off into water of far greater than diving depth before it was found.

Greek sponge divers also found a wreck at Mahdia in the Bay of Tunis in 1907. An entire temple was aboard, as well as many bronzes. Greek trawlermen—fishermen who drag nets along the bottom in water somewhat deeper than that divers work—brought up parts of bronze statues from Cape Artemision in 1926. An expedition was begun to recover other statues (including the larger-than-life bronze of a naked Zeus casting a spear—a replica of which stands at the entrance to the United Nations building in New York), but the water was too deep for the diving technology of the day, and after several serious accidents the site was abandoned and perhaps lost.

These finds of truly great statuary were, of course, tremendously exciting to the art world; but they were of less significance to classical archeologists, who did not feel that random finds of ships carrying art objects were very important in defining societies and cultures. It is also possible that much of the early undersea work was ignored because of its technical crudity in mapping, dating, and processing of finds. Peter Throckmorton[10] quotes a Turkish official in the Department of Antiquities in 1960 as saying: ". . . underwater archaeology is not very important or interesting. We archaeologists are interested in the culture of the people, not in minor details of ship construction, which is of interest only to a few specialists."

The beginnings of recent activity in salvaging old ships from shallow water seems to have begun with Jacques Y. Cousteau, who organized a team of divers in 1952 to excavate a wreck at Gran Congloué, not far from Marseilles. It was an exciting adventure, widely reported in the press and enhanced somewhat by Cousteau's colorful descriptions and motion pictures. In this exca-

10. P. Throckmorton, personal communication.

vation, which followed an undersea slope downward as deep as 50 meters, the Cousteau group pioneered many techniques that have since been greatly refined, including the use (for archeological purposes) of television, air lift, and the Aqualung.[11] The excavators worked in the glare of publicity and could not be expected to foresee many technical problems of wreck excavation. Looking back on the work, it now seems possible that there were at the site two wrecks, one above the other, which got confused. This does not seem quite as serious now as it did at the time, when that site was thought to be unique.

The Congloué operation was a good beginning in the sense that it attracted the world's attention to a new source of historical material that has since revealed a great deal about ancient times. That excavation was followed by a flurry of artifact hunting by scuba divers, mostly along the French Mediterranean coast. These hunts were characterized by confusion of purpose and numerous diving accidents. In the early days, divers were not archeologists, nor vice versa. As wrecks were destroyed in the search for artifacts, few careful drawings were made and most of the data were lost forever. But in 1958, Throckmorton and, shortly afterward, George Bass, decided that classically correct underwater excavations were technically feasible and that a Bronze-Age wreck off Cape Geledonia, Turkey, was a good place to attempt one. This very successful archeological work, in which the first precision, three-dimensional surveys of ship and cargo were made, has been described in detail by Bass.[12] It was followed by a similar work by Throckmorton[8] in Italy and Greece, and Katzev[13] in Cyprus; thus a technology for precise excavation of wrecks in shallow water has been established.

However, the depth at which such work is practicable is set by the usual physiological problems of diving. Even today, in doing extended work deeper than, say, 50 meters, there is a real chance of diver-archeologists getting the bends or having other kinds of diving trouble. True, it is possible to put down living quarters and remain at depth for weeks at a time, to use complex mixed-gas apparatus, or to use a saturated diving system, in which one lives under pressure at the surface and is carried to and from the depths in a pressure chamber. Doubtless these methods will eventually be used and the archeological diving work will be extended as deep as perhaps 200 meters; but the costs and problems increase greatly, while the archeological advantages do not improve at the same rate.

Bass and others successfully used side-looking sonar for detecting wrecks off the coast of Turkey in water depths to 130 meters; they identified the targets with television.[14] Bass also used the small submarine *Asherah*, equipped with stereophotographic mapping cameras, to take overlapping photographs from which accurate maps could be made. Although further development of the equipment was said to be required, these techniques were effective. However, they are not known to be in use at the present time. Throckmorton and others have used bottom-penetrating sonar to find wrecks beneath harbor muds.

So much for the history of undersea work and the present state of work in shallow water. The operations in deep water will be quite different.

WHERE TO LOOK FOR SUNKEN SHIPS

There will be an essential difference in the

11. J. Y. Cousteau, *Nat. Geogr.*, January 1954, p. 125.
12. G. F. Bass and P. Throckmorton, *Archaeology* 14, No. 2 (1961), p. 728; G. F. Bass, *Expedition* 3, No. 2 (Winter 1961), p. 131.
13. M. L. Katzev, *Nat. Geogr.*, June 1970, p. 841.

14. G. F. Bass, *A Diversified Program for the Study of Shallow Water Searching and Mapping Techniques* (University of Pennsylvania Museum, Philadelphia, 1968); M. S. McGehee, B. P. Luydendyk, D. E. Boegman, *Location of Ancient Roman Shipwreck by Modern Acoustic Techniques* (Report No. MPL-U-98/67, Marine Physical Laboratory, Univ. of California, San Diego, 1967).

means of finding deep-water wrecks. Wrecks in shallow water have generally been found accidentally by sponge divers, trawlers, or sport divers. Throckmorton, who probably has located more wrecks in the Mediterranean than anyone else, has relied largely on talking to sponge divers in waterfront bars. He pleasantly engages them in Turkish, Greek, or Italian over auzo or raki and eventually turns the conversation to "old pots in the sea"—amphorae. Every sponge diver in the Mediterranean knows where there are a few mounds of such pots marking the site of an old wreck. Eventually, they tell him and he records the location. He has personally checked many of these, and his charts show hundreds of sites.

In deep water there is no similar mechanism for getting a lead on a promising wreck. Instead, it will be necessary to rely on statistics that are based on knowledge of the ancient ports and population centers, the kinds of cargoes and location of trade routes, and the nature of the old ships and the way they were sailed. Probably the last of these is the most important, since the mariners of old did not sail in straight lines between ports as a modern ship would. The ships were small and they sailed between protected bays, sometimes stopping to avoid the strong afternoon winds. They would minimize runs in the open sea by sailing within a few miles of the coast, anchoring during the calm, and sailing with the favoring winds as much as possible. Often the outward voyage would follow a different route from the return, and both might change seasonally or stop entirely during the months of November to March because of the increased risk during bad weather. The size of the ship and the way it was rigged, as well as the destination of its cargo, also influenced the route.

The story of what is known about the ancient ships is well told in the works of Lionel Casson[15] and others, but there is also

15. L. Casson, *Ships and Seamanship in the Ancient World* (Princeton Univ. Press, Princeton, N.J., 1971).

much that is not known about the many varieties of ships and cargoes over a period of several thousand years.

A very brief review of the history of Mediterranean shipping before the Christian era is appropriate. This was an age when empires were formed, established their colonies, collapsed, and were replaced. Because rough terrain and unsettled political situations made land travel very difficult and risky, there may have been more ships afloat in some periods than there are now. The objective is to find places along well-traveled sea lanes where ships would have troubles that would cause them to sink in deep water. Then a search for sunken ships can be conducted where the statistics are more favorable.

Egyptian ships were trading along the eastern Mediterranean at least as early as 2500 B.C., one of their first imports being cedar from Lebanon for shipbuilding. Minoan Crete was the first great sea power in the Mediterranean. In the years 1800 to 1500 B.C., its ships explored the Aegean and Black seas and pioneered the trade routes, as far west as Sicily, that were destined to last for centuries. By 1500 B.C., the Mycenaeans controlled the seas, giving way in about 1200 B.C., to the Phoenicians, who extended the trade routes into the western basin and built colonies at least as far west as Cadiz, which is beyond Gibraltar. Their sailing technology was remarkable, and some people think that they may have sailed on across the Atlantic. On their return eastward, they carried metals—tin, silver, lead, and iron—mainly from the mines of Spain.

Gradually, by about 800 B.C., dominance of the sea shifted to Greek shippers, who dealt more in such bulk commodities as olive oil, wine, wheat, pottery, and hides. The Greeks founded hundreds of colonies and linked these to the homeland by ship traffic; thus, by 500 B.C., Athens (Piraeus) was a major trade center. Its influence gave way to that of Rhodes and Alexandria, until, in the century before Christ, the Romans dominated the sea (Figure 6-23). Presumably all

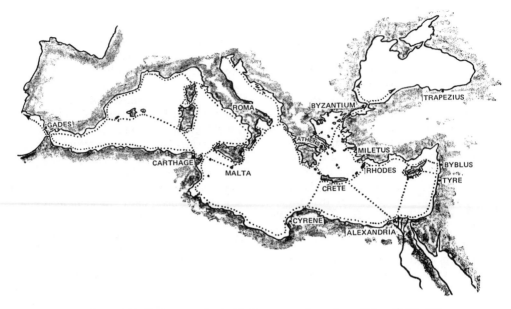

Figure 6-23. Trade routes in the Mediterranean, from about 500 to 200 B.C.[24]

of the countries around the sea had ships that were involved to some extent in trading, piracy, and intermittent wars. The point is that there were a great many different kinds of ships at sea, manned by crews of different nations, during the millennium B.C. The routes they followed remained virtually unchanged for thousands of years.

WHY SHIPS SANK IN DEEP WATER

Why would a ship suddenly sink in deep water—especially considering the fact that the routes they followed were generally near shore and that risks were minimized by anchoring in protected waters when the weather was bad? There were accidents of many kinds. The weather in the Mediterranean is known for sudden, violent change, and the old ships could not respond rapidly or sail into the wind very well. In the Aegean, they feared the meltime, a northerly, afternoon wind that has gusts to gale force and that dies down again, to leave the sea calm by

midnight.[16] It would be very easy for a ship to be blown well offshore and sunk in a sudden squall. In the Adriatic these northerly gales are called boras, and they reach force 10.[17] In the late fall and winter there is the sirocco, which can blow at a steady force 6 for a week, with gusts to force 8 or more. In 1966, a modern steamer went down in one, with a loss of 166 people. These violent and unexpected winds are known to have been the principal cause of losses of sailing ships in the eastern Mediterranean in the 19th century A.D.—and they doubtles were in previous centuries as well.

The ancient ships were generally small (less than 40 meters long) and made of wooden planks held edge to edge, caravel

16. H. M. Denham, *The Aegean* (Murray, London, 1963), p. 63.
17. Force 10 is a whole gale with winds of 50 knots; it can make (fully developed) waves 7 to 10 meters high. Force 6 is a strong breeze with winds of 25 knots; it can make (fully developed) waves 3 to 4 meters high.

style, with mortised joints. If the planks came apart, the weight of cargo and ballast would take a ship down in a few minutes. This kind of accident might be aggravated by a shift of cargo, a weakening of the wood by marine borers, or by one large wave. Generally the ships did not have much freeboard, and many were not completely decked. In heavy weather they could take water over the sides and quickly founder or capsize.

Pirates were the scourge of shipping for many hundreds of years; they doubtless sank many of the vessels they caught. Fire was also a frequent cause of loss—coals would fall from the galley stove as the ship pitched, or a carelessly held torch would start a blaze. Some nations had strict safety laws about fire on shipboard. In later centuries, there were several occasions when whole fleets were lost by fire. Besides the other hazards, warships had a chance of being rammed in battle or being scuttled after capture.

One concludes that ships sank in ancient times for the same reasons they did in the 19th century. Because the old ships were smaller, less well built and rigged, and sailed by less well-trained crews, the statistics must have been worse. Throckmorton[8, p. 34] points out that, in the years from 1964 to 1869, 10,000 insured merchant ships were lost—1000 without a trace. Of the 372 British naval ships lost by mishap between 1793 and 1850, nearly half were lost by running onto unmarked shoals, and 78 of them, or 21 per cent foundered at sea, usually in deep water with all hands. He also notes that on a single reef at Yassi Ada he and Bass found more than 15 wrecks, ranging from 3rd century B.C. to the 1930's.

The number of ships involved in commercial traffic during the millenium B.C. can be used to estimate the number of wrecks lying in deep water. Yalouris[18] believes there were over 300 active ports by the 4th century B.C., with an average capacity of 40 ships each, and that half of these ships were at sea. If so, the standing crop of ships was

18. F. Yalouris, unpublished data.

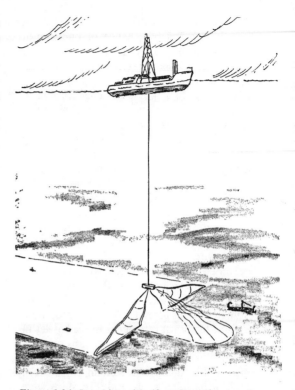

Figure 6-24. Searching with *Alcoa Seaprobe*. A pod containing both a forward and a side-looking sonar, as well as television, is attached to the tip of a weighted pipe. Together the sonars sweep a pathway about 400 meters wide at a speed of about 1 meter every 2 seconds.

about 12,000. This number increased markedly in Roman times, although it was much less at the beginning; therefore, a conservative estimate would probably be that there were, on the average, 6000 ships in trade over the 1000-year period. If one judges the average life of a merchant ship to be 40 years, some 150,000 ships were built.

Being guided by the statistics of ship losses during the 1800's, one can guess that half eventually retired safely and the other half were lost on voyages. If, of those lost, 20 per cent foundered, then there are 15,000 ships down offshore—many in deep water. If the total length of the trade routes in the eastern Mediterranean were 6000 nautical miles (1 nautical mile = 1.9 kilometers) and the average width 10 nautical miles, there is a reasonable chance of one ship in every 4 square nautical miles of bottom along those routes. By selecting specific areas were ships were known to get into trouble, it is possible to improve the chances considerably. There are numerous high-probability sites in the straits between islands, between islands and mainland, and at major "jumping off points," where ships left the security of nearby land and set out into open waters. Such sites exist between the Peloponnesus and Crete, east of Crete to Rhodes and Turkey, between Italy and Yugoslavia, between Tunis and Sicily, around Malta, near Gibraltar, and along the coasts of Lebanon, Israel, and Cyprus.

Somewhat more is known about the losses of military ships in the same period, and there are more fixed points to guide estimates in this case. There are literally hundreds of historical references to sea battles in which astonishing numbers of ships were involved. For example, at the battle of Economus in 255 B.C., 250 Roman ships faced 200 Carthaginian ships; only 16 ships were lost in the battle (which Rome won), but in a storm off Camarina shortly afterward, 250 of the remaining ships were wrecked. When Agrippa fought Sextus at Naulochus in 42 B.C., 600 ships were engaged. When Anthony and Cleopatra met Augustus at Actium in 37 B.C., about 100 ships of the 900 involved were lost.

The life-span of a trireme is estimated at 20 to 30 years,[18] if it were not destroyed by warfare or storm. Thus a force of 300 ships would be maintained by building ships at the rate of 10 to 15 a year. A law of Themistocles decreed that 20 ships should be constructed each year, in order to ensure replacement of the old ships. We do not know how long it was enforced, but the total number of warships built in Athens in the 5th and 4th centuries B.C. was around 1200. Considering the number of states with fleets, the size of the fleets, the rapidity of the replacement, and the life expectancy and survival chances of the ships, I estimate that

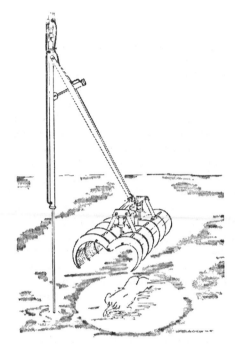

Figure 6-25. The exploratory tongs are guided by television and powered by a sea water hydraulic system. They are supported by an adjustable arm and are held steady by a pin 10 meters long. The pin can be driven into the bottom when the object to be recovered is seen. Tongs open to about 3 by 3 meters and can lift 5 metric tons. This system can perform delicate operations of lowering, rotating, and closing. It retracts through the center well into the ship.

25,000 warships may have been built before the time of Christ.

If one figures battle losses at 10 per cent and losses at sea for another 10 per cent [Casson (reference 3, p. 157) says the Romans lost four ships to weather for every one to enemy action], then there may have been 5000 warships sunk—many of them in deep water. The locations of the battles are often known within a few miles.

Although I have used the millennium B.C. in this discussion of ship losses, it is obvious that the sinkings continued, at least through the age of sail, at about the same rates. The ships were larger, better built, and easier to sail, but the storms and wars and accidents went on. Ships that sank much later than the ancient ones may be even more interesting. For example, the Battle of Lepanto was fought not far from Actium 1608 years after Anthony and Augustus fought there; another 100-odd ships went down, this time with a very different array of artifacts. Sunken ships have been accumulating in the Mediterranean for 4000 years. Sorting through them on the bottom and identifying those of interest to archeological salvors may be a bit of a problem.

THE SHIP AND ITS EQUIPMENT

Having decided that very old ships and their cargoes were among the most valuable objects to be found on the sea floor, and knowing the limitations of other possible methods, I proposed in 1962 the search and recovery system described here.[19] With few changes it has been developed into the *Alcoa Seaprobe.*

Alcoa Seaprobe is like a drilling ship with a center well and is propelled and maneuvered by two Voith-Schneider vertical-axis propellers, fore and aft. It is 75 meters long, 17 meters beam, and draws about 4 meters. The ship is all electric, with a power plant forward; it has quarters for 50 persons aft. (It is the world's largest aluminum structure—constructed mainly of 5456 aluminum, to demonstrate the excellent marine properties of that material.) The ship is steered by means of a console on the bridge, which controls the propellers and makes it possible to exert thrust in any direction almost instantaneously. Thus it can be dynamically positioned[20] manually or, with certain navigational inputs, automatically.

Above the 4 by 12 meter center well, there is an aluminum derrick that can lift up to 400,000 kilograms. The available working load of the derrick and draw works will be 200,000 kilograms (200 metric tons, with a safety factor of 2).

The derrick and pipe-handling system is capable of lowering or recovering drill pipe at an average rate of about 0.25 meter per second, to a depth of 6000 meters (although only about 2400 meters of pipe will normally be aboard). Depending on the operational situation and the depth, the lower end of the pipe will be weighted with up to 20 metric tons of drill collars. The drill pipe ordinarily carried is 11.5 centimeters in diameter, in 10-meter joints; it is handled as pairs, or "doubles," 20 meters long. It can be raised or lowered with precision a centimeter at a time, or rapidly at several meters a second (if necessary to avoid obstacles). The pipe can be rotated slowly to change the orientation of the instrument pod or tools at its lower tip.[21]

The information and control cable that connects the instrument pod to the scientific control center is attached to the outside of the pipe by special clips at 20-meter intervals. Within that cable, sonar and television signals come up coaxial conductors; power and instructions go down adjacent pairs. The information is recorded on paper recorders and on television tape; because precise position and time are recorded marginally at short intervals, it is possible to refind objects on the bottom.

The principal searching instruments are two sonars mounted in the pod; one scans directly ahead, and the other sees the bottom on each side at right angles to the motion of the pod (Figure 6-24). The side-looking sonar has a frequency of 177.5 kilo-

19. W. Bascom, U.S. Patent No. 3,215,976 (2 November 1965).
20. Dynamic positioning is a phrase first used by the author in 1960 to describe the process of equipping a ship with multiple propellers whose net thrust can be controlled from a central console, enabling the ship to hold position in deep water, relative to a fixed marker, in spite of winds, waves, and currents.

21. In addition to search and recovery, the *Alcoa Seaprobe* is capable of drilling for geological cores in deep water, generally following the method devised by the author and his associates in 1960 (W. Bascom, Ed., *Experimental Drilling in Deep Water* Publication No. 914, National Academy of Sciences, Washington, D.C., 1961).

hertz and a beam width of 0.3 degree. It sends and receives sound pulses in such a fashion that objects that project above the bottom (such as a rock or a ship) will better reflect the sound and produce a darker spot on the record. Because the sea floor immediately beyond an object is in the sonic shadow, the record of that area remains white. As the ship (and the pod) moves, the line-by-line record of a succession of pings builds a sonic picture of the bottom; thus, the size and position of any object can be determined within a few meters, as can the approximate height of the object. The resolution is such that, at a distance of 200 meters, the sonar can distinguish between two objects that are more than 1 meter apart.

Since the principal information obtained is a white shadow, it is evident that the sonar beam must strike the bottom at a substantial angle (there are no shadows near the nadir). This means that a pathway immediately beneath the pod must be searched by the forward-scanning sonar. This sonar works in a similar fashion; but, because it scans an area rather than a thin slice of bottom, it is not practical to record this information. Instead, any objects seen on its plan-presentation scope are noted for time and position; later they are plotted by hand on the record from the side-looking sonar.

The necessary precise navigation will be done in one of several ways, depending on the distance from land, the depth of the water, and the size of the area to be searched. For short-range work in depths to 2000 meters when an area of up to 50 square nautical miles is to be searched, the following method will ordinarily be used. Two taut-moored, subsurface buoys will be installed some 10 miles apart outside the search area; a surface buoy with radar transponder will be tied to each. Under most circumstances, the movement of a transponder buoy relative to a point on the surface above its anchor will be less than 20 meters. The ship will then obtain its position relative to these fixed buoys by ranging on

them with radar. The digitized distance to each buoy will be recorded at 1-minute intervals, which are also coded automatically on the margin of the recordings. The maximum uncertainty of position will be about 40 meters (half the length of the ship).

SEARCHING, INSPECTION, AND RECOVERY

When the ship arrives at a site selected for search, the transponder buoys are installed and the area is surveyed with an ordinary echo sounder, in order that the search can be conducted approximately along contour lines. The instrument pod is lowered on the weighted pipe until it is about 60 meters above the bottom. Then the ship begins to move along the planned course at a speed of about 0.8 meter per second. At this height above the bottom, the sonar scans a pathway about 400 meters wide. At this velocity, it searches an area of about 20,000 square meters per minute, a square kilometer every hour, and a square nautical mile every 4 hours. Allowing for an overlap in pathways of 50 meters, to make up for the navigational uncertainty, and allowing time for turns and adjusting pipe length, the average time needed to search 1 square nautical mile will be about 6 hours.

Doubtless, many objects will be detected in each square nautical mile searched. These will be given priorities for inspection, based on experience in interpreting such records. Ships that rest upright on the bottom should be relatively easy to identify as ships. Deciding whether or not they are valuable as salvage targets is a more difficult problem.

The pod that holds the sonars also holds the television camera and lights, but the sonar height of 60 meters above the bottom is beyond the range of the television (about 15 meters is the maximum range in clear water).

Therefore, when a visual inspection is to be made of the objects found with sonar, the pod is lowered until it is a few meters above the bottom. The ship navigates by radar back to the priority site, and a more local-ized search begins. While the ship holds posi-tion, the tip of the pipe is maneuvered by means of a propeller just above the pod. This propeller, driven by sea water pumped down the pipe, is capable of moving the pod up to 50 meters in any direction (depending on depth and pipe weight). Watchers on the ship will look for and attempt to identify the object and, if it is a ship, to determine the approximate kind and age. All wrecks would be carefully photographed for what-ever information they contain.

If the ship is worth close inspection, the next step would be to retract the pod and replace it with the exploratory tongs shown in Figure 6-25. The tongs are a grasping tool guided by television. They are supported by a movable arm that extends out from a cylinder containing a 10-meter-long, hydrau-lically driven pin. This pin is used to fasten the tip of the pipe to the bottom adjacent to the wreck, and to give a fixed point about which the tongs can be rotated. The tongs can thus make slow, sure movements, in spite of small movements of the ship above. Watchers on *Alcoa Seaprobe* can wait for the dust to settle without the ship's drifting off course. Each of the tong tips has a pipe through which clear sea water is pumped to wash sediment away so that a better inspec-tion can be made. Objects weighing less than 5 metric tons can be grasped and lifted with considerable delicacy. Specific objects of art or items of cargo can be recovered with these tongs and withdrawn through the cen-ter well into the ship.

When an entire ship suitable for recovery is located, the super-tongs will be towed to the site. These tongs would weigh about 50 metric tons and would be supported on the surface by their own pontoons, whose buoy-ancy can be adjusted. On site, the tongs may be slung beneath the ship, secured to the tip of the pipe, and lowered. As with the explor-atory tongs, these will have television equip-ment mounted so that those on the ship can see the object to be seized. When fully closed, these tongs will surround a cylinder

about 100 meters in diameter and 12 meters long. They are closed by a hydraulic system worked by leverage, which exerts increasing-ly greater force, up to about 10 metric tons per double tine (60 metric tons for six double tines). This should be sufficient to penetrate the mud beneath a hull and to close without touching the old ship.

The dimensions were selected with the sizes of ancient ships in mind. Figure 6-26 shows the super-tongs holding a small Phoe-nician-like trading ship of 1000 B.C. The ship is about 16 meters long and 7 meters beam. (Half full of mud, this would ap-proach the weight-lifting capacity of the *Alcoa Seaprobe*.) Depending somewhat on the amount of mud and cargo in the hull, the remaining structural strength of the wood,[22] and the shape, ships up to the rated capacity of 200 metric tons could be lifted. Much longer ships would be neatly cut into two or more pieces with a hydraulically driven chain saw, and the lifting process re-peated. These pieces can be reassembled later, probably with less difficulty than the fragments of ships found in shallow water to date.

When the super-tongs have been retracted to a position just below the ship, the *Alcoa Seaprobe* will slowly move into quiet, shal-low water and set the burden on the sea floor. There the mud, ballast, and cargo can be removed by divers, thus lightening and strengthening the hull for its final move.

A special museum barge (Figure 6-27) would then be brought in, and an ordinary, large capacity marine derrick will lift tongs and hull from the sea bottom to the barge's tank, which is a few meters deep and filled with water. The super-tongs are removed

22. R. Marx, personal communication. Marx found the complete lower ribs of a 300-year-old galleon covered with sand in shallow water. The oak ribs were 20 centimeters square; beneath the outermost 1 centimeter, they "looked like new" and were in a condition so good that furniture could be made of them. He notes that this is unusual but not unique, and that the Spanish wreck he found in deep water[6] seems to be equally sound.

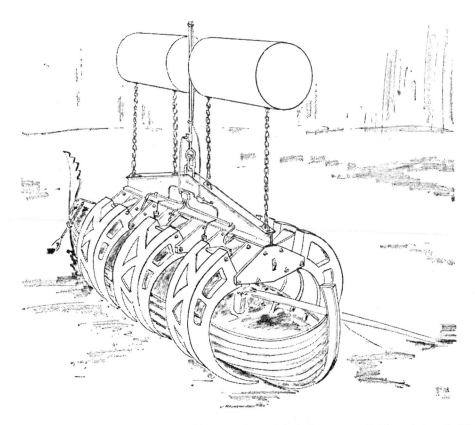

Figure 6-26. The super-tongs shown are lifting a small vessel from deep water. Guidance is by television; closing power comes from sea water pumped down the pipe at high pressures. These tongs (which weigh 50 metric tons) would be towed to the salvage site on their own flotation and used in a nearly neutral condition of buoyancy.

Figure 6-27. A museum barge could be constructed to support a tank in which an ancient ship could be carefully rebuilt underwater and displayed. A crane that can lift 200 metric tons (and that is available around most major ports) would be used to lift the super-tongs and the ship from shallow sea water into the tank and then return the tongs to the sea.

(and possibly sent back for the rest of the hull), and the archeologists can complete the reconstruction underwater, with controlled conditions of light and chemicals. The ancient ship will be visible to the public, but will remain submerged in a controlled environment. Because the barge would be mobile, it can be conveniently moved about the world for display.

LEGAL CONSIDERATIONS

The objects being considered for salvage are clearly part of the general heritage of our civilization and should eventually become the property of the world's great museums. Since the countries and kingdoms that once owned these ships are long gone, it may not be easy to say who owns them, and it may not even be possible to determine what flag the ship sailed under. However, in most cases the following general principle will apply.

Many countries subscribe to those Law of the Sea conventions that specify a 3-mile zone of territorial waters (measured from a baseline drawn between promontories) and, beyond that, a contiguous zone to 12 nautical miles. The laws of most countries would prohibit ships of a foreign power from conducting salvage operations within these zones without specific permits.

Beyond the 12-mile limit are the high seas, where a vessel may generally be said to possess the right to conduct scientific research and to search for sunken property. Private vessels drive the right of the country whose flag they fly.

A coastal state may lay special claims to the sea floor on the continental shelf (whose limit is generally considered to be 200 meters). This kind of claim is intended to protect a state's natural resources to the limits of exploitability, but sunken property can hardly be considered a natural resource. Since the depths of greatest interest for *Alcoa Seaprobe* are well outside both the 12-mile limit and the 200-meter contour, it would seem that the ship will clearly be operating on the high seas. This would be true under U.S. law, off our own coast.

However, as a matter of scientific courtesy, one might invite archeologists from nearby states to participate. In some cases, the ship may be invited into territorial waters to perform search and recovery in the interest of international science or for a local government. Depending on the nature of the material salvaged and its origin, it should be possible to negotiate a reasonable arrangement for distributing the finds among the appropriate museums. The unique capability of the *Alcoa Seaprobe* makes it likely that several of the countries around the Mediterranean will be interested in cooperative archeological ventures.

CONCLUSION

Obviously this kind of work is expensive; so is any work in modern science, compared to what it cost a few years ago. But the results will justify the expense. Salvaging one ancient ship from deep water will probably cost no more, for example, than salvaging the *Maine,* or the *Vasa,* or the Roskilde Viking ships. The daily cost of *Alcoa Seaprobe* will be about half that of an ordinary tanker, destroyer, or deep-sea drilling ship. The preservation and maintenance of the ship salvaged will be a fraction of similar costs for the *Texas* or the *Queen Mary.*[23] It

23. The *Maine,* sunk in 1898, was salvaged from the floor of Havana harbor in 1911; the *Vasa,* sunk in 1628, was raised from Stockholm harbor in 1961; the Roskilde Viking ships, sunk about 1400, were raised from a Danish harbor in 1962. The cost of each of these was several million dollars. Daily charter rates for a medium-sized tanker of 50,000 tons are now about $20,000; rates for destoyer, if computed on the same basis as those for a civilian ship, would be about $22,000 per day, plus depreciation and shore support. The day rate of the *Glomar Challenger* is about $16,000. The battleship *Texas,* now on public display at San Jacinto battlefield, was made into a presentable monument for a cost of about $1 million. The *Queen Mary,* now a convention center-museum at Long Beach, California, has a cost over $50 million to date.

is quite possible that a museum of ancient ships can be made self-supporting. Certainly it will attract the general public as well as the scholars, the tourists as well as the natives.

Who can say what a Roman trireme, or a Phoenician trader, or a Cyprean pirate ship is worth in terms of a better understanding of history? By these standards, it would seem that our civilization can afford to recover those ancient ships that can provide insights into the living and working conditions of another civilization, especially when they are

24. A. A. M. Van der Hyden and H. H. Scullard, Eds., *Atlas of the Classical World* (Nelson, London, 1959).

not discoverable by other means. A new intellectual adventure that is understandable by the public and that generates interest and excitement can contribute to the appreciation and support of science as a whole.

SUMMARY

There is reason to believe that some old wooden ships on the deep-sea floor have survived for thousands of years without much change. They will not be covered with much sediment, and it will be possible to find them using new searching techniques. These are embodied in the system of the *Alcoa Seaprobe,* which is also equipped to identify and raise old ships.

7

ESTUARINE AND MARINE POLLUTION

The Role of Man in Estuarine Processes

L. EUGENE CRONIN   1967

INTRODUCTION

Vulnerability to human influence is a characteristic of estuaries. They lie in proximity to man's terrestrial habitat, produce large quantities of his food supply, and are doorways between the oceans and the land masses. Each receives the impact of many human activities throughout an entire watershed, and many are subjected to the most intensive levels of use applied to any marine water areas. It is, therefore, appropriate to identify and consider the past effects of man on the fundamental processes in estuaries and to contemplate future beneficial and detrimental influences on these fascinating, complex, and important waters.

Man's historical development has been closely linked with the estuaries. Humans have always exhibited a natural affinity for water and these bays and river mouths often present unique advantages. They are semienclosed and therefore provide natural harbors; they are effective nutrient traps and therefore are rich in food; they connect the oceans and the inland rivers so that they are natural transportation centers; and their often high rates of flux and flush permit disposal of great quantities of waste. The history of exploration, colonization, and settlement of the coasts of the North American continent illustrates the use of estuaries in the development of new populations and cultures.

The effect of human activity on these estuaries was probably unimportant prior to about 1850, and was limited to the effects of silt erosion from agricultural areas and the disposal of human wastes. The enormous expansion during the last century in industrial activity, production and use of power, diversity of manufactured materials, transportation, fishing intensity, and human population have all placed diverse and increasing pressures on these waters. They all affect the processes of the estuaries and their capacity for future use.

The uses of estuaries are described by

Dr. L. Eugene Cronin is director and research professor with the Natural Resources Institute and director of the Chesapeake Biological Laboratory at the University of Maryland. He was former commissioner of the U.S. Department of Health, Education, and Welfare Secretary's Commission on Pesticides and their Relationship to Environmental Health. He is currently on the DDT Advisory Committee of the Environmental Protection Agency. He has written and helped write over fifty articles.

From *Estuaries* (Lauff, G. H., ed.), A.A.A.S., Pub. No. 83, pp. 667-689, 1967. Reprinted with light editing and by permission of the author and the American Association for the Advancement of Science. Copyright 1967 by the American Association for the Advancement of Science.

others in this volume, but additional examples can be added:

1. The Rhine River, according to Bolomey (1959), is heavily polluted by the 40 million people who live along its course, as well as by the great industrial centers like the Ruhr and the Saar. The need to eliminate this pollution is so great that an international commission has been established. The estuarine areas near Rotterdam receive the net product.

2. In Baltimore Harbor, near the head of Chesapeake Bay, the equivalent of about 400 tons of concentrated sulfuric acid is released daily from a single large industrial operation (Stroup *et al.*, 1961).

3. The city of Washington, D. C., uses the upper Potomac Estuary as the final stage in its sewage treatment process (Auld, 1964). Brehmer (1964) has pointed out that this is a present average daily addition of 22,700 pounds of phosphorous and 68,100 pounds of nitrogen. This annual release of eight million pounds of phosphorous and 25 million pounds of nitrogen will nearly double by the year 2000, as treatment plant effluent increases from 200 million gallons a day to 360 million gallons a day. Brehmer further emphasizes that present knowledge of the effects of this pollution is insufficient to predict the effects on the estuary or even to establish rational limits.

4. Of the ten largest metropolitan areas in the world, seven border estuarine areas (New York, Tokyo, London, Shanghai, Buenos Aires, Osaka, and Los Angeles). They contain over 55 million people and enormous industrial activity. One-third of the population of the United States lives and works close to estuaries.

5. Bulkheading, dredging, and filling to create waterfront real estate have already permanently changed the nature of some estuaries. The Branch of River Basin Studies of the U. S. Department of the Interior reviews and comments on federal projects which would affect fish and wildlife resources. They have reported on 426 different projects in five years which would affect coastal areas. They expect to start 100 more projects this year, 15 for flood control, 20 for navigation, 20 for beach erosion or hurricane protection projects, and 50 for private proposals.

6. Without waiting for inexpensive nuclear energy, man has begun to engineer vast estuarine changes. The 2,700 km^2 Zuider Zee has been enclosed for 32 years, essentially fresh for 27 years, and largely converted to sub-sea-level agricultural land. Holland has also begun to transform salt and brackish waters into lakes in the great Delta Project of southwest Holland (Vaas, 1963).

7. More than a billion pounds of biocidal chemicals have been used in the United States, and Butler *et al.* (1962) point out that many of these substances and their oxidation products reach the coastal waters and the estuaries. Cottam (1960) reported that 35 million pounds of arsenical salts, 45 million pounds of copper sulfate, six million pounds of organic phosphates, and 130 million pounds of chlorinated insecticides and fungicides were used in one year. Six thousand brand-name biocidal materials are now available and almost unrestricted in use. Rachal Carson (1962) has cited specific and vivid examples of the results of application of control chemicals in the areas of the Miramichi in Canada, in the Indian River of Florida, in New Jersey, and at other sites.

These brief examples of the effect of human activities are further evidenced by experience in various parts of the world. They serve as reminders, not as an adequate review.

The future will bring very rapid increase in all the present uses of estuaries, and entirely new pressures and modifications are already taking shape: (1) California is studying alternate plans for damming, diverting, filling, and vastly modifying large areas in the Sacramento-San Joaquin region and elsewhere (Jones *et al.*, 1963). (2) There is a well-advanced plan to divert the waters of the Sabine River, the Neches, the Trinity, the Brazos, the Colorado, and the Nueces from east Texas to west Texas. No water would escape to the coastal estuaries (Thompson, 1961).

The development of inexpensive sources of power in unprecedented quantities may make these seem like mere practice sessions. In this compilation to assemble and assess our knowledge, constructive perspective can be gained by reviewing the effects of man on estuaries.

This review is presented to single out the most significant estuarine processes which man affects, to offer summary and partial assessment of the location and nature of human effects, to suggest the present benefits and losses from these effects, and to consider the future role of man in the estuaries.

No attempt will be made here to stress their importance in human welfare, or the economic and social values which they provide. The enormous diversity of systems, which results in highly significant individuality, is assumed. So, too, is the inherent dynamism of most estuaries. We must consider generalized effects by specific examples, and the net effects superimposed on vigorous rhythmic systems.

There are no physical, chemical, or biological processes unique to the estuary, but many are typical of this complex and distinctive mixture of sea and river. In the cases and discussions which follow, primary attention will be given to these questions:

1. What physical, chemical, and biological processes are unusually significant in the estuary and may be modified by man?

2. How have human activities affected these processes beyond the normal range of variation present in the virgin estuary?

3. What are the possibilities for future management of estuarine processes for optimal achievement of human values from estuaries?

The literature of estuaries contains many research reports, administrative summaries, and discussions dealing with specific problems, and many of these will be cited later. No previous general review was located, but several contributions have been especially

helpful and stimulating. They include the series of reports on estuarine hydrography by Pritchard (1951, 1952, 1955, 1959a, b, 1960); on flushing and biological effects by Ketchum (1950, 1951a, b, c, 1954); on the grand-scale effects of enclosing the Zuider Zee (Havinga, 1935, 1936, 1941); on biology of pollution by Hynes (1960); on biological aspects of estuaries by Hedgpeth (1957); on physical and chemical aspects by Emery and Stevenson (1957); Mansueti (1961) on the nature of man's effects; Nelson (1947) on enrichment; H. T. Odum and his students (1958, 1962) on estuarine processes; Rounsefell (1963) on the choices of management objectives; Sykes (1965) on multiple usage; and E. P. Odum (1961b) on imaginative and constructive approaches to new potentials. Baughman's valuable bibliography (1948) contains many pertinent annotations. Sverdrup *et al.* (1942) continues to serve as a splendid general reference.

The topics presented here are numerous and diverse, ranging from upland erosion to invertebrate toxicity and estuarine hydrography. The pertinent literature is, therefore, represented by illustrative examples rather than by exhaustive inclusion.

Certain processes have been chosen for emphasis here because they meet two criteria: (1) each is significant in many or most estuaries, at least of the coastal plain type; (2) each is now subjected to substantial (i.e., beyond normal range) modification by man.

The physical attributes which will receive emphasis are salinity, temperature, river flow, and basin shape.

Chemical modifications which will be discussed include the addition of biocides, nutrient chemicals, pulp mill wastes, and certain exotics.

Geologically, only silt and siltation are considered.

Among the highly varied modifications of biological processes, human predation (more commonly called fishing!) and the introduction of new species have been chosen as illustrations.

The lists could be very long, but these

may serve to summarize present knowledge and provide guidelines for the future.

ACTIVITIES IN THE WATERSHED

Since the estuary is the recipient of effects from changes throughout the watershed, a review of the pertinent human activities along the contributory waterways can be helpful. They vary and are important in the estuarine processes.

Modification of River Flow

Many human activities affect the quantity of inflow of fresh water, its temporal distribution, and its contents. River flow can be reduced, especially by diversion of river water for human consumption (Nelson, 1960; Jones et al., 1963; Ketchum, unpublished ms.; Thompson, 1961), for the vast increases in artificial irrigation of agricultural land (Mansueti, 1961), and by the intentional or accidental use of spillways or breaks in levees (Gunter, 1952, 1956). Conversely, flow can be significantly increased in the basins receiving the diversion. More frequently, increase in the total annual output is the result of denuding the watershed by removing vegetation and by other activities that decrease the insoak and subsurface retention. This is especially vivid in the paved urban areas (Renn, 1956) and along highways where as much as 30 acres per mile is paved or carefully sloped to maximize runoff. These also increase the flashiness of rivers, with greater flooding in high-flow periods and drought in low-flow seasons. Counteracting forces do exist, however, in the increasing number of small and large dams, many of which are specifically designed for moderation of the river flow and long-term release, and in improved general conservation practices.

Gross Effects

The gross estuarine effects of changed river flow are rather well understood, al-though more subtle effects have infinite local variation. The most pervasive effect is on the general hydrographic structure and behavior of the estuary. River flow is a prime factor in the determination of salinity distribution in the estuary (Ketchum, 1951a, c) and of the vertical and horizontal physical structure of the estuary (Pritchard, 1955). Pritchard has shown that increased river flow converts a homogeneous-type estuary through moderate stratification to strong and persistent stratification. Cronin et al. (1962) showed that this conversion occurs in the Delaware, following runoff variation. Pritchard (personal communication) has provided additional evidence of the power of flow. During most summers, the deeper waters of the central Chesapeake are severely depleted of oxygen as they are transported up-estuary. Stratification of cooler, saltier, denser deep water under warmer, fresher, lighter surface water may be very strong, and an increase in river flow during this period enhances stratification, enlarges the area of depleted oxygen, and may do extensive damage to estuarine organisms (Carpenter and Cargo, 1957).

More specific estuarine effects of change in river flow have been shown by Beaven (1946), Gunter and Hall (1963), Nelson (1960), and by others. Beaven documented the control by the Susquehanna River over salinity in the upper Chesapeake Bay, and provided excellent evidence that all major oyster mortalities recorded for the upper Bay from 1907 to 1946 were associated with and probably the direct result of periods of high runoff of the river.

Biological Effects

The intermittent controlled release of fresh water from Lake Okechobee to the St. Lucie Estuary in Florida has probably enhanced the fisheries by nutrient supply, and may benefit croaker, mullet, anchovy, and menhaden, according to Gunter and Hall (1963). A better pattern of release could be of increased benefit to the fisheries.

Nelson (1931), in reviewing the arguments relating to diversion of Delaware River water to New York City in the Hudson River basin, cited many estimates of the ecological and biological effects. As an example, he predicted an up-bay movement of oyster drills. *Urosalpinx cinerea,* to invade several excellent oyster seed beds previously protected by low salinities. In more general terms, salinity is known to limit the distribution of oysters and many other estuarine species (Gunter, 1955; Galtsoff, 1960; Korringa, 1952). Davis (1958) and others have demonstrated from laboratory experiments that the eggs and larvae of clams, *Mercenaria* (*Venus*) *mercenaria,* and oysters (*Crassostrea virginica*) have optimal salinity requirements for development and growth, and the optimal salinity for the eggs of the oyster may be governed by the degree of salinity at which the parent oysters develop gonads.

Additional biological effects of flow change are known. Diversion may disturb the migratory patterns of fish. Gaussle and Kelley (1963) found that flow reversal in the San Joaquin River, because of exportation of water through a power plant, has apparently affected salmon runs, presumably because "home stream" water was not present to stimulate ascent and spawning. The degree of dilution affects the decrease of bacteria in polluted estuaries, although Ketchum *et al.* (1952) found this to be a much smaller factor than the bactericidal effect of sea water. Ketchum (1954) also showed that the vigor of estuarine circulation, which is greatly affected by river flow, determines the reproductive rate necessary for maintenance of plankton populations. Pritchard (1951) suggested that the upstream movement of deep water, also affected by river flow, may transport young croaker, *Micropogon undulatus,* from the spawning ground (off Chesapeake Bay) upstream at 0.2-0.4 knots, or 130 miles in less than 20 days. Our later experience in the Bay suggests that this is a primary and essential method of dispersion for the young of weakfish (*Cynoscion regalis*), spot (*Leiostomus xanthurus*), blue

crab (*Callinectes sapidus*), and other species. Pritchard has also shown (1952) that similar movements may provide oyster larvae for the greatest seed oyster area of the world, in James River of Virginia. Bousefield (1955) vividly related such flow-dominated circulation to the distribution of barnacles in the Miramichi Estuary. Odum and Wilson (1962) expressed the possibility that bays with little flushing may develop higher productivity and more effective regeneration of nutrients. Low river flow would obviously favor these effects.

Chemical Effects

Flow modification also affects the chemical content of waters entering the estuarine system. Carpenter (1957) found that the calcium content of the Susquehanna is broadly, but imprecisely, related to flow. Renn (unpublished), in preliminary scanning of the distribution of ABS (alkyl benzene sulfonate), principal ingredient of the nonbiodegradable detergents, observed the effective reduction caused by dilution. Laundry outfalls showed levels up to 140 ppm ABS, which was reduced by dilution, absorption, blow-off of foam, and degradation (about 5 per cent per week) to 0.2-0.3 ppm in rivers and estuaries. Agricultural or silvicultural pesticides would be similarly affected.

Siltation

Siltation in estuaries is caused by both natural and human factors. Deforestation, flashing runoff, and poor agricultural practices contribute. Burt (1955) showed that river discharge affects the distribution of the inorganic suspended load in the Chesapeake, as many others have seen from other estuaries. Wolman *et al.* (1957) calculated that 60 million cubic feet of silt per year are deposited in the estuary of the Potomac River near Washington. Mansueti (1961) estimated that half of the former upper estuary spawning areas for fish and shellfish beds for oysters have been destroyed or shifted downstream by sediments in the Chesapeake. He also pointed out the progressive filling of deeper channels by such silt. Gunter (1956) summarized the soil transport of the Mississippi River as 730 million tons of soil per year into the Gulf of Mexico—38 thousand acres, three feet deep. He deplores the narrow canalization of this transport, which formerly overflowed the basin.

Human countermeasures are again available, but not well utilized, in this country at least. Soil conservation is progressing. Small watershed dams and larger reservoirs also retain silt effectively.

Enrichment

Organic enrichment of tributaries is an ancient problem. In direct relation to estuaries, Jeffries (1962) has described the nutritional contribution to Raritan Bay of nitrate-nitrogen and phosphate from sewage in the Raritan River and the resultant dense phytoplankton blooms. Renn (1956) cites human organic loading as being 0.4 lb/person/day. The quantities reaching the estuary will obviously vary greatly with distance, river flow, loading, and many other factors, so that generalizations here are inappropriate.

Dams and Barriers

Physical barriers in tributary streams create very special and important effects. Hynes (1960) and Mansueti (1961) pointed out that dams absolutely block anadromous fish migrations and may eliminate important runs unless fishways are provided. A series of dams on the Susquehanna has cut off migrations of the white shad (*Alosa sapidissima*) and reduced the spawning area of the striped bass (*Roccus saxatilis*). Others have blocked herrings and other species. Whitney (1961) pointed out that, in addition to physical blockage, dams create reservoirs which differ greatly in circulation, temperature, and currents from the original stream, so that anadromous fish may not be successful even if they pass the wall. Fishways also introduce an artificial but effective factor for genetic selection. In addition, reservoir water can become depleted of oxygen, modified in temperature, and changed in other ways (Whaley, 1960). The common practice of drawing relatively deep water through turbines can release undesirable conditions into the upper estuary. An aged fisherman pointed out other modifications in the upper Chesapeake. Before dams, the upper Bay was subject to annual ice scour alternating with heavy siltation. Dams prevent the ice movement, trap the silt, and produce much greater stability of the bottom. Heavy vegetation grows where it once could not.

Selected Cases

To illustrate the range and magnitude of watershed effects on estuaries, four vivid examples, past and future, will be briefly described. They are the leveeing of the lower Mississippi River, operation of the Bonnet Carré Spillway, plans for water management of the Potomac River, and diversion from the Delaware River to the Hudson River.

Levees. Gunter (1952, 1953, 1956, 1957a) describes the development of levees in the lower portion of the 1,257,000 square mile Mississippi basin (Figure 7-1). First construction was in 1717, and a broader program was inaugurated in 1735 by private interests. The State of Louisiana assumed responsibility in 1828, the Mississippi Valley Commission in 1879, and the U. S. Army Corps of Engineers in 1928. Levees are now up to 35 feet high. Among the many results, runoff is faster and peakier; velocity increase transports more silt; alluviation, sedimentation, and flooding of the swamps, marshes, and estuaries have virtually ceased; and enormous quantities of silt are directly deposited in the Gulf of Mexico. These have brought serious, and usually detrimental, changes to the estuarine areas of Louisiana. Drainage of nutrients from land is reduced; salinity is increased and stabilized; island erosion is increased; and bays may move inland. Species

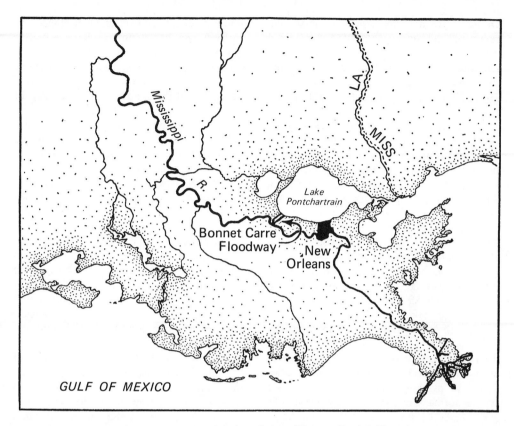

Figure 7-1. Lower Mississippi River Basin, showing relation of Bonnet Carré Spillway to the estuarine areas bordering the Gulf of Mexico.

appear to be responding to these changes and some oyster reefs are disappearing.

Bonnet Carré Spillway. In the same system, a great spillway 7,700 feet wide, was completed in 1932 (Figure 7-1). It was designed to protect New Orleans from Mississippi River floods, and can carry 250 thousand cubic feet per second (cfs). It diverts water into Lake Pontchartrain, thence through Lake Borgne and Mississippi Sound, and eventually to the Gulf of Mexico. Gunter (1953) describes the results of openings in 1937, 1945, and 1950. Since all of the receiving area is estuarine, the effects are of interest here. In Lake Pontchartrain, motile organisms are driven out, and many non-motile forms are killed by low salinities. A small area is covered by mud. Most or all of the oysters in certain beds are destroyed, with lower loss over a wider area, although oyster pests and predators are killed out. Nutrient is added to the area, estimated in 1950 as 40 thousand tons. Following return to normal salinities, unusually great production of shrimp and other marine life is observed. In Gunter's opinion, the total beneficial economic effect out-weighs the partial oyster mortalities which occur in some years.

The Potomac. The Potomac is a very flashy river. With an average flow at Washington, D. C., of 11,000 cfs, it has an observed range from 800 cfs to 500,000 cfs.

Droughts, floods, heavy siltation, massive enrichment, and increasing needs for sustained water supply all present problems in its control and use. Many partial or complete answers have been proposed (Wolman *et al.*, 1957), but the most recent suggestion is a dramatic example of efforts to manage a river and control its effects on an estuary.

The U. S. Army Corps of Engineers has developed four alternate plans, with primary interest in flood control, recreation, water supply, and water quality control (U. S. Army Engineer District, Baltimore, N. Atl. Div., 1963). The one recommended would include establishment of 418 headwater small reservoirs, 16 major reservoirs, 3 small flood control projects, and a general program of land management and conservation. The cost is estimated at $498 million, and it could be completed about 2010; the plan is now being reviewed.

Estuarine effects have been only partially estimated. Augmentation of low summer and fall flow is the obvious, and reduction of spring maxima is inevitable. Uncontrolled low-flow averages about 1,100 cfs, and would normally be expected to decline in 50 years to 800 cfs because of upstream irrigation and other consumptive uses. If the plan is effected, minimal flow would be about 4,700 cfs, which is 42 per cent of the annual average. Specific projection of all the effects of this change on the dynamics, populations, and character of the Potomac Estuary offers challenging problems. It has not even been attempted. It is already clear, however, that the upper estuary, at least, would be significantly and permanently different, especially during summer months. Physical circulation, nutrient availability, and biotic communities would all be significantly modified.

The Delaware diversion. New York City, on the Hudson River, has long sought to obtain water from the Delaware River watershed. Delaware Bay, principally through the State of New Jersey, fought such diversion in 1929-1931 and 1952-1954, primarily on the grounds the estuarine productivity

would be decreased. Nelson championed this position and has summarized the estuarine argument (1960). Despite evidence which was more precise than usual, the Supreme Court allowed a diversion of 1,240 cfs (800 mgd). Part of this is already in effect. Part of the pattern provided will guarantee minimal flows at Trenton during low-flow periods. Nelson argued that the minimal flows would be ineffective in combating oyster drills, that drills would penetrate farther on to productive seed beds with higher average salinities, and that the nutrient loss would be especially damaging.

Ketchum was asked to predict the effects, and provided estimates of channel salinities under normal flows and under diversion with controlled summer minima. The near future will test his precision, but his predictions provide an excellent example of the nature of the effects of intentional flow modification on estuarine conditions (Ketchum, unpublished).

The river in 1952 had a mean flow of 11,700 cfs, with usual lows near 2,000 cfs and highs near 29,000 cfs. With stipulated augmentation of flow at Trenton, and diversion of 1,240 cfs, Ketchum predicts that mean flow will be reduced from 11,700 to 10,530 cfs, low flow increased from 2,157 to 2,893 cfs, and high flow reduced from 28,900 to 21,950 cfs. Salinity will increase at mean flow a maximum of 0.85 $^{\circ}/_{\circ\circ}$, decrease at low flow a maximum of 1.08 $^{\circ}/_{\circ\circ}$, and increase at high flow a maximum of 3.2 $^{\circ}/_{\circ\circ}$. Range would thus be reduced, with a maximum at the center of the estuary, where annual range of 16.46 $^{\circ}/_{\circ\circ}$ would become 12.85 $^{\circ}/_{\circ\circ}$, a difference of 4.61 $^{\circ}/_{\circ\circ}$. Isohalines would be moved upstream, except during augmented flow. Biologically, the important conditions which limit the viability or success of populations are generally the extreme conditions to which the populations are exposed. The proposed diversion should, therefore, be beneficial to the populations within the estuary.

At the hearings, other biologists expressed broad concern about the detrimental effects, but were unable to make and substantiate specific predictions of those effects.

It is interesting to compare these projections with the report of Segerstraale (1951) that average changes of 0.5-0.75 °/$_{oo}$ produced significant changes in the effective distribution of estuarine species.

Apparently, no one has made an effort to predict the effects of the addition of 1,240 cfs of water to the Hudson Estuary, perhaps because no one objects to the addition, or because the lower Hudson at New York City seems to be beyond reclamation.

ACTIVITIES IN THE ESTUARY

Physical Processes

With increasing frequency and intensity, human activities are changing the physical conditions and processes in estuaries.

Thermal Addition

One dramatic and growing group of effects arises from the addition of heat on a continuous basis. The most important sources have been generating plants for electricity, using great quantities of water to cool condensers. Research on thermal effects in fresh water has been extensive, permitting summary by Ingram and Towne (1959), Hynes (1960), and others for the fluvial situation, where physical effects and the biological sequence produced are reasonably predictable. This is not so for estuaries, where tidal flux and the presence of a different community of animal and plant species introduce complicating factors. Pannell *et al.* (1962) reported observations which illustrate the difficulties of predicting effects, and the types of effects which may occur. Specific predictions were used in designing the release of 26 million gallons per hour with a rise from intake to outfall of 12°F. The prediction was that warmed water would spread thinly on the surface. It was observed, however, that an area 1,000 feet in

radius is warmed at least 5°, but most of it is between 3 and 9 feet below the surface, and the plume swings with tidal action.

It was predicted that little heat would be absorbed by the main body of the estuary, and observed that virtually all heat dispersion is by mixing. The most interesting observation was that most warmed water remained below cooler but fresh water and above cooler salt water. The interplay of salinity and temperature produced this unexpected vertical series, with salinity dominant.

Biologically, the heat appears to have extended the local breeding season of the boring gribble, *Limnoria,* and increased the incidence of the shipworm, *Teredo.* These sessile forms were affected, but no significant change in zooplankton was detected. Other effects remain to be evaluated.

In the United States, a valuable experiment may be provided by the construction of a 670 MW steam generating plant on the Patuxent Estuary, a tributary of the Chesapeake. The plant is now under construction and will warm 500,000 gallons per minute by 11.5°F in warm months and half that quantity by 23°F in winter. This volume is about 50 per cent more than the average freshwater input of the watershed above this point, so that the estuary will be used as a tidal cooling lagoon. Fortunately, intensive studies are preceding and following the construction of the plant. The Chesapeake Biological Laboratory has established a research team and developed cooperation from about 40 scientists in a dozen institutions and agencies for pre and post-operative studies of circulation, salinity, oxygen and thermal distribution, phytoplankton abundance and productivity, zooplankton distribution, bacterial density, fouling rates and species, benthic community composition, fish egg and larval distribution, adult fish distribution and migration, crab abundance, and other aspects of the area. Perhaps this will permit improved prediction and rational regulation of such activities. Present projections of the need for electric power call for demand to increase in this country from 60 billion gallons per day (bgd) in 1955, to 131 bgd in 1975, to 200 bgd in the 1980's (Picton, 1956). Against these figures, consider the estimates of total dependable surface runoff, 385 bgd at present increased to 630 bdg by 1975 or 1980. Power demand in this country and many others grows even more rapidly than populations, and increasing pressure to use estuaries for cooling is inevitable.

Otto Kinne at Helgoland has recently provided an excellent and valuable review of the effects of temperature on marine and brackish-water animals (1963) as part one of a survey of the effects of both temperature and salinity. The biological effects of thermal change, affecting all chemical and biological rates and processes, are profound. Kinne devotes portions of his review to temperature tolerance and lethal limits, effects on metabolism and activity, reproductive success, distribution, organism size, meristic characters and shape, and biotic adaptation to temperature.

Comprehension of these effects will be invaluable in efforts to prevent estuarine damage or to utilize heat to obtain optimal benefits.

Changed Salinity

Human activities in the estuary occasionally affect salinity. Examples are provided by pumpage of large volumes of fresh water into the estuary (Wolman *et al.,* 1957) or by engineering changes affecting the fundamental pattern of circulation, such as major modifications of channel depth.

The great Dutch conversion of an estuarine area of the Zuider Zee to a freshwater lake was completed in 1932, when an enclosure dike was completed (Figure 7-2), cutting off 2700 km^2 of rich, warm, shoal water with a salinity of 10-15 $^\circ/_{oo}$ in the inner portion and 15-25 $^\circ/_{oo}$ in the outer portion. The results, descirbed and discussed by Havinga (1935, 1936, 1941, 1949, 1959), show a vivid example of man's impact.

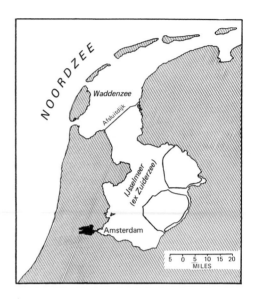

Figure 7-2. The old Zuider Zee, cut off by an enclosure dike or "Afsluitdijk" to convert an estuary to a lake and dry land.

Salinity decreased as the flow of the IJssel River continued while ocean water was almost completely prevented from entering (Figure 7-3). Stability was attained about 1937, and subsequent salt intrusion is limited to the water near two locks in the dike. The mero-estuarine species which had provided a large fishery (herrings, anchovy, and shrimp) and various non-commercial species were quickly and vastly reduced. Herring

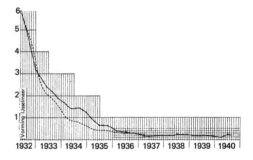

Figure 7-3. Chlorinity in the IJsselmeer following enclosure. The broken line indicates the calculated chlorinity level. The solid line shows the observed level (after Havinga, 1941).

were seen in great numbers at the dike, but were unable to reach the low-salinity water. Anchovy sought a different estuarine condition, relatively high temperature, and were equally frustrated. Stenohaline marine and polyhaline organisms were killed, and eventually their predators starved. With varying periods of endurance, the edible mussel, *Mytilus edulis;* the soft-shell clam, *Mya arenaria;* the green crab, *Carcinus maenas;* and other forms (*Cardium edule, Tellina balthica, Corophium* sp., *Heteropanope tridentata, Gobius minutus,* and others) died out. Eventually, the predators dependent on intolerant forms also disappeared.

Most of the euryhaline forms passed out of existence more slowly in the IJsselmeer, but some survived. *Neomysis vulgaris* was highly successful in the new environment, and provides much of the food for larger species. Limnetic species spread slowly but

effectively; the motile forms achieved wide dispersion.

As a result, the older fishery has been entirely replaced by freshwater production of pikeperch, bream, roach, and others. A striking survivor is the eel, *Anguilla vulgaris,* which thrives as the principal commercial species. It is aided by nighttime locking of young elvers into the lake when they appear along the dike.

Havinga points out that the result is faunistic poverty, marked by adaptations and intrusions, although total human values have been greatly enhanced. Fish production of 16 million kilograms per year was replaced by 1959 with 7 million kg of fish plus 70 million kg equivalent of pork on the reclaimed land. This may treble when all the potential *polder* reclamation is completed.

The Netherlands has embarked on a second great program for modifying brackish-

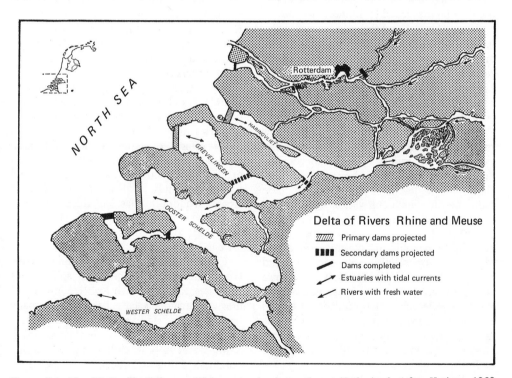

Figure 7-4. The "Delta Plan" for modifying estuaries in southwest Netherlands (after Havinga, 1959, modified to show works completed by mid-1964).

water areas (Figure 7-4). Primarily for the control of highly destructive floods, they have begun the "Delta Plan" at the mouths of the Rhine, Meuse, and Scheldt Rivers, to be completed in 1978. Three arms of the sea will be cut off and a series of additional dams will modify flow and convert salt and brackish areas to fresh water. Thorough hydrographic study before, during, and after construction is being accompanied by intensive and extensive biological surveys (Hartog, 1963; Vaas, 1963). Present predictions of the effects are stimulating and instructive, in view of Dutch experience in other locations. Tidal action will cease for most of the region, and be reduced throughout the area. Salinity will decrease rapidly, modified by leaching of salt from *polders,* and is not expected to fall below 0.3 $^\circ/_{oo}$. Stenohaline marine organisms will perish, although euryhaline species will persist for a long period and a brackish-water population will develop, then be succeeded by limnetic species.

Hartog has defined the euhalinicum, polyhalinicum, mesohalinicum, oligohalinicum, and freshwater conditions under the present regime, and published on some of the Amphipoda. Other studies are covering Hirundinea, Isopoda, Gastropoda, Gammaridae, turbellarians, fish, plankton, and vegetation.

Human safety will be gained, but it is most regrettable that part of Holland's important mussel industry and all of her Zeeland oyster industry may be sacrificed. Korringa (1958) described the threat to these industries, in which cultivation and management have been brought to a high and intensive level. He pointed out that it might theoretically be possible to determine the allowable range of salinity, silt content, sand transport, current velocity, plankton, and other requirements and seek new oyster areas, but he was not optimistic. Artificial culture of seed would be required, and this is still a difficult technique, without demonstrated economic justification on a large scale.

Modifications of Basins

The shape of the basin of an estuary has many effects on hydrographic dynamics and, as a result, on other processes. Pritchard has expressed the effect of modification (1955) by pointing out that conversion from marked stratification to vertical homogeneity is favored by increasing the width and opposed by increasing the depth. Shore erosion, siltation, channel dredging and spoil deposition would all have local effects. Skyes (1965) has noted that the establishment of cities on the shore is usually followed by an expanding pattern of reclamation, fills, causeways, and bridges, permanently altering the entire area. The premium which is placed on waterfront residences creates dramatic changes in basins by a pattern of dredging and filling for housing and for industry. Thompson (1961) summarized the usual estuarine effects as reduction in the water area; denudation of the bottom as fill is removed; modification of currents and tidal exchange; alteration in salinity, temperature, and perhaps oxygen content; and sediment dispersion.

Boca Ciega Bay, on the west coast of Florida, is a clear example of the present and future magnitude of these changes (Figure 7-5). Hutton *et al.* (1956) felt that the combined effects of sedimentation and basin change in Boca Ciega Bay might be to reduce or eliminate fishing, destroy breeding and nursery grounds, and create low oxygen areas. Woodburn has provided a guide (1963) for reviewing proposals to change shorelines, and suggests principles to be followed.

Comprehension of the relationships between the shape of the basin and various processes can be put to highly constructive uses. Scale models have been used in many parts of the world to study physical patterns of circulation and to test new possibilities. Simmons (1959) is enthusiastic about the potentials of using models in estuarine re-

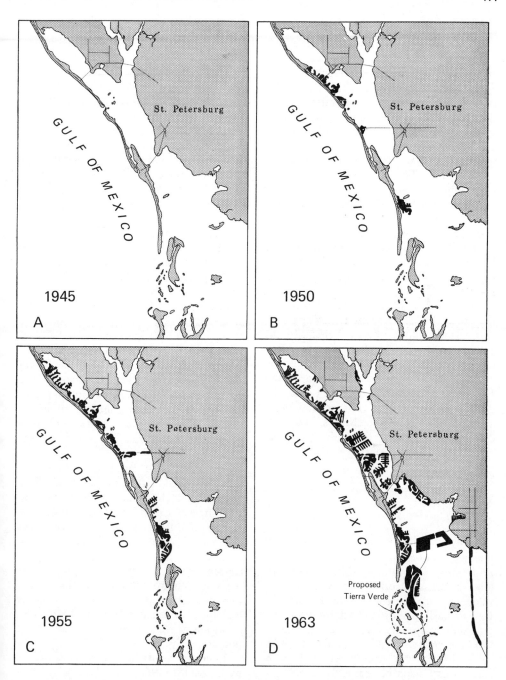

Figure 7-5. Progressive development of land by dredging and filling in Boca Ciega Bay near St. Petersburg, Florida. (Courtesy Mr. James Sykes and the U.S. Bureau of Sport Fisheries and Wildlife).

search. By using such techniques as single slug and continuous release, methylene blue chloride and other dyes, and artificial roughening, he is confident that models can produce accurate integration of the tidal, density, and freshwater forces that affect dispersion, dilution, and flushing of introduced materials. However, present estuarine models usually cannot scale non-conservative, or interacting, or time-related materials and processes. Pritchard (1960) has called attention to possible limitations in model studies of processes involving diffusion phenomena. Present estuarine models include the Delaware Estuary, San Francisco Bay, Puget Sound, Thames Estuary, New York Harbor, Matagordo Bay, Galveston Bay, Narragansett Bay, and others. Additional models are planned.

Man has created a rather special change in the "shape" of estuaries by removing a portion of the land mass between them. In this country an intercoastal waterway allows water, aquatic organisms, and people to move freely between bays along the entire East Coast and much of the Gulf Coast. The great canal systems of Europe and other areas provide similar interchanges. The greatest effects will probably involve biological exchanges and extensions.

Chemical Methods

Biocides

Many chemical compounds inhibit vital biological processes. Most compounds do so in excessive concentrations. In recent years, great industrial effort has gone into the development of compounds which could be used in small quantities to interfere with life processes in specific target species. The results are variously called weedicides, herbicides, bactericides, fungicides, and pesticides. Miss Carson has correctly termed the group "biocides" because of their fundamental effect. They destroy life. Advantages and disadvantages from their use are both

very great. These, and toxic substances never intended for use in control work, are present in estuaries in increasing amounts.

The U. S. Fish and Wildlife Service has been especially concerned with biocidal effects, and very valuable work has been done on two broad problems. What are the effects of these materials on estuarine species other than the original target? Can biocides be used constructively in the management of estuarine organisms? This research provides valuable guidance. It is, however, most regrettable that there has as yet been little support for research on the fundamental physiological mechanism of biocides, and on the effects of known toxins on the broad spectrum of non-commercial species.

Davis investigated the effects of 31 pesticides on the eggs and larvae of the oyster and clam (1961). He reported great variation in toxicity within each group of chemicals, ranging from 90 per cent mortality of oysters in 0.05 ppm DDT to improved growth of clams in 5 ppm of lindane. Several apparently beneficial materials may have reduced bacterial action. Reduction in growth rate was identified as a useful indicator of toxicity. Davis felt that it will eventually be possible to select materials for highly specific control.

The effect of toxins on growth was more fully developed by Butler *et al.* (1962). They observed that change in growth rate provides a sensitive bioassay technique, allowing detection of differences in one or two days. Chlordane, heptachlor, and rotenone were observed to be inhibitory within 24 hours in concentrations as low as 0.01 ppm. All of the common agricultural pesticides tested were toxic, but there was some indication that they are released from internal storage in organisms when environmental concentration drops.

In efforts to utilize biocides constructively, Loosanoff (1960) and his associates have screened hundreds of compounds in a search for effective methods for controlling green crabs (*Carcinus maenas*) and other arthropod predators on molluscan shellfish. They identified a large number of effective compounds, which can also kill shrimp, prawns, copepods, and other crustaceans. Perhaps it is suggestive of the complexity and difficulties of this field to note that my own professional interest in the blue crab (*Callinectes sapidus*) and the copepods of the estuaries makes me regard this new knowledge with conflicting feelings. Loosanoff also reported that many of the materials tested adversely affect molluscs. 1.0 ppm of DDT caused the death of oyster larvae, and 0.025 ppm interfered with growth. He suggested combinations which might be effective chemical barriers to minimize the invasion of shellfish beds by arthropods, and urged thorough evaluation of every such possibility prior to use.

Additional efforts in the control of undesired species have been reported recently. Hanks (1963) learned that baitfish could be soaked in lindane solution to reduce green crab populations and reduce crab immigration. Lindsay (1963) tested several materials for the control of ghost shrimp, *Callianassa* sp., and Japanese drills, *Ocinebra japonica,* in Puget Sound. Shrimp could be controlled and drills could be prevented from entering experimental plots, but Lindsay urged that the methods be avoided until long-term effects and public health problems could be fully understood. The use of DDT on oyster cultch to control barnacles (*Elminius*) was tested for the effects on oysters (Waugh and Ansel, 1956). This control technique had been discovered by Loosanoff and tested by others. The set of oysters doubled, initial growth was somewhat inhibited but followed by excellent growth, to reach 40 per cent greater size than controls in 2½ months.

With reference to plants, estuarine research on biocides is extremely limited. 2,4-D was effective in killing Eurasian milfoil in the upper Chesapeake Bay (Rawls, 1964), and Beaven *et al.* (1962) showed that the effective concentration had no acute effects on crabs, oysters, clams, and fish. A secon-

dary mortality occurred, however, when the dense mat of killed plants decayed on the bottom, producing anaerobic and toxic conditions.

An excellent summary of the results of research on pesticides has been prepared by Butler and Springer (1963). It brings together the present knowledge pertinent to coastal waters, including a number of individual reports which must be omitted in the present general discussion. Pesticides have been developed on a significant scale only since World War II, and research on effects still lags far behind the rapid growth of this economically valuable field of products. The review includes the results of laboratory and field studies on phytoplankton, crustaceans, molluscs, fish, reptiles, birds, and mammals. It is abundantly clear that there is significant variation between species, between chemical families of compounds, and between individual compounds. Irregularity in the results of field tests suggests the reasonable probability that actual toxicity might be affected by temperature, silt, flushing rates, salinity, and other environmental factors.

These authors also comment constructively on the special problems related to pesticide applicate. Chronic toxicity may be more subtle than acute toxicity, but as devastating. Acquired resistance has been demonstrated for some vertebrates. The specific capacities of organisms to concentrate pesticides are poorly known, but may be extremely important because oysters concentrated 96 per cent of available DDT within two days and retained much of this for substantial periods. The possibilities of recycling and build-up were noted, citing freshwater experience in which original applications of 0.014-0.02 ppm of TDE (DDD) eventually produced concentration of 2,500 ppm in fish and 1,600 ppm in fish-eating birds. The eventual use of chemicals which are specific for target species and non-toxic to all others is cited as the most promising future avenue of effort.

During March, 1964, the British Government ordered withdrawal of the chlorinated hydrocarbons, aldrin and dieldrin, from use. Action was apparently based on evidence of accumulative contamination, and also on some uncertainty of the physiological human effects. Also in March, 1964, the U. S. Public Health Service and the State of Louisiana established endrin used for cane borers as the probable cause of the killing of 10 million fish in the Mississippi River basin, including brackish waters of the Gulf of Mexico. As compared with environmental concentrations, fish blood showed a thousandfold increase, and fatty tissue a tenthousandfold accumulation.

In the estuarine studies which have been conducted, these chlorinated hydrocarbons are all dangerous to important estuarine species, but so are many others. Revelations of damage and restrictions on use are not yet ended.

Intensified attention to pesticides should not obscure the ever-present possibilities of other toxic elements and compounds. The paucity of such research in estuaries is impressive in view of the great industrial growth of recent years, and the consequent threat to inshore waters. Alexander *et al.* (1936) illustrated one type of effect in that low levels of cyanide showed linear increase in toxicity with temperature, but nearly logarithmic increase in toxicity with decrease in oxygen concentration. Olson *et al.* (1941) and Davis (1948) explained the effects of flocculent copper as pollution on diatoms, and learned that precipitating particles carry some diatoms from the water. This would decrease productivity, although Davis found no evidence of substantial loss. Galtsoff (1960) observed that the precipitate can be dealt with by oysters, but suspects that it may be harmful to larvae. He also notes the absorption and storage of copper, mercury, lead, and arsenic by oysters and other bivalves.

Ideally, research on the effects of chemical additions to the estuary should precede or parallel industrial development, new brand production, and chemical process modification. Regrettably, this is not the present practice. Therefore, new damage and danger remain as continuous and growing threats in coastal waters.

Nutrient Chemicals

Human wastes, or their degradation products, are universally placed in the river, the estuary, or the sea (see Koch, 1959, for a comprehensive European summary), and provide a continuous source of nitrogen and phosphorous in various combinations, plus a variable and imperfectly understood mixture of many substances. The great human preference for concentrating near estuaries makes the impact even greater at that site. Nitrogen and phosphorous are essential for photosynthetic elaboration, and availability sometimes limits production, so that they must receive principal discussion.

Before examining some of the effects of human waste, it is of interest to note that other potential mixtures of waste are added to estuaries in significant quantities. Ryther (1954) and others of the staff of the Woods Hole Oceanographic Institution (Ryther *et al.,* 1958a; Guillard *et al.,* 1960) have studied the remarkable growth of algae in Great South Bay and Moriches Bay, on Long Island, New York. Broad study of hydrography, chemistry, and biology has demonstrated that heavy pollution from surrounding duck farms, with high organic nitrogen but unusually low N:P ratio, combines with the topography and hydrography of these bays to yield a very dense population of small algae, dominated by *Nannochloris atomus* and a species of *Stichococcus.* These have seriously reduced the oyster-producing capacity of these estuaries by unbalanced overenrichment.

Effects of nutrient addition, rather than tables of the prevailing quantities, are of interest here. Hynes (1960) has provided a valuable condensation of knowledge of the fluvial and lacustrine effects of organic pollution, and demonstrates that processes are rather well established (see also Ingram and

Towne, 1959). Of about 45 pages on the biological effects of organic matter, Hynes was able to devote only two to estuaries because "there is little detailed information on the biological consequences of estuarine pollution." He cites tidal activity and rapid environmental changes as complicating factors. However, the general pattern of dense algal blooms, often accompanied by esthetically undesirable appearance and odor is very familiar (Bartsch, 1961).

Rational attempts must be made to understand these effects, in view of the certain growth of coastal populations. E. P. Odum (1961a) provides a constructive point of departure when he recognizes that the addition of large amounts of organic waste into natural waters creates a new ecosystem, and that an effective response to it is most likely to be found in studying the fundamental principles and processes involved. He includes a helpful analysis of the energy flow involved and the nutrient cycling which occurs, and suggests avenues of attack when the results are inconsistent with human wishes. Krause (1961) and others have outlined the fundamentals of algal physiology, and have urged a basic approach to these highly practical problems.

The distinctive problems and potentials of estuarine enrichment have already received attention. Mansueti (1961) stresses the widespread estuarine tendency to act as a nutrient trap, with tidal cycles and slow flushing to enhance the availability of nutrients for chemical or biological use. Pritchard (1959b) reviewed the physical mechanisms affecting the distribution of conservative materials introduced, and showed that: (a) oscillatory tidal motion produces longitudinal spread; (b) non-tidal dispersion will respond to the density of the material and the pattern of net motion in the estuary (vertically homogeneous, two-layer, three-layer, etc.); (c) net flushing will occur, and usually be predominantly along the right side of the estuary; and (d) entrapment by lateral indentures can provide important reservoirs. He predicted that computation may eventually be feasible for predicting the distribution of conservative contaminants, but that non-conservative materials will be much more difficult. I would add that biological processes and sequences in estuaries rarely permit such broad-scale projections with useful accuracy as yet, although they merit continued effort.

Artificial enrichment has been essayed, but it is expensive because of the cost of fertilizers and of accompanying research. Pratt (1949) found that the addition of superphosphate could elevate phosphate levels in a salt pond, although not by calculated amounts. Repeated addition held levels well above pre-treatment concentrations. Nitrate additions, on the other hand, produced short spurts to predicted levels, but these could not be held. A net increase in the standing crop was obtained, and it lasted for a significant period. Earlier large-scale study of changes in a fertilized sea-loch by Raymont (1949) and others showed that plankton production increased, bottom fauna was enhanced, and fish production eventually responded favorably. Use of sewage for enrichment of freshwater ponds is an ancient technique.

Development of practical methods for profitable management of the enormous quantities of fertilizing materials available to estuaries offers a challenging field for further research and for contribution to human welfare. At present, these vast quantities are released in undirected patterns, without using their great potential for profitable selected improvement.

The specific effects at sewage outfalls into estuaries also need further study. Through the tidal cycle, the release plume will swing upstream and downstream, and cover a central area twice or continuously. Effects on planktonic and nektonic species may be ephemeral, and the best record of effect is often made on the benthic community. The affected zones are oval, circular, or semi-circular, in contrast with the more linear fluvial effects noted by Gaufin and Tarzwell (1956), Hynes (1960), and Ingram and

Towne (1959). Estuarine patterns have been observed by Blegvad (1932), Fraser (1932), Filice (1954a, b), Reish (1959a, b), and McNulty (1961). Successive zones away from the outfall may include an area of sludge or soft muck with no macroinvertebrates; a poor zone which may have characteristic species present, sometimes dwarfed; a relatively rich zone with heavy populations of molluscs, worms, diatoms, and other species; and normal communities. Effects of local circulation, substrate, salinity, and other variables would produce almost infinite variations. In broader terms, McNulty saw a small damaged area surrounded by a much larger enriched area, and could distinguish indicator organisms and communities for both. Galtsoff (1956), however, reported that sludge from domestic sewers has almost completely smothered the formerly productive shellfish beds in the vicinity of large cities, including New York, Boston, and Norfolk. Hynes cited the reports of Pentelow (1955) that sewage causes severe deoxygenation in some estuaries like the Thames to promote a condition which blocks the upstream and downstream migration of sea trout and salmon so that these runs have become extinct. There is need for additional research on the effects of sewage—and much room for improvement in handling it.

Radioactive Wastes

Radioactive wastes are entering coastal waters in increasing quantities from such activities as research, munitions, industry, and medicine. The estuaries are likely to receive, and retain, greater concentrations than the oceans. These are of concern because they may cycle and recycle until they enter human food supplies in significant quantities, and they may also affect the genetic structure of aquatic organisms.

Chipman (1958, 1960) has cited the complexity of obtaining complete answers, since the fate of radionuclides is dependent on many physical, chemical, and biological processes. Pritchard (1960) stated that the principal factors involved will usually be the form of the contaminant, dilution, advective transport, turbulent diffusion, uptake on sediments, and extraction by the biota from solution or from the sediments. Chipman and his associates (Chipman et al., 1958) reported on a series of experiments designed to establish uptake rate of various probable radionuclides by common species and to track these materials through major food chains. Of all the suggested fates, Donaldson (1960) found biological activity to be the most important factor in the distribution and localization of radioactive products at Eniwetok. He found no evidence of biological effects of these products, probably because competition quickly eliminates injured individuals. The applicability of these generalizations to estuaries has not been tested. Extreme local situations may be dangerous and important, but apparently the total effect is not yet great.

Pulp Wastes

The wastes from plants that process wood into cellulose products by various processes are released into many estuarine areas. A great deal has been written which contains many conflicting observations and opinions. As an example, in 1947 Galtsoff et al. were satisfied that pulp-mill waste was the principal cause of decline of the productivity of oyster bars in the York River of Virginia. In a very different area, Waldichuk (1959) described a variety of releases in British Columbia, and noted that chemical and biological degradation may or may not be present, depending largely on the mechanics of release. The report offered by Gunter and McKee to the Washington Pollution Control Commission in 1960 appears to contain a rational summary: (1) sulfite waste is exceedingly complex, varying with species of wood, treatment, and digestion chemicals; (2) effects have been erratic and mixed, with investigational results including both stimulation and depression of the biota; (3) physical dispersion and dilution are important aids in disposal; (4) interim regulations should be established; and (5) research is essential to study the effects of the liquor and methods of rendering it non-destructive.

Exotic Chemical Effects

Wherever large quantities of chemicals are placed in estuaries, they cause change. Each case is local and specific. Some special instances are, however, potentially instructive and important.

Stroup et al. (1961) commented on the effects of adding large quantities of acid to the carbonate-buffered waters of Baltimore Harbor. They found that pH is decreased, and the partial pressure of CO_2 is increased. This increase in CO_2 tension may be significantly favorable to photosynthesis but unfavorable to the fauna.

Galtsoff (1960) emphasized the capacity of many organisms to concentrate elements and compounds from the environment. Bivalve molluscs, for example, absorb copper, mercury, lead, and arsenic near industrial areas. Oysters, clams, and scallops can contain concentrations of zinc over 100 thousand times that of surrounding waters (Chipman et al., 1958).

Oil, motor exhaust fumes, ship bilge, unusual industrial wastes, garbage dumping, and a limitless variety of special chemical additions exist, and require comprehension and control where they are important.

Pollution-control efforts are universal, but extremely variable. It is improbable, however, that understanding and effectively controlling chemical pollution will ever catch up in the race against new products and processes. In most cases, regulations are created after damage occurs rather than with intelligent foresight.

Geological Processes

Man has directly influenced geological processes, principally by changing silt production and distribution. Within the estuary, this is usually the result of shoreline con-struction (Sykes, 1965), dy[...]ter, 1957b), cutting of wate[...] canals, or certain specialized fishin[...] tions such as hydraulic dredging fo[...] shell clams (Manning, 1957). Manning d[...] scribed the Maryland gear, developed since 1950; it jets a trench in the bottom about 30 inches wide and up to 18 inches deep. Heavy materials drop rapidly, but clams and other coarse materials are conveyed by belt to the surface; the fine silts and clays are dispersed, usually to no more than 50 feet on either side of the cut. Most material returns to the trench.

Effects of silt handling are usually relatively local, although Hellier and Kornicker reported in 1962 that hydraulic canal dredging deposited silt as deep as 27 cm, and as far as 0.5 miles from the dredge. Many factors affect dispersion. Bartsch (1960) has identified the sources and results of silt in fresh water. Turbidity obviously interferes with light transmission and photosynthesis, and usually increases oxygen demand (Odum and Wilson, 1962). After heavier materials settle quickly, turbidities rarely exceed the natural levels caused by wind. Galtsoff (1960) included sedimentation as one of the negative factors in the environment of the oyster, since even the capacity of this remarkably silt-tolerant species can be overcome by smothering loads. Although severe special cases exist Gunter's comment seems to have general application—deleterious effects are real but localized, and nutrient release may offset the damage done.

Biological Processes

All the modifying activities considered in earlier sections affect biological processes. They operate through effects on photosynthetic production, nutrient cycling, changed food supply, altered activity patterns, direct maiming or killing, and many other pathways in the ecosystem. None of these effects of man are understood fully for estuarine areas. Our basic knowledge of effect on biological processes is limited and most of our

und the econom-
Two areas of hu-
c evidence that we
ncing these effects
sical and chemical
he transplanting of
ic communities are
review.

417

Human Predation (Fishing)

In the Chesapeake Bay and the immediate coastal waters, about 500 million pounds of aquatic organisms are stripped annually from the total estuary-dependent biota by human predation. Sykes (1964) reports 1,104,000,000 pounds in 1960 for the north portion of the Gulf of Mexico. This is a very specialized predation, which is highly selective by species and by size; it is seasonal in its effects, and does not return the captured nutrients directly into the system. Not all the effects on the biological processes are yet known, but many can be cited or suggested.

The species preyed upon by humans in the estuaries of the Middle Atlantic coast of North America can be grouped into three categories: wild resident species, wild transient species, and cultured species. Man's predation has different effects on each of these groups, but there are several general effects which apply to all.

The wild resident species, which spend virtually all their life history in the estuary, were, obviously, present in the estuary prior to man's intensive use. Consider the oyster bed communities, soft shell clams, and striped bass of the Chesapeake: early harvest was light, with inefficient gear; power boats, nylon nets, better dredges, navigational aids, increased experience, and more efficient predatory techniques have placed increased pressure on these and other residents. Four known and possible changes could be made. First, the structures of the community can be substantially altered: for example, all utilized oyster beds could be changed in species composition, size, distribution for oysters,

and physical structure. Secondly, the total abundance and distribution of the prey species can be substantially changed. (Maryland's oyster catch is about 10 per cent of earlier levels, despite gear improvement and high price. Many old beds are barren, without oysters or their usual associates.) The size and age composition of prey species can be modified, which appears to have occurred in striped bass, where human predation is heaviest on young fish. A fourth possibility is that growth rates, distribution, and spawning success might be enhanced. (Soft shell clams may be growing more rapidly and densely on worked beds than on undisturbed areas.)

The transient wild species like shad, herrings, etc., which move into or through the estuary to breed or to feed, are exposed to intensive inshore predation for only part of their lives, but it may be under highly vulnerable circumstances. A principle of fishery biology was expressed by an experienced fisherman when he told me that "Fish are fairly safe from overfishing unless we can get at them in a bottleneck that they must pass through. Then, we can really murder them." Although increased fishing efficiency also contributes to the decreases, the bottlenecks, like migration to limited spawning areas, offer the greatest opportunities for the depletion of stocks by overfishing and for a significant genetic selection. The sturgeon appears to be depleted, for example, but critical studies are lacking; it is rare in this area, and extirpation is imaginable. Significant genetic selection can be effected by size limits, net-mesh regulations, and controlled seasons if the entire stock is regularly exposed to selective predation. Fishways interposed on essential spawning migrations may select to favor strains which can pass ladders.

The cultured and semi-cultured species are particularly vulnerable to change. Complete culture, involving production of young or seed from selected parents and supervision until harvesting, apparently is not now possible for any estuarine species on a commercial scale. *Crassostrea virginica, Ostrea edulis, Mytilus edulis, Chanos chanos,* and *Penaeus setiferus* are all partially cultured, and these and others will probably be fully controlled in the future. Several important steps can be taken on behalf of these species, including provision of effective substrate for larvae of oysters and clams, transportation and concentration in favorable growing areas, partial or complete protection from other predators, feeding, and intensive and thorough harvesting. All of these are violent modifications of the natural biological processes of the estuary, and may result in improved growth rate, survival and condition of the cultured species. Oysters on planted cultch, for example, when transplanted to good growing areas where food is more plentiful and natural predators absent, show improved survival and growth. Genetic selection may also result, as in the case of oysters which may have been modified by culture and by very wide-scale transplanting along the entire Atlantic coast—the potentials in this field will be enormously enhanced when artificial breeding becomes practicable. Still another result may be increased parasitic and natural predator damage; planted oyster beds can be wiped out by cow-nose rays, oyster drills decimate seed beds, and the intensive populations in planted beds may be more severely parasitized by microparasites.

The capture and permanent removal of large quantities of any of these species is in itself a radical interruption of biological processes. Under undisturbed conditions, each animal would compete, die, be consumed and digested, pass into the nutrient sequence, and continue as recycled elements and compounds. The only net losses to the system would come from flushing and sedimentation. When we extract large chunks of organized organic materials, it seems likely that the total production of the prey species is probably increased because the removal of some organisms provides space and food for their competitors; a substantial quantity of organic material leaves the estuary, at least temporarily, although much may return as sewage and industrial waste; species compet-

ing with the prey species are favored; and a broad spectrum of modifications affects the unused species, including those that feed the species directly captured, its parasites, natural predators, and the rest of the ecosystem.

Far too little attention has been directed to research on these effects of human activity in the estuary. Many of the possibilities have not been tested. The speculative comments here are only indications of the actions and reactions that are modified, and suggest some challenging avenues for future investigation.

Artificial Introduction of Species

Each species attains a distribution which balances its needs and its environment. Distributions normally change slowly, accompanied by constant adjustments in that balance. Man, however, has often violently disrupted this leisurely pattern by transplanting species to new areas. This has sometimes been done intentionally in estuaries to increase or improve yields of food, and it has often been done inadvertently, carrying species along with transplants, on ship hulls, or by carelessness.

Great benefit is possible from introductions. The oyster fishery on the west coast of North America depends primarily on seed transplanted each year from Japan. Four hundred thirty-five striped bass (1879-81) and 15 thousand white shad fry (1871) were carried laboriously from the East Coast to the West Coast; both are widely established, with substantial benefits and no known damage to the receiving waters.

On the other side of the economic balance are the parasites, predators, and competitors of species which have food value. The Portugese oyster, *Gryphea (Crassostrea) angulata,* was introduced to the coast of France, where it gradually drove out some of the superior *Ostrea edulis* (Galtsoff, 1946). The voracious screw borer, *Urosalpinx cinerea,* one of the worst of the oyster drills, was accidentally carried into English waters with American oysters, and is a serious and extensive predator (Korringa, 1952). *Urosalpinx* has also been taken to the Puget Sound area, where it joined another immigrant, *Tritonalia japonica,* an oriental species which is considered to be the most destructive drill of the area (Korringa, 1952). The extensive transplantations of oyster stocks among waters of the Atlantic and Gulf Coasts have been suggested as suspected mechanisms for introducing to new areas the fungus, *Dermocystidium marinum,* the microparasite called MSX, and other parasites. Oyster competitors have been observed to create serious problems in new waters. A mudworm, *Polydora ciliata,* was introduced into Australia about 1870 (Nelson, 1946). It changed the industry, forcing oyster culture off the bottoms and onto stakes or stone slabs. A small slipper limpet, *Crepidula fornicata,* was taken to Europe from America. It grows to giant size there and has spread over many areas, especially in Holland and England (Korringa, 1952). It increased so vigorously that it actually threatened to replace the Dutch oyster. Korringa believes that it is a space competitor, harbors a serious shell disease as it decays, and destroys great numbers of oyster larvae.

The barnacle, *Elminius modestus,* accidentally brought from the Southern Hemisphere to England (Korringa, 1952), competes vigorously with oyster larvae for setting space and probably destroys larvae. As reported earlier, DDT applications appear to be useful in its control. The shell of an oyster provides a habitat for a remarkable variety of protozoans, snails, and the eggs and spores of other species. It is possible that the transplantation of oysters, oyster shells, and seed has modified the distribution of more aquatic species than any other human activity.

The drama and problems associated with artificial introductions in estuaries are not, however, limited to invertebrates. Like the starling and the English sparrow on land,

carp, goldfish (*Carassius auratus*), the walleye or yellow pike perch (*Stizostedion vitreum*), catfish, and others have entered the Middle Atlantic area and other waters (Mansueti, 1961). The effects are not yet measured, but the carp and goldfish are regarded with serious concern.

Water chestnut, *Trapa natans,* was imported into the United States as a handsome ornamental plant. Accidental release near Washington, D. C. produced, within ten years, beds covering 10,000 acres (Rawls, 1964). It blocked navigation, provided a breeding site for mosquitoes, and produced devilish "caltrops" or hard-spined seed cases. Expensive mowing, hand-picking and chemical treatment have reduced it to a controlled threat.

Eurasian watermilfoil, *Myriophyllum spicatum,* is widely distributed in Europe, Asia, and Africa, where it is a modest member of the flora. In Chesapeake Bay, however, it has recently become a serious menace to many interests, blocking navigation, preventing boating and swimming, interfering with seafood harvesting, increasing siltation, and encouraging mosquitoes. It thrives over a wide salinity range from 0 $^{\circ}/_{\circ\circ}$ to 15 $^{\circ}/_{\circ\circ}$, and can tolerate 20 $^{\circ}/_{\circ\circ}$, reproduces effectively by fragmentation, and survives in all depths less than about 9 feet. At least 100 thousand acres are infested, and new tributaries are invaded each year. Control has been effective on an expensive local basis, applying 2,4-D in clay pellets. This, in turn, opens serious questions about the dangers of the control method. Beaven *et al.* (1962) showed that standard applications have no acute effects on clams, crabs, oysters, or some fish, but the possibilities of chronic effects, residues, accumulation, and human intake are not yet resolved. Perhaps all this unfinished story began with the emptying of a fish bowl containing this attractive plant.

These varied instances of introduction without comprehension of natural controls and without effective restraint convey their own point. One need only add that each is likely to be an irreversible act, with permanent effects.

ACTIVITIES IN THE OCEAN

The sea is still beyond man's control, and the strenuous efforts he makes to change his terrestrial environment still seem puny in comparison with oceanic forces. He can, however, successfully block the sea from its tributary bays, with results that are important to the estuary.

Pritchard and many others have stressed the power and importance of tidal currents, which provide the energy for horizontal translation and for mixing throughout the estuary. He showed in 1955, for instance, that increased tidal velocity tends to convert a stratified estuary to vertical homogeneity. When ocean barriers are constructed or removed, the change will have far greater importance than merely reducing or encouraging the intrusion of salt and of marine species. Ryther *et al.* (1958), in considering the algal problem in Great South Bay and Moriches Bay, attributed improvement in the entire area to the reopening of Moriches Inlet to restore effective exchange with the sea.

Hundreds, perhaps thousands, of coastal inlets and harbors have been modified by engineering efforts to stabilize, improve, or protect them for various human endeavors. Groins, breakwaters, channel dredging, bulkheading, and filling all change the natural patterns and processes. Each inlet is a specific and local case, and the concept that engineering changes should always be preceded by thorough consideration of all the physical, chemical, and biological results to be effected should be encouraged. This has not been the usual sequence, and most inlet engineering appears to have been single-purpose modification. Improved comprehension of the total effects of altering relations between ocean and estuary is desirable and will be increasingly valuable.

Important special problems appear when nearshore ocean waters are used for waste disposal. Wastes can be translocated into the estuaries by such mechanisms as the regular migrations of anadromous fish and the great inflow of oceanic water with tidal currents. Ketchum, in an unpublished paper, reviewed pertinent data for one area, the mouth of Delaware Bay. He concluded that discharge would be undesirable at any site within five miles of the coast, but showed the great variation in waste-receiving capacity at different locations. Further, he stressed the necessity for specific on-site studies prior to selection of waste sites. Pritchard (1960) effectively outlined the steps necessary in evaluating sites for disposing of radioactive wastes. He also stressed that this inshore environment comes into more intimate contact with man than any other marine waters. Therefore, it is most likely to receive wastes and, simultaneously, most in need of protection from excess wastes. He discussed many of the factors which influence the fate of these wastes. As understanding of estuarine processes grows, the probabilities of irreparable and massive damage from ignorance in waste disposal should decline.

ON THE RESILIENCY OF ESTUARIES

The reading necessary for the preparation of this review has deepened and clarified my personal concern with the future welfare of estuaries. It is clear that many destructive forces are being applied widely, that pressures are increasing at a very rapid rate, but that intelligent planning and control of estuarine changes are rare. Fundamental and practical understanding of estuaries is now increasing at an impressive rate, but this growth appears to be dangerously slow in comparison with the increase in disruption of estuarine systems.

Several of the characteristic physical, chemical, and biological features of the estuary provide an interesting and valuable resistance to change. Scars often heal quickly. The factors aiding resiliency have not yet been adequately investigated, but several illustrative and stimulating examples are known:

1. The vigor of the rhythmic and turbulent circulation pattern continuously and endogenously renews the supply of water, food, larvae, and other essential elements to any small damaged area. This aids in recovery and protects long-term net stability patterns.

2. The substantial buffering capacity of estuaries, usually operating through the carbonate system, is another element which resists changes imposed on estuaries. It is not so great as the buffering capacity of the open ocean, but it is greater than most rivers, and is enormously important in the estuaries where pollution is received.

3. Exogenous renewal is also normally continuous, because estuaries receive continuous input from rivers and from the ocean. Since the river-sourced and ocean-sourced populations are substantial components of the estuarine biomass, the addition of organisms from these sources is important in normal estuarine sequences and in recovery from damaging or toxic change.

4. Many species have biological characteristics which provide special advantages in estuarine survival. These characteristics usually protect the species against the natural violence of estuaries, and they are often helpful in resisting external forces, like man.

Additional examples of contributors to estuarine resiliency include the oyster, the blue crab and the striped bass. The oyster, for instance, has been cited by Galtsoff (1960) for great tolerance to temperature (0°C to about 35°C), salinity (about 5 $\,^\circ/_{oo}$ to 35 $\,^\circ/_{oo}$), and for its remarkable ability to hermetically seal the valves of its shell to isolate the animal from unfavorable conditions for as long as three weeks. Nelson (1938) described the oyster's complex and effective mechanisms for dealing with the high silt content of coastal waters. The blue

crab uses the net upstream flow of deep waters in estuaries to provide annual redistribution of juveniles to all the tributaries and upsteam areas of the Chesapeake Bay. Mansueti (1961) stressed the remarkable resiliency of the striped bass in surviving the increased pressures and damages in the Chesapeake system. He suggested that the specific gravity of the eggs of his species may be crucially important to survival, since the semi-buoyant eggs released in nearly fresh water are buoyed by turbulence as the embryos develop, and protected from silt smothering. When the larvae emerge, they have been carried downstream by net surface flow to waters of higher salinity, past the zone of maximum turbidity, so that they can feed. A last example can be drawn from the copepod populations of estuaries. The mechanisms which permit maintenance of large plantonic populations in these turbulent and flushed systems are not yet fully understood but appear to be complex and effective. They may include vertical migrations (which could move populations downstream at night and upstream by day), reservoirs in marginal areas, and other attributes or patterns of individual or population behavior (Rogers, 1940; Bousefield, 1955; Barlow, 1955; Cronin et al., 1962).

These resilient forces, and the others which certainly exist, are welcome allies in the efforts to achieve optimal balance between man's effects on estuaries and their capacities.

THE FUTURE ROLE OF MAN

Man's past effects on estuaries have been poorly and incompletely planned, unimaginative, and frequently destructive. In view of the many important uses served by these waters, and the size of the growing pressures on them, it is imperative that a new major human force be utilized in the future—the force of intelligent management. It will require application of many kinds of tools and techniques, ranging from original fundamental oceanographic research to regulatory changes and public education. The yields can be great indeed. Some of the approaches necessary for intelligent management are now apparent, and it is perhaps timely to summarize them and the achievement which could be gained.

The Tools for Management

Research

This review has illuminated the need for basic research at every possible level to identify the significant estuarine processes, quantify relationships, and resolve many complexities to predictable patterns. Despite the visible and valuable increase in attention to the estuaries in recent years, most of the controlling parameters are yet incompletely and inadequately comprehended.

Among the physical processes, only gross estimates can be provided for the patterns of flushing, circulation, vertical mixing, and diffusion, and for such specific factors as light absorption, sedimentation rates, and location, current direction and velocity, and salt distribution. The valuable contributions of Ketchum, Pritchard, and others permit some of these estimates to be made under certain highly specific or broadly generalized conditions, but each new estuary observed is still likely to require an extensive descriptive survey, trial-and-error fitting of computed relationships, and a relatively long period of correction by well-designed local observation before useful precise predictions can be made. Field studies of interesting phenomena, increased use and refinement of physical and mathematical models, and controlled laboratory studies are all urgently needed.

Similar urgency exists for growth in chemical and geological comprehensions. The fate of the most abundant natural terrigenous elements and compounds is partly known but the chemical sequences involving rarer materials are almost untouched. Others have reviewed, with greater detail and higher competence, the present state of chemical knowledge of estuaries, but it is relevant here to point out that such knowledge appears to be completely insufficient for understanding and intelligent judgment of many present and probable human activities with chemical implications. These include the introduction of vast quantities of complex industrial chemical mixtures, the release of stored materials from bottom sediments by dredging, the pouring of sewage chemicals into the system, and other violent modifications. As usual, practical resolution of the problems involved will be almost totally dependent on fundamental comprehension.

The insufficiency of present biological knowledge in estuaries is equally evident. Indeed, it is rather shocking, since the areas have been so heavily utilized for food production for centuries. The oyster, for example, is considered to be the most studied of invertebrate animals, and yet there are important unanswered questions on nutrition, genetics, general physiology, reproduction, and even the morphology of the group. Similar and greater ignorance exists about all other estuarine species. Principal attention has always been focused on the relatively small numbers of economically useful species, and there is urgent need for more knowledge of phytoplankton, zooplankton, a wide variety of benthic organisms, and the large array of vertebrates and invertebrates which are part of the dynamic estuarine biota, but which are not consumed, sold, or otherwise directly exploited by humans. Interrelationships between and among species must be illuminated, the interplay between organisms and the estuarine environment is but vaguely known, and full understanding of estuarine biology lies many years of research ahead.

Only passing and superficial attention can be given here to other research needs which appear to exist in the field of economics and social and political science. Intelligent management of estuaries will require a valid estimation of the economic results of various alternate routes of action, and will require modification of existing patterns of human activity. Perhaps research is not the necessary addition which will make this feasible, but there do appear to be major unanswered problems. Evaluation of recreational use has been notoriously difficult, especially in the non-consumptive fields like beauty, cleanliness, and personal refreshment. In social and political areas, the total effects of modifying the strongly independent behavior of fishermen, for instance, appear to be complex and difficult to comprehend.

Public agencies carry almost all the cost of estuarine research, since the public interests are so large and diverse. This is totally appropriate for basic research on the processes and principles of estuarine phenomena, and equally fitting for uses which directly serve public fisheries, waste disposal, recreation, and similar interests. There is, however, another potential and proper source for the substantial supporting funds required. When use of the estuaries for financial profit is desired, the cost of research on the effect of use should often be placed where the profit will be realized. An industry which proposes to save the cost of alternative methods of waste disposal by placing waste into an estuary (or other public holding) should pay the costs of research necessary for evaluation of the proposal. Under many circumstances, these funds might be supplemented by public funds, since the findings will be useful in other problems of interest to the public. Many examples exist of such research support, without control, by the using industries, but the policy of increased assignment of research costs to the profit maker appears to be equitable and to be necessary as a source of needed funds.

Pragmatic Experiments

Direct practical experiments in management, in which gross and relatively uncomprehending efforts are made to achieve identified objectives, appear to offer substantial and valuable assistance toward the intelligent management of estuaries. These are usually far less desirable and effective than efforts which can be properly based upon adequate understanding of the prin-

ciples involved, but several special circumstances argue for serious consideration and use of some pragmatic studies.

The pressures on estuaries are so great now and the threats are so urgent that there is not always time to develop comprehension. The rate of increase in use is so much greater than the rate of increase in knowledge that shortcuts are sometimes fully justified. The alternative appears to be destruction of many areas while we await the results of lengthy and complex research.

When a substantial change in an estuary cannot be deferred (usually because the opposing arguments are excessively weak in data and evidence), it can often be turned to constructive use in terms of improved understanding. Every possible effort should be made to capitalize on these forced but usable opportunities. Suggestions for research expenditures often face resistance, especially on a project regarded as a *fait accompli*, but adequate and competent study before and after a change can be valuable. No normal research budget can afford the cost of enormous diversion, or pollutant release, or engineering change, or thermal addition, or wide-scale pesticide application, which political decisions may require. Thorough pre- and post-treatment study, based on statistically sound sampling techniques, and paired when possible with another untreated estuary, can be of great value. It is essential that these studies meet several criteria, however. The pre-study and post-study periods must be long enough to distinguish the effects of change from normal (and often great) variations. Skilled professional guidance must be provided and not entrusted to inexperienced staffs. Interpretations must be based on objective criteria, free from adverse political pressures.

Under such circumstances, valuable knowledge can often be obtained.

Administration

The operating control of the modification of estuaries is dispersed among state and federal legislative bodies, executive agencies, and the public will. The mechanisms and experience of the groups cannot be discussed here, but recent vivid examples are interesting and may be stimulating.

In Massachusetts, recent state legislation requires review and prior approval of all proposed modifications of marshland, shores, and coastal lands. This law overrides the privileges of private ownership, in recognition of the effects of such modifications on the public interest. Other states, towns, and small governmental bodies are essaying similar careful management.

Maryland established a committee of its Board of Natural Resources to consider the problem of disposing of the large quantities of sediment removed in channel creation, improvement, and maintenance. The Board has approved and effectuated a change in policy from simple dumping in deep water to creation of useful diked land whenever feasible. Sometimes expensive and destructive waste can be converted into constructive, and even profitable, real estate.

At the federal level, mention has already been made of the legal requirement that the Branch of River Basin Studies of the U.S. Department of the Interior must review and comment on proposed federal projects which would affect fish and wildlife resources. This includes estuaries, but review and comment is not a substitute for proper review and control—which would include study of all the effects of such projects.

Many other governments have initiated similar efforts to control and to manage rationally the violent changes which are occuring. Local conditions are as varied as the estuaries themselves, but examples of good management do exist, and they merit consideration and emulation.

The Yield

The benefits which may be derived from proper management of coastal areas are great enough to merit the costs. The gains appear to include:

1. Substantial growth in basic knowledge of many significant physical, chemical, geological, and biological processes as the result of the major research efforts essential for such management. Paralleling this would be an important increase in the training of competent scientists in these fields.

2. Vast improvement in the ability to predict the effects of important proposed engineering or management changes. Such predictions could be made in terms of organic production, dollar value, and other criteria.

3. Development of rational and balanced objectives, based upon the real capacities and limitations of various estuarine systems. The possible alternatives would be comprehended and choices would be made for good reasons. As an interesting example of conflict, Rounsefell (1963) wishes to make maximum use of the high potentials of estuaries as nursery areas for commercially useful fish, and takes rather violent exception to suggestions by H. T. Odum and Wilson (1962) of techniques which would maximize photosynthetic production in estuaries. In this case and others, there may be genuine conflicts of interest. However, it seems more pertinent to emphasize that we do not now understand enough about the ultimate results of photosynthetic increase or about how to single out and achieve the optimal balance in use of the factors which determine fish density.

4. The greatest benefits to be gained lie in achieving positive and ultimate uses of estuaries as replacement for negative resistance to change, temporizing half-measures, and short-term patchwork in management. Balance by informed decision could be reached among uses of estuaries for food production, waste disposal, recreation, navigation, esthetic enjoyment, and research. Without any assumption of unerring wisdom, there is hope for remarkable improvement in long-term use of these areas.

On Positive Thinking

Much of this review has described the destructive, or at least uncomprehended, estuarine changes through man's efforts. However, the cited literature also contains suggestions for the positive and profitable manipulation of rivers, bays, marshes, and other coastal areas. Some of these proposals are well supported by experiments or field evidence, and others will require rigorous investigation prior to acceptance and improvement. The following suggestions do not by any means exhaust these ideas, but they indicate the directions and vitality of some of the recognized possibilities.

Chemical additions could protect or enrich estuaries, if they were used intelligently to offset undesirable conditions or to supplement limiting elements. Hynes (1960) has pointed out that the most outstanding problem in disposing of sewage is the appalling waste of nutrients. Brehmer (1964) noted that the District of Columbia area spends $2.1 million annually in the unmanaged release of $3.2 million worth of fertilizing materials. Constructive techniques are not fully developed, but Føyn (1959), for instance, has learned to remove phosphorous from wastes entering the Oslo Fjord by electrolytic precipitation. E. P. Odum (1961a) pointed out six ways by which phytoplankton blooms might be controlled. No problem here is likely to be beyond the capacity of industrial ingenuity, if economic incentives become sufficiently high. An interesting example may be provided in the case of the detergents. Widespread public reaction against visible suds and reported detergent residues in natural waters produced serious possibilities of strict prohibitive legislation and regulations. Pertinent industries have made massive investment in the discovery and development of "biodegradable" and other degradable cleansing compounds, and now promise that the problem will be effectively solved. If the competitive interest of such industrial giants can be focused on other estuarine problems, magnificent achievements may result.

Thermal additions might be constructively used. Spawning of all species and photosynthetic production are controlled by temperature, and offer tantalizing opportunities

for management. Huntsman (1950) suggested the use of warmed water in controlling the movement of fish to concentrate populations, or for other purposes. The present increase in attention to research on many thermal effects should suggest some desirable possibilities. Imaginative engineering and improved biological understanding might be fruitful partners in achieving useful hot spots.

Improved races and species can be selected and introduced. Galtsoff (1956) has pointed out some of the variation that exists among races of oysters and the possibility of selection for desirable characteristics under intensive cultivation. Provenance research has been so widely successful in forestry and agriculture that its potentials are beyond question. The successful introduction of new species presents greater difficulties and requires very thorough preparation and, perhaps, a share of good luck. Korringa (1952) has, however, pointed out that success might sometimes be encouraged and accelerated by simultaneous importation of natural control mechanisms. The literature contains many records of introduction, with both successes and failures, available to guide future efforts. A peculiar blending of conservative concern and imaginative daring may be necessary in improving upon present races and species.

Management—the intentional modification of the factors determining production by a species—can be carried to higher levels than have yet been achieved in estuaries. Maximum culture of oysters, clams, and other usable herbivores may offer the greatest potentials, since they are nutritionally supported near the broad base of the food chains. For instance, Glude (1951) outlined the sequence of seed production, predator control, mortality prevention, fattening, and controlled harvest which might increase oyster production substantially. Similar concepts of population comprehension and management should eventually be possible, not only for the species of fish and shellfish directly utilized, but also for the supporting zooplankton species, phytoplankton, and all the necessary parts of the complex biota of estuaries.

The gross ecology of the estuary can be manipulated to advantage. Present and probable engineering capacities are so great as to require new thinking about potential use of these capacities as well as the more usual effort to resist their application. It may, in some circumstances, be found desirable to store and release river water; divert huge volumes; radically alter channels, currents, and tides; or in other ways introduce major alterations. Those who oppose such suggestions are often expressing fear of the unpredictable consequences, and might alter their position if sufficient knowledge existed to permit accurate prediction and evaluation of all the results.

RECOMMENDATIONS

1. Every level of estuarine research should be supported and speeded to the maximum rate consistent with competence. Basic research in this challenging, complex, and available marine environment is critically needed. Applied research and pragmatic "experiments" must also be fully utilized to prevent the uses and abuses from destroying these areas. The development of research programs should be aggressive and opportunistic.

2. The full force of intelligent management should be brought to bear in man's role in estuarine processes. Research, public education, and wise management are vital to the optimal future use of estuaries, with appropriate and balanced policies and practices applied in each estuarine system.

3. Selected estuaries, as "typical" as possible, should be set aside in the public interest to provide undisturbed research centers. These must be protected from the creeping exploitation that has eventually destroyed the character and value of many estuaries, and reserved for esthetic enjoyment, for comparison with utilized estuaries, and for the many kinds of research which would yield increased comprehension without damage to the system studied.

Author's Note: This paper is Contribution No. 269, Natural Resources Institute of the University of Maryland.

Valuable assistance was received from the following individuals who sought out and sent me original information on estuaries, or material which is difficult to obtain: James E. Sykes and Seton H. Thompson of the U.S. Fish and Wildlife Service, Clair Idyll of the University of Miami, Robert Ingle of the Florida State Board of Conservation, Gordon Gunter of the Gulf Coast Research Laboratory, P. Korringa of the Netherlands Institute for Fisheries Research, and J. E. G. Raymont of the University of Southampton. Their deep interest in the problems discussed here is especially welcome. Mrs. A. J. Mansueti and Arie de Kok of the Chesapeake Biological Laboratory provided illustrations and translations which have been valuable. Mrs. Leone H. Williams performed the essential typing chores with competence, and she and others provided constructive editorial review.

LITERATURE CITED

Alexander, W. B., B. A. Southgate, and R. Bassindale, 1936. Survey of the River Tees. II. The estuary—chemical and biological. *J. Marine Biol. Assoc U. K.* 717-724.

Auld, D. V., 1964. Waste disposal and water supply. In *Problems of the Potomac Estuary;* pp. 13-18. Proc. Interstate Comm. Potomac River Basin, Washington, D.C.

Barlow, J. P., 1955. Physical and biological processes determining the distribution of zooplankton in a tidal estuary. *Biol. Bull.* **109**(2), 211-225.

Bartsch, A. F., 1960. Settleable solids, turbidity, and light penetration as factors affecting water quality. In *Biological problems in water pollution. Trans. 1959 Seminar, Robert A. Taft Sanitary Eng. Center Tech. Rept. W60-3,* 118-127.

Bartsch, A. F., 1961. Introduced eutrophication, a growing water resource problem. In *Algae and Metropolitan Wastes. Trans. 1960 Seminar, Robert A. Taft Sanitary Eng. Center Tech. Rept., W61-3* 6-9.

Baughman, J. L., 1948. *An annotated bibliography of oysters with pertinent material on mussels and other shellfish and an appendix on pollution.* Texas A. and M. Res. Found., Agr. and Mechanical College of Texas, College Station, Texas.

Beaven, G. F., 1946. Effect of Susquehanna River stream flow on Chesapeake Bay salinities, and history of past oyster mortalities on upper Bay bars. *Third Ann. Rept., Maryland Bd. Nat. Resources*, 123-133.

Beaven, G. F., C. K, Rawls, and G. E. Beckett, 1962. Field observations upon estuarine animals exposed to 2,4-D. *Proc. N. E. Weed Contr. Conf.* **16**, 449-458.

Blegvad, H., 1932. Investigations of the bottom fauna at outfalls of drains in the Sound. *Rept. Danish Biol. Sta.* **37**, 5-20.

Bolomey, J. G. W., 1959. Effect of the Rhine on Netherlands beaches. *Proc. First Intern. Conf. on Waste Disposal in the Marine Environment*, 164-174.

Bousefield, E. L., 1955. Ecological control of the occurrence of barnacles in the Miramichi estuary. *Bull. Natl. Museum Can.* **137**, 1-69.

Brehmer, M. L., 1964. Nutrient enrichment in the Potomac estuary. In *Problems of the Potomac Estuary;* pp. 47-50. Proc. Interstate Comm., Potomac River Basin, Washington, D.C.

Burt, W. V., 1955. Distribution of suspended materials in Chesapeake Bay. *J. Marine Res.* **14**(1), 47-62.

Butler, P. A., and P. F. Springer, 1963. Pesticides—A new factor in coastal environments. *Trans. 28th N. Am. Wildlife and Natl. Resources Conf., 378-390.*

Butler, P. A., A. J. Wilson, Jr., and A. J. Rick, 1962. Effect of pesticides on oysters. *Proc. Natl. Shellfisheries Assoc.* **51**, 23-32.

Carpenter, J. H., 1957. A study of some major cations in natural waters. *Chesapeake Bay Inst. Tech. Rept.,* **15**, 1-75.

Carpenter, J. H., and D. G. Cargo, 1957. Oxygen requirement and mortality of the blue crab in the Chesapeake Bay. *Chesapeake Bay Inst. Tech. Rept.* **13**, 1-22.

Carson, R., 1962. *Silent Spring.* Houghton Mifflin Co., Boston.

Chipman, W. A., 1958. Biological accumulation of radioactive materials. *Proc. First Ann. Texas Conf. Utilization Atomic Energy, Misc. Publ. Texas Eng. Exp. Sta.,* 36-41.

Chipman, W. A., 1960. Accumulation of radioactive pollutants by marine organisms and its relation to fisheries. In *Biological Problems in Water Pollution. Trans. 1959 Seminar, Robert A. Taft Sanitary Eng. Center Tech. Rept., W60-3,* 8-14.

Chipman, W. A., T. R. Rice, and T. J. Price, 1958. Uptake and accumulation of radioactive zinc by marine plankton, fish, and shellfish. *U.S. Fish and Wildlife Serv., Fishery Bull.* **135**, 279-292.

Cottam, C., 1960. A conservationist's view of the new insecticides. In *Biological problems in Water Pollution. Trans. 1959 Seminar, Robert A. Taft Sanitary Eng. Center W60-3,* 42-45.

Cronin, L. E., J. C. Daiber, and E. M. Hulburt, 1962. Quantitative seasonal aspects of zooplankton in the Delaware estuary. *Chesapeake Sci.* **3**(2), 63-93.

Davis, C. C., 1948. Studies of the effects of industrial pollution in the lower Patapsco River area. II. The effects of copperas pollution in plankton. *Maryland Dept. Res. and Education, Chesapeake Biol. Lab. Publ.* **72**, 1-12.

Davis, H. C., 1958. Survival and growth of clam and oyster larvae at different salinities. *Biol. Bull.* **114**(3), 296-307.

Davis, H. C., 1961. Effects of some pesticides on eggs and larvae of oysters, *Crassostrea virginica,* and clams, *Venus mercenaria. Com. Fisheries Rev.* **23** (12), 8-23.

Donaldson, L., 1960. Radiobiological studies at the Eniwetok test site and adjacent areas of the western Pacific. In *Biological problems in water pollution. Trans. 1959 Seminar, Robert A. Taft Sanitary Eng. Center Tech. Rept., W60-3:* 1-7.

Emery, K. O., and R. E. Stevenson, 1957. Estuaries and lagoons. I. Physical and chemical characteristics. *Geol. Soc. Am., Mem. 67* **1**, 673-749.

Filice, F. P., 1954a. An ecological survey of the Castro Creek area in San Pablo Bay. *Wasman J. Biol.* **12**(1), 1-24.

Filice, F. P., 1954b. A study of some factors affecting the bottom fauna of a portion of the San Francisco Bay estuary. *Wasman J. Biol.* **12**(3), 257-292.

Føyn, E., 1959. Chemical and biological aspects of sewage discharge in inner Oslofjord. *Proc. First Intern. Conf. Waste Disposal in the Marine Environment:* 279-284.

Fraser, J. H., 1932. Observations on the fauna and constituents of an estuarine mud in a polluted area. *J. Marine Biol. Assoc.* **18**(1), 69-84.

Galtsoff, P. S., 1946. Comment on the importation of foreign shellfish. *Com. Rept., Natl. Shellfisheries Assoc.*

Galtsoff, P. S., 1956. Ecological changes affecting the productivity of oyster grounds. *Trans. 21st N. Am. Wildlife Conf.* 408-419.

Galtsoff, P. S., 1960. Environmental requirements of oysters in relation to pollution. In *Biological Problems in Water Pollution. Trans. 1959 Seminar, Robert A. Taft Sanitary Eng. Center Tech. Rept., W60-3,* 128-133.

Galtsoff, P. S., W. A. Chipman, Jr., J. B. Engle, and H. N. Calderwood, 1947. Ecological and physiological studies of the effect of sulfate pulp mill wastes on oysters in the York River, Virginia. *U. S. Fish and Wildlife Serv., Fishery Bull.* **51**, 58-186.

Gaufin, A. R., and C. M. Tarzwell, 1956. Aquatic macroinvertebrate communities as indicators of organic pollution in Lytle Creek. *Sewage Ind. Wastes* **28**, 906-924.

Gaussle, D., and D. W. Kelley, 1963. The effect of flow reversal on salmon. *Ann. Rept. Delta Fish and Wildlife Protection Study. Append. A, A1-A16.*

Glude, J. B., 1951. The effect of man on shellfish populations. *Trans. 16th N. Am. Wildlife Conf.,* 397-403.

Guillard, R. R. L., R. F. Vaccaro, N. Corwin, and S. A. M. Conover, 1960. Report on a survey of the chemistry, biology, and hydrography of Great South Bay and Moriches Bay, conducted during July and September 1959, for the townships of Islip and Brookhaven, New York. *Woods Hole Oceanog. Inst. Ref.* 60-15.

Gunter, G., 1953. Historical changes in the Mississippi River and the adjacent marine environment. *Univ. Texas, Publ. Inst. Marine Sci.* **2**(2), 119-139.

Gunter, G., 1953. The relationship of the Bonnet Carré Spillway to oyster beds in Mississippi Sound and the "Louisiana Marsh" with a report on the 1950 opening. *Univ. Texas, Publ. Inst. Marine Sci.* **3**(1), 17-71.

Gunter, G., 1955. Mortality of oysters and abundance of certain associates as related to salinity. *Ecology* **36**, 601-605.

Gunter, G., 1956. Land, water, wildlife and flood control in the Mississippi Valley. *Proc. Louisiana Acad. Sci.* **19**, 5-11.

Gunter, G., 1957a. Wildlife and flood control in the Mississippi Valley. *Trans. 22nd N. Am. Wildlife Conf.* 189-196.

Gunter, G., 1957b. How does siltation affect fish production? *Natl. Fisherman* **38**(3), 18-19.

Gunter, G., and G. E. Hall, 1963. Biological investigations of the St. Lucie estuary (Florida) in connection with Lake Okeechobee discharges through the St. Lucie Canal. *Gulf Res. Repts.* **1**(5), 189-207.

Gunter, G., and J. E. McKee, 1960. A report on oysters, *Ostrea lurida* and *Crassostrea gigas,* and sulfite waste liquors. *Spec. Consultants Rept. to Pollution Control Comm., State of Washington,* February 1960. Olympia, Washington.

Hanks, R. W., 1963. Chemical control of the green crab, *Carcinus maenas* (L.). *Proc. Natl. Shellfisheries Assoc.* **52**, 75-86.

Hartog, C. Den., 1963. The amphipods of the deltaic region of the Rivers Rhine, Meuse and Scheldt in relation to the hydrography of the area. I. Introduction and hydrography. *Neth. J. of Sea Res.* **2**(1), 29-39.

Havinga, B., 1935. Het Zuiderzeegebied in zijn huidige phase van ontwikkeling. *Vakblad voor Biologen* **17** (4), 64-73.

Havinga, B., 1936. De veranderingen in den hydrographischen toestand en in de macrofauna van Zuiderzee en IJselmeer gedurende de jaren 1931-1935. *Mededeelingen van de Zuiderzee-Commissie* **4**, 1-26.

Havinga, B., 1941. De veranderingen in den hydrographischen toestand en in de macrofauna van het IJsselmeer gedurende de jaren 1936-1940. *Mededeelingen van de Zuiderzee-Commissie* **5**, 1-18.

Havinga, B., 1949. The enclosing of the Zuyder Zee and its effect on fisheries. *U. N. Sci. Conf. Conserv. and Util. of Resources. (Wildlife 1(a)/5; Water 7(c)/3).*

Havinga, B., 1959. Artificial transformation of salt and brackish water into fresh water lakes in the Netherlands and possibilities for biological investigations. *Arch. Oceanog. Limnol.* **10**, 47-52. (Suppl.)

Hedgpeth, J. W., 1957. Estuaries and lagoons. II. Biological aspects. In *Treatise on marine ecology and paleoecology;* vol. 1, Ecology. *Geol. Soc. Am. Mem.* **67**, 693-729.

Hellier, T. R., Jr., and L. S. Kornicker, 1962. Sedimentation from a hydraulic dredge in a bay. *Univ. Texas, Publ. Inst. Marine Sci.* **8**, 212-215.

Huntsman, A. G., 1950. Population dynamics in estuaries. In *A Symposium on Estuarine Ecology.* Atl. Estuarine Res. Soc. Yorktown, Virginia.

Hutton, R. F., B. Eldred, K. D. Woodburn, and R. M. Ingle, 1956. The ecology of Boca Ciega Bay with special reference to dredging and filling operations. *Florida Marine Lab., Tech. Ser.* **17**, 1-86.

Hynes, H. B. N., 1960. *The biology of polluted waters.* Liverpool Univ. Press, Liverpool.

Ingram, W. M., and W. W. Towne, 1959. Stream life below industrial outfalls. *Public Health Repts.,* **74**(12), 1059-1070.

Jeffries, H. P., 1962. Environmental characteristics of Raritan Bay, a polluted estuary. *Limnol. Oceanog.* 7, 21-31.

Jones, R. L., D. W. Kelley, and L. W. Owen, 1963. Delta fish and wildlife protection study. *Ann. Rept. No. 2.* Resources Agr. Calif., Sacramento, California.

Ketchum, B. H., 1950. The exchanges of fresh and salt waters in tidal estuaries. *Proc. Colloq. Flushing of Estuaries, Office of Naval Res.* 1-23.

Ketchum, B. H., 1951a. The flushing of tidal estuaries. *Sewage Ind. Wastes* 23(2), 198-209.

Ketchum, B. H., 1951b. The dispersion and fate of pollution discharged into tidal waters, and the viability of enteric bacteria in the sea. Progress Rept. Res., Grant No. 1249(C2), Period Dec. 1, 1949 to Feb. 1, 1951. *Woods Hole Oceanog. Inst. Ref. No. 5-11.*

Ketchum, B. H., 1951c. The exchanges of fresh and salt waters in tidal estuaries. *J. Marine Res.* **10**, 18-38.

Ketchum, B. H., 1954. Relation between circulation and planktonic populations in estuaries. *Ecology* **35**, 191-200.

Ketchum, B. H., J. C. Ayers, and R. F. Vaccaro, 1952. Processes contributing to the decrease of coliform bacteria in a tidal estuary. *Ecology* 33(2), 247-258.

Kinne, O., 1963. The effects of temperature and salinity on marine and brackish water animals. I. Temperature. *Oceanog. Marine Bio. Ann. Rev.* **1**, 301-340.

Koch, P., 1959. Discharge of waste into the sea in European coastal areas. *Proc. First Intern. Conf. Waste Disposal in the Marine Environment* 122-130.

Korringa, P., 1952. Recent advances in oyster biology. *Quart. Rev. Biol.* **27**, 266-303, 339-365.

Korringa, P., 1958. The Netherlands shellfish industry and the Delta plan. *Intern. Council Explor. Sea, C. M., Shellfish Comm. No. 60.*

Krause, R., 1961. Fundamental characteristics of algal physiology. In *Algae and Metropolitan Wastes. Trans. 1960 Seminar, Robert A. Taft Sanitary Eng. Center Tech. Rep., W61-3, 40-47.*

Lindsay, C. E., 1963. Pesticide tests in the marine environment in the State of Washington. *Proc. Natl. Shellfisheries Assoc.* **52**, 87-97.

Loosanoff, V. L., 1960. Some effects of pesticides on marine arthropods and mollusks. In *Biological Problems in Water Pollution. 1959 Seminar, Robert A. Taft Sanitary Eng. Center Tech. Rept., W60-3,* 89-93.

Manning, J. H., 1957. The Maryland soft shell clam industry and its effects on tidewater resources. *Maryland Dept. Res. and Education Resource Study Rept., No. 11.*

Mansueti, R. J., 1961. Effects of civilization on striped bass and other estuarine biota in Chesapeake Bay and tributaries. *Proc. Gulf Caribbean Fisheries Inst., November 1961,* 110-136.

McNulty, J. K., 1961. Ecological effects of sewage pollution in Biscayne Bay, Florida: Sediments and the distribution of benthic and fouling macroorganisms.

Bull. Marine Sci., Gulf Caribbean **11**(3), 393-447.

Nelson, T. C., 1931. Expert testimony in *New Jersey vs. New York et al.;* Master hearings, U. S. Supreme Court, 283 U.S. 336, No. 16, original. Argued April 13-15, 1931.

Nelson, T. C., 1938. The feeding mechanism of the oyster. I. On the pallium and the branchial chambers of *Ostrea virginica, O. edulis,* and *O. angulata,* with comparisons with other species of the genus. *J. Morph.* **63**, 1-61.

Nelson, T. C., 1946. Comment on the importation of foreign shellfish. *Com. Rept. Natl. Shellfisheries Assoc.*

Nelson, T. C., 1947. Some contributions from the land in determining conditions of life in the sea. *Ecol. Monographs* **17**, 337-346.

Nelson, T. C., 1960. Some aspects of pollution, parasitism and inlet restriction in three New Jersey estuaries. In *Biological Problems in Water Pollution. Trans. 1959 Seminar, Robert A. Taft Sanitary Eng. Center Tech. Rept.,* W60-3, 203-211.

Odum, E. P., 1961a. Factors which regulate primary productivity and heterotrophic utilization in the ecosystem. In *Algae and metropolitan wastes. Trans. 1960 Seminar, Robert A. Taft Sanitary Eng. Center Tech. Rept.,* W61-3, 65-71.

Odum, E. P., 1961b. The role of tidal marshes in estuarine production. *The Conservationist* **15**(6), 12-15.

Odum, H. T., 1959. Analysis of diurnal oxygen curves for the assay of reaeration rates and metabolism in polluted marine bays. *Proc. First Intern. Conf. Waste Disposal in the Marine Environment,* 547-555.

Odum, H. T., and C. M. Hoskin, 1958. Comparative studies on the metabolism of marine waters. *Univ. Texas, Publ. Inst. Marine Sci.* **5**, 16-46.

Odum, H. T., and R. F. Wilson, 1962. Further studies on reaeration and metabolism of Texas bays. *Univ. Texas, Publ. Inst. Marine Sci.* **8**, 23-55.

Olson, R. A., H. F. Brust, and W. L. Tressler, 1941. Studies of the effects of industrial pollution in the Lower Patapsco area. I. Curtis Bay region, 1941. *Chesapeake Biol. Lab. Publ.* **43**, 1-40.

Pannell, J. P. M., A. E. Johnson, and J. E. G. Raymont, 1962. An investigation into the effects of warmed water from Marchwood power station into Southampton water. *Proc. Inst. Civil Engrs.* **23**, 35-62.

Pentelow, F. T. K., 1955. Pollution and fishes. *Verhandl. Intern. Ver. Limnol.* **12**, 768-771.

Picton, W. L., 1956. Water use in the United States, 1900-1975. Water and Sewerage Industrial and Utilities Div., *U. S. Dept. Commerce, Bus. Serv. Bull., 136,* Washington, D.C.

Pratt, D. M., 1949. Experiments on the fertilization of a salt water pond. *J. Marine Res.* **8**, 36-59.

Pritchard, D. W., 1951. The physical hydrography of estuaries and some applications to biological problems. *Trans. 16th N. Am. Wildlife Conf.,* 365-375.

Pritchard, D. W., 1952. The physical structure, circulation, and mixing in a coastal plain estuary. *Chesapeake Bay Inst., The Johns Hopkins Univ. Tech. Rept. 3, Ref. 52-2.*

Pritchard, D. W., 1955, Estuarine circulation patterns. *Proc. Am. Soc. Civil Engrs.* **81**, 1-11.

Pritchard, D. W., 1959a. Computation of the longitudinal salinity distribution in the Delaware estuary for various degrees of river inflow regulation. *Chesapeake Bay Inst., The Johns Hopkins Univ. Tech. Rept. 18, Ref. 59-3.*

Pritchard, D. W., 1959b. The movement and mixing of contaminants in tidal estuaries. *Proc. First Intern. Conf. Waste Disposal Marine Environment* 512-525.

Pritchard, D. W., 1960. Problems related to disposal of radioactive wastes in estuarine and coastal waters. In *Biological Problems in Water Pollution. Trans. 1959 Seminar, Robert A. Taft Sanitary Eng. Center Tech. Rept.,* W60-3, 22-32.

Rawls, C. K., 1964. Aquatic plant nuisances. In *Problems of the Potomac Estuary;* pp. 51-56. Proc. Interstate Comm. Potomac River Basin. Washington, D. C.

Raymont, J. E. G., 1949. Further observations on changes in the bottom fauna of a fertilized sea loch. *J. Marine Biol. Assoc.* **27**(1), 9-19.

Reish, D. J., 1959a. An ecological study of pollution in Los Angeles-Long Beach Harbors, California. *Occasional Paper, Allen Hancock Foundation* **22**, 1-119.

Reish, D. J., 1959b. The use of marine invertebrates as indicators of water quality. *Proc. First Intern. Conf. Waste Disposal in the Marine Environment*, 91-103.

Renn, C. E., 1956. Man as a factor in the coastal environment. *Trans. 21st N. Am. Wildlife Conf.*, 470-473.

Rogers, H. M., 1940. Occurrence and retention of plankton within the estuary. *J. Fisheries Res. Bd. Can.* **5**(2), 164-171.

Rounsefell, G. A., 1963. Realism in the management of estuaries. *Marine Resources Bull. No. 1, Alabama Marine Res. Lab.* Dauphin Island, Alabama.

Ryther, J. H., 1954. The ecology of phytoplankton blooms in Moriches Bay and Great South Bay, Long Island, New York. *Biol. Bull.* **106**, 198-209.

Ryther, J. H., R. F. Vaccaro, E. M. Hulburt, C. S. Yentsch, and R. R. L. Guillard, 1958a. Report on a survey of the chemistry biology and hydrography of Great South Bay and Moriches Bay conducted during June and September, 1958, for the townships of Islip and Brookhaven, Long Island, New York. *Woods Hole Oceanog. Inst., Ref., No. 58-57.*

Ryther, J. H., C. S. Yentsch, E. M. Hulburt, and R. F. Vaccaro, 1958b. Dynamics of a diatom bloom. *Biol. Bull.* **115**, 257-268.

Segerstraale, S. G., 1951. The recent increase in salinity off the coast of Finland and its influence upon the fauna. *J. Conseil Intern. Exploration Mer.*, *17*, 103-110.

Simmons, H. B., 1959. Application and limitations of estuary models in pollution analyses. *Proc. First Intern. Conf. Waste Disposal in the Marine Environment*, 540-546.

Stroup, E. D., D. W. Pritchard, and J. H. Carpenter, 1961. Final report, Baltimore Harbor study. *Chesapeake Bay Inst., The Johns Hopkins Univ. Tech. Rept. 26, Ref. 61-5.*

Sverdrup, H. U., M. W. Johnson, and R. H. Fleming, 1942. *The Oceans.* Prentice-Hall, Inc., Englewood Cliffs, New Jersey.

Sykes, J. E., 1964. Requirements of Gulf and South Atlantic estuarine research. *Proc. 16th Ann. Session, Gulf and Caribbean Fish. Inst., 1963*, 113-120.

Sykes, J. E., 1965. Multiple utilization of Gulf Coast estuaries. *Proc. Southeast Game and Fish Comm., 17th Ann. Conf. 1963*, 323-326.

Thompson, S. H., 1961. What is happening to our estuaries? *Trans. 26th N. Am. Wildlife and Nat. Resources Conf.*, 318-322.

U. S. Army Engineer District, Baltimore, N. Atlantic Div., 1963. *Potomac River Basin Report, Summary*, 1-43.

Vaas, K. F., 1963. Annual report of the Delta Division of the Hydrobiological Institute of the Royal Netherlands Academy of Sciences for the years 1960 and 1961. *Neth. J. Sea Res.* **2**(1), 68-76.

Waldichuk, M., 1959. Effects of pulp and papermill wastes on the marine environment. *2nd Seminar Biol. Problems in Water Pollution, Robert A. Taft Sanitary Eng. Center.*

Waugh, G. D., and A. Ansell, 1956. The effect, on oyster spatfall, of controlling barnacle settlement with DDT. *Annals Applied Biology* **44**, 619-625.

Whaley, R. C., 1960. Physical and chemical limnology of Conowingo Reservoir. *Chesapeake Bay Inst., The Johns Hopkins Univ. Tech. Rept. 20, Data Rept. 32, Ref. 60-2*, 1-140.

Whitney, R. R., 1961. The Susquehanna fishery study, 1957-1960: A report of a study on the desirability and feasibility of passing fish at Conowingo Dam. *Maryland Dept. Res. and Education Contrib., 169*, 1-81.

Wolman, A., J. C. Geyer, and E. E. Pyatt, 1957. A clean Potomac River in the Washington metropolitan area. *Interstate Comm. Potomac River Basin, October 1957*, Washington, D. C.

Woodburn, K. D., 1963. A guide to the conservation of shorelines, submerged bottoms and saltwaters with special reference to bulkhead lines, dredging and filling. *Marine Lab., Florida Board Cons., Educ. Bull.* **14**, 1-8.

Control of Estuarine Pollution

JEROME B. GILBERT   RONALD B. ROBIE*   1971

INTRODUCTION

In the area of environmental concern, there is a growing awareness that nearly every one of man's activities affects the environment. Our history has shown that many seemingly innocuous decisions which do not have immediate adverse effects have proven to be damaging over extended periods of time. The present condition of our nation's estuaries serves as a glaring example of this reality.

Estuaries[1] are one of the nation's most

important assets. They are utilized for a wide range of commercial, industrial, and recreational activities while simultaneously serving a vital role in the natural cycles of fish, animal and plant life.

Because of the natural mixing of fresh and salt waters, the estuarine environment produces a wide variety of living organisms, from microscopic species to large numbers of fish and shellfish, birds, and mammals. Many species, such as clams and oysters, spend their entire life cycles in the estuaries. Others, particularly shrimp, migrate from the sea to estuarine nursery areas. In these rich waters, they grow to sub-adult size before returning to the sea to complete their life cycles. The anadromous species, such as salmon and striped bass, pass through the estuaries to their spawning grounds farther upstream, and the young return through the estuaries to the ocean. At least two-thirds of the animal populations in the oceans spend an essential portion of their life cycle in

* The authors are indebted to James Wernecke for his research assistance. The views expressed herein, however, are those of the authors and not of the State of California.

1. For the purposes of this article an estuary will be defined as a body of water which has a free connection to the sea and within which sea water is mixed with fresh water derived from land drainage. See Pritchard, *What is an Estuary: Physical Viewpoint*, in Estuaries (G. Lauff ed. 1967).

Jerome B. Gilbert is a civil engineer and consultant with offices in Sacramento, California; he also teaches at the University of California at Davis. He was executive officer of the State Water Resources Control Board in California from 1969 to 1972.

Ronald B. Robie is a lawyer and adjunct professor at the University of the Pacific (McGeorge School of Law) and has been a member of the State Water Resources Control Board in California since 1969. Before that, he was a consultant to the Assembly Water Committee, California Legislature.

From *Natural Resources Journal*, Vol. 11, No. 2, pp. 256-273, 1971. Reprinted with authors' revisions and light editing and by permission from the authors and *Natural Resources Journal* published by the University of New Mexico School of Law, Albuquerque, New Mexico.

estuarine waters or are dependent on species that do. Innumerable waterfowl and shore-birds depend on the plant and animal organisms of the coastal zone for their food. Many winter and nest in these waters.

The base for all animal life in estuaries is the abundant variety of plant growth, from mangroves to eelgrass and algae. They are supported by the mixing and flushing action of the tides and the organic nutrients which collect to produce the rich bottoms and wet-lands.[2]

Estuaries' role in the support of fish and wildlife is inconsistent with their intensive use by man. This inconsistency extends to both competition between resource use and resource protection. Rivers bring accumulations of municipal and industrial waste[3] and urban runoff adds fertilizers and nutrients. Excessive siltation from upstream land use practices and reclamation activities of adjacent land owners have resulted in the filling of extensive water areas.[4] Upstream diversions may change the position of the fresh water-salt water interface in the estuarine zone, thereby affecting fish and wildlife habitats.[5]

Concurrent with public determination to

end pollution of the environment has been rising concern over the fate of our nation's estuaries[6]—concern that ranges in direction from the serious effects of such pollutants as chlorinated hydrocarbons[7] to the increasingly demonstrated adverse effects of numerous toxic substances upon the estuarine environment.[8] Due to the complex nature of the estuarine environment,[9] and the fact estuarine areas are population centers,[10] there is a tendency to suggest they must be considered and managed as individual environmental units. But, estuaries are not "the problem." It is the watershed, it is regional, it is metropolitan. Estuaries depend on land and water management; they are assaulted from all sides by organic and sediment im-

6. Estuarine problems have been the subject of a number of Congressional Committee Hearings. See, *Hearings on the Nation's Estuaries: San Francisco Bay and Delta, California Before the Subcommittee on Conservation and Natural Resources of the House Committee on Government Operations,* 91st Cong., 1st Sess., (1969); *Hearings on the National Oceanographic Program Before the Subcommittee on Oceanography of the House Committee on Merchant Marine and Fisheries,* 91st Cong., 1st Sess., pts 1 & 2 (1969). Estuaries have also been the subject of several recent federal studies. See U.S. Dept. of the Interior, National Estuary Study (1970), and U.S. Fed. Water Pollution Control Admin., National Estuarine Pollution Study (1968).

7. Chlorinated hydrocarbon pesticides are a group of insecticides that contain at least carbon, hydrogen and chlorine. In general, they are persistent in the environment, have an affinity for fatty tissue and are toxic to numerous insects. Examples are DDT, Dieldrin, Endrin, Chlordane and Toxaphene.

8. Manufacturing processes are becoming more complex creating greater amounts of exotic wastes potentially toxic to humans and aquatic life; and the effects of current levels of such substances as cadmium, lead and mercury are still not fully understood. Council Report, *supra* note 2, at 52.

9. The estuarine environment is continually in a state of change. Salt and fresh water concentrations are subject to any variation in the level of fresh water input. In turn, fresh water input is determined by upstream use, seasonal variation in the weather, and variances in the year to year precipitation levels. *See* U.S. Fed. Water Pollution Control Admin., Marine Biology and Pollution Ecology Training Manual, at C23-1, (1970).

10. Eight of our most populous metropolitan areas are located in estuarine zones and Great Lakes areas, as are 15 of the largest U.S. cities. National Estuary Study, *supra* note 4, at 50.

2. U.S. President's Council on Environmental Quality, First Annual Report on Environmental Quality 176 (1970) [hereinafter cited as Council Report].

3. For example, a limited investigation of pesticides undertaken as part of the San Francisco Bay-Delta Water Quality Control Program found that between 10,000 and 20,000 pounds of chlorinated hydrocarbons entered the Bay-Delta system in 1965. The data indicated that from twenty to forty per cent of the chlorinated hydrocarbons entering the system were discharged in municipal and industrial wastes. Kaiser Engineers and Assoc. Firms, Final Report to the State of California, San Francisco Bay-Delta Water Quality Control Program, at II-8, XII-23 (1969).

4. A recent study rated 62 per cent of California estuaries as severely modified by landfill activity. U.S. Department of the Interior, Fish and Wildlife Service, 1 National Estuary Study 25 (1970).

5. Migrating birds, anadromous fish, shellfish and a wide variety of aquatic life depend upon a sometimes delicate balance for their survival. An alteration in the chemical or physical characteristics of environmental zones can severely alter the number and variety of species.

balances, toxic substance accumulation, and salinity intrusion.[11]

Water pollution control efforts until recently were designed primarily to protect the quality of water used for consumptive purposes, and since estuarine waters are not generally sources of domestic water, the control of estuarine pollution has lagged behind the control of pollution in entirely fresh water areas.[12] A number of other factors have contributed to the estuaries' falling behind in the race for environmental protection and enhancement, including problems regarding the effectiveness of pollution control efforts as well as difficulties in efforts to measure pollution in an estuarine environment. This article will attempt to shed light on some of these problems.

THE LEGAL BASIS FOR CONTROL

The scientific problems associated with water quality control in estuaries, which are discussed elsewhere in this article, are more than matched by the problems caused by the nation's intricate governmental systems and by the political values relating to estuarine management. Governmental responsibility is divided between federal, state and local jurisdictions.[13] Various laws dealing with estuarine management are often times conflicting, and unsettled public use rights,[14] disputed titles and overlapping provisions of law make difficult the orderly administration of our estuarine resources. This section will discuss the estuarine control activities of the several levels of government.

Federal Government

Constitutional Basis of Authority

The federal role in estuarine management is supported by a number of broad congressional grants of authority. Because estuarine areas are often extensively involved in commerce, the "Commerce Power"[15] affords the federal government its most significant basis from which to regulate estuarine-related activities. The Federal Water Pollution Control Act,[16] which applies to interstate and coastal waters,[17] and the regulation of navigation, principally by the United States Army Corps of Engineers,[18] are the primary federal activities based on this power.

Under the "Property Power,"[19] the federal government exercises influence in estuaries through control of property owned by the United States. Similarly, under the General Welfare Clause,[20] the United States, through the Department of the Interior's

---

11. Salinity intrusion extends to the movement of saltwater into groundwater basins as well as up streams that flow into the estuary. For a more complete discussion of this problem see Gindler and Holburt, *Water Salinity Problems: Approaches to Legal and Engineering Solutions,* 9 Natural Resources J. 329 (1969).
12. President's Commission on Marine Science, Engineering and Resources, Our Nation and the Sea, A Plan for National Action 74 (1969).
13. U.S. Dept. of Interior, Fish and Wildlife Service, 2 National Estuary Study 212 (1970).
14. This is a cause of confusion to many states. Recently, the California Supreme Court held that historic use of shoreline areas is to be a major consideration in determining public use. The Court also stated that the courts should encourage public use of shoreline areas whenever that can be done consistently with the federal Constitution. Gion v. Santa Cruz, 2 Cal. App.3d 29, 465 P.2d 50, (1970). For a discussion of the subject see Sax,

The Public Trust Doctrine in Natural Resources Law: Effective Judicial Intervention, 68 Mich. L. Rev. 473 (1970).
15. U.S. Const. art. I, § 8, para. 3—to regulate commerce with foreign nations and among the several states. Extended to include activities "affecting commerce." See County of Mobile v. Kimball, 102 U.S. 691 (1881).
16. Water Pollution Control Act, 62 Stat. 1155 (1948), *as amended* 33 U.S.C. 1151 et seq. [hereinafter the Federal Water Pollution Control Act in its amended form is cited as F.W.P.C.A.].
17. F.W.P.C.A. § 10(a) "The pollution of interstate or navigable waters ... shall be subject to abatement as provided by this Act."
18. In 1935 Congress provided generally that investigations and improvements of rivers, harbors and other waterways shall be under the jurisdiction and prosecuted by the Department of the Army under the direction of its Secretary and supervision of the Chief of Engineers. Act of Aug. 30, 1935, § 1, 49 Stat. 1028, 33 U.S.C. 540.
19. U.S. Const. art. IV § 3, para. 2.
20. U.S. Const. art. I, § 8, para. 1.

Bureau of Reclamation, constructs water storage projects on the tributaries of estuaries. The operation of these projects can have a significant impact upon the estuaries and the land and water resources that surround them.[21]

The "Treaty Power"[22] provides an interesting basis for authority inasmuch as the United States is a party to two major multilateral treaties relating to pollution of the sea by oil[23] and to a number of treaties and agreements with Mexico and Canada that affect the quality and use of boundary waters.[24]

Administrative Regulation

With constitutionally vested powers and the broad interpretation thereof forming a base, Congress has assigned authority for a number of estuarine-related activities to various federal agencies.[25]

The Department of the Interior has major administrative responsibilities in the estuarine zone. The Estuarine Areas Act of 1968[26] gives the Secretary of the Interior review authority over federal development activities affecting estuarine resources. Furthermore, numerous agencies within the Department have administrative responsibilities ranging from the study and protection of wildlife to the development of water resources that will eventually affect the estuary.

The recently formed Environmental Protection Agency,[27] which houses the principal federal regulatory functions in the environmental field, is significantly involved in estuarine management. The Agency's Water Quality Office oversees the establishment and enforcement of federal water quality standards for interstate and coastal waters. It also administers grant programs to assist states and public agencies in the administration of water quality programs, research, and construction of water quality control facilities.[28] The construction of treatment facilities with federal financial assistance has been the major factor in the upgrading of the quality of waste discharges to estuarine waters in many parts of the country.[29]

21. For example, the massive Central Valley Project in California, undertaken as a reclamation project, significantly affects the flow of water through the Sacramento/San Joanquin Delta and the full effect of this project upon fish and wildlife and water quality is not yet known. See United States v. Gerlach Livestock Co., 339 U.S. 725 (1950).
22. U.S. Const. art. II, § 2, para. 2. Treaties made under the authority of the United States shall be the supreme law of the land. U.S. Const. art. VI, para. 2.
23. The International Convention for the Prevention of Pollution of the Seas by Oil. 12 U.S.T. 2989 (1961). The United States Convention of the High Seas, [1962] 13 U.S.T. 2313. Recently, President Nixon made a proposal for a new treaty that would, among other things, protect the ocean from pollution. Wkly. Comp. Presidential Docs., May 25, 1970, 677-678.
24. The International Boundary and Water Commission, United States and Mexico. Rio Grande, Colo., and Tijuana Treaty, 59 Stat. 1219, T.S. No. 994. The Boundary Waters Treaty, 1909, 36 Stat. 2448, T.S. No. 548, authorized the creation of the International Joint Commission of the United States and Canada.
25. This has resulted in some confusion between agencies. Activities of one agency will often conflict or cancel the efforts of another. See *Federal Pollution Attack Gains Steam, But Long-Term Outlook Remains Cloudy*, 2 Government Executive 50-52 (1970).

26. 16 U.S.C. § § 1221-26 (1968). The Act authorized a general study and inventory of estuaries (See § 1222), and requires all federal agencies in planning for water and land resource use and development to give consideration to estuaries, their natural resources, and their importance for commercial and industrial developments (See § 1224).
27. See Reorganization Plan No. 3 (H.R. Doc. No. 91-364) Operative Dec. 2, 1970. The Agency has assumed responsibilities formerly held by the Atomic Energy Commission regulating radioactivity from nuclear installations, which often are or will be located adjacent to or within estuaries. In addition, the Agency has also assumed responsibilities formerly held by the Dept. of Health, Education and Welfare and administered through the Environmental Health Service including functions exercised by: The National Air Pollution Control Administration, the Environmental Control Administration and the Air Quality Advisory Board, also the functions in regard to establishing tolerances for pesticide chemicals and the functions of the Bureaus of: Solid Waste Management, Water Hygiene, and Radiological Health.
28. F.W.P.C.A. § § 6-8.
29. The nation's municipal waste-handling systems show an investment of $880 million for the year

The Defense Department has considerable influence in estuarine areas due to the presence of a number of military installations in these areas and the active role the Army Corps of Engineers has played through its civil works program.

Through activities of the Coast Guard, the Department of Transportation also performs a number of service activities directed at the beneficial use of estuarine waters. The Coast Guard is charged with the enforcement of federal laws in the navigable waters of the United States, and with the maintenance and operation of aids to navigation.[30]

Many federal agencies and laws also indirectly affect estuarine areas.[31] For example, a recently enacted provision of the Federal Water Pollution Control Act[32] adds the consideration of environmental factors to the existing statutory authority of many federal agencies, the most prominent of which is the Corps of Engineers.[33] Furthermore, considerable federal control is exercised over state and local actions through the review responsibilities in assorted federal grant programs other than those contained in the Federal Water Pollution Control Act.[34]

State and Local Government

Generally, state and local governments have the most direct authority in estuarine areas. The most substantial basis for their regulation of estuarine activities falls under the so-called police power.[35] This power supports state water quality regulation and land-use controls.

Many states have delegated significant authority[36] in estuarine management and land use to local government,[37] and in some cases these local controls are protected from state legislative interference by so-called "home rule" provisions under which municipal affairs or matters not of statewide significance are constitutionally protected powers of local government.[38]

1969. This amount, however, did little more than cover replacement and growth needs developed in the same year. Total investment requirements will conservatively amount to $10 billion over the years 1970-74 if all existing deficiencies are corrected and no new deficiencies incurred. U.S. Department of the Interior, Federal Water Pollution Control Administration, The Economics of Clear Water, Summary Report 5 (1970). Recent estimates prepared during Congressional action on amendments to the Federal Water Pollution Control Act range from $15-25 billion between 1972 and 1977.
30. See 14 USC., Ch. 5 (Supp. III), 33 U.S.C., Ch. 7, 33 U.S.C. 157. Also, F.W.P.C.A. § 13 provides that the secretary of the department in which the Coast Guard is operating is charged with the enforcement of federal standards in respect to the control of sewage from vessels.
31. For instance, the U.S. Forest Service, Department of Agriculture, manages the forestry aspect of watershed protection and 12 national forests involve lands that drain directly into estuarine areas. U.S. Department of the Interior, 3 National Estuarine Pollution Study, at V-27 (1969). Presently, under the Environmental Policy Act of 1969 (Pub. L. No. 91-190, Jan. 1, 1970), all federal agencies are required to submit reports regarding the environmental impact of their proposed actions. See § 102.
32. F.W.P.C.A. § 21(b). Applicants for a federal license or permit to conduct any activity that may result in a discharge into the navigable waters of the United States are required to submit a certification from the state in which the discharge will originate to the effect that activity will not violate applicable water quality standards.

33. Also, F.W.P.C.A. § 21(a) requires each federal agency having jurisdiction over any real property, a facility, or engaged in any federal public works project to insure compliance with applicable water quality standards.
34. A number of grant programs administered by the Departments of the Interior, Defense, Agriculture, and Housing and Urban Development directly affect estuarine management. For example, see California Assembly Committee on Water, Handbook of Federal and State Programs of Financial Assistance for Water Development (1972 ed.).
35. The inherent and plenary power in states over persons and property which enables the people to prohibit all things inimical to comfort, safety, health, and welfare of society. Drysdale v. Prudden, 195 N.C. 722, 143 S.E. 530, 536 (1928).
36. Generally, local government authority stems entirely from powers delegated by the parent state.
37. Notably, a few states have maintained or reasserted their land-use powers. See Hawaii Rev. Laws § 205-10 (1968) and [6] Me. Rev. Stat. Ann. Tit. 12 § 681-89 (Supp. 1970).
38. Article XI, § § 6 and 8(j) of California's Constitution gives charter cities the power to "make and enforce all laws and regulations in respect to municipal affairs, subject only to the restrictions and limitations provided in their several charters."

Generally, state activities in estuarine management are concentrated in state water pollution control agencies.[39] For the most part, these pollution control efforts have been designed to regulate municipal and industrial waste discharges.

There has been a trend toward establishing special purpose governmental agencies devoted to specialized problems affecting estuaries or a particular estuary. One of the most notable examples of the latter was the establishment in California of the San Francisco Bay Conservation and Development Commission.[40] The basic function of the Commission is to control the filling of San Francisco Bay, an activity which has already reduced the surface area of the Bay from 700 square miles to slightly over 400 square miles. The remarkable support the Commission received from the California public was shown in 1969 when the California Legislature made the Commission permanent and expanded its jurisdiction to include all the territory located between the shoreline of San Francisco Bay and a line 100 feet landward.[41]

Conflicts and Limitations

Federal—State

Much of the active disagreement between state and federal governments concerns the ownership of submerged lands.[42] In 1953, Congress attempted to resolve this conflict with passage of the Submerged Lands Act,[43] which placed title in the coastal states to the submerged lands within their boundaries, defined their seaward boundaries as extending three geographical miles from the coastline (three marine leagues into the Gulf of Mexico), and placed these lands and their resources under applicable state law.[44] However, because coastlines consist of numerous indentations and extensions and because many coastal states have developed and extended authority over coastal islands, the definition and design of coastline boundary standards and the seaward extension of state authority are still the subject of debate.[45]

President Nixon's treaty proposal of May 23, 1970[46] includes a recommendation that would establish a 12-mile territorial sea. This proposal could have a direct effect upon the regulation of estuarine pollution. Pollution of the sea is subject to the action of the tide, currents and winds. Discharges into the sea will, in many instances, float toward the coastal zone and result in degradation of the estuarine environment just as discharges into the coastal zone will have an effect on the sea. Because the President's proposal would extend United States' jurisdiction over a 12-mile area, pollution standards and regulations that include estuarine considerations could be initiated. However, the questions of

---

39. A discussion of those rights, remedies and defenses relating to water quality is presented in 3 Gindler, Waters and Water Rights 37-195 (1967). For a summary of state water pollution control agencies see Hines, *Nor Any Drop To Drink: Public Regulation of Water Quality*, 52 Iowa L. Rev. 186 (1966-67). However, the California Water Code provides that, "In acting upon applications to appropriate water, the board shall consider water quality control plans which have been established . . . and may subject such appropriations to such terms and conditions as it finds are necessary to carry out such plans." (Sec. 1258) West. Supp. 1970. Thus water quality considerations are an integral part of water rights administration.
40. Created in 1965 by the McAteer-Petris Act Cal. Govt. Code § § 66600-66653 [West 1966], the Commission is charged with responsibility for preparing "a comprehensive and enforceable plan for the conservation of the water of the bay and the development of its shoreline." For background on the Commission and its activities see: Committee on Government Operations, Protecting America's Estuaries: The San Francisco Bay-Delta 7-12 (1970).
41. Cal. Govt. Code § 66610(b) (West Suppl 1970).

42. See United States v. California, 332 U.S. 19 (1947); also United States v. Louisiana, 394 U.S. 11 (1969).
43. 43 U.S.C. § § 1301-15 (1953).
44. *Id.* § 1311.
45. See United States v. California, 381 U.S. 139 (1965); also United States v. Louisiana, 394 U.S. 11 (1969).
46. See The International Convention for the Prevention of Pollution of the Seas by Oil, *supra* note 23.

authority allocations between the federal and state governments would still be unresolved. Proposed legislation was introduced in the 91st Congress providing for state planning jurisdiction over offshore areas in which the concerned states have a legitimate interest.[47]

Another conflict has been the scope of regulation which the federal government could exercise under the Federal Water Pollution Control Act. Estuarine waters are covered by the Act because they are defined as "interstate waters."[48] The Act provides that state water quality standards for estuaries are subject to federal approval,[49] and if they are unsatisfactory the federal government may impose standards.[50]

In 1970 acting under Executive order 11574 (December 23, 1970), the U.S. Army Corps of Engineers began to require waste discharge permits of all U.S. industries discharging to navigable waters and their tributaries. This directly duplicated the programs of many states including California, Michigan, New York and others. But it added uniformity on a national basis potentially strengthening some weak state and local efforts. The statutory authority for this program is the Rivers and Harbors Act of 1899 (33 U.S.C. 401-413; Sec. 407 is referred to as the "Refuse Act"). The act is not designed to cope with modern pollution problems and Congress appears ready to supplant the whole Corps effort (few permits had been issued by the end of 1971) with a comprehensive national regulation program administered by EPA and the states. See S.

2770 and HR 11896, 92nd Congress, 1st Session.

State-Local

Probably of more concern from a practical standpoint than federal/state conflicts are the self-imposed limitations on state ability to regulate estuarine areas. For example, despite the "public trust doctrine,"[51] many coastal states have transferred ownership of submerged lands to private individuals or local government[52] with the result that the most direct state basis for regulating the use of these lands has been lost. Only a few states have provided comprehensive state regulations covering coastal activity and development, including the placing of structures.[53] However, as with any exercise of the police power, a governmental agency must be able to meet changing judicial interpretations as to what constitutes the taking of property without just compensation.[54] This question was raised during the consideration of legislation to extend the San Francisco Bay Conservation and Development Commission's authority to include the regulation of structures on the shoreline.[55]

51. Under this doctrine title to the tidelands is held in trust by the state to be used by the people. The state is obligated to protect the public rights of navigation, fishing and commerce. See Pollard's Lessee v. Hagon, 44 U.S. (3 How.) 212,229 (1844). See also Council Report, *supra* note 2 at 176.
52. Exceptions include Hawaii, Texas and Alaska which own their estuarine zones. National Estuarine Pollution Study, *supra* note 31, at V-133-34.
53. Massachusetts, Connecticut and North Carolina have wetlands protection laws while Hawaii, Wisconsin, and to some extent Oregon have exercised statewide powers over the contiguous dry lands. Council Report, *supra* note 2, at 178. In 1972, several attempts were made to provide for state regulation of California's coastal activities, but they failed to gain legislative approval. See California Senate Bills. Also California Assembly Bills.
54. U.S. Const. amend. V commands that "private property [shall not] be taken for public use, without just compensation." The problem of what constitutes a compensable taking of property has long been a source of confusion to scholars and courts.
55. San Francisco Bay Conservation and Development Commission, San Francisco Bay Plan 3-4, 37-38 (1969).

47. S.2802, S.3183 and S.3460, 91st Cong. (1969-70). Only one of these bills would cover the proposed 12-mile territorial sea extension (S.3183 which defines coastal zone as extending "seaward to the outer limit of the United States territorial sea"). All of these bills reserve in the Federal Government the right to review and approve the states' planning or operating programs for their coastal zones.
48. 6 U.S. Dept. of the Interior, Fish and Wildlife Service, National Estuary Study, at E-2 (1970).
49. F.W.P.C.A. § 10.
50. *Id.*

Maine's Wetlands Act,[56] which was designed to protect the ecology of coastal areas, recently failed to meet the test of substantive due process.[57]

Problems are also created by the frequent reluctance of local governments to establish and implement land-use plans which maximize environmental protection in estuarine areas. A local government's most important legal tools in this regard are zoning and taxation. However, local agencies are frequently restricted by a relatively small tax base and are thus prevented from the purchase of estuarine areas. This situation also tends to encourage local planning bodies to accept industrial development (which increases employment and tax revenues) at the expense of environmental protection. Frequently, even when estuarine areas are zoned for protection of environmental values, special-use allocations or subsequent rezoning for industrial and commercial activities result in degradation of the environment.[58] Clearly, local government has given priority to its tax base and lip service to conservation. It would appear that significant state or federal financial assistance to local government will be necessary if local efforts at limiting land use in estuarine areas are to be successful. As an alternative, special tax treatment of the lands involved could be considered.

In addition to being limited by financial and special interest pressures, local government is frequently limited in its ability to manage an estuary because of areal limitations in governmental jurisdiction. Typically, an estuarine area is under the jurisdiction of numerous cities, counties and other special-purpose governmental agencies which frequently have differing tax bases, powers and estuary-related priorities. Under the best of circumstances, even a limited degree of coordination among local government concerns in estuarine areas is difficult to ob-

tain.[59] More difficulty arises when an estuary involves more than one state. The northeastern United States has witnessed the development of several compacts designed to administer a cooperative multi-state effort of estuary protection. Of these, the Tri-State [60] Compact and the Delaware River Basin Compact[61] have the broadest range of activities in relation to estuarine water quality. Although these compacts recognize the need for state-level administrative cooperation, studies have indicated their effectiveness has been limited and that many of the signatory states continue to provide administrative controls outside the framework of their respective compacts.[62]

California: A Case in Point

The San Francisco Bay-Delta Estuary, located at the confluence of the Sacramento and San Joaquin rivers at the head of Suisun and San Francisco Bays, is the most important estuary in the state. The Central Valley of California, which comprises nearly 40 per cent of the state's total area, is tributary to the Delta and more than 5.75 million people reside in the adjacent counties.[63]

California has a comprehensive and broadly interpreted[64] state water quality control act[65] designed to protect the quality

56. Me. Rev. Stats. Ann. Tit. 12 § § 4701-09 (1970).
57. See Maine v. Johnson, 265 A.2d 711 (1970).
58. National Estuarine Pollution Study, *supra* note 31, at V-147, V-155.

59. This situation prompted the Planning and Conservation League of California to comment: "The odds against a thousand city governments regulating an end to boosterism in the coastal zone of California are roughly equivalent to the classic probability of a million monkeys pecking away at typewriters and someday producing 'Hamlet.' " The Riverside Press Enterprise, Sept. 20, 1970.
60. New Jersey, New York and Connecticut.
61. Delaware, New Jersey, New York, Pennsylvania and the U.S.
62. National Estuarine Pollution Study, *supra* note 31, at V-201.
63. Final Report to the State of California, San Francisco Bay Delta Water Quality Program, *supra* note 3, at XIV-1 to XIV-35.
64. See 26 Op. Cal. Att'y Gen. 88 (1956); 27 Op. Cal. Att'y Gen. 482 (1956) and 43 Op. Cal. Att'y Gen. 302 (1964).
65. The Porter-Cologne Water Quality Control Act, Cal. Water Code § § 13000-13951 (West Supp. 1970).

of state waters[66] from the discharge of waste[67] from all sources.[68] Administration of water quality control is carried out by a five-member, full-time State Water Resources Control Board,[69] and nine, nine-member[70] decentralized regional boards which act on an area wide basis. Supervision, budgetary review, approval of regional water quality plans, resolution of disputes between regional boards,[71] and appeal responsibility are placed in the State Board.[72]

Other state agencies which affect the San Francisco Bay-Delta Estuary include the State Lands Commission, custodian of approximately 634,653 acres of state-owned land, a large portion of which is tide and submerged lands in the estuarine zones,[73] the Department of Water Resources, which is concerned primarily with water resources investigations and the development of fresh water supplies;[74] the Department of Fish and Game, which has enforcement authority concerning fish kills and is the State's scientific arm for investigating the effects of water pollution on marine life;[75] the Department of Health, which regulates public health aspects of water use;[76] and the De-

partment of Conservation's Divisions of Forestry, Mines and Geology, Oil and Gas.[77] Each of these agencies exerts considerable influence in regard to management and maintenance of the water quality in the estuarine area.

Much of California's fresh water supply is concentrated in the northern part of the State and involves the extensive watershed of the San Francisco Bay-Delta Estuary. Diversions of this water supply from the Estuary to other portions of the State by the Federal Central Valley Project[78] and the State Water Project[79] affect the extent of salinity intrusion in the Delta which in turn affects the ecology.[80] Delta water users are protected by statutory provisions requiring the State Water Project, in coordination with

---

66. Cal. Water Code § 13000 (West Supp. 1970).
67. "Waste" includes sewage and any and all other waste substances . . . associated with human habitation, or of human or animal origin, or from any producing, manufacturing, or processing operation of whatever nature. Cal. Water Code § 13050(d) (West Supp. 1970).
68. See for example, State Water Resources Control Board Resolution 70-23, Aug. 6, 1970 (Cal.).
69. Cal. Water Code § § 174-188.5 (West Supp. 1970).
70. Cal. Water Code § § 13200-13207 (West Supp. 1970).
71. Cal. Water Code § 13320(d) (West Supp. 1970).
72. Cal. Water Code § § 13168, 13320(a) (West Supp. 1970).
73. The Commission has exclusive jurisdiction over all ungranted tidelands and submerged lands owned by the state including the authority to lease or otherwise dispose of such lands. Cal. Pub. Res. Code § 6301 (West 1956).
74. Cal. Water Code § 150 (West Supp. 1956).
75. The provision most used by the department in its enforcement activities is Cal. Fish & Game Code § 5650 (West 1968). See also § 5652.
76. The department is given responsibility for the maintenance of pure water for domestic use [Cal.

Health & Safety Code § 203 (West Supp. 1956)], the authority to revoke permits issued to any person supplying water for domestic use (§ 4011), and also the authority to regulate the disposal of many wastes (§ § 4401, 4400).
77. See Cal. Pub. Res. Code § § 630-647, 2002-2322, 3001-3234 (West Supp. 1970).
78. The Federal Central Valley Project was authorized in 1935 under provisions of the Emergency Relief Act as a reclamation project. It is a multipurpose development to supply water for irrigation, municipal, industrial, and other uses, improve navigation on the Sacramento River, control floods in the Central Valley, and produce hydroelectric energy. It includes 16 dams and some 900 miles of conduits, tunnels, and canals.

Major reservoirs include Lake Shasta on the Sacramento River, Folsom Lake and Auburn Reservoir on the American River, Millerton Lake on the San Joaquin River, and New Melones Reservoir on the Stanislaus River. Major aqueduct systems are the Delta-Mendota Canal, Friant-Kern Canal, Madera Canal, and Corning Canal. Other key features are the San Felipe Division, Trinity Division, and San Luis Division.
79. The California State Water Project is a multipurpose water development that conserves and distributes water, produces electrical energy and provides flood control, recreation, and fish and wildlife enhancement. The initial facilities of the Project—now 95 per cent completed or under construction—include 18 reservoirs, 15 pumping plants, 5 powerplants and 580 miles of aqueducts. Parts of the Project have been in service since 1962; water deliveries will be made from the southern terminus in 1972.
80. See California State Water Resources Control Board water rights Decision 1379 (July 28, 1971) for a discussion of these factors in relation to the State Water Project and Central Valley Project.

the Federal Central Valley Project to provide salinity control and an adequate water supply[81] and also by the State's recognition of the sensitive interrelationship between water quality and water quantity as expressed through water rights allocation.[82] California law recognizes both the riparian[83] and appropriative[84] doctrines of allocation of water resources. Under this latter doctrine, permits for appropriation of water are issued by the State Water Resources Control Board.[85]

The development and implementation of land-use planning programs in the San Francisco Bay-Delta Estuary is typical of the confusion and difficulties which arise on the local level. The San Francisco Bay-Delta Estuary consists of 12 counties, 104 cities and numerous limited-purpose special districts which have powers affecting the environment of the estuary;[86] there is no single- or

multi-purpose agency covering the entire estuary. The San Francisco Bay Conservation and Development Commission[87] is the only agency approximating areawide jurisdiction.

Even the readily isolated problem of waste disposal is fragmented among many separate jurisdictions. Although a three-year comprehensive study of pollution of the San Francisco Bay-Delta area recommended in 1969 that an areawide agency be established to handle waste disposal in the area the initial response was negative both in the Legislature and within the area itself. However, in 1971 the Legislature finally created the Bay Area Sewer Services Agency and more detailed studies of subregional parts of the Bay area were underway. But the determination to justify piecemeal solutions to pollution on technical, financial, and, most importantly, local autonomy grounds remains, despite a clear state and national interest in this estuary. The critical problem of achieving local action for environmental protection remains.[88]

PLANNING FOR WATER RESOURCES MANAGEMENT

Failure to provide nationwide guidance of land use has complicated the problem. Local agencies empowered to decide how land is used have continued to operate within their

81. Cal. Water Code § § 12202-05, 12220 (West Supp. 1970).

82. In California the Water Resources Control Board has the authority to approve appropriations by storage of water to be released for the purpose of protecting or enhancing the quality of other waters put to beneficial use [Cal. Water Code § 1242.5 (West Supp. 1970)], to take into account the amounts of water needed to remain in the source for the protection of beneficial uses, including any uses specified to be protected in any relevant water quality control plan [Cal. Water Code § 1243.5 (West Supp. 1970)], and to institute necessary court actions to adjudicate rights or to further the physical solutions necessary for the protection of the quality of groundwater [Cal. Water Code § 2100 (West Supp. 1970)]. For a discussion of water quality and water rights see Robie, *Relationship Between Water Quality and Water Rights*, Contemporary Developments in Water Law 73-83 (Water Resources Symposium No. 4, C. Johnson and Lewis S. eds. University of Texas 1970).

83. Under this doctrine the law recognizes that each riparian owner has a right to the reasonable use of water on land riparian to a watercourse. It is a judically oriented common law system concerning the rights of one riparian vis-a-vis other riparians.

84. The doctrine of prior appropriation states that the first in time to use the water beneficially is the first in right.

85. Cal. Water Code § 1250 (West Supp. 1970).

86. More than 275 local public entities in the 12-county study area perform functions related to the provisions of water or sewer service. Of the 104 cities located in the 12-county Bay-Delta area, 68

provide sewer service and 48 provide water service. Sewerage functions are performed by 155 public entities in the 12 counties. In addition to the 68 cities, 44 entities operate pursuant to the Sanitary District Act. The remaining 18 entities operate pursuant to one of 10 other acts which meet the specific needs of the service area. A total of 103 public entities provide domestic water service in the 12 counties. In addition to the 48 cities which provide water service, there are 40 districts which operate pursuant to the county water district law and 15 other entities providing water service under nine special district acts. There are 17 special flood control and water conservation districts in the 12-county Bay-Delta area. Final Report to the State of California, San Francisco Bay-Delta Water Quality Control Program, *supra* note 63, at 2.

87. See Cal. Govt. Code, note 40 *supra*.

88. California Stats. 1971, Ch. 909. See also State Water Resources Control Board *Clean Water For San Francisco Bay*, January 1971.

narrow areas of authority while ignoring the regionwide results of their fragmented decisions.[89] Only recently has there been a general realization that communities were neglecting long-term resource protection to achieve short-run improvements in the tax base or economic situation. This last-minute awareness has dramatized the need for proper land-use planning techniques that can insure a predictable rate and direction of development compatible with environmental goals.[90]

Water Quality Controls

To a limited extent, water quality controls have been used to indirectly fill the void created by the lack of adequate land-use plans. For instance, in recent years there has been an acceleration in the planning and construction of waste treatment facilities on an areawide basis.[91] Because of the absence of land-use planning, such water quality planning may be subject to criticism as accomplishing only a limited purpose. Appropriate predictions of land use and consideration of other environmental factors must necessarily supplement water quality plans. Recent federal regulations emphasize the land-use planning responsibilities expected from water quality management by requiring area-wide planning as a prerequisite to federal construction grants.[92]

Adequacy of Water Quality Controls

The past few years have seen significant

progress in the development and implementation of state water pollution control programs[93] through programs of indirect or direct control of waste disposal.[94] However, the effectiveness of existing and proposed waste treatment facilities is being questioned.[95] In addition there is no general acceptance in the scientific community of new planning needs, particularly as they relate to waste treatment facility design. Although it is technically simple and relatively inexpensive to reduce the oxygen-demanding characteristics of waste, it is more difficult to reduce the wide range of toxic chemicals, heavy metals, and nutrients that are discharged from most types of today's treatment plants.

Also evident are the limitations inherent in current waste discharge regulations. For example, the traditional methods of measuring pollution[96] are no longer adequate. They do not consider problems of toxicity or the long-term cumulative effects (such as increased productivity[97]) of the discharge of

89. Council Report, *supra* note 2, at 184.
90. There has been some federal recognition of the need to coordinate land use planning activities with environmental considerations. See The Natural Land Use Policy Act, S. 3354, 91st Cong. (1970) which calls for ecological factors to be used as criteria in land use planning. For a discussion see Caldwell, *The Ecosystem as a Criterion for Public Land Policy*, 10 Natural Resources J. 203 (1970).
91. For example, the regional systems in San Diego, Seattle and Toronto were forerunners in the construction to accommodate areawide considerations.
92. See 35 Fed. Reg. 10756 (1970).

93. Council Report, *supra* note 2, at 50.
94. These include such measures or the construction of public waste treatment facilities, judicial or administrative orders requiring dischargers to either cease or correct deficiencies, and tax incentives awarded industries to bring their discharges into compliance with acceptable standards.
95. In the last few years, communities around the nation have invested large sums in improvement of waste treatment facilities but in many streams the degree of treatment needed is far beyond the technical capability of existing or proposed facilities.
  There are many types of pollutants that cannot be effectively controlled by treatment such as pesticides and products that contain phosphates. Both of these pollution sources and other similar products must be controlled at the source and new federal authority is needed to assure rapid elimination of dangerous products from the market. See Final Report, San Francisco Bay-Delta Water Quality Control Program, *supra* note 3, at XX-31 to XX-40 and X-1 to X-10.
96. Traditional water pollution parameters include Biochemical Oxygen Demand (BOD), Chemical Oxygen Demand (COD), suspended solids and coliform counts. These parameters measure the oxygen depleting characteristics, the particulate matter content and the numbers of coliform bacteria respectively in wastewaters or the receiving water.
97. Production can be defined as the total amount of cellular organic matter that is formed within a certain time from the raw material nutrients sup-

nutrients into confined portions of the estuary.

Certain rigid governmental policies also tend to complicate the problem and may in some cases increase the already existing problems of estuarine productivity.[98]

The discharge of heated wastes, primarily industrial and power plant cooling water, provides another threat to the estuarine environment. Small increases in the temperature can have serious effects, particularly in estuaries that support anadromous fish runs.[99] Protection from this source of pollution requires either the elimination of heated waste discharges or their rigid control.

Problems in maintenance of water quality in the estuary also encompass the difficulties involved by depletion of freshwater supplies caused by upstream diversions and storage projects. This is a critical problem in the coastal areas of the arid west, and it is becoming an increasingly significant problem in the estuaries of the Eastern United States where rainfall is more evenly divided throughout the year. Water stored primarily to meet consumptive purposes can be released in natural channels to meet environmental demands as well as the water supply needs for domestic, agricultural, and industrial uses.[100] But this would mean substantially increased magnitude and scope of water quality planning efforts, fully coordinated with planning for the protection and development of other natural resources. Considerations such as these suggest that protection of our nation's waters should depend less upon programs limited to the regulation of waste discharges and more upon management programs which include water quantity, water quality, and land-use controls.

A CHOICE FOR THE FUTURE

The ability to provide effective environmental management programs depends upon the implementation of comprehensive development plans supported by the enforcement of land-use controls. Existing regulatory efforts of federal, state, and local government can reduce pollution loads of waters and contribute to the protection and enhancement of our nation's natural assets. But, until the use of land is controlled within a framework of areawide or statewide planning programs (in some instances, this must be multi-state), these efforts cannot prevent the continuing degradation of the total environment.

The complex nature of estuaries demonstrates the need for programs that can regulate pollution control activities at their source. Estuaries are an inseparable part of an upstream watershed. Any upstream development will have a dramatic impact on the estuaries' sensitive environmental characteristics. Controls must extend not only to waste discharges but also to the development activities in the total watershed or basin of which the estuary is an integral part.

Since it may be many years before adequate plans can be developed, it is essential that existing and fragmented regulatory and planning efforts be simplified and consolidated. The wave of environmental concern

plied. In aquatic terminology, "Production" or "Productivity" usually expresses the rate of algal growth in the body of water. This is often referred to as "algal primary productivity." See Calif. State Water Quality Control Board, Eutrophication—A Review, Pub. No. 34, (1967).
98. For example, the Federal Government supports the concept that secondary treatment (85% BOD removal) be provided for all communities (with limited exceptions) regardless of individual local water conditions. See proposed rule at 35 Fed. Reg. 8942 (1970).
99. In his presentation entitled "Research on Thermal Pollution Report on the Columbia River and Estuary" presented at the Annual Pacific Marine Fisheries Commission meeting held at Coeur D'Alene, Idaho, Nov. 21, 1968, George R. Snyder reported that anadramous fish have been blocked in the Okanogan River, Washington by high water temperatures and that temperature blocks to fish migration have been observed near the confluence of the Snake and Columbia Rivers.
100. McCullough and Vayder, Delta-Suisun Bay Water Quality and Hydraulic Study, Journal of the Sanitary Engineering Division 801-27 (Proceedings of the American Society of Civil Engineers, Oct. 1968).

has the capability of generating environmental bureaucracy of unprecedented proportions. Many federal and state agencies have strengthened their role in enforcing pollution standards, but they often compete with each other to do the most environmental good. As a result of this interagency competition many massive projects which might further degrade the environment and those projects that are needed to correct environmental damage are caught up in a web of paperwork, hearings and controversy.

The federal role in estuaries should be limited to technical support and financial assistance administered by one central agency. Although the creation of the Environmental Protection Agency is a positive step in this direction, residual power in other federal agencies[101] creates external conflicts. Because the federal government is too far removed from the geographically and politically scattered estuaries, quality control can be most effectively exercised at the state level.[102] Municipal governments on the other hand have limited financial resources, and their attempts at controlling the

extensive estuarine area are often faced with a wealth of private interests that frustrate conservation-related regulations. The states in partnership with local governments have both the legal basis and the administrative ability to provide the means by which the estuarine environment can be protected and enhanced. The federal government could by example and by deed greatly alleviate estuarine pollution. The Corps of Engineers through control of the placement of polluted dredging spoils and the Navy through control and removal of toxic painting wastes and shipboard sewage could end damages from what will constitute the last remaining sources of pollution in many estuaries.

Recent examples in California have shown that estuaries can be effectively managed.[103] Single-minded, single-purpose programs that attempt to separate them from the total environment and assign responsibility for their management to the federal government will result in nothing more than a continuation of the present situation. State controlled management programs based on land-use planning and consideration of the total effect on the environment can be administered within a framework of cooperation between state and local government to produce maximum protection and enhancement of the estuary.

101. For instance, the Department of Housing and Urban Development is involved with land-use planning, The Environmental Protection Agency is concerned with envrionmental controls, and the Corps of Engineers and Bureau of Reclamation is concerned with public works projects.
102. The unanimous conclusion of three federal studies was that responsibility for the management of estuaries should reside with the states. See National Estuarine Pollution Study, *supra* note 31, at V-259; The National Estuary Study, *supra* note 4, at 73; and Our Nation and the Sea, A Plan for National Action, *supra* note 12, at 8.

103. Except for the persistent problem of vessel waste pollution, San Diego Bay is a clean bay as a result of an areawide waste treatment and disposal system. Water quality control programs have substantially improved the quality of the Los Angeles Harbor and the variety and number of marine organisms are increasing in San Francisco Bay.

Ocean Pollution Seen from Rafts

THOR HEYERDAHL   1971

There are times when an observation is made by someone who is looking for something else. This was the case when the present speaker experimented with prehistoric types of watercraft to travel back into man's past, and yet stumbled upon three practical observations which have a bearing upon man's future:

1. The ocean is not endless.
2. There is no such thing as territorial waters for more than days at a time.
3. Pollution caused by man has already reached the farthest section of the world ocean.

It may seem superfluous to state that the ocean is not endless, something the world has known since Columbus crossed the Atlantic in 1492. Yet I dare insist that this fact has not sufficiently penetrated our minds, we all subconsciously act as if the ocean has horizons running into the endless blue sky. When we build our city sewers in pipes far enough into the sea, when we dump poisonous refuse outside territorial borders, we feel we dispose of it for ever in a boundless abyss. We have known for centuries that our planet has no edges and that the oceans interlock in a never-ending curve around the world, but perhaps it is this uninterrupted curve that gives us the feeling of endlessness, this feeling that the ocean somehow continues to curve into space. From all continents we keep on sending our refuse into the presumably endless ocean almost with the impression that we send it away into space. Rarely do we stop and think of the fact that the ocean is nothing but a very big lake, landlocked if we go far enough in any direction. Other than being the largest body of water on earth, its main distinction from other lakes is that they usually have an out-

Dr. Thor Heyerdahl is an internationally known ethnologist and explorer. He is perhaps best known for his *Kon-Tiki* expedition on which six Norwegian scientists drifted westward on a model of a prehistoric Inca balsa-wood raft from Peru to the Polynesian atoll Raroia in the Tuamotu Archipelago—a distance of 4300 miles. This expedition tested the theory that some inhabitants of these islands came from South America. More recently, his *Ra I* and *II* expeditions gained worldwide attention when his international crew of seven men traveled in papyrus rafts from Africa to the Caribbean Islands. These expeditions tested the theory that the pyramid builders of the Western Hemisphere, particularly in Mexico and Central America, were influenced by the pyramid builders in Egypt.

From a statement read before the United States Senate Committee on Ocean and Atmosphere on November 8, 1971.

let to carry away excessive natural solutions and pollution, whereas the ocean has none. Through a worldwide, non-stop flow, all the excess waste and refuse that run from lakes and land assemble in the ocean, and only clean water evaporates back into the atmosphere. There were days not far back when our ancestors would laugh at the idea that man could pollute and kill a lake so big that nobody could see across it. Today Lake Erie is only one of a long series of lakes destroyed by man in the most different parts of the world. Place ten Lake Eries end to end and they span the ocean from Africa to America. True, the ocean is deeper than any lake, but we all know that due to photosynthesis the bulk of life is restricted to the thin upper layer, and we also know that an estimated 90 per cent of all marine life happens to be on the continental shelves which represent only 10 per cent of the total ocean area. Add to this that if half a dozen towns send their refuse into Lake Erie, all the cities, all the farmlands, all the rivers and ships of the world channel their refuse into the ocean, directly or in a roundabout way. No wonder then, that a time has come when even the world ocean has begun to become visibly polluted.

This discovery, which was first forced upon me while drifting at surface level in the mid-Atlantic in 1969, helped to open my own eyes to the fact that the ocean has its limits, and the closer one gets to know it the more easily this can be perceived. When we rush across it with engine-driven craft we feel that it is thanks to the modern traveling speed that the continents seem to be not so immensely far apart. But when you place yourself on a primitive raft and find that, entirely without engine or modern means of propulsion, you drift across the largest oceans in a matter of weeks, then you realize that you made it, not because of modern technique but quite simply because the ocean is not at all endless. With a speed slightly faster than that of average surface pollution, I crossed the Pacific from South America to Polynesia on nine balsa logs in

1947, and, on bundles of reeds, from Africa to the Caribbean Islands almost twice within a year in 1969-70. Each of these oversea voyages on an aboriginal type of watercraft was intended as an eye-opener for fellow anthropologists who, like the average layman, have retained the universal concept of the ocean as an endless waste, unsurmountable by pre-Columbian craft because of its boundless dimensions. This concept is wrong, and we run the risk of harming ourselves dearly unless we abandon this medieval concept of the endless sea and accept the fact that the ocean itself is nothing more than a big, salt lake, limited in extent and vulnerable as all the smaller bodies of water.

A second dangerous illusion equally hard-to-die is the image of territorial water. We draw a line parallel to the coast, three miles, ten miles, or a hundred miles off shore, and declare the inside as territorial water. There is no such thing as territorial water, the ocean is in constant motion, like the air. We can draw a line on the ocean floor and lay claims to the static land on the bottom, but the body of water above it is as independant of the map as is the atmosphere above dry land: wind and currents disregard any national border lines. Refuse dumped inside Peruvian territorial waters equals refuse dumped around the shores of Polynesia; refuse dumped inside Moroccan territorial waters equals refuse dumped in the Caribbean Sea. Any liquid piped into the port of Safi in Morocco, and any buoyant material dumped outside the local breakwaters just where our papyrus bundle-boat was let adrift, will run along as on a river straight to tropic America where some will wash against the beaches and some will move on up along the east coast of the United States. Moroccan territorial waters in a matter of weeks or months become American territorial waters, with all the good and evil this may involve. The salt sea is a common human heritage, we can divide the ocean floor between us, but we shall for ever be deemed to share the common water which rotates like soup in a boiling kettle: the spices one nation puts in will be tasted by all the consumers.

Only when we abandon the almost superstitious awe for the immensity of the sea, and the misconception of coastal water as a stagnant body, can we fully understand what is happening when visible pollution is scattered the full length of the North Atlantic surface current which flows perpetually from northwest Africa to tropical America. This entire span of the ocean, from continent to continent, contain among other modern refuse an immeasurable quantity of small drifting oil clots. They were accidentally noticed during the crossing with the papyrus raft-ship *Ra I* in 1969 and deliberately surveyed and sampled the next year during the crossing with *Ra II*.

In fact, in organizing our marine experiment with the first papyrus vessel ever to be tested at sea in modern times, our expedition group was initially unprepared for pollution studies. The objectives of the enterprise were to investigate the sea-going abilities and geographical range of the oldest type of watercraft used by man's earliest civilizations in the Mediterranean world as well as in Mexico and Peru, and furthermore to test the effects of multi-national cooperation in cramped quarters and under stress. We were seven men from seven nations on *Ra I* and eight from eight nations on *Ra II*. At sea, however, early in the voyage of *Ra I*, pollution observations were forced upon all the expedition members by the conspicuous presence of oil clots and undoubtedly also because of our own proximity to the ocean surface and slow progress through the water.

Departing from the Moroccan port of Safi on the northwest coast of Africa on May 25th, 1969, the seven men on board *Ra I* became aware of traveling in polluted water for the first time on June 6th, at 24°38'N and 17°06'W, or about a hundred miles (160 km) off the coast of Mauritania. The sea was now rolling calmly and we noticed the surface to be densely scattered with brownish to pitch-black lumps of asphalt-like material

as big as gravel and floating at close intervals on and just below the surface. The clots were drifting with the surface current in our direction, but benefiting more from the tradewinds we moved considerably faster, averaging a speed of about 2 to 2.5 knots. The local current speed is about 0.5 knots. Knowing that our reed-vessel was near the circum-African shipping lane, we climbed the mast and began to scout for ships, being convinced that we had entered the wake of some nearby oil tanker that had just cleaned its tanks. No ship was seen. On June 8th, having advanced about a hundred miles farther to the southwest, we found ourselves again sailing through similarly polluted water, still without any ship in sight. The following day we sailed into an area of the ocean where the same flotsam included pieces of larger size, some appearing as thick, black flakes of irregular shape up to 5-6 inches in diameter. The local ocean water itself gradually turned into an opaque and greyish-green color instead of being transparent and clear blue; it was recorded in the expedition journal as resembling harbor water at the outlet of city sewers.

Although sporadic lumps were noted, no specific entry was made in the expedition log until June 30th, when our position was at 15°45'N and 35°08'W, that is virtually in the mid-Atlantic with Africa and America almost at the same distance. Here once more we suddenly entered an area so polluted that we had to be attentive in washing ourselves or dipping our toothbrushes into the water, to avoid the seemingly endless quantities of oil clots of sizes ranging from that of a grain of rice to that of a sandwich.

Ra I covered 2700 nautical miles (*ca* 5000 km) in 54 days, and on July 15th and 16th, shortly before abandoning the test vessel we found ourselves again in the same general type of polluted water. Our position was now 13°32'N and 47°20'W, or some six hundred miles (960 km) east of the island of Barbados and slightly closer to the mainland coast of South America. Many of the clots had an eroded or pitted surface, and small barnacles as well as algae were occasionally seen growing on them.

Some samples were collected and at the end of the voyage delivered with a brief report to the Permanent Norwegian Delegation at the United Nations. Although no deliberate or preconceived observations were made, the voyage with *Ra I* resulted in the involuntary recording of six day's traveling through visibly strongly polluted water in the course of eight weeks of trans-Atlantic sailing. Thus, more than 10 per cent of the surface water traversed by *Ra I* was visibly polluted by a rich flotsam of nonorganic material of rather homogeneous appearance and undoubtedly resulting from modern commercial activity.

Our report to the United Nations in 1969 aroused a general interest, not least among scientists and shipping authorities, and prepared for what we might again encounter, we decided to keep a systematic record of daily observations when we embarked on the voyage of *Ra II* the following year. *Ra II* was again launched outside the breakwaters of the same ancient port of Safi in Morocco, this time on May 17th, 1970. As the water along the west coast of Africa and in the latitudes where we were to undertake the Atlantic crossing is not at all stagnant, but moves toward America with a speed of 0.5 knots or more, it is clear that we did not voyage through the same surface water this second time. In fact, the surface water observed by us from *Ra I* had been displaced more than four thousand miles during the year that had passed between the departures of the two consecutive raft expeditions. In other words, the water which we traversed along the African coast in May 1969, had long since deposited its flotsam along the Caribbean shores or else carried it into the initial part of the Gulf Stream, by the time we embarked on the second voyage in 1970. Correspondingly, the water seen around us as we abandoned *Ra I* short of Barbados in July 1969, would this subsequent year be on its return flow with the Gulf Stream back across the North Atlantic, heading for Europe. Nothing could impede this eternal circulation of ocean water, westward near the Equator and eastward in the far north, caused by the rotation of the earth itself. Thus the pollution we saw during *Ra II* was wholly independent of anything we observed on *Ra I*.

During our experiment with *Ra II,* in addition to the regular entries in the expedition log, a special pollution record was kept by Madani Ait Ouhanni, who also at reasonable intervals collected samples of the asphalt-like clots which, toward the end of the voyage, were handed to the United Nations' research vessel *Calamar* for subsequent transfer to the Norwegian UN Delegation. The samples were taken by means of a fine-meshed dip-net. It should be noted that in the rippled seas oil-clots were difficult to detect unless washed on board or drifting past very close to our papyrus deck. Only when the wave surface was smooth, or the floating objects were of conspicuous size, was it possible to detect and record pollutants passing more than six or eight feet away from the *Ra.* Thus, the considerable quantities of oil clots and other floating refuse which were found to float close alongside our papyrus bundles reflect the true dimensions of the problem is estimated in a broader geographical scope. It should also be noted that the route followed by *Ra II* was straighter and somewhat more northerly than that of *Ra I* which constantly broke the rudders and was forced on a drift voyage down beyond the latitude of the Cape Verde Islands.

On the background of these facts, it is disheartening to report that drifting oil clots were observed forty out of the fifty-seven days it took *Ra II* to cross from Safi to Barbados. This is 72 per cent of the traveling time spent in water where oil clots could be seen. From May 17, 1970 when we left the port in Morocco (at 32°20'N and 9°20'W) until and including June 28 when we had reached 15°54'N and 45°56'W, we recorded oil pollution on forty days out of forty-three. On the three days when pelagic oil lumps were not seen, Ouhanni's entries in the pollution record state that the sea was too rough for proper observation. It may thus be safely assumed that the 2407 nautical miles (4350 km) covered by *Ra II* during the initial 43 days of its voyage represented an uninterrupted stretch of polluted surface water, the degree of visible pollution varying from slight to very grave. It is slightly encouraging to note, however, that with the exception of some sporadic lumps observed on July 30th, no record of such particles was made during the remaining 700 miles to Barbados. This curious fact should not delude us through, since this was the very area where we noted extreme pollution the previous year. Also, on our arrival in Barbados, the owner of our east coast hotel reported that oil clots were sometimes so common on this beach that it was a problem to keep carpets clean from lumps that had stuck to his clients' feet.

Perhaps the sudden disappearance of oil clots in front of the Caribbean Islands during the 1970 crossing can be ascribed to a temporal irregularity in the local movement of water. The disappearance of the clots coincided with the sudden arrival of feeders from northbound branches of the South Equatorial Current, which were noticed both in our own displacement and simultaneously indicated by sudden changes in water temperature. Nevertheless, although the seemingly ever-present oil lumps disappeared this time, plastic containers and other imperishable manmade objects were observed sporadically until the last day of our crossing.

The average extent of oil pollution recorded during the voyage of *Ra II* amounted to lumps of asphalt-like material the size of finger-tips or smaller, scattered far apart in otherwise clean water. There would be days when only a very few such lumps could be seen from sunrise to sunset, whereas in exceptional cases the water was so polluted that a bucket could not be filled without some floating clots being caught at the same time.

The first very seriously polluted water was entered by *Ra II* four days after departure, on May 21st, at 29°26'N and 11°40'W, about 100 nautical miles off the African coast before we entered the passage between the Canary Islands and Morocco. From early that morning until the evening of the following day, *Ra II* was drifting very slowly through calm water that was thickly polluted by clusters of solidified oil lumps commonly of the size of prunes or even potatoes. Many of these lumps were dark-brown, mousy, and pitted, more or less covered by marine growth, whereas others were smooth and black, with the appearance of being quite fresh. For a duration of two days, the surface water, containing large quantities of these lumps, was also covered intermittently by a shallow white foam such as develops from soap or synthetic washing powder, while occasionally the ocean's surface was even shining in rainbow-colors as from gasoline. The sea was smooth and a vast quantity of dead coelenterates could be seen for considerable distances on both sides of our track. The expedition journal recorded that "the degree of pollution is shocking."

The following week only sporadic lumps were noticed, until on May 29th, at 25°43'N and 16°23'W, when our records again show that "the pollution is terrible." During the previous night, oil lumps, of which the biggest were the size of a large fist, had been washed on board during darkness, to remain as the water filtered through the papyrus as through the fringes of whalebone. Barnacles, marine worms, crustaceans, and sometimes bird feathers, were found attached to the oil lumps. The high degree of pollution was this time witnessed for three consecutive days, when swimming inevitably meant colliding with the sticky clots. On May 31st, at 25°00'N and 17°07'W, the expedition journal has the following entry: "An incredible quantity of shell-covered asphalt lumps today, big as horse-droppings and in clusters everywhere. One plastic bottle and one metal oil-can also observed, plus a large cluster of greenish rope, and nylon-like material besides a wooden box and a carton. It is shocking to see how the Atlantic is getting polluted by Man." No ships were seen in the vicinity.

The next entry into seriously polluted waters was on June 16th. At 18°26'N and 34°28'W, again virtually in the mid-Atlantic, the surface of the waves and as far as we could see below contained endless quantities of large and small oil lumps.

Ra II completed its Atlantic crossing on July 12th, 1970, landing on Barbados after covering 3270 nautical miles (*ca* 6100 km) in fifty-seven days. Although pelagic oil clots represented the most consistently recurring type of visible pollution during the two *Ra* voyages, it should be made clear that other debris from man of a rather heterogeneous kind was also common, even where oil was absent. Thus, in 1970, pollution in the form of plastic containers, metal cans, glass bottles, nylon objects, and other perishable and nonperishable products of man, representing refuse from ships and shores, passed close by the sides of our raft at intervals from the day of departure to the day of landing.

This was in marked contrast to our experience during the voyage of the raft *Kon-Tiki* two decades earlier. A noted aspect of that voyage, which then took place in the Pacific, was that not a single oil clot, in fact not a single sign of Man's activities, was seen during the 4300 miles crossing. From the day we left Callao in Peru until we landed on Raroia atoll in Polynesia 101 days later, we were constantly impressed by the perfect purity of the sea. The first trace of other human beings observed was the wreck of an old sailing vessel thrown up on the reef where we landed. Although, in fact, the contrast refers to two different oceans, the currents rotate between them and the difference between observations in 1947 and 1970 is so marked that it probably has some bearing on the rapidity with which we pollute the sea.

Through the State Department of Norway, a meeting was arranged between representatives from different scientific institutions and the oil industry who were invited to discuss an analytic program for the oil clot samples collected by the *Ra II* expedition. The analytic program was designed to determine whether the samples represented crude or refined oil, and also to estimate the origin, whether it could be leakage from drill, scattered oil from a single wrecked super-tanker like *Torrey Canyon,* oil from marine organisms, or mixed discharges from many different vessels. The analytic work was carried out by the Central Institute for Industrial Research in Oslo, and their findings can be summarized as follows:

The results of the infrared spectras show that the samples consist mainly of saturated hydrocarbons or mineral oil. Some samples seem to contain compounds from decomposed crude oil or heavy fuel oil. Vegetable and animal oils are apparently absent. According to the results of a gas chromatographic analysis, the saturated hydrocarbons were normal paraffins (n-paraffins) with 14 to 40 carbon atoms in each molecule with maximum around 20 and 30 carbon atoms. Such n-paraffins are generally, but not exclusively, the major fractions in mineral oil from the U.S.A. and North Africa. The samples showed a wide range in their contents of nickel and vanadium which indicate that they have derived from geographically different sources. In short, the conclusion was that the countless oil clots drifting about from continent to continent represent crude oil pollutants not from one leakage or one wreck, but from different sources. We are hardly far off then if we suspect the major part of the oil clots to be the scattered refuse from the numerous tankers which daily discard their ballast water at sea before entering a port of loading.

It was not an objective of the *Ra* expeditions to draw biological or ecological conclusions from our observations. Our aim is merely to call attention to observations that were virtually forced upon us by our prolonged proximity to the surface of the sea. Yet, one cannot refrain from certain deductions. Clearly, the time has passed when ocean pollution was a mere offense to human aesthetics because the surf throws oil and rubbish upon the holiday-makers' beaches. Much has been written about the tendency of oil molecules to expand in thin layers over wide areas of water, thus impeding the photosynthesis needed by the oxygen-producing phytoplankton. Those of us who sat on the two *RA*s observing fishes, large and small, nibbling at any floating particle wonder how the almost ever-present oil clots can avoid affecting the metabolism of the marine fauna and flora; not least the filter-feeding fishes and whales which swim with open mouth and, like the reed-bundles of *Ra,* let the water sieve through whereas plankton and oil clots alike get stuck in gills, whalebones or intestines. Small fish may get wise to the presence in their own element of unpaletable oil clots, but larger marine species have no way of gaping over plankton without taking in nonorganic material floating alongside as well. In addition, the oil lumps examined showed that they very frequently provided a foothold for live organisms which ride along as a sort of bait attracting the attention of bypassing fish. I am referring here to the fact that small *Cirripedia,* or edible barnacles (identified as *Lepas pectinata*) were very commonly sitting in regular clusters on the lumps. Various edible crustaceans were also frequently found clinging to the lumps, notably an isopod (*Idothea metallica*) and a small pelagic crab (*Planes minutus*). Marine worms hid in the pitted surface, and the shell of a tiny dead cuttlefish (*Spirula spirula*) was found in one sample.

In closing, I may be permitted a personal remark. A much more far-reaching study than our improvised sampling will be needed before we can judge the durability and effects of this steadily increasing flotsam of oil and debris. Perhaps bacterial activity and disintegration will finally sink or efface the oil from the ocean's surface, but certainly not before a large percentage is washed up against the continental and island shores. Having first personally witnessed the almost uninterrupted host of clots rotating about in the mid-ocean, I have subsequently visited some shorelines of the three continents bordering on the land-locked Mediterranean Sea and found a belt ranging in color from grey to black along the waterline of cliffs exposed to the polluted surf. In certain areas, like on the otherwise attractive island of Malta, it is as if the entire coastline to a height of six or eight feet above water-level has been smeared by a black impregnation. Where the invisible marine paintbrush has been at work there is no sign of life, neither algae nor molluscs, crustaceans or any other marine species naturally at home on such rocks. The coastal cliffs and reefs represent, as we know, a major breeding place for pelagic plankton and a necessary stepping stone in the life cycle of a great many of the species of paramount importance to Man.

I stress again, there are few things as illusive as the concept of territorial waters. What others dump at sea will come to your shores, and what you dump at home will travel abroad irrespective of national legislation. We must start at national level, but we must quickly move on to international agreements if we shall be able to protect our common ocean for future generations.

Scientific Aspects of the Oil Spill Problem

MAX BLUMER   1971

THE EXTENT OF MARINE OIL POLLUTION

Oil pollution is the almost inevitable consequence of our dependence on an oil-based technology. The use of a natural resource without losses is nearly impossible and environmental pollution occurs through intentional disposal or through inadvertent losses in production, transportation, refining and use. How large is the oil influx to the ocean? The washing of cargo tanks at sea, according to the director of Shell International, Marine Ltd.[1] had the potential in 1967 of introducing 2.8 million tons into the ocean, assuming that no use was made of the Load on Top (LOT) technique. With the increase in ocean oil transport from 1967 to 1970 this potential has grown to 6 million tons. The LOT technique is not being applied to one quarter of the oil tonnage moved by tankers; conse-

quently, these vessels introduce about 1.5 million tons of oil into the sea. The limitations of the LOT technique have been described by E. S. Dillon[2]: the technique is not always used even if the equipment exists, the equipment may be inadequate, shore receiving facilities may be lacking and principal limitations lie in the formation of emulsions in heavy seas or with heavy crude oils. Insufficient time may be available for the separation of the emulsion or the oil water interface may not be readily recognized. In addition the most toxic components of oil are also readily soluble in water and their disposal into the ocean could be avoided only if clean ballasting were substituted for the LOT technique. For these reasons it is estimated that the present practices in tanker ballasting introduce about 3 mil-

1. Statement by J. H. Kirby, quoted by J. R. Wiggins, Washington Post, March 15, 1970.

2. Dillon, E. Scott, "Ship Construction and Operation Standards for Oil Pollution Abatement," presented to a Conference on Ocean Oil Spills, held by the NATO Committee on Challenges of Modern Society, Brussels, November 2-6, 1970.

Dr. Max Blumer is an organic geochemist and senior scientist with Woods Hole Oceanographic Institution, Massachusetts. He studies natural hydrocarbons and pigments.

From *Environmental Affairs*, Vol. 1, No. 1, pp. 54-73, 1971. Reprinted with light editing and by permission of the author and the Environmental Law Center of the Boston College Law School. This paper was presented to a Conference on Ocean Oil Spills, held by the NATO Committee on Challenges of Modern Society, Brussels, November 2-6, 1970. Contribution No. 2616 of the Woods Hole Oceanographic Institution.

lion tons of petroleum into the ocean. The pumping of bilges by vessels other than tankers contributes another 500,000 tons.[3] In addition, in-port losses from collisions and during loading and unloading contribute an estimated 1 million tons.[4]

Oil enters the ocean from many other sources whose magnitude is much less readily assessed. Among these are accidents on the high seas (*Torrey Canyon*) or near shore, outside of harbors (West Falmouth, Mass.), losses during exploration (oil based drilling mud) and production (Santa Barbara, Gulf of Mexico), in storage (submarine storage tanks) and in pipeline breaks, and spent marine lubricants and incompletely burned fuels. A major contribution may come from untreated domestic and industrial wastes; it is estimated that nearly 2 million tons of used lubricating oil is unaccounted for each year in the United States alone, and, a significant portion of this reaches our coastal waters.[5,6]

Thus, the total annual oil influx to the ocean lies probably between 5 and 10 million tons. A more accurate assessment of the oil pollution of the oceans and of the relative contribution of different oils to the different marine environments is urgently needed. Such an assessment might well lie within the role of the NATO Committee on Challenges of the Modern Society.

With the anticipated increase in foreign and domestic oil production, with increased oil transport and with the shift of production to more hazardous regions (Alaska, continental shelf, deep ocean), we can expect a rapid increase of the spillage rate and of the oil influx to the ocean. Floating masses of

crude oil ("tar") are now commonly encountered on the oceans and crude oil is present on most beaches. Oil occurs in the stomach of surface feeding fishes[7] and finely dispersed hydrocarbons occur in marine plants (e.g. sargassum[8]) and in the fat of fish and shellfish.[6,9a,b] Hydrocarbons from a relatively small and restricted oil spill in the coastal waters of Massachusetts, U.S.A., have spread, nine months after the accident to an area occupying 5000 acres (20 km^2) offshore and 500 acres (2 km^2) in tidal rivers and marshes. The effect on the natural populations in this area has been catastrophic. The full extent of the coverage of the ocean bottom by petroleum hydrocarbons is unknown; chemical analyses are scarce or nonexistent.

EVALUATION OF THE THREAT

Oil: Immediate Toxicity

All crude oils and all oil fractions except highly purified and pure materials are poisonous to all marine organisms. This is not a new finding. The wreck of the *Tampico* in Baja, California, Mexico (1957) "created a situation where a completely natural area was almost totally destroyed suddenly on a large scale. . . . Among the dead species were lobsters, abalone, sea urchins, starfish, mussels, clams and hosts of smaller forms."[10] Similarly, the spill of fuel oil in West Falmouth, Massachusetts, U.S.A., has virtually extinguished life in a productive coastal and intertidal area, with a complete kill extend-

3. Statement by C. Cortelyou, Mobil Oil Company, quoted by W. D. Smith, The *New York Times*, April 19, 1970.
4. Blumer, M., "Oil Pollution of the Ocean," in *Oil on the Sea*, D. P. Hoult, ed., Plenum Press, 1969.
5. Anon., "Final Report of the Task Force on Used Oil Disposal," American Petroleum Institute, New York, N.Y., 1970.
6. Murphy, T. A., "Environmental Effects of Oil Pollution," Paper presented to the Session on Oil Pollution Control, American Society of Civil Engineers, Boston, Mass., July 13, 1970.

7. Horn, M. H., Teal, J. H. and Backus, R. H., "Petroleum Lumps on the Surface of the Sea," *Science* 168, 245, 1970.
8. Youngblood, W. W. and Blumer, M., unpublished data, 1970.
9a. Blumer, M., Souza, G., and Sass, J., "Hydrocarbon Pollution of Edible Shellfish by an Oil Spill," *Marine Biology* 5, 195-202, 1970.
9b. Blumer, M., Testimony before the Conservation and Natural Resources Subcommittee, Washington, D.C., July 22, 1970.
10. North, W. J., "Tampico, a Study of Destruction and Restoration," *Sea Frontiers* 13, 212-217, 1967.

ing over all phyla represented in that habitat (Hampson and Sanders[11] and unpublished data). Toxicity is immediate and leads to death within minutes or hours.[12]

Principally responsible for this immediate toxicity are three complex fractions. The *low boiling saturated hydrocarbons* have, until quite recently, been considered harmless to the marine environment. It has now been found that this fraction, which is rather readily soluble in sea water, produces at low concentration anaesthesia and narcosis and at greater concentration cell damage and death in a wide variety of lower animals; it may be especially damaging to the young forms of marine life.[13] The *low boiling aromatic hydrocarbons* are the most immediately toxic fraction. Benzene, toluene and xylene are acute poisons for man as well as for other organisms; naphthalene and phenanthrene are even more toxic to fishes than benzene, toluene and xylene.[14] These hydrocarbons and substituted one-, two-, and three-ring hydrocarbons of similar toxicity are abundant in all oils and most, especially the lower boiling, oil products. Low boiling aromatics are even more water soluble than the saturates and can kill marine organisms either by direct contact or through contact with dilute solutions. *Olefinic hydrocarbons,* intermediate in structure and properties, and probably in toxicity, between saturated and aromatic hydrocarbons are absent in crude oil but occur in refining products (*e.g.,* gasoline and cracked products) and are in part responsible for their immediate toxicity.

Numerous other components of crude oils are toxic. Among those named by Speers and Whitehead,[15] cresols, xylenols, naphthols, quinoline and substituted quinolines and pyridines and hydroxybenzoquinolines are of special concern here because of their great toxicity and their solubility in water. It is unfortunate that statements which disclaim this established toxicity are still being circulated. Simpson[16] claimed that "there is no evidence that oil spilt round the British Isles has ever killed any of these (mussels, cockles, winkles, oysters, shrimps, lobsters, crabs) shellfish." It was obvious when this statement was made that such animals were indeed killed by the accident of the *Torrey Canyon* as well as by earlier accidents; work since then has confirmed the earlier investigation. In addition, this statement, by its emphasis only on the adult forms, implies wrongly that juvenile forms were also unaffected.

Oil and Cancer

The higher boiling crude oil fractions are rich in multiring aromatic compounds. It was at one time thought that only a few of these compounds, mainly 3,4-benzopyrene, were capable of inducing cancer. As R. A. Dean[17] of British Petroleum Company stated, "no 3,4-benzopyrene has been detected in any crude oil . . .. [I]t therefore seems that the risk to the health of a member of the public by spillage of oil at sea is probably far less than that which he normally encounters by eating the foods he enjoys." However, at the time this statement was made, carcinogenic fractions containing 1,2-benzanthracene and alkylbenzanthracenes had already been isolated by Car-

11. Hampson, G. R., and Sanders, H. L., "Local Oil Spill," *Oceanus* 15, 8-10, 1969.
12. Sanders, H. L., Testimony before the Conservation and Natural Resources Subcommittee, Washington, D.C., July 22, 1970.
13. Goldacre, R. J., "The Effects of Detergents and Oils on the Cell Membrane," Suppl. to Vol. 2 of Field Studies, Field Studies Council London, 131-137, 1968.
14. Wilber, C. G., *The Biological Aspects of Water Pollution,* Charles C. Thomas, Publisher, Springfield, Ill., 1969.

15. Speers, G. C. and Whitehead, E. V., "Crude Petroleum," in *Organic Geochemistry,* Eglinton, G. and Murphy, M. R. J., eds., Springer, Berlin, 638-675, 1969.
16. Simpson, A. C., "Oil, Emulsifiers and Commercial Shell Fish," Suppl. to Vol. 2 of Field Studies, Field Studies Council, London, 91-98, 1968.
17. Dean, R. A., "The Chemistry of Crude Oils in Relation to their Spillage on the Sea," Suppl. to Vol. 2 of Field Studies, Field Studies Council, London, 1-6, 1968.

ruthers, Stewart and Watkins[18] and it was known that "biological tests have shown that the extracts obtained from high-boiling fractions of the Kuwait oil . . . (method) . . . are carcinogenic." Further, "Benzanthracene derivatives, however, are evidently not the only type of carcinogen in the oil. . . ." In 1968, the year when Dean claimed the absence of the powerful carcinogen 3,4 benzopyrene in crude oil, this hydrocarbon was isolated in crude oil from Libya, Venezuela, and the Persian Gulf.[19] The amounts measured were between 450 and 1800 milligrams per ton of the crude oil.

Thus, we know that chemicals responsible for cancer in animals and man occur in petroleum. The causation of cancer in man by crude oil and oil products was observed some years ago, when a high incidence of skin cancer in some refinery personnel was observed. The cause was traced to prolonged skin contact by these persons with petroleum and with refinery products. Better plant design and education, aimed at preventing the contact, have since reduced or eliminated this hazard.[20] However, these incidents have demonstrated that oil and oil products can cause cancer in man, and have supported the conclusions based on the finding of known carcinogens in oil. These references and a general knowledge of the composition of crude oils suggest that all crude oils and all oil products containing high boiling aromatic hydrocarbons should be viewed as potential cancer inducers.

Safeguards in plant operations protect the public from this hazard. However, when oil is spilled into the environment we loose control over it and should again be concerned about the possible public health hazard from cancer-causing chemicals in the oil. We have shown that marine organisms ingest and re-

18. Carruthers, W., Stewart, H. N. M. and Watkins, D. A. M., "1,2-Benzanthracene Derivatives in a Kuwait Mineral Oil," *Nature* 213, 691-692, 1967.
19. Graef, W. and Winter, C., "3,4 Benzopyrene in Erdoel," *Arch. Hyg.* 152/4, 289-293, 1968.
20. Eckardt, R. E., "Cancer Prevention in the Petroleum Industry," *Int. J. Cancer* 3, 656-661, 1967.

tain hydrocarbons to which they are exposed. These are transferred to and retained by predators. In this way even animals that were not directly exposed to a spill can become polluted by eating contaminated chemicals. This has severe implications for commercial fisheries and for human health. It suggests that marketing and eating of oil contaminated fish and shellfish at the very least increases the body burden of carcinogenic chemicals and may constitute a public health hazard.

Other questions suggest themselves: Floating masses of crude oil now cover all oceans and are being washed up on shores. It has been thought that such stranded lumps are of little consequence ecologically. It has been shown that such lumps, even after considerable weathering, still contain nearly the full range of hydrocarbons of the original crude oil, extending in boiling point as low as $100°C$. Thus such lumps still contain some of the immediately toxic lower boiling hydrocarbons. In addition, the oil lumps contain all of the potentially carcinogenic material in the 300-500° boiling fraction. The presence of oil lumps ("tar") or finely dispersed oil on recreational beaches may well constitute a severe public health hazard, through continued skin contact.

Low Level Effects of Oil Pollution

The short-term toxicity of crude oil and of oil products and their carcinogenic properties are fairly well understood. In contrast to this we are rather ignorant about the long term and low level effects of oil pollution. These may well be far more serious and long lasting than the more obvious short term effects. Let us look at low level interference of oil pollution with the marine ecology.

Many biological processes which are important for the survival of marine organisms and which occupy key positions in their life processes are mediated by extremely low concentration of chemical messengers in the sea water. We have demonstrated that marine predators are attracted to their prey by organic compounds at concentrations be-low the part per billion level.[21] Such chemical attraction—and in a similar way repulsion—plays a role in the finding of food, the escape from predators, in homing of many commercially important species of fishes, in the selection of habitats and in sex attraction. There is good reason to believe that pollution interferes with these processes in two ways, by blocking the taste receptors and by mimicking for natural stimuli. The latter leads to false response. Those crude oil fractions likely to interfere with such processes are the high boiling saturated and aromatic hydrocarbons and the full range of the olefinic hydrocarbons. It is obvious that a very simple—and seemingly innocuous—interference at extremely low concentration levels may have a disastrous effect on the survival of any marine species and on many other species to which it is tied by the marine food chain.

Research in this critical area is urgently needed. The experience with DDT has shown that low level effects are unpredictable and may suddenly become an ecological threat of unanticipated magnitude.

The Persistence of Oil in the Environment

Hydrocarbons are among the most persistent organic chemicals in the marine environment. It has been demonstrated that hydrocarbons are transferred from prey to predator and that they may be retained in organisms for long time periods, if not for life. Thus, a coastal spill near Cape Cod, Massachusetts, U.S.A., has led to the pollution of shellfish by fuel oil. Transplanting of the shellfish to clean water does not remove the hydrocarbons from the tissues. Oil may contaminate organisms not only at the time of the spill; hydrocarbon-loaded sediments continue to be a source of pollution for many months after the accident.

Oil, though lighter than water, does not remain at the sea surface alone; storms, or the uptake by organisms or minerals, sink the oil. Oil at the sea bottom has been found after the accidents of the *Torrey Canyon,* at Santa Barbara, and near Cape Cod. Clay minerals with absorbed organic matter are an excellent adsorbent for hydrocarbons; they retain oil and may transport it to areas distant from the primary spill. Thus, ten months after the accident at Cape Cod, the pollution of the bottom sediments covers an area that is much larger than that immediately after the spill. In sediments, especially if they are anaerobic, oil is stable for long time periods. Indeed, it is a key fact of organic geochemistry that hydrocarbons in anaerobic recent sediments survive for millions of years until they eventually contribute to the formation of petroleum.

COUNTERMEASURES

Compared to the number and size of accidents and disasters the present countermeasures are inadequate. Thus, in spite of considerable improvement in skimming efficiency since the Santa Barbara accident, only 10% of the oil spilled from the Chevron well in the Gulf of Mexico was recovered.[22] From an ecological point of view this gain is nearly meaningless. While we may remain hopeful that the gross esthetic damage from oil spills may be avoided in the future, there is no reason to be hopeful that existing or planned countermeasures will eliminate the biological impact of oil pollution.

The most immediately toxic fractions of oil and oil products are soluble in sea water; therefore, biological damage will occur at the very moment of the accident. Water currents will immediately spread the toxic plume of dissolved oil components and, if the accident occurs in inshore waters, the whole water column will be poisoned even if

21. Whittle, K. J. and Blumer, M., "Chemotaxis in Starfish, Symposium on Organic Chemistry of Natural Waters," University of Alaska, Fairbanks, Alaska, 1968 (in press).

22. Wayland, R. G., Federal Regulations and Pollution Controls on the U.S. Offshore Oil Industry, this conference.

the bulk of the oil floats on the surface. The speed with which the oil dissolves is increased by agitation, and in storms the oil will partly emulsify and will then present a much larger surface area to the water; consequently, the toxic fractions dissolve more rapidly and reach higher concentrations. From the point of view of avoiding the immediate biological effect of oil spills, countermeasures are completely effective only if *all of the oil is recovered immediately* after the spill. *The technology to achieve this goal does not exist.*

Oil spills damage many coastal and marine values: water fowl, fisheries, and recreational resources; they lead to increased erosion; they diminish the water quality and may threaten human life or property through fire hazard. A judicious choice has to be made in each case: which—if any—of the existing but imperfect countermeasures to apply to minimize the overall damage or the damage to the most valuable resources. Guidelines for the use of countermeasures, especially of chemical countermeasures, exist[23] and are being improved.[24] Some comments on the ecological effects and desirability of the existing countermeasures appear appropriate.

Detergents and Dispersants

The toxic, solvent-based detergents which did so much damage in the clean-up after the *Torrey Canyon* accident are presently only in limited use. However, so-called "nontoxic dispersants" have been developed. The term "nontoxic" is misleading; these chemicals may be nontoxic to a limited number of often quite resistant test organisms but they are rarely tested in their effects upon a wide spectrum of marine organisms including their juvenile forms, preferably in their normal habitat. Further, in actual use all dispersant-oil mixtures are severely toxic, because of the inherent toxicity of the oil, and bacterial degradation of "nontoxic" detergents may lead to toxic breakdown products.

The effect of a dispersant is to lower the surface tension of the oil to a point where it will disperse in the form of small droplets. It is recommended that the breakup of the oil slick be aided by agitation, natural or mechanical. Thus, the purpose of the detergent is essentially a cosmetic one. However, the recommendation to apply dispersants is often made in disregard of their ecological effects. Instead of removing the oil, dispersants push the oil actively into the marine environment; because of the finer degree of dispersion, the immediately toxic fraction dissolves rapidly and reaches a higher concentration in the sea water than it would if natural dispersal were allowed. The long term poisons (e.g. the carcinogens) are made available to and are ingested by marine filter feeders, and they can eventually return to man incorporated into the food he recovers from the ocean.

For these reasons I feel that the use of dispersants is unacceptable, inshore or offshore, except under special circumstances, *e.g.*, extreme fire hazard from spillage of gasoline, as outlined in the Contingency Plan for Oil Spills, Federal Water Quality Administration, 1969.[23,24]

Physical Sinking

Sinking has been recommended. "The long term effects on marine life will not be as disastrous as previously envisaged. Sinking of oil may result in the mobile bottom dwellers moving to new locations for several years; however, conditions may return to normal as the oil decays."[25] Again, these conclusions

23. Contingency Plan for Spills of Oil and Other Hazardous Materials in New England, U.S. Dept. Interior, Federal Water Quality Administration, Draft, 1969.
24. Schedule of Dispersants and Other Chemicals to Treat Oil Spills, May 15, 1970, Interim Schedule, Federal Water Quality Administration, 1970.

25. Little, A. D., Inc., "Combating Pollution Created by Oil Spills," Report to the Dept. of Transportation, U.S. Coast Guard, Vol. 1: Methods, p. 71386 (R), June 30, 1969.

disregard our present knowledge of the effect of oil spills.

Sunken oil will kill the bottom faunas rapidly, before most mobile dwellers have time to move away. The sessile forms of commercial importance (oysters, scallops, etc.) will be killed and other mobile organisms (lobsters) may be attracted into the direction of the spill where the exposure will contaminate or kill them. The persistent fraction of the oil which is not readily attacked by bacteria contains the long term poisons, *e.g.,* the carcinogens, and they will remain on the sea bottom for very long periods of time. Exposure to these compounds may damage organisms or render them unfit for human nutrition even after the area has been repopulated.

The bacterial degradation of sunken oil requires much oxygen. As a result, sediments loaded with oil become anaerobic and bacterial degradation and reworking of the sediments by aerobic benthic organisms is arrested. It is one of the key principles of organic geochemistry that hydrocarbons in anaerobic sediments persist for million of years. Similarly, sunken oil will remain; it will slow down the resettlement of the polluted area; and it may constitute a source for the pollution of the water column and of fisheries resources for a long time after the original accident.

For these reasons I believe that sinking of oil is unacceptable in the productive coastal and offshore regions. Before we apply this technique to the deep ocean with its limited oxygen supply and its fragile faunas we should gather more information about the interplay of the deep marine life with the commercial species of shallower waters.

Combustion

Burning the oil through the addition of wicks or oxidants appears more attractive from the point of view of avoiding biological damage than dispersion and sinking. However, it will be effective only if burning can start immediately after a spill. For complete combusion, the entire spill must ... by the combustion promoters, since ... will not extend to the untreated are ... practice, in stormy conditions, this may ... impossible to achieve.

Mechanical Containment and Removal

Containment and removal appear ideal from the point of avoiding biological damage. However, they can be effective only if applied immediately after the accident. Under severe weather conditions floating booms and barriers are ineffective. Booms were applied during the West Falmouth oil spill; however, the biological damage in the sealed-off harbors was severe and was caused probably by the oil which bypassed the booms in solution in sea water and in the form of wind-dispersed droplets.

Bacterial Degradation

Hydrocarbons in the sea are naturally degraded by marine microorganisms. Many hope to make this the basis of an oil removal technology through bacterial seeding and fertilization of oil slicks. However, great obstacles and many unknowns stand in the way of the application of this attractive idea.

No single microbial species will degrade any whole crude oil; bacteria are highly selective and complete degradation requires many different bacterial species. Bacterial oxidation of hydrocarbons produces many intermediates which may be more toxic than the hydrocarbons; therefore, organisms are also required that will further attack the hydrocarbon decomposition products.

Hydrocarbons and other compounds in crude oil may be bacteriostatic or bacteriocidal; this may reduce the rate of degradation, where it is most urgently needed. The fraction of crude oil that is most readily attacked by bacteria is the least toxic one, the normal paraffins; the toxic aromatic hydrocarbons, especially the carcinogenic polynuclear aromatics, are not rapidly attacked.

ment in bacterial oil
he complete oxida-
e oil requires all the
),000 gallons of air
herefore, oxidation
ere the oxygen con-
previous pollution
tion may cause ad-
ditional ecological damage through oxygen depletion.

Cost Effectiveness

The high value of fisheries resources, which exceeds that of the oil recovery from the sea, and the importance of marine proteins for human nutrition demand that cost effectiveness analysis of oil spill countermeasures consider the cost of direct and indirect ecological damage. It is disappointing that existing studies completely neglect to consider these real values.[17] A similarly one-sided approach would be, for instance, a demand by fisheries concerns that all marine oil production and shipping be terminated, since it clearly interferes with fisheries interests.

We must start to realize that we are paying for the damage to the environment, especially if the damage is as tangible as that of oil pollution to fisheries resources and to recreation. Experience has shown that cleaning up a polluted aquatic environment is much more expensive than it would have been to keep the environment clean from the beginning.[26] In terms of minimizing the environmental damage, spill prevention will produce far greater returns than cleanup— and we believe that this relationship will hold in a *realistic* analysis of the overall cost effectiveness of prevention or cleanup costs.

THE RISK OF MARINE OIL POLLUTION

The Risk to Marine Life

Our knowledge of crude oil composition and

of the effects of petroleum on marine organisms in the laboratory and in the marine environment force the conclusion that petroleum and petroleum products are toxic to most or all marine organisms. Petroleum hydrocarbons are persistent poisons. They enter the marine food chain, they are stabilized in the lipids of marine organisms, and they are transferred from prey to predator. The persistence is especially severe for the most poisonous compounds of oil; most of these do not normally occur in organisms and natural pathways for their biodegradation are missing.

Pollution with crude oil and oil fractions *damages the marine ecology* through different effects:

1. Direct kill of organisms through coating and asphyxiation.[27]

2. Direct kill through contact poisoning of organisms.

3. Direct kill through exposure to the water soluble toxic components of oil at some distance in space and time from the accident.

4. Destruction of the generally more sensitive juvenile forms of organisms.

5. Destruction of the food sources of higher species.

6. Incorporation of sublethal amounts of oil and oil products into organisms resulting in reduced resistance to infection and other stresses (the principal cause of death in birds surviving the immediate exposure to oil[28]).

7. Incorporation of carcinogenic and potentially mutagenic chemicals into marine organisms.

8. Low level effects that may interrupt any of the numerous events necessary for the propagation of marine species and for the survival of those species which stand higher in the marine food web.

26. Ketchum, B. H., *Biological Effects of Pollution of Estuaries and Coastal Waters,* Boston Univ. Press, 1970 (in press).

27. Arthur D. R., "The Biological Problems of Littoral Pollution by Oil and Emulsifiers—a Summing up," Suppl. to Vol. 2 of Field Studies, Field Studies Council, London, 159-164, 1968.

28. Beer, J. V., "Post-Mortem Findings in Oiled Auks during Attempted Rehabilitation," Suppl. to Vol. 2 of Field Studies, Field Studies Council, London, 123-129, 1968.

The degree of toxicity of oil to marine organisms and the mode of action are fairly well understood. On the other hand, we are still far from understanding the effect of the existing and increasing oil pollution on the marine ecology on a large, especially world-wide, scale.

Few, if any, comprehensive studies of the effects of oil spills on the marine ecology have been undertaken. Petroleum and petro-leum products are toxic *chemicals;* the long term biological effect of oil and its persis-tence cannot be studied without chemical analyses. Unfortunately, chemical analysis has not been used to support such studies in the past and conclusions on the persistence of oil in the environment have been arrived at solely by visual inspection. This is not sufficient; a sediment can be uninhabitable to marine bottom organisms because of the presence of finely divided oil, but the oil may not be visually evident. Marine foods may be polluted by petroleum and may be hazardous to man but neither taste nor visual observation may disclose the presence of the toxic hydrocarbons.

A coordinated biological and chemical study of the long-term effect and fate of a coastal oil spill in West Falmouth, Massa-chusetts, U.S.A. has shown that even a rela-tively low boiling, soluble and volatile oil persists and damages the ecology for many months after the spill. In this instance about 650 tons of #2 fuel oil were accidentally discharged into the coastal waters off the Massachusetts coast. I wish to summarize our present findings of the effect of this accident.

Persistence and Spread of the Pollution[9a,b,29]

Oil from the accident has been incorporated into the sediments of the tidal rivers and

marshes and into the offshore sediments, down to 42 feet, the greatest water depth in the sea. The fuel oil is still present in inshore and offshore sediments, eight months after the accident. The pollution has been spread-ing on the sea bottom and now covers at least 5000 acres offshore and 500 acres of marshes and tidal rivers. This is a much larger area than that affected immediately after the accident. Bacterial degradation of the oil is slow; degradation is still negligible in the most heavily polluted areas and the more rapid degradation in outlying, less af-fected, areas has been reversed by the influx of less degraded oil from the more polluted regions. The kill of bottom plants and ani-mals has reduced the stability of marshland and sea bottom; increased erosion results and may be responsible for the spread of the pollution along the sea bottom.

Bacterial degradation first attacks the least toxic hydrocarbons. The hydrocarbons remaining in the sediments are now more toxic on an equal weight basis than imme-diately after the spill. Oil has penetrated the marshes to a depth of at least 1-2 feet; bacterial degradation within the marsh sedi-ment is still negligible eight months after the accident.

Biological Effects of the Pollution[11,12]

Where oil can be detected in the sediments there has been a kill of animals; in the most polluted areas the kill has been almost total. Control stations outside the area contain normal, healthy bottom faunas. The kill as-sociated with the presence of oil is detected down to the maximum water depth in the area. A massive, immediate kill occurred off-shore during the first few days after the

29. This and the next two sections of the paper were written nine months after the West Falmouth oil spill. The following reports, giving the status after two years, are now available:
  (a) The Persistence and Degradation of Spilled Fuel Oil. *Science.* (1972) 176, 1120-1122.

  (b) The West Falmouth Oil Spill. I. Biology. Howard L. Sanders, J. Frederick Grassle, and George R. Hampson. WHOI 72-20.
  (c) The West Falmouth Oil Spill. II. Chemistry. Data Available in November, 1971. M. Blu-mer and J. Sass. WHOI 72-19.
These reports are available from the National Tech-nical Information Service, Springfield, Va. 22151.

accident. Affected were a wide range of fish, shellfish, worms, crabs and other crustaceans and invertebrates. Bottom living fishes and lobsters were killed and washed up on the beaches. Trawls in 10 feet of water showed 95% of the animals dead and many still dying. The bottom sediments contained many dead clams, crustaceans and snails. Fish, crabs, shellfish and invertebrates were killed in the tidal Wild Harbor River; and in the most heavily polluted locations of the river almost no animals have survived.

The affected areas have not been repopulated, nine months after the accident. Mussels that survived last year's spill as juveniles have developed almost no eggs and sperm.

Effect on Commercial Shellfish Values[9a,b]

Oil from the spill was incorporated into oysters, scallops, softshell clams and quahaugs. As a result, the area had to be closed to the taking of shellfish.

The 1970 crop of shellfish is as heavily contaminated as was last year's crop. Closure will have to be maintained at least through this second year and will have to be extended to areas more distant from the spill than last year. Oysters that were removed from the polluted area and that were maintained in clean water for as long as 6 months retained the oil without change in composition or quantity. Thus, once contaminated, shellfish cannot cleanse themselves of oil pollution.

The tidal Wild Harbor River, a productive shellfish area of about 22 acres, contains an estimated 4 tons of the fuel oil. This amount has destroyed the shellfish harvest for two years. The severe biological damage to the area and the slow rate of biodegradation of the oil suggest that the productivity will be ruined for a longer time.

Some have commented to us that the effects measured in the West Falmouth oil spill are not representative of those from a crude oil spill and that #2 fuel oil is more toxic than petroleum. However, the fuel oil is a typical refinery product that is involved in marine shipping and in many marine spillages; also, the fuel oil is a part of petroleum and as such it is contained within petroleum. Therefore, its effect is typical, both for unrefined oil and for refinery products. In terms of chemical composition crude oils span a wide range; many lighter crude oils have a composition very similar to those of the fuel oils and their toxicity and environmental danger corresponds respectively. However, many crude oils contain more of the persistent, long term poisons, including the carcinogens, than the fuel oils. Therefore, crude oils can be expected to have even more serious long term effects than the lower boiling fuel oils.

The pollution of fisheries resources in the West Falmouth oil spill is independent of the molecular size of the hydrocarbons; the oil taken up reflects exactly the boiling point distribution of the spilled oil. Thus, spills by other oils of different boiling point distributions can be expected to destroy fisheries resources in the same manner.

We believe that the environmental hazard of oil and oil products has been widely underestimated, because of the lack of thorough and extended investigations. The toxicity and persistence of the oil and the destruction of the fisheries resources observed in West Falmouth are typical for the effects of marine oil pollution.

The Risk to Human Use of Marine Resources

The destruction of marine organisms, of their habitats and food sources directly affects man and his intent to utilize marine proteins for the nutrition of an expanding population. However, the presence in oil of toxic and carcinogenic compounds combined with the persistence of hydrocarbons in the marine food chain poses an even more direct threat to human health. The magnitude of this problem is difficult to assess at this time. Our knowledge of the occurrence of carcinogens in oil is recent and their relative concentrations have been measured in very few oils. Also, our understanding of the fate of hydrocarbons, especially of carcinogens, in the marine food chain needs to be expanded.

Methods for the analysis of fisheries products for the presence of hazardous hydrocarbons exist and are relatively simple and the analyses are inexpensive. In spite of this no public laboratory in the United States—and probably in the world—can routinely perform such analysis for public health authorities. There is increasing evidence that fish and shellfish have been and are now being marketed which are hazardous from a public health point of view. Taste tests, which are commonly used to test for the presence of oil pollutants in fish or shellfish, are inconclusive. Only a small fraction of petroleum has a pronounced odor; this may be lost while the more harmful long term poisons are retained. Boiling or frying may remove the odor but will not eliminate the toxicity.

The Risk to the Recreational use of Marine Resources

The presence of petroleum, petroleum products and petroleum residue ("tar," "beach tar") is now common on most recreational beaches. Toxic hydrocarbons contained in crude oil can pass through the barrier of the human skin and the prolonged skin contact with carcinogenic hydrocarbons constitutes a public health hazard. Intense solar radiation is known to be one of the contributing factors for skin cancer. The presence of carcinogens in beach tar may increase the risk to the public in a situation where a severe stress from solar radiation already exists.

The Risk to Water Utilization

Many of the toxic petroleum hydrocarbons are also water soluble. Water treatment plants, especially those using distillation, may transfer or concentrate the steam-volatile toxic hydrocarbons into the refined water streams, especially if dissolved hydrocarbons are present in the feed streams or if

particulate oil finds its way into the plant intake.

CONCLUSIONS

1. Oil and oil products must be recognized as poisons that damage the marine ecology and that are dangerous to man. Fisheries resources are destroyed through direct kill of commercially valuable species, through sub-lethal damage and through the destruction of food sources. Fisheries products that are contaminated by oil must be considered as a public health hazard.

2. Only crude estimates exist of the extent of marine oil pollution. We need surveys that can assess the influx of petroleum and petroleum products into the ocean. They should be world-wide and special attention should be paid to the productive regions of the ocean; data are needed on the oil influx from tankers and non-tanker vessels, on losses in ports, on offshore and inshore accidents from shipping, exploration and production and on the influx of oil from domestic and industrial wastes.

3. The marine ecology is changing rapidly in many areas as a result of man's activities. We need to establish baseline information on composition and densities of marine faunas and floras and on the hydrocarbon levels and concentrations encountered in marine organisms, sediments and in the water masses.

4. All precautions must be taken to prevent oil spills. Prevention measures must be aimed at eliminating human error, at the present time the principal cause of oil spills.

5. Spill prevention must be backed by effective surveillance and law enforcement. *In terms of cost effectiveness spill prevention is far superior to cleanup.*

6. Perfection and further extension of the use of the Load on Top methods is promising as a first step in reduction of the oil pollution from tankers. The effectiveness of the technique should be more closely assessed and improvements are necessary in interface detection, separation and measure-

ment of hydrocarbon content in the effluent, both in the dispersed and dissolved state. On a longer time scale, clean ballast techniques should supersede the Load on Top technique.

7. The impact of oil pollution on marine organisms and on sources of human food from the ocean has been underestimated because of the lack of coordinated chemical and biological investigations. Studies of the effect of oil spills on organisms in different geographic and climatic regions are needed. The persistence of hydrocarbon pollution in sea water, sediments and organisms should be studied.

8. Research is urgently needed on the low-level and long term effects of oil pollution. Does oil pollution interfere with feeding and life processes at concentrations below those where effects are immediately measured? Are hydrocarbons concentrated in the marine food chain?

9. Carcinogens have been isolated for crude oil but additional efforts are needed to define further the concentrations and types of carcinogens in different curde oils and oil products.

10. The public health hazard from oil derived carcinogens must be studied. What are the levels of oil derived carcinogens ingested by man and how wide is the exposure of the population? How much does this increase the present body burden with carcinogens? Is there direct evidence for the causation of cancer in man by petroleum and petroleum products outside of oil refinery operations?

11. Public laboratories must be established for the analysis of fisheries products for toxic and carcinogenic chemicals derived from oil and oil products, and tolerance levels will have to be set.

12. The ocean has a limited tolerance for hydrocarbon pollution. The tolerance varies with the composition of the hydrocarbons and is different in different regions and in different ecological sub-systems. The tolerance of the water column may be greater than that of the sediments and of organisms.

An assessment of this inherent tolerance is necessary to determine the maximum pollution load that can be imposed on the environment.

13. Countermeasures which remove the oil from the environment reduce the ecological impact and danger to fisheries resources. All efforts should be aimed at the most rapid and complete removal since the extent of the biological damage increases with extended exposure of the oil to sea water.

14. Countermeasures that introduce the entire, undegraded oil into the environment should be used only as a last resort in situations such as those outlined in the Contingency Plan of the Federal Water Quality Administration, involving extreme hazard to a major segment of a vulnerable species of waterfowl or to prevent hazard to life and limb or substantial hazard of fire to property. Even in those cases assessment of the long term ecological hazard must enter into the decision whether to use these countermeasures (detergents, dispersants, sinking agents).

15. As other countermeasures become more effective, the use of detergents, dispersants and sinking agents should be further curtailed or abolished.

16. Efforts to intensify the natural bacterial degradation of oil in the environment appear promising and should be supported by basic research and development.

17. Ecological damage and damage to fisheries resources are direct consequences of oil spills. In the future, the cost of oil leases should include a fee for environmental protection.

18. Environmental protection funds derived from oil leases should be used to accomplish the necessary research and education in the oil pollution field.

ACKNOWLEDGMENTS

The author expresses his gratitude for continued support to the National Science Foundation, to the Office of Naval Research and to the Federal Water Quality Administration.

8

OCEAN LAWS
AND MANAGEMENT

Ecology, Law, and the "Marine Revolution"

CARLETON RAY   1970

INTRODUCTION

Man has not yet solved the age-old paradox upon which his civilizations have many times foundered; namely, that high population numbers with high cultural levels demand high environmental productivity, yet exploitation of Nature produces environmental destruction and ecological collapse. When the numbers of humans will come to exceed the total carrying capacity on Earth, as is already the case in many nations, no one can say; but if Man does not learn the lessons of history, there is no doubt that this catastrophic situation will occur relatively soon. The survival of Man, or anyway of civilization as we now know it, will surely depend upon how he handles this challenge.

There are two dominant features of the marine part of this challenge: first, the development of international law with enforcement for exploitation of the sea, and, second, the development of ecosystem-based conservation practices. The latter includes the cessation of existing destructive practices, the assessment of marine environments relative to the carrying capacity of Earth for Man, and the creation of marine parks, sanctuaries, and control areas for research. These ecological aspects have to date been attacked in a piecemeal fashion. Ultimately, the answers will depend upon value judgements about what sort of a world we choose to live in. The late Fairfield Osborn (1953) asked: "Is the purpose of our civilization really to see how much the earth and human spirit can sustain?"

This paper considers biology and law as they reflect upon what we may call the Marine Revolution. Biology and law require different approaches. The body of law by which we exercise control and responsibility is of Man's creation. It should reflect common sense and be capable of rational alteration. Natural phenomena may make no "sense" at all, and their complexities are

Dr. G. Carleton Ray is an associate professor with the Department of Pathobiology, School of Hygiene and Public Health at Johns Hopkins University, Maryland. He is director of the International Biological Program on Marine Mammals and serves on the National Academy of Sciences' Panel on Marine Aquatic Life and Wildlife. Physiology and acoustical mechanisms of seals and walruses are his research interests. He has published over fifty technical and popular articles.

From *Biological Conservation*, Vol. 3, No. 1, pp. 7-17, 1970. Reprinted with author's revisions and light editing and by permission of the author and *Biological Conservation*, Elsevier Publishing Company, Ltd., England.

infinite. It has been stated that the ecosystem is not only more complex than we think it is; it is also more complex than we can think. The ecologist can rarely be definitive. He often experiences great difficulty in explaining, even to some fellow scientists and especially to engineers and technicians, the real nature of the ecological crisis. Ehrlich's (1969) "Eco-catastrophe" sounds to many like alarmist stuff, yet it has a fundamental basis of perception.

To a great extent we are slaves of our own history. The *laissez faire* spirit of exploitation, the goal of economic growth, Man's socio-religious beliefs which separate him from Nature, and the conflict and case-history methods of law make little sense when applied to the environment. The emerging "Marine Revolution" poses to those concepts a challenge which magnifies the importance of the sea far beyond its resource value. The wide recognition that this is so is reflected by the numbers of recent symposia and reports on the exploration, use, and legal régimes of the sea. Unfortunately, meetings of the American Bar Association and the Marine Technology Society, among others, have been composed almost entirely of industry representatives, lawyers, and a scattering of government officials, naval personnel, and fisheries biolgists—the last mostly representing mission-oriented government agencies or industry. Marine ecologists have been virtually absent!

In spite of this, the intensifying debate has produced the beginnings of workable ideas. The ecosystem approach may be just over the horizon, the greatest present need being for marine ecologists to make their voices heard. If consideration for the ecosystem be added to the debate, it is possible that non-destructive and cooperative exploitation on an international basis will result, and perhaps then marine ecosystems will not suffer further.

THE MARINE REVOLUTION

Man's massive entry into "inner space" initiates what we are calling the Marine Revolution. It is resulting in increased resource utilization and new régimes for law, politics, and socio-economics, as Man investigates, uses, and hopefully will conserve, that three-quarters of the world's surface which has been mostly foreign to him.

Agricultural and Industrial Revolutions

Some thousands of years ago, Man began to grow his own food. This change from the hunter-gatherer to the agriculturalist comprized the Agricultural Revolution. It led to the diversity of occupations which marks present urban culture. The Agricultural Revolution produced more food in a more accessible form than was available to the hunter-gatherer. Food, which presumably had been a limiting factor, was limiting no more. The carrying capacity of land for humans rose and the population grew accordingly.

The Industrial Revolution has been going on for the last two centuries or more. It has been marked by the growth of science and technology, by increased resource-use, and by expanded diversity and efficiency of human skills. It has meant a turning away from the agricultural way of life to an increasingly urbanized and "artificial" one. It once again increased the carrying capacity of the land for human beings and led to a spectacular decrease in death control without concomitant birth control. Most significantly of all, the Industrial Revolution, in its greed for resources, has produced environmental destruction at an astounding and dangerous pace. Forests have been cut, land has been eroded and stripped, bays have been polluted and filled, and the result of all of these and other activities has been to lower the long-term carrying capacity of land for future human populations, notwithstanding the temporary increase which technology has made possible. Such environmental wastage makes our wish to provide a better life for our children seem to be sheer hypocracy.

The Marine Revolution

Thus does Man turn to the seas which become increasingly vital for his resources. However, the Marine Revolution is not totally a consequence of the exhaustion of the land. Man also turns to the sea as it lies before him in the form of a challenge which he is now becoming technologically able to accept. "Products are sold on an open world market that cares nothing about the origin of the material; one competes only against price" (Bascom, 1966).

Thus, we accept the challenge of the sea, being not a little starry-eyed over our technology. But we must remind ourselves that Man remains a hunter-gatherer in the oceans; in only an insignificant few places does he farm the sea. This contrast between developing technology and the inadequacies of cultural and legal frameworks for regulation is a characteristic of "revolution."

The Marine Revolution is, to my mind, quite as important a development as the previous Agricultural and Industrial Revolutions. It is no more obvious on a day-to-day basis than the Agricultural and Industrial Revolutions were in their time. Future Man will clearly see this Revolution as his inner-space logistics and utilization increase.

ECOSYSTEMS AND HOMEOSTASIS

The ecosystem is the fundamental functional unit of the natural world. It is comprised of all the living and non-living components of an environment and the totality of their interrelationships. An ecosystem has properties of self-sustainment. Solar energy must be added, but nutrients and other materials are recycled. Examples are a lake, a forest, an estuary, and a coral reef.

Carrying Capacity, Limiting Factors, and Synergisms

Carrying capacity may be defined as the number of individuals of a species within a particular ecosystem beyond which no major increase in numbers may occur. It fluctuates about an equilibrium level and may change seasonally or even daily. It is regulated according to Liebig's "law" of the minimum and Shelford's "law" of tolerance, which together state that the presence or abundance of an organism locally is determined by the amounts of critical materials available or by the local levels of environmental factors such as salinity or temperature.

It is typical of ecology that "laws" are easy to state but difficult to prove. A major reason for this is *synergism;* that is, environmental factors often act together to produce effects which are different quantitatively or qualitatively from the effects expected separately or additively. Carrying capacity and limiting factors apply to all living things. The foolish assumption is that technology may negate them for Man. Technology cannot alter ecological laws, though it can redirect utilization in limited ways.

Productivity

Productivity is determined by turnover rate. The standing crop or biomass is a poor indicator of this, as it tells little about how often materials are recycled. Plants absorb about one per cent of solar energy for photosynthesis. An examination of trophic levels from these producers to primary, secondary, or tertiary consumers, reveals that each step involves about a 90 per cent loss of energy. Thus, food-chains are usually short and each trophic level shows much lower total production than its predecessor.

Nutrients, unlike energy, are recycled. The biogeochemical cycles of gases, salts, and minerals, are most efficient in complex ecosystems. Man can occasionally increase productivity through the addition of substances which once were limiting. More often, his "making the desert bloom" fails in the long run through failure to recognize the interrelationships of these cycles.

Primary   productivity   varies   widely.

Deserts and the waters of the deep oceans, which together cover most of the Earth, produce less than one gram of dry organic matter per square meter per day. Grasslands, waters over the continental shelf, and marginal agriculture produce 0.5 to 3 gm; moist forests and agriculture produce 3 to 10 gm; estuaries, inshore seas, and intensive agriculture produce 10 to 25 gm (Odum, 1959).

Owing to their large total productive area and volume, the seas contain more living material than the land supports. However, Man's utilization is at a higher trophic level in the sea: land = sun → grass → cow; sea = sun → phytoplankton → zooplankton → primary carnivore (e.g. herring) → secondary carnivore (e.g. tunny). The seas contain a much greater total diversity of life in terms of classes of animals than does the land, but owing to the lower oxygen content of water than air, the seas are dominated by animals of lower metabolic rate, but higher ecological efficiency than birds and mammals. Lastly, the sea provides a more stable environment than the land; in it, the "weather" is mild and the productive season is long. For all these reasons, marine productivity is not equivalent to that of land.

Homeostasis, Simplification, and Pollution

Homeostatis defines the "balance of nature." All ecosystems depend upon recycling for sustainment and upon complexity for stability. These involve intricate mechanisms analogous to (but more complex than) the heat-producing, dissipating, and conserving mechanisms which regulate human body temperature. Ecosystems are never perfectly balanced, but homeostatic mechanisms give them recuperative power which, when exceeded, leads to breakdown; the eutrophication of Lake Erie is a classic example of such excess.

A major part of homeostasis lies in complexity which insures both productivity and stability, and also has aesthetic value for Man (Elton, 1958; Dasmann, 1968). Man is a simplifier of ecosystems and thus reduces their recuperative power. The many forms of pollution are the most serious stresses in this regard. Historically, Man has depended upon maximum homeostatic capacities of the environment to endure pollution; but in simplifying and polluting at the same time, he attacks with a two-edged sword.

Is the ocean too large to disrupt? I think not. According to the Task Force on Environmental Health and Related Problems (1967), the American people and their environment are being exposed to half-a-million different alien substances with 20,000 new ones being added each year. Some of these go to sea. For instance, pesticides have been distributed throughout the world's oceans through the vectors of air and precipitation (Frost, 1969). Polikarpov (1966) suggests that radionuclide pollution of the seas may already be at a dangerous level for some organisms. Hedgpeth (in press) remarks that our standards for waste disposal are anthropocentric and that laboratory tests on pollutants are "interesting, but possibly academic as far as the real world is concerned"—in other words, waste-level standards set for Man are not necessarily those which ecosystems will tolerate.

MAN'S USE OF THE SEA

Only recently has Man begun to explore the sea throughout its three dimensions. The first extensive exploration of the deep sea was in 1873-76 by *H.M.S. Challenger.* Not quite a century later, Man has visited the ocean's deepest place in a research submarine and knows that all marine waters are capable of supporting life.

The Marine Revolution consists of five major aspects, which are related to, but by no means coincidental with, its dominating challenges mentioned in the Introduction. These aspects are: fisheries, minerals and mining, military interests, science and technology, and conservation and recreation. Emery (1966) gives world values of marine resources in 1964 as follows: biological—US

6.4×10^9; geological—US 3.6×10^9; and chemical—US 1.3×10^9. Biological resources will always be the most valuable, even if surpassed economically, for Man cannot exist without them, and they are largely renewable.

Fisheries

Fisheries remain the most difficult aspect of international law of the sea. This is due mainly to the fact that most commercially important marine animals move and cannot be claimed. It is ludicrous to discover that certain benthic organisms are, in fact, classified as "minerals" under the Convention of the Continental Shelf. In some cases it is of advantage to the exploiter that they should be so classified, an instance being the Alaska king crab (Oda, 1968); in other cases the reverse is true, instances including some shrimps (Neblett, 1966). Fisheries resources include various algae, plankton, shellfish, fishes, turtles, and mammals (Walford, 1958); but, as has been pointed out above, Man's utilization represents only a fraction of total marine productivity.

Over-utilization continues to dominate fisheries, especially off-shore ones. Clark (1967) states that Japanese long-lining accounted for almost a million billfishes in 1965. Even larger quantities of tunny were taken. Evidence is accumulating that such utilization cannot be sustained. Perhaps even more serious than overfishing is inshore habitat destruction. Over two-thirds of all commercial and sport fishes of the eastern United States depend upon inshore environments at some critical time of their life-cycle. The most effective way to extirpate a species is by environmental disruption, and this is being done inshore at a rapid pace.

Consideration of energetics lead many to propose exploitation at lower trophic levels. Complex size/metabolic factors and fishing efficiency strongly indicate, however, that higher-order consumers are more effective fishermen and converters of energy than Man is. A total "plankton" fishery should be considered as a last, and none too satisfactory, resort. Those who have taste-tested swordfish and plankton might agree! The choice, however, should not be between swordfish and plankton; given proper management, we could have both.

The concept of "yield" is vital biologically and legally. Fisheries biologists have emphasized the asymptotic attainment of maximum biomass through controlled utilization. Such a yield may or may not conform to economic efficiency or to local market value—hence the preference of "optimum" over "maximum" yield (Crutchfield, 1968).

W. M. Chapman (1966) states an exploitive point of view: "When the fishing effort has increased beyond the point of maximum sustainable yield, the fishing can ordinarily be permitted to expand without serious damage to the resource." He ignores Allee's principle (Odum, 1959), which is that density is in itself a limiting factor for population growth and survival. Relative abundance of the species in a community is a contributor to homeostasis. Thus, it is biologically most sound to change population size as little as possible in natural systems.

Christy (1966) considers broader aspects of utilization: ". . . somehow or other it will be necessary to limit the number of fishermen that can participate in a fishery. Such limitations can be achieved only by further restricting the 'freedom of the seas'; and this clearly raises questions about the meaning of this freedom and about the distribution of wealth." This approach appears to me more susceptible to ecological application than Chapman's more narrowly-stated views.

Aquaculture presents different sorts of problems from hunter-gathering, and may be the dominant provider of the future. Aquaculture is a major concern of the U.S. Sea Grant Program (Abel, 1968). Ryther and Bardach (1968) and Bardach and Ryther (1968) review aquaculture and make the point that it will be carried out largely along coasts—exactly the areas currently most stressed at the hand of Man. To reconstitute coastal environments, or to fertilize them artifically, is difficult or impossible. The key to aquaculture is clearly the maintenance of natural productivity.

Minerals and Mining

Reading in this field often leaves one impressed with the viewpoint that somehow we are slaves to "economic growth." Close (1968) speaks of "the care and feeding of a gigantic industrial complex." One hopes that only a segment of industry would speak so carelessly, but it does appear true that an awareness of ecology and a willingness to exploit the non-living resources at little or no expense to the living are indeed rare. If mineral exploitation continues by sea as it has by land, the predicatable results are frightening to contemplate. Strip mining is one parallel example.

Mero (1966, 1968), Luce (1968), and Young (1968), review the diversity of mineral resources in the sea. Inshore mineral exploitation is already heavy, but a consensus exists that only a few minerals, such as oil and gas, are currently feasible of exploitation. This is evidently based upon the lack of a favorable legal and economic climate, not upon the lack of technological capability. Further, it is not true that exploitation will progress from shallower to deeper water, any such progress being a function of the resource sought (Wilkey, 1969).

Off-shore mineral production in 1968 was 6 per cent of the world total and of it oil and gas accounted for 84 per cent (Economic Associates, 1968). In 1965, 16 per cent of the free world's oil was produced off-shore, the result of the work of 325 rigs which have drilled many thousands of wells (Dozier, 1966); oil has been produced from wells in as much as 104 m of water (Wilkey, 1969), and exploratory drilling was carried out in 1968 in the Gulf of Mexico in over 3600 m. At any one time, about 30,000,000 tons of oil are at sea in tankers. From U.S. off-shore wells alone the production of oil has so far been 2×10^9 barrels,* and of gas

* 1 barrel = ca 200 liters; 1 cubic meter = ca 30 cubic feet.

5.5×10^{12} ft^3,* at an investment of U.S. $6 thousand million, and with the ultimate potential of 15-35 thousand million barrels of oil and 90-170 $\times 10^{12}$ ft^3 of gas (Nelson and Burk, 1966). The massive pollution potential of the oil industry has been previewed by the tragic *Torrey Canyon* and Santa Barbara disasters. We can be certain that these episodes are not the last of their kind, and probably there will be far bigger ones.

Military Interests

Military activities in the oceans are shrouded in secrecy. It would, for instance, be interesting to know what the degree of radionuclide pollution is from Soviet and U.S. nuclear-powered submarines. Both Harlow (1966) and Hearn (1968) give as the U.S. Navy's viewpoint the contention that maximum freedom to use all dimensions of the sea must be maintained in order to exploit naval strength to the fullest in the best national interest. I think it fair to state that such a position is shared by the military of other major powers. The effect is to raise a serious obstacle to internationalization, to expanded territorial management, and to peaceful use of the sea-floor.

It is difficult for me to understand why putting the sea-bed under a "peaceful purposes only" treaty, as has already been done for outer space and Antarctica, is not in the "best national interst." Evidently, military influence was a major factor in preventing that principle from being accepted at the 1967 United Nations debate on the subject (Eichelberger, 1968). As yet the sea-bed is not much utilized militarily, though the waters over the floor of the sea certainly are. Thus, it is particularly disturbing to read that "military strategists . . . have been looking for better ways to put the sea to use for the purposes of national defense" (New York Times, 1969).

It must be pointed out that military interests are not necessarily contrary to fishing or mineral exploitation. In any case, international progress on these last should not be held up by conflicts with the military authorities.

Science and Technology

The United States, among other nations, is heavily committed to marine exploration, science, development, and conservation. Reports on the highest level are numerous, including: Interagency Committee on Oceanography (1963, 1967); National Academy of Sciences (1964, 1967, 1969); Panel on Oceanography, President's Science Advisory Committee (1966); National Council on Marine Resources and Engineering Development (1967, 1968a, 1968b); and the Commission on Marine Science, Engineering and Resources (1969).

The last-mentioned, the so-called Stratton Commission Report, departs courageously from—while also building upon—the baseline established by its predecessors and is no doubt that most significant of them all. It is broadly ecological and international in nature, and recommends a U.S. National Oceanographic and Atmospheric Agency for centralization of U.S. research, exploration, data collection, and education. Further, it proposes an International Registry Authority for ocean claims—with régimes for ocean bottoms, a delineated continental shelf, and an intermediate zone. The Commission also stresses optimal use of coastlines on a long-term basis in which industry, water quality, and aquaculture would be regulated under Federal law to guard against deterioration of the inshore marine environment. A useful review of this Report, including both the pros and the cons, is provided by the Program of Policy Studies in Science and Technology (1969).

Looking not at reports, but at budgets, produces some dismay. Ocean Science News (1969) states the current U.S. Federal commitment to marine matters to be $528 million per annum, of which only $150.6 million is in basic and applied research, $143 million being in national security—and this in the very year of Man's travel to the moon and continued development of supersonic

transport! The overall oceanic budget has grown 22 per cent since 1968, when Economic Associates, Inc. (1968) remarked: "what remains to be pointed out is the very low level of Federal expenditure on ... resources and their environment, compared with Federal oceanologic programs in general and, decidedly so, with the Federal effort in such a field as outer space."

The International Biological Programme's Marine Productivity section deserves mention. The IBP theme of "The biological basis of productivity and human welfare" is ideally suited to the needs of Man during the initial period of the Marine Revolution. However, at the current level of funding (only U.S. $7 million for all U.S. IBP sections in fiscal year 1970), it is certain that IBP cannot fulfil its goals.

Conservation and Recreation

To many, conservation and recreation involve inter alia the establishment of parks, sanctuaries, and control areas for research (Ray, 1961, 1965, 1966, 1968; V. J. Chapman, 1968; Randall, 1969). However, conservation and recreation must not be confined to protected areas. Both must principally be concerned with the maintenance of ecosystem homeostasis on a worldwide basis, and this is a large order indeed.

The concepts of conservation have been developed for terrestrial environments and are only vaguely applicable to the sea. The oceans together occupy a vastly larger part of the biosphere than does the land, and they are more continuous. The sea's rate of change, its biotic complexity, and our ignorance of its three-dimensional hydrosphere, are of a different order of magnitude from their counterparts on the more familiar land. For both land and sea, modern conservationists have become less concerned with the placing of "fences" about sea or landscape, valuable as protective measures are, than with an ecological concept of the total ecosystem of which Man forms a part. A good basis of conservation policy exists for land and, in part, for inshore seas. For the high seas, this is not the case.

LEGAL RÉGIME OF THE SEA

Ultimately, Man's marine activities of all kinds must be legally regulated. Griffin (1967) states: "To a large extent, a period of legal conjecture is ending." The problem is "... to evolve policies and a legal régime which will maximize all beneficial uses of ocean space.... Under no circumstances, we believe, must we ever allow the prospects of rich harvest and mineral wealth to create a new form of colonial competition among the maritime nations." A contrary view is that of Ely (1967a): "Above all, we should not now cede to any international agency whatsoever the power to veto American exploration of areas of the deep sea which are presently open to American initiative. We can give away later what we now keep, but the converse is sadly false." Ely (1967b, 1968) later extended these views.

Basically, the argument concerns whether the sea and sea-floor are *res nullius* (belonging to no one but subject to claim) or *res communis* (property of the world community).

Eichelberger (1968) puts the matter another way when he says: "Either [the sea] opens up another threat of conflict or another area of cooperation." Of course, the argument is not so simple. As Friedham (1966) and Belman (1968) point out, traditional law of the sea is imperfect, but there is legitimate hesitancy towards creating new modes when our experience with the sea and our ignorance of its resources are both still great.

Historical Background

In 1609, Grotius wrote *Mare Liberum* as a challenge to national jurisdiction of areas of ocean. This brief for the Dutch Government was directed toward breaking the Portuguese monopoly of the East Indies spice trade. Gradually, and in partial response to struggles for supremacy between Britain and Spain, the principle of "freedom of the high seas" was accepted.

The concept of a territorial sea was born when Bijnkershoek wrote *De Comino Maris* in 1702. A territorial width of three nautical miles (*ca* 6 km) has been attributed to the distance of a cannon-ball shot, but the range of cannon at the time was only a single nautical mile. Probably the three-mile limit began with a British instruction to her Ambassadors, in 1672, that control should be exercised one marine league (= 3 nautical miles) from shore (Weber, 1966). Three nautical miles was never adopted as a limit universally; claims of up to 12 such miles (*ca* 24 km) have always been valid.

A Convention of 1884 sustained all states' rights to lay cable on the deep sea-floor; but it was not until the Treaty of Paria, between Britain and Venezuela in 1942, and the Truman Proclamation of 1945, that any state claimed jurisdiction and control over any part of the sea-floor. By its important action, the United States effectively laid claim to an area of shelf larger than Alaska and Texas combined.

Three-and-a-half centuries of precedent thus led to recognition of the following zones: (1) internal waters and bays within the control of the coastal state; (2) territorial sea under the control of the coastal state; (3) continental shelf over which the coastal state might claim control; (4) contiguous zones for special purposes; (5) the high seas, held to be *res communis;* and (6) the deep sea-floor, held to be *res nullius.* New technology for ocean research and exploitation after World War II indicated obvious conflict under this system.

The International Law Commission had been created in 1947 under the United Nations. It proposed in 1956 that a Conference on Law of the Sea be held. This occurred in 1958 at Geneva and adopted four Conventions as follows:

(1) Territorial Sea and the Contiguous Zone: ratified 10 September 1964. This Convention confirmed the control of the coastal state over all resources within a territorial sea. In addition, the coastal state might declare control over a contiguous zone

for security, fishery, fiscal, immigration, or sanitary purposes, but not to interfere with the right of innocent passage. The width of the fisheries zone is still undecided. Of 91 coastal states, 49 declare 12 nautical miles, 17 declare more than 12, 10 declare between 3 and 12, and 15 declare 3 miles in 1966 (Oda, 1968). A narrow territorial sea is favored by military interests and by states with international fishing fleets; Japan is the only major fishing nation which adheres to three miles. A wide territorial sea is favored by states wishing to protect a coastal fishery. Obviously, the U.S. has been in a delicate position and only recently declared 12 nautical miles to be the width of its fisheries zone, the territorial width remaining 3.

(2) High Seas: ratified 30 September 1962. This includes all waters outside territorial ones and declares freedoms of navigation, overflight, fishing, and the laying of submarine cables and pipelines. Also included are regulations on piracy and pollution.

(3) Continental Shelf: ratified 10 June 1964. This Convention is mainly concerned with the sea-floor and does not include the water lying above. It has already been pointed out that certain living resources are included. The most serious contention concerns the extent of the shelf, which is defined in the Convention as extending: ". . . to the sea-bed and subsoil of the submarine area adjacent to the coast, but outside the area of the territorial sea, to a depth of 200 meters or, beyond that limit, to where the depth of the superjacent waters admit of the exploitation of the natural resources of the said areas." Two schools of thought prevail here. One contends that as this Convention is entitled "Continental Shelf," the sea bottom beyond its geographic limits of about 200 m depth is not included. The other contends that the exploitability provision defines a "juridical shelf" which could include the slope or even the whole ocean bottom. It should be kept in mind that the shelf area is a huge one; without the slope it comprises 10×10^6 mi^2 (about 28 $\times 10^6$ km^2), which is equal to 20 per cent of the total land area on Earth (Mero, 1966, 1968). An excellent review of the problem is that of Tubman (1966).

(4) Fishing and Conservation of Living Resources of the High Seas: ratified 20 March 1966. This remains the most controversial of the Conventions, being the only one which did not more or less standardize a body of existing custom but which contained genuine innovation. The problem that one non-cooperating state could vitiate fishery conservation efforts was a major reason for calling the Geneva Conference. This Convention "virtually forces consideration of the need for conservation of a fish stock by all participating nations if only one (or an adjacent coastal state) insists on it," but "it says nothing about the principles to be followed, nor, more fundamentally, about the objectives sought" (Crutchfield, 1968). It does not treat allocations or provide more than case-by-case consideration of conservation.

Prognosis

Christy (1968) outlines four approaches to the developing law of the sea. The "wait and see" approach leaves exploitation to chance. Support for wait-and-see comes in part from proponents of case law who heed the dictum of Oliver Wendell Holmes: "The life of the law is not logic, but experience." Additional support accrues from those who note our lack of knowledge and experience in the sea.

The second approach is that of the "national lake." The obstacle here is that the division of the sea would be highly inequitable. The U.S.S.R. would get little, whereas tiny oceanic islands would gain title to huge territories.

The "flag" approach is the third. It is supported mainly by mineral and military interests of powerful nations. Burke (1966a, 1966b, 1968, 1969), McDougal (1968), and Wilkey (1969), all defend this point of view, emphasizing traditional processes of mineral claim on and under a sea-bed held to be res nullius. Some are willing to make concessions on an international registry or towards cooperation in pollution and security. On the other hand, Young (1968), Krueger (1968), and Eichelberger (1968), hasten to point out that the flag approach is but a form of neocolonialism which would rapidly lead to a gold-rush. Nor does the flag approach, with its unavoidable competitive nature, make much sense ecologically.

The United Nations has shown its resolve by a series of resolutions. One of 31 December 1968, designated Resolution 2467A-2467D (XXIII), includes the following points: (1) promotion of international cooperation; (2) exploitation for the benefit of mankind; (3) prevention of pollution; (4) desirability of peaceful use of the sea-bed; and (5) endorsement of an International Decade of Ocean Exploration.

I find it impossible to argue against any of these goals, and equally impossible to see an alternative to internationalism in achieving any of them. Precedents of treaties on Antarctica and outer space exist though both Young (1968) and Eichelberger (1968) point out that the ocean floor is not *tabula rasa* (i.e. a "blank slate") as were in some senses both Antarctica and outer space. However, they do not point out that virtually all of Antarctica was under territorial claim, and that nuclear testing and exploration had been carried on in outer space before those treaties were signed. Both treaties involved a yielding of claims and nullifications of military interests. It is difficult to sea why such yielding could not also take place for the sea-floor, the superjacent waters, and even some sections of shelf. One thing is certain; under no reasonable circumstances would the exploiter lose by international control. All that might ensue would be more efficient utilization and a cleaner sea.

Gargantuan problems exist with regard to internationalism. Burke (1966*a*, 1966*b*, 1968, 1969), Alexander (1966), and Griffin (1967), review the problems of disarmament, bilateral and multilateral agreements, the extent of off-shore claims, scientific freedom in research, and many others. Burke (1969), particularly, examines difficulties in applying the Stratton Commission Report. However, one should not be deterred from a path simply because it is stony.

CONCLUSION

The sea lies today like a huge plum which Man is ready to pluck, but toward which he gropes in quandary. this paper emphasizes the application of ecology to this Marine Revolution. We see that historically we have grown to treat the sea as the land—exploitively and as a "frontier" to be conquered. There is no longer room for doubt that this is a collision course and that the "conquest" of Nature threatens Man's existence as a species with high "culture."

Much as we might wish it so, the sea is not a placebo for our destruction of the land. The very existence of Conventions on the sea are cause for optimism and proof of awareness of the need for change. To the international lawyers belongs most of the credit. However, there persist such items as the "house" lawyer's fear of loss of proprietary rights, the industrialist's fear of loss of claim, and the fisherman's fear of loss of *laissez faire* exploitation. Many maintain that we do not yet know enough about the sea, nor do we have sufficient experience with it, to change our *modus operandi*. Nevertheless, one must agree with Belman (1968): "If law awaits developments, it loses the ability to shape them."

The ecosystem principle must serve as the overriding guide for shaping our future resource use. We simply do not dare exceed limits of homeostasis in the sea. Ripley (1966) states: "The basic problem therefore is to acquire sufficient knowledge about or ecosystems to provide feedback controls essential to homeostasis." It is true that we do not as yet have all the knowledge we might desire, but it is also true that we know enough now to be able intelligently to monitor our actions. We *can* assume that every one of our actions puts some stress on the environment. We *can* put aside expediency, tradition, and false economic idols. We *can* negate flimsy and obsolescent national boundaries. We *can* shift the burden of proof for ecological damage from the plaintiff-community to the defendant-exploiter. The problem is not the ability to change; it is the desire and necessary understanding.

A new brand of environmental biologist must become increasingly involved in the Marine Revolution; without him, no purely political or legal solution will suffice. Non-biologists, even lay conservationists, have too rarely shown comprehension of the complexities of the living world and they are not equipped to deal with the sophistication of ecosystem ecology. However, the biologists have been largely unwilling to commit themselves. Darling (1967) has pinpointed part of the problem: "... public policy has to be ahead of public consensus ... ecology and conservation can move surely into the hurly-burly without losing scholarly integrity, a course most of us must be prepared to follow...." Biology must to a new degree achieve interaction with politics and the law. Scientific integrity must be defended and this is not in conflict with a willingness to "stick one's neck out."

There is apparently no end in sight either to Man's reproductive potential or to his infinite conceit that he shall inherit the (still productive?) Earth. Yet there is a limit to the sea as to the land. The uniqueness of the Marine Revolution lies in part in the fact that Man is recognizing the limits of the Earth as he is developing exploitation of its most remote and unknown region—the oceans and seas. It also lies in the fact that the oceans' and seas' uncertain ownership forces Man at last to consider alternatives to provincialism and nationalism. Indeed it may be said that the Marine Revolution, for the first time in Man's history, ties survival with international cooperation.

ACKNOWLEDGMENTS

To the following I owe thanks for helpful comments on this paper: Eugenie Clark, University of Maryland; Raymond F. Dasmann, Conservation Founcation; Sidney R. Galler, Smithsonian Institution; Roger M. Herriott, The Johns Hopkins University; A. Starker Leopold, University of California (Berkeley); Nicholas Polunin, *Biological Conservation*; John E. Randall, Bishop Museum, Honolulu, Hawaii; George W. Ray, Jr., private lawyer; Frank M. Potter, Jr., Environmental Clearinghouse; Charles H. Southwick, The Johns Hopkins University; Richard Young, private lawyer. I am indebted to all these persons, for their very constructive criticism.

REFERENCES

Abel, Robert B. (1968). A history of federal involvement in marine sciences: emergence of the National Sea Grant Program. *Natural Resources Lawyer* 1, 105-14.

Alexander, Lewis M. (1966). Offshore claims of the world. Mimeo. for *Law of the Sea Institute*. University of Rhode Island, June-July, 8 pp.

Bardach, John E. and Ryther, John H. (1968). *The Status and Potential of Aquaculture. Vol. II: Particularly Fish Culture.* Amer. Inst. Biol. Sci., Clearinghouse for Fed. Sci. & Tech. Information, Springfield, Virginia, 225 pp., illustr.

Bascom, Willard (1966). Mining in the sea. Mimeo. for *Law of the Sea Institute,* University of Rhode Island, June-July, 1 p.

Belman, Murray J. (1968). The role of the State Department in formulating federal policy regarding marine resources. *Natural Resources Lawyer* 1, 14-22.

Burke, William T. (1966a). Technological development and the law of the sea. Mimeo. for *Law of the Sea Institute,* University of Rhode Island, June-July, 18 pp.

Burke, William T. (1966b). Legal aspects of ocean exploitation—status and outlook. Pp. 1-23 in *Exploiting the Ocean. Trans. Second Annual Mar. Tech. Soc. Conf. & Exhibit.,* Mar. Tech. Soc., Washington, D.C.

Burke, William T. (1968). A negative view of a proposal for United Nations ownership of ocean mineral resources. *Natural Resources Lawyer* 1, 42-62.

Burke, William T. (1969). Law, Science and the ocean. Mimeo. for *Law of the Sea Institute,* University of Rhode Island, August, 34 pp.

Chapman, V. J. (1968). Underwater reserves and parks. *Biol. Conserv.* **1**, 53.

Chapman, Wilbert M. (1966). Fishery resources in offshore waters. Mimeo for *Law of the Sea Institute,* University of Rhode Island, June-July, 18 pp.

Christy, Francis T., Jr. (1966). The distribution of the seas' fisheries wealth. Mimeo. for *Law of the Sea Institute,* University of Rhode Island, June-July, 14 pp.

Christy, Francis T., Jr. (1968). Alternative régimes for the marine resources underlying the high seas. *Natural Resources Lawyer* **1**, 63-77.

Clark, Eugenie (1967). The need for conservation in the sea. *Oryx* **9**, 151-3.

Close, Frederick, J. (1968). *The Sleeping Giant. An Industrial Viewpoint on the Potential of Oceanography.* Mar. Tech. Soc., Washington, D. C. Reprinted by Alcoa, Pittsburgh, 12 pp.

Commission on Marine Science, Engineering and Resources (1969). *Our Nation and the Sea.* US Government Printing Office, Washington, D. C., xi + 305 pp.

Crutchfield, James A. (1968). The convention on fishing and living resources of the high seas. *Natural Resources Lawyer* **1**, 114-24.

Darling, F. Fraser (1967). A wider environment of ecology and conservation. *Daedalus* **96**, 1003-19.

Dasmann, Raymond F. (1968). *A Different Kind of Country.* Macmillan, New York, and Collier-Macmillan, London, viii + 276 pp., illustr.

Dozier, J. R. (1966). Offshore oil and gas operations present and future. Mimeo. for *Law of the Sea Institute,* University of Rhode Island, June-July, 11 pp.

Economic Associates, Inc. (1968). *The Economic Potential of the Mineral and Botanical Resources of the US Continental Shelf and Slope.* A Study Prepared for the National Council on Marine Resources and Engineering Development, Clearinghouse for Fed. Sci. & Tech. Information, Springfield, Virginia, 520 pp.

Ehrlich, Paul (1969). Eco-catastrophe. Reprinted from *Ramparts,* Int. Planned Par-

enthood Fed., New York, September, 5 pp.

Eichelberger, Clark M. (1968). A case for the administration of mineral resources underlying the high seas by the United Nations. *Natural Resources Lawyer* **1**, 85-94.

Elton, Charles S. (1958). *The Ecology of Invasions by Animals and Plants.* Methuen, London, and John Wiley, New York, 181 pp., illustr.

Ely, Northcutt (1967a). American policy options in the development of undersea mineral resources. Mimeo. for *90th Ann. Meeting Amer. Bar Assn,* Honolulu, August, 10 pp.

Ely, Northcutt (1967b). The administration of mineral resources underlying the high seas. Mimeo. for *Amer. Bar Assn National Institute on Marine Resources,* Long Beach, California, June, 12 pp.

Ely, Northcutt (1968). A case for the administration of mineral resources underlying the high seas by national interests. *Natural Resources Lawyer* **1**, 78-84.

Emery, K. O. (1966). Geological methods for locating mineral deposits on the ocean floor. Pp. 24-43 in *Exploiting the Ocean. Trans. Second Ann. Marine Tech. Soc. Conf. & Exhibit.,* Mar. Tech. Soc., Washington D. C.

Friedham, Robert L. (1966). Conflict over law: voting behavior at the United Nations Law of the Sea Conference. Mimeo. for *Law of the Sea Institute,* University of Rhode Island, June-July, 16 pp.

Frost, Justin (1969). Earth, air, water. *Environment* **11**, 14-33.

Griffin, William L. (1967). The emerging law of ocean space. *Int. Lawyer* **1**, 548-87.

Harlow, Bruce A. (1966). Territorial sea concept. Mimeo. for *Law of the Sea Institute,* University of Rhode Island, 1 p.

Hearn, Wilfred A. (1968). The role of the United States Navy in the formulation of federal policy regarding the sea. *Natural Resources Lawyer* **1**, 23-31.

Hedgpeth, Joel (in press). Atomic waste disposal in the sea: an ecological dilemma. In *The Careless Technology* (Ed. M. Taghi Farvar & John Milton). Natural History Press, Doubleday, New York.

Interagency Committee on Oceanography (1963). *Oceanography, the Ten Years*

Ahead. ICO Pamphlet No. 10, Federal Council for Science and Technology, Washington D. C., 54 pp.

Interagency Committee on Oceanography (1967). *National Oceanographic Program*. ICO Pamphlet No. 24, Federal Council on Science and Technology, iv + 107 pp.

Krueger, Robert B. (1968). The Convention on the Continental Shelf and the need for its revision and some comments regarding the régime for the lands beyond. *Natural Resources Lawyer* 1, 1-18.

Luce, Charles F. (1968). The development of ocean minerals and law of the sea. *Natural Resources Lawyer* 1, 29-35.

McDougal, Myers S. (1968). Revision on the Geneva Convention on the law of the sea (Comments). *Natural Resources Lawyer* 1, 19-28.

Mero, John L. (1966). Review of mineral values on and under the ocean floor. Pp. 61-78 in *Exploiting the Ocean. Trans. Second Ann. Mar. Tech. Soc. Conf. & Exhibit.*, Mar. Tech. Soc., Washington, D. C.

Mero, John L. (1968). Mineral deposits in the sea. *Natural Resources Lawyer* 1, 130-7.

National Academy of Sciences (1964). *Economic Benefits from Oceanographic Research*. NAS Publ. 1228, 50 pp.

National Academy of Sciences (1967). *Oceanography 1966: Achievements and Opportunities*. NAS Publ. 1492, 183 pp.

National Academy of Sciences (1969). *An Oceanic Quest: The International Decade of Ocean Exploration*. NAS Publ. 1709, 115 pp.

National Council on Marine Resources and Engineering Development (1967). *Marine Science Affairs—A Year of Transition*. US Govt. Printing Office, Washington, D. C., v + 157 pp.

National Council on Marine Resources and Engineering Development (1968a). *International Decade of Ocean Exploration*. US Govt. Printing Office, Washington, D. C., i + 7 pp.

National Council on Marine Resources and Engineering Development (1968b). *Marine Science Affairs—A Year of Plans and Progress*. US Govt. Printing Office, Washington, D. C., xiii + 228 pp.

Neblett, William R. (1966). A fishery view of recent law of the sea conferences. Mimeo. for *Law of the Sea Institute*, University of Rhode Island, June-July, 15 pp.

Nelson, T. W. and Burk, C. A. (1966). Petroleum resources of the continental margins of the United States. Pp. 116-33 in *Exploiting the Ocean. Trans. Second Ann. Mar. Tech. Soc. Conf. & Exhibit.*, Mar. Tech. Soc., Washington, D. C.

New York Times (1969). Seabed potential for arms studied, 8 Oct.

Ocean Science News (1969). 1970 federal ocean market. *Ocean Science News* 11, 2-3.

Oda, Shigeru (1968). The Geneva Conventions on law of the sea: some suggestions for their revision. *Natural Resources Lawyer* 1, 103-13.

Odum, Eugene P. (1959). *Fundamentals of Ecology*, second edn. W. B. Saunders, Philadelphia & London, xvii + 546 pp., illustr.

Osborn, Fairfield (1953). *The Limits of the Earth*. Little, Brown, Boston, x + 238 pp.

Panel on Oceanography, President's Science Advisory Committee (1966). *Effective Use of the Sea*. US Govt Printing Office, Washington, D. C., xv + 144 pp.

Polikarpov, G. G. (1966). *Radioecology of Aquatic Organisms*. Reinhold, New York, xxviii + 314 pp.

Program of Policy Studies in Science and Technology (1969). A critical review of the Marine Science Commission Report. The George Washington Univ. and the Mar. Tech. Soc. Law Committee, Washington, D. C., 114 pp.

Randall, John E. (1969). Conservation in the sea. *Oryx* 10, 31-38.

Ray, Carleton (1961). Marine Preserves for ecological research. Pp. 323-8 in *Trans. 26th No. Amer. Wildlife & Nat. Res. Conf.*, Wildlife Man. Inst., Washington, D. C.

Ray, Carleton (1965). The scientific need for shallow-water marine sanctuaries. Pp. 83-98 in *Symp. on Sci. Use of Natural Areas*, XVI Int. Cong. Zool. Reprinted by *Field Research Projects, Natural Area Studies*, No. 2, Coconut Grove, Fla., x + 103 pp.

Ray, Carleton (1966). Inshore marine conservation. Pp. 77-87 in *First World Conference on National Parks*. US Govt Printing Office, Washington, D. C., xxxiv + 471 pp.

Ray, Carleton (1968). *Marine Parks for Tanzania.* Conservation Foundation, Washington, D. C., 47 pp., illustr.

Ripley, W. Dillon (1966). The future of environmental improvement. Reprinted by *The Graduate School Press,* from *Environmental Improvement: Air, Water and Soil,* US Dept Agriculture, Washington, D. C., pp. 85-93.

Ryther, John H. and Bardach, John E. (1968). *The Status and Potential of Aquaculture. Vol. 1: Particularly Invertebrates and Algal Culture.* Amer. Inst. Biol. Sci., Clearinghouse for Fed. Sci. & Tech. Infor., Springfield, Virginia, 261 pp.

Task Force on Environmental Health and Related Problems (1967). *A Strategy for a Liveable Environment.* Rept to the Sec. of Health, Education & Welfare, US Govt Printing Office, Washington, D. C., xxi + 90 pp.

Tubman, William C. (1966). The legal status of minerals located on or beneath the sea floor beyond the continental shelf. Pp. 379-404 in *Exploiting the Ocean. Trans. Second Ann. Mar. Tech. Soc. Conf. & Exhibit.,* Mar. Tech. Soc., Washington, D. C.

Walford, Lionel A. (1958). *Living Resources of the Sea.* Ronald Press, New York, xi + 321 pp.

Weber, Alban (1966). Our newest frontier: the sea bottom. Some legal aspects of the continental shelf status. Pp. 405-11 in *Exploiting the Ocean. Trans. Second Ann. Mar. Tech. Soc. Conf. & Exhibit.,* Mar. Tech. Soc., Washington, D. C.

Wilkey, Malcolm R. (1969). The role of private industry in the deep ocean. *Symp. on Private Industry Abroad.* Southwestern Legal Foundation, Dallas, Texas, June. Reprinted by Mathew Bender, Washington, D. C., 40 pp.

Young, Richard (1968). The legal régime of the deep sea floor. *Amer. Jour. Int. Law* **62,** 641-53.

National Jurisdiction and the Use of the Sea

LEWIS M. ALEXANDER 1968

JURISDICTIONAL ZONES OF THE SEA

The limits of jurisdiction of the United States differ with respect to each of the three physical environments of the nation. On land, jurisdiction ends at our boundaries with Canada and Mexico. In space, international law does not yet recognize any specific limits to national sovereignty, although there is acknowledged to be a zone of outer space well away from the earth's surface to which a country's jurisdiction does not extend. In the sea there are national boundaries for different purposes, and beyond these boundaries various forms of control may still prevail. To the United States, as to other maritime countries, the nature and extent of its jurisdiction in the marine environment is of prime importance because of the effects this jurisdiction has on our opportunities for use of the sea's resources.[1]

1. The problems of utilizing the marine environment led to the formation in 1966 of the National

The term "resource" is used here to refer both to goods and to services from the sea. Marine resources differ in several respects from those of the land. One difference is the three-dimensional nature of the sea itself, permitting activities to take place simultaneously on the surface and within the water, as well as on and under the seabed. A second difference is the high mobility of many of the marine resources, and of the water medium itself. Finally, within a country's land boundaries, political control over resources is *exclusive* in the sense that no other nation has rights to them; in the sea, however,

Council on Marine Resources and Engineering Development, and in early 1967 of the Commission on Marine Science, Engineering and Resources—the latter body charged with recommending to the President and to Congress a national oceanographic program that will meet present and future national needs. The author is currently serving as a member of the Commission staff. Nothing in this article should be construed in any way as reflecting opinions of the Marine Science Commission or of the Marine Council.

Dr. Lewis M. Alexander is chairman and professor of the Department of Geography at the University of Rhode Island. He is the executive director of the Law of the Sea Institute and the former deputy director of the President's Commission on Marine Science, Engineering and Resources. His research interests include political geography and law of the sea studies. He has published and edited over thirty-five books and articles.

From *Natural Resources Journal*, Vol. 8, No. 3, pp. 373-400, 1968. Reprinted with light editing and by permission from the author and *Natural Resources Journal* published by the University of New Mexico School of Law, Albuquerque, New Mexico.

rights to resources may be shared or "common property" in nature, thus adding new dimensions to jurisdictional problems.[2] This article discusses the nature and extent of our offshore jurisdiction in terms of the major uses we make of the marine environment.

The ocean areas of the globe have traditionally been divided into two basic components: the high seas which are free to the use of all countries, and the marginal seas over which coastal states exercise jurisdiction. Events in recent decades have served to complicate this division. Questions of national jurisdiction have been extended to the seabed and subsoil of the oceans, and to the airspace above them. New technologies and new uses of the marine environment have brought with them new pressures for extensions of national controls well away from the coast. These events have been augmented by political trends—the conflict of ideologies, the achievement of independence by many countries, the pressures of population on world food supplies—with the result that concern is growing in many quarters that the traditional freedom of the high seas will be gradually eroded away by national claims. These claims, in turn, rest on the real or imagined needs that governments feel they have for protection in their offshore waters from the competition of foreigners.

United States policy on jurisdictional matters in the marine environment rests on two principal bases: the international regulations contained in the four Geneva Conventions, and the bilateral and multilateral agreements which this nation has made with other states concerning specific uses of the sea. The four Conventions were adopted at the 1958 Geneva Law of the Sea Conference, to which 82 countries sent delegations.[3] These Conventions have been ratified

by a sufficient number of governments so that they are now in effect.[4]

The Conventions provide for the partitioning of the oceans into five jurisdictional zones. First is the zone of *internal waters,* including certain bays, estuaries, and other adjacent waters, over which a coastal state exercises complete sovereignty. Seaward of this is the *territorial sea,* over which the sovereignty of the coastal state is limited only by the right of innocent passage by foreign vessels. There is no agreement among states on a standard breadth for the territorial sea. In late 1967, of 92 states which claimed a definite breadth for their territorial waters, 28 (including the United States) claimed three miles, 33 (including the Soviet Union) claimed twelve miles, and 24 adopted breadths between three and twelve miles. The remaining 7 had territorial claims in excess of twelve miles.[5]

Beyond territorial limits is the *contiguous zone,* in which coastal states may exercise the control necessary to prevent and punish infringements of their customs, fiscal, immigration or sanitary regulations.[6] The contiguous zone may not extend more than twelve miles seaward of the base-line from which the breadth of the territorial sea is measured; thus, countries which claim twelve miles as their territorial limit are not entitled to an additional contiguous zone. Equally important is the fact that many countries, including the United States, now consider their contiguous zone to be also an exclusive fisheries area in which foreign fishing is prohibit-

---

2. A shared resource may belong to two or more nations, but a common property resource falls within the purview of no particular nation, and thus all countries are free to share in its exploitation.
3. The four Conventions were: Convention on the Territorial Sea and the Contiguous Zone, Convention on the High Seas, Convention on Fishing and

Conservation of the Living Resources of the High Seas, and Convention on the Continental Shelf. See, United Nations Conference on the Law of the Sea, *Off. Rec.,* U.N. Doc. A/C 13/1-43 (1958).
4. By the end of 1967, thirty-three states had ratified the Territorial Sea Convention, forty had ratified the High Seas Convention, twenty-five had ratified the Fishing Convention, and thirty-six had ratified the Convention on the Continental Shelf.
5. See the complete table of offshore claims in Alexander, *Geography and the Law of the Sea,* 58 Annals, Assn'n of Am. Geographers 177 (1968).
6. Controls exist over these infringements only if they are committed within the coastal state's territory or territorial sea.

ed except by special agreement.[7] Some 28 states with territorial breadths of less than twelve miles have extra-territorial fisheries zones out to the twelve-mile limit.[8]

A fourth jurisdictional zone is the *continental shelf,* the shallow platform extending out from the land for varying distances beneath the sea. Three physical variables here are (1) that the depth at which the gentle incline of the shelf breaks to a more precipitous slope varies considerably from place to place, (2) that the distance from shore at which this "edge" of the shelf occurs also varies greatly throughout the world, and (3) that in some localities there are deep basins or canyons in the shelf, seaward of which are additional shallow areas. The 1958 Geneva delegates, seeking agreement on identifying the shelf, defined it as "the seabed and subsoil of the submarine areas adjacent to the coast but outside the area of the territorial sea, to a depth of 200 meters or, beyond that limit, to where the depth of the superjacent waters admits of the exploitation of the national resources of the said areas." This definition also applies to submarine areas adjoining the coasts of islands. Over its continental shelf, the coastal state exercises sovereign rights for the purpose of exploring and exploiting its natural resources.

The fifth zone constitutes the *high seas,* which are open to all nations. Specifically, countries are guaranteed the following four freedoms on the high seas: freedom of navigation; freedom of fishing; freedom to lay submarine cables and pipelines; and freedom

to fly over the high seas. In addition, there are other freedoms that are "recognized by the general principles of international law." Nowhere in the Geneva Conventions is reference made to the legal status of the seabed and subsoil underlying the high seas beyond the limits of the continental shelf.

Although the Conventions represent a decisive forward step in the formulation of the international law of the sea there still remain many problems of ambiguity. Some matters were left unresolved by the Geneva Conventions. Others, although covered by the Convention articles, are subject to wide differences of interpretation. In still other cases, it may in time prove wise to seek revision of certain articles in the light of changing technological or economic conditions. These ambiguities will be discussed in terms of specific uses which the United States makes of the sea.

USES OF THE SEA

In considering the major uses made of the sea by the United States we start by noting the principal categories of uses for which jurisdictional problems involving foreigners may be important. These categories are transportation, commercial fishing, mining, scientific research, and military operations. Each will be treated separately in terms of uncertainties regarding national jurisdiction.

Transportation

Although freedom of navigation on, and flight over, the high seas is guaranteed by the Geneva High Seas Convention, limitations may arise with respect to marginal waters off a nation's coastline. The three sets of conditions under which this might occur are with respect to the internal waters of a foreign state, to its territorial waters, and to international straits.

Since innocent passage by foreign vessels is not guaranteed through a state's internal waters a question arises as to the limits a country may claim to such waters. The

7. Under an Agreement of November, 1967, with the U.S.S.R., Soviet vessels are permitted to fish within a designated area in the 3-12 mile zone off the New Jersey coast between January 1 and April 1 of each year. In addition, both Soviet and Japanese fishing vessels are permitted under certain conditions to fish between 3 and 12 miles off portions of the Alaskan coast. See the February 13, 1967, Agreement with the Soviet Union, and the May 9, 1967, Agreement with Japan.

8. Combining these 28 with the 33 countries claiming a twelve-mile territorial sea brings to 61 the number of countries (out of a total of 111 independent states bordering on the sea) which prohibit foreigners from fishing within twelve miles of their shores.

Geneva articles are quite precise on the delimitation of the baseline marking the outer limits of internal waters with respect to such matters as islands, low-tide elevations, permanent harbor works, and the closing line to be used across the mouths of bays.[9] The articles are less definite on the special circumstances surrounding the delimitation of straight baselines along the coast. Article 4 of the Territorial Sea Convention states: "In localities where the coastline is deeply indented and cut into, or if there is a fringe of islands along the coast in its immediate vicinity, the method of straight baselines joining appropriate points may be employed in drawing the baseline from which the breadth of the territorial sea is measured."[10] Still, there is considerable room for debate on what the Article authorizes in terms of specific baselines. If a coastal state follows a liberal interpretation in delimiting its offshore boundaries, it may find itself challenged by other interested countries; such a challenge might eventually go to the International Court of Justice, as occurred in the Anglo-Norwegian Case.[11]

The Geneva delegates failed to agree on two related matters: the definition of "historic bays," and the delimitation of straight baselines in the case of archipelagos. Some coastal states have claimed that certain nearshore waters have traditionally been treated as a part of the national domain, and, while they cannot be closed off as internal under the delimitation systems spelled out in the Conventions, they should nevertheless be recognized as coming within that state's sovereignty.[12] A variation of this is the suggested "closed sea" regimes, under which jurisdiction over an extensive water body, such as the Black or Baltic Seas, would rest with the states which border on them. In either case freedom of navigation in marginal waters would be threatened.

The delimitation of straight baselines about archipelagos was the subject of considerable debate, but no decisions, at the 1958 Geneva Conference.[13] The rationale for such delimitation appears to rest on the fact that the inter-island waters form the connecting links among a multi-island state, and that therefore they should be treated as internal. Straight baselines have already been delimited about Indonesia and the Philippines, and within these baselines the waters are classed as internal. Among other recently-independent countries are the Maldives and Western Samoa, both of which consist of island groups, and which may in time seek straight baseline regimes to close off inter-island waters.

Within its territorial waters a coastal state is entitled to suspend the innocent passage of foreign ships "if such suspension is essential for the protection of its security."[14]

9. A closing line, for example, could be drawn across the mouth of Cape Cod Bay, where the distance from Plymouth to Provincetown is less than 24 miles. But this would not be true in the case of Bristol Bay, Alaska, where the closing line is 160 miles.

10. The waters enclosed by these baselines are internal in nature. However, Article 5(2) of the Territorial Sea Convention states "where the establishment of a straight baseline . . . has the effect of enclosing as internal waters areas which previously had been considered as part of the territorial sea or of the high seas, a right of innocent passage . . . shall exist in those waters." 3 *Off. Rec.*, A/C. 13/39 (1958).

11. The United Kingdom challenged the Norwegian method for delimiting straight baseline along its coasts in a case which eventually was decided in favor of Norway. See Evensen, *The Anglo-Norwegian Fisheries Case and its Legal Consequences*, 46 Am. J. Int'l L., 609 (1952).

12. In 1963 the State of Alaska sought unsuccessfully to close off the waters of Bristol Bay on the grounds of its "historic" nature. See Arctic Maid Fisheries, Inc. v. State, Sup. Ct. of Alas., No. 316. *Dismissed per stipulation*, (1963). For general discussions of the historic bay problem see L. J. Bouchez, The Regime of Bays in International Law (1964), and M. P. Strohl, The International Law of Bays (1963).

13. A clear discussion of this problem is in Evensen, *Certain Legal Aspects Concerning the Delimitation of the Territorial Waters of Archipelagos*, in 1 U.N. Conf. on the Law of the Sea, *supra*, note 3. See also Sorensen, *The Territorial Sea of Archipelagos*, 315, Questions of International Law: Presented to J. P. A. Francois on the Occasion of his Seventieth Birthday (1959).

14. Freedom of navigation by American merchant vessels in foreign territorial waters has not to date been impeded by any nation.

Article 16 of the Territorial Sea Convention goes on to note that such suspension must be temporary in nature, apply to specified areas of the territorial sea, and be "without discrimination against foreign ships." In fact, such details may be meaningless, as in the case of the blanket prohibitions which have long been in effect against Israeli shipping within the territorial waters of the United Arab Republic. In a world of political turbulence and expanding territorial claims this right of suspension of foreign shipping may in time prove to be an extremely troublesome issue.

The case for international straits is also a complex one. Article 16 of the Territorial Sea Convention provides: "There shall be no suspension of the innocent passage of foreign ships through straits which are used for international navigation between one part of the high seas and another part of the high seas or the territorial sea of a foreign State." The article, however, fails to identify the criteria for a strait "used for international navigation," nor, of course, does it attempt to define "innocent passage."[15] How long, for example, need a strait be traversed by foreign ships in order to qualify as one which is used for "international navigation"? If the country which controls the strait in question feels that armed attack by its neighbor is imminent, must it view with impunity the import of oil and other supplies by its neighbor?[16] And what if one state has re-

search vessels off another's coasts and these vessels seek innocent passage through coastal straits? The acquisition of oceanographic data may be regarded as inimical with innocent passage.

The use of international straits is closely tied to the question of territorial limits. Obviously the number of straits lying entirely within coastal states' territorial waters is far greater under a universal twelve-mile territorial regime than would be the case for three or six miles.[17] Freedom of overflight is also involved here, for there may or may not be a strip of high seas waters extending through the straits.

Still another transportation problem is that of the "genuine link" requirement in shipping registry which holds that a state "must effectively exercise its jurisdiction and control in administrative, technical and social matters over ships flying its flag." Many vessels, operating under "flags of convenience" are owned by nationals of one state but registered in another.[18] One important question is whether a state may challenge the practices of another country in granting nationality to vessels on the grounds that the necessary "genuine link" between flag and ownership does not

---

15. In the summer of 1967 the Soviet Union refused permission for passage of U.S. Coast Guard vessels "Edisto" and "East Wind" through the 23½-mile wide Vilkitsky Straits south of Severnaya Zemlya, connecting the Kara and Laptev Seas. The Soviet action was apparently based on the contention that these were warships and that advance notification was required of such ships. Although the Soviets gave no reason for their refusal one interpretation might be that since the vessel was carrying out oceanographic research in the waters north of the U.S.S.R., its purpose in seeking passage was not "innocent" but rather to seek data on the nature of the inter-island passage.

16. The dispute in the spring of 1967 over the right of Israeli shipping to use the Strait of Tiran into the Gulf of Aqaba pointed up some of the difficulties surrounding Article 16, including the

difficulty of enforcing its provisions. For a background of this dispute see Selak, *A Consideration of the Legal Status of the Gulf of Aquaba*, 52 Am. J. Int'l L. 660 (1958), and *Passage Through the Suez Canal of Israel-Bound Cargo and Israel Ships* 51 J. Int'l L. 530 (1957). Also Gross, *Geneva Conference on the Law of the Sea and the Right of Innocent Passage through the Gulf of Aquaba* 53 J. Int'l L. 564 (1959).

17. Kennedy, *A Brief Geographical and Hydrographical Study of Straits which Constitute Routes for International Traffic*, in U.N. Conf. on the Law of the Sea 114, *supra* note 3. See also Table III, Widths of Selected Straits and Channels in Sovereignty of the Sea, U.S. Department of State, Geographic Bulletin No. 3 (1965).

18. See B. A. Boczek, Flags of Convenience (1962). This problem is closely associated with that of pollution by ships, either within or beyond a coastal state's territorial limits. In the case of the March, 1967, *Torrey Canyon* disaster off southwestern England the vessel was registered in Liberia but owned by an American company and chartered to a British concern.

exist.[19] The United States is particularly vulnerable in this respect since it is the principal country making use of "flags of convenience" arrangements.

Commercial Fishing

A second use of the sea by the United States is for fishing. In this use the industry has a political importance—both domestically and internationally—well in excess of its relative economic significance to the nation's economy. The value of the domestic catch in 1966 was $454 million, while that of fishery processed products was $1.2 billion,[20] as against a total gross national product for that year of $740 billion.

The nation's commercial catch has remained rather stable over the past thirty years, fluctuating between 4.3 and 5.4 billion pounds. During the same period the total world fish catch has more than trebled, resulting in a decline of the United States' position from second to sixth among the fishing nations. Conversely, the demand for fisheries products in the United States has been growing steadily, so that by 1966 some 65 per cent of our fishery products (valued at $724 million) was imported.[21]

Despite the failure of the fishing industry to expand the volume of catch, its interests remain of critical importance to the United States for the following reasons: (1) fishing activities on the high seas carry with them implications of national security and pres-

tige; (2) there are strong psychological and political connotations in the industry; (3) there are international implications in "food from the sea" programs which the United States has inaugurated. The security/prestige factor of commercial fisheries may be seen in the training of seamen provided for by the industry, the knowledge of the marine environment gained from fishing operations, the use of fishing vessels in wartime, and, from the standpoint of prestige, the negative values provided by the appearance of large technologically-efficient foreign fishing vessels which compete for the catch with smaller American vessels in waters close to our coasts. The psychological/political element is evident in the traditional appeal of our fisheries operations and in the role of congressmen from Alaska, Washington, Massachusetts and other states in seeking to protect the interests of American coastal fishermen. The international implications center on the expectations in many countries of developing the untapped food potential in the sea for their expanding populations. [22]

The gap between myth and reality may at times be a broad one, but the facts are: first, that the United States has the potential for an economically-viable fishing industry; and second, that such an industry, to flourish, must have access to the fisheries resources beyond the present national limits. The twin problems should be attacked simultaneously; without the one the other loses much of its rationale.

The bulk of the United States' catch

19. A discussion of this "genuine link" problem is given in McDougal and Burke, The Public Order of the Oceans: A Contemporary International Law of the Sea 1008-1141 (1962). The authors feel that "Registration alone . . . recommends itself . . . as the appropriate 'link' in attribution of national character, between state and ship."

20. Data for the domestic catch is from Fisheries of the United States, 1966 (Washington: United States Department of the Interior, Fish and Wildlife Service, Bureau of Commercial Fisheries, C.F.S. 4400, 1967).

21. Balanced against this figure were exports valued at $85 million. In 1966 we imported 53 per cent of our edible fishery products, and 75 per cent of industrial fishery products (used for fish meal, fish oil, etc.).

22. Much has been written on the potential of the sea. See, for example, Schaefer, The Potential Harvest of the Sea, 94 Transactions of the Am. Fisheries Soc'y 923 (1965) and Chapman, Food Production from the Sea and the Nutritional Requirements of the World, Conference on Law, Organization and Security in the Use of the Ocean (1967). In contrast to the present world harvest of some 66 million metric tons, estimates of the fisheries potential of the sea range from 200 million to 2 billion tons. It has been calculated that a doubling of the current world catch, if this increase were processed into Fish Protein Concentrate, would make up the animal protein deficiencies for half the world's population.

comes from the coastal waters. Some 63 per cent of the average annual fishery by volume in the 1959-1963 period was taken within twelve miles of the American coast, representing 60 per cent of the total catch value.[23] Another 17 per cent by volume, and 25 per cent by value was taken in United States coastal waters beyond the twelve-mile limits, while 9 per cent by volume and 15 per cent by value was taken on the high seas beyond our coastal waters. It has been estimated that on a sustained yield basis some four times the present catch could be harvested within twelve miles of the United States coast without substantially changing present fishing methods.[24] Much of the increase would be in species not now economically valuable to the American fishermen.[25] And, the catch in coastal waters beyond the twelve-mile limit could potentially amount to ten times the quantity United States fishermen are now harvesting from these areas, although, again, reliance would have to be put on new types of fisheries.[26]

The pursuit of economic efficiency in the fishing industry has been amply discussed by Crutchfield, Christy and Scott, Pontecorvo, and others.[27] Among the steps advocated for efficiency are limitations to entry, repeal of state laws which support inefficient use, and removal of the bans on the use of foreign vessels by United States fishermen. The fishing industry, according to some proponents, is plagued by too many small units, a lack of investment capital, and diversity of interests among harvesters, processors, and marketers of fisheries products. Because of internal divisions of interest the industry cannot speak with one voice in Congress. Much of its equipment is outdated, and the industry is plagued by lack of federal support for such programs as gear or technology improvements, or the development of fish protein concentrate.

Not all the commercial fishing enterprise ie experiencing economic difficulties. The shrimp, tuna, and king crab fisheries are flourishing, but in each case there is a unique aspect of the resource availability.[28] The salmon and halibut stocks are protected by treaty against most foreign competition;[29] menhaden, oyster, and blue crab resources exist for the most part within the twelve-mile limit. But the haddock, cod, flounder, whiting, ocean perch, hake, scallop and other fisheries are open to substantial harvesting by foreign vessels, and in practically all the American fisheries, economists argue, there has been too great an effort applied in terms of the total returns available to the fishermen.[30]

---

23. Statistics supplied by the Bureau of Commercial Fisheries.
24. Statement by Donald L. McKernan, then Director of the Bureau of Commercial Fisheries, Fish Protein Concentrate Hearings on S.2720 Before the Committee on Commerce, 89th Cong., 2d Sess. 44 (1966).
25. Among these are anchovies, blue crab and oysters.
26. *Supra* note 24. Among under-utilized species beyond the twelve-mile limit are thread herring, hake, jack mackerel, and shrimp.
27. Biological and Economic Aspects of Fisheries Management (J. Crutchfield ed. 1959); F. Christy and A. Scott, The Common Wealth in Ocean Fisheries (1965); Pontecorvo, *Regulation in the North American Lobster Fishery*, in Economic Effect of Fishery Regulation 239 (R. Hamlisch ed. 1962).

28. The king crab is considered by the U.S. Government as a natural resource of the shelf over which this nation has exclusive exploitation rights (although agreements with Japan and the Soviet Union permit these nations a certain level of catch each year in the Gulf of Alaska). There have been no serious overfishing or foreign competition problems to date in the shrimp industry. The Eastern Pacific tuna is also largely dominated by the U.S., although here annual quotas have had to be established for the yellowfin tuna. See Kask, *Present Arrangements for Fishery Exploitation*, The Law of the Sea: The Future of the Sea's Resources 56 (1968).
29. The 1952 International North Pacific Fisheries Convention, signed by Japan, Canada, and the United States, provides for Japanese abstention from fishing for salmon east of Longitude 175°W., and for halibut and herring in certain waters between this longitude and the North American coast. For an excellent discussion of this and other North Pacific fisheries matters, see *North Pacific Fisheries Symposium*, 43 Wash. L. Rev. (1967).
30. Economists agree that where there is no control over the number of fishermen, the total cost of any fishery tends to match the total revenues, so that any sharable profit becomes dissipated. See

The United States is involved in nine international commissions and conventions designed to regulate various fisheries in the interests of conservation. It has bilateral agreements with Canada, Mexico, Japan, and the Soviet Union regarding details of fisheries operations off the United States coasts; through United Nations agencies and other international bodies we participate in the growing number of regional and world-wide organizations designed to research and administer various aspects of the world fisheries resources. But the persistent question remains: Is the United States doing enough to uphold its long-term interests in fisheries exploitation?[31]

These interests might be thought of in terms of (1) strengthening the economic structure of our domestic industry, (2) assuring the industry a reasonable share of the catch beyond our national limits, and (3) making available greater amounts of the sea's food potential for developing nations. The first goal may involve a substantial rehabilitation of much of the domestic industry and the introduction of limited-entry provisions into many of the fisheries. The third objective can probably not be realized through the use of fish harvested by Americans for the hungry nations since this fish is too expensive. Rather, the interested nations themselves should be encouraged to harvest, process, and distribute internally (or among nearby developing countries) the fisheries products. The second goal concerns us here since it involves extension of jurisdiction beyond the twelve-mile limit.

There are several ways in which new

authority over coastal fisheries could be established. We could extend our territorial limits out to one or two hundred miles from shore and deny foreigners the right to fish in these waters, except perhaps under special conditions. Alternatively, we could extend our exclusive fisheries—rather than territorial—limits out to this distance, or to the edge of the continental shelf; again, the freedom of foreigners to fish these waters would be proscribed.[32] In the case of an exclusive fisheries zone, the "historic rights" of foreigners to fish these waters could be recognized.[33] Still another procedure would be to seek the right to enforce unilateral conservation restrictions beyond the twelve-mile limit. These restrictions might or might not discriminate against foreign fishermen.[34]

Suggestions have been made that the high seas fisheries be placed under international control, on a worldwide or a regional

Crutchfield, *Over-capitalization of the Fishing Effort*, The Law of the Sea: The Future of the Sea's Resources 23 (1968).

31. The problem of defining our interests is complicated by the possibility that the nation might be better served by importing even higher percentages of our fishery needs, and allowing uneconomic segments of the domestic fishing industry to gradually disappear. But most persons acquainted with the industry still feel that the national interest would best be served by attempting to aid and strengthen the industry, rather than to abandon it.

32. According to some writers an extension of national sovereignty would be involved in the proclamation of a territorial or exclusive-fishery zone. In the latter case, certain restrictions on the movement of ships, aircraft, or submersibles would not apply. But little serious thought has actually been given to the legal distinctions between the two types of zones.

33. There is much diversity of opinion on the definition of "historic fishing rights." How, for example, would a coastal state's historic rights (as in the case of some New England groundfisheries) equate with the rights of foreign fishermen? Definitions of historic rights were included in the final American-Canadian Proposal at 1960 Geneva Law of the Sea Conference and in the 1964 European Fisheries Convention. For the former, see Second United Nations' Conference on the Law of the Sea: *Off. Rec.* U.N. Doc. A/C 19/8 (1960). A description of the Fisheries Convention is in *Developments in the Law of the Sea, 1958-64*, British Institute of Int'l and Comparative L. (1965).

34. Two points are involved in the problem of unilateral conservation measures proclaimed by the coastal state. One is that the measures may apply to stocks which foreigners are more actively exploiting than the coastal state's fishermen (i.e., tuna off Peru), and the foreigners may have no recourse for presenting their arguments that the conservation measures are without scientific justification. A second point is that the coastal state may deliberately restrict foreign efforts (through licensing, quotas, or other means) while not imposing commensurate restraints on its own nationals beyond the twelve-mile limit. The variations in these types of discriminatory practices are almost limitless.

basis.[35] Recourse might be made to the 1958 Geneva Convention on Fishing, to which the United States is a party. Under this Convention, if a country feels a coastal stock beyond territorial limits is threatened by overfishing, it has the right unilaterally to invoke conservation measures which are applicable to both domestic and foreign fishing, as long as those measures do not discriminate against foreign fishermen.[36] The machinery set up by the Geneva Fisheries Convention has never been implemented, nor, according to some experts, is it likely to be put to the test in the near future.[37]

Care should be taken to distinguish between conservation agreements which limit the total catch without affecting the question of who gets how much of the combined harvest, and allocation systems which guarantee states a specified share of the total catch. Some "national quota" schemes would reserve a fixed amount of the catch under any circumstances to the coastal state, presumably on the basis of propinquity; other schemes would determine the coastal states's rights to high seas fisheries near its shores on the basis of long-standing investments, need, or economic dependence on the sea.[38]

Operationally, the United States, to date, has depended on a series of bilateral and multilateral agreements to provide protection against excess foreign competition in its offshore fisheries. Three problems exist with respect to such agreements. First, since the agreements are voluntary in nature, a large measure of "give and take" may be involved, thus diminishing the advantages to be gained from the protection which they seek to provide. A second problem is that most such agreements are short-lived, requiring frequent renegotiations and concessions. A third is that such agreements are binding only on signatory states; all other states are free to ignore their provisions.

In the light of these difficulties might it not be wiser for the United States to seek international recognition of this nation's "special rights" to certain fisheries resources in its coastal waters beyond the twelve-mile limit? Such rights might be based on our long-term investments in coastal fisheries, on the economic dependence of certain Northeastern and Northwestern areas (particularly Alaska) on these fisheries, or simply on the concept that a coastal nation has some general rights against excessive fishing by foreigners in its coastal waters. One system for recognizing these rights would be the assignment of quotas of the total maximum sustainable yield of valuable species to the coastal state and to foreign countries.[39]

35. A principal rationale for this proposal is contained in F. Christy and A. Scott, *supra* note 27. Because of the growing problems of competition and of economic waste they argue that only through international management of high seas fisheries can rational exploitation of these resources be brought about. See also Christy, *The Distribution of the Sea's Wealth in Fisheries*, in The Law of the Sea: Offshore Boundaries and Zones 106 (L. Alexander ed. 1967).

36. Article 7 of the Fisheries Convention permits a coastal state to adopt unilateral measures of conservation in any area of the high seas adjacent to its territorial sea, but laces this with many provisions, among them that there is urgent need for the conservation measures "based on appropriate scientific findings" and that affected foreign fishermen have the right to appeal to a special arbitral commission.

37. Herrington notes the limited membership of the Convention (twenty-five out of a possible 130 or more states), the lack of provisions to handle the impact of massive long-range fishing on developed coastal fisheries, the absence of international enforcement provisions, and the need to accelerate action on imperative conservation measures. Herrington, *The Future of the Geneva Convention on Fishing and the Conservation of the Living Resources of the Sea*, 62, The Law of the Sea: The Future of the Sea's Resources 62 (1968). Also, *The Convention on Fisheries and the Conservation of Living Resources: Accomplishments of the 1958 Geneva Conference*, 26, *supra* note 35.

38. The case for "special rights" is described in Garcia-Amador, The Exploitation and Conservation of the Resources of the Sea (1963). See also the Resolution, Special Situations Relating to Coastal Fisheries, adopted at the 1958 Geneva Law of the Sea Conference.

39. The validity of the maximum sustainable yield concept has been hotly debated. Theoretically, there is some point in the extent of a fishery where the annual harvest, together with the mortality due to natural predators, combine to limit the species

When each participating country has reached the limit of its assigned quota it would cease fishing that particular species until the following year. This system already exists between the United States and Canada, and between the Soviet Union and Japan, with respect to Pacific salmon.

Crutchfield, in discussing this national fishery.[40] The costs to the coastal state of surveillance and enforcement of such a program may be high. An additional problem is that of acceptability of such a regime by all interested parties. What would countries as the Soviet Union, Japan, and Poland, for example, have to gain from the adoption of a national quota system in the fisheries off the United States coasts, if this guaranteed the United States as large, or larger, share of valuable species than it now obtains? One answer, Crutchfield suggests, might lie in "possible trade-offs, not only in fisheries but with respect to other interests."[41]

Three possible approaches oppose any type of unilateral extension by the United States of its fisheries jurisdiction beyond the twelve-mile limits. A first would hold that not enough now is known about our coastal fisheries to justify specific regulations; what we need now is more fisheries research, not more restrictions.[42] A second argument would be that such extension might jeopardize the interests of the shrimp and tuna industries which fish close to the shores of other nations, and whose value of catch stood first and third respectively among United States fisheries in 1966.[43] Finally, it could be argued that seaward extensions of fisheries jurisdictions throughout the world could lead to serious underutilization of stocks, since many countries might not permit foreigners to harvest underexploited fisheries within the proscribed zones.

Political pressures in the United States may well force the government in coming years to take action against the increasing number of foreign vessels operating off our coasts. If ad hoc concessions and agreements prove inadequate to the task of protecting what the coastal fishing industry feels to be its legitimate rights, then some unilateral actions may be considered. The establishment of a national quota system for selected species is probably the least drastic of such actions. These quotas would apply to the total catch of the affected species, both within and beyond the twelve-mile limit.

How the quotas would be determined is, of course, a focal problem in this system. For some fisheries, such as Pacific salmon and halibut, the United States and Canada have successfully pursued an "abstention" arrangement under which they alone share in the harvest, on the basis of past conservation measures.[44] But the American share of the

to merely reproducing its number year after year. But Crutchfield, among others feels that maximum sustainable yield "is a thoroughly ambiguous term" and "that the function relating the yield of organic marine products to harvesting efforts does not necessarily reach a peak and thereafter decline as the rate of exploitation increases." Crutchfield, *Zones of National Interest: Convention on Fishing and Living Resources of the High Seas,* 1967 A.B.A. Symposium on Marine Resources (in press). For Crutchfield, and other economists, a more meaningful limit is the point of maximum net economic yield, that is, the point where the difference between total costs of production and total revenue is greatest. (But see Schafer's counter arguments, The Law of the Sea: The Future of the Sea's Resources 128 (1968).
40. See his excellent article, *Management of the North Pacific Fisheries: Economic Objectives and Issues* 43 Wash. L. Rev. 283 (1967).
41. *Ibid.* at 307.
42. See, e.g., Chapman, *Fishery Resources in Offshore Waters* 87, *supra* note 35.

43. The 200-mile claims of Peru, Chile and Ecuador have long plagued the American tuna fishermen. United States protests have been of no avail and the vessels' owners are now buying licenses to fish within the 200-mile limit. Any extension of American fisheries claims might not only encourage these countries but also Caribbean nations (off whose coasts shrimp vessels fish) to extend their jurisdictional claims.
44. The "abstention" principle is based on the following three conditions: (1) evidence indicating that more extensive exploitation of a stock will not provide a substantial increase in yield; (2) the exploitation of the stock is limited or otherwise regulated for the purpose of maintaining or increasing its maximum sustainable productivity; (3) the stock is the subject of extensive scientific study designed to discover whether it is being fully utilized, and the conditions necessary for maintaining its maximum sustained productivity.

catch of Atlantic haddock, cod, and other groundfish is declining in the face of foreign activities. Our historic rights here, and in the hake, saury, and pollock fisheries, are less well-established. The allocation of national quotas would presumably come from international negotiations among the interested countries. Tortuous as this process might be, the alternatives could be worse: depletion and possible destruction of the resources; gradual "freeze-out" of American fishermen beyond the twelve-mile limit; or unilateral extension of United States claims to exclusive fisheries well out into the oceans.[45]

Mining

The principal jurisdictional problems related to mining on and beneath the floor of the sea involve the determination of the outer limits of the continental shelf, and the possible regimes to be adopted for the deep ocean floor beyond these limits. As noted earlier the Continental Shelf Convention defines the shelf as being "adjacent to the coast" and extending seaward to the isobath of 200 meters (656 feet) or beyond "to where the depth of the superjacent waters admits of the exploitation of the natural resources." This definition tends to be open-ended, depending on the use of the terms "exploitation" and "adjacent." Existing technology would permit the exploitation of minerals from the floor of the sea in at least 6,000 feet of water and probably more. But the costs of such operations are prohibitive; thus, the limits of exploitability (at least for the United States) are economic in nature, rather than technological.

Assuming (1) that the term "exploitability" has a commercial connotation, (2) that the depths to which jurisdiction extend for

one state also apply to all others, (3) that the areas to be exploited must be "adjacent" to the claimant's coast, and (4) that the framers of the 1958 Convention presumed that there was *some* ultimate limit to the continental shelf, and, if we accept the proposition that commercial exploitation must actually be taking place beyond the 200-meter isobath in order for countries to extend their claims to the shelf on the grounds of exploitability, then the problem of jurisidiction in deep waters may be some years away. The deepest commercial recovery of oil off the United States is some 340 feet of water. No other nation is exploiting oil, gas, or hard minerals at these depths. American companies are carrying out exploratory drilling beyond the 200-meter isobath, and floating rigs are reportedly capable of drilling in 1,300 feet of water, but exploration is not the same thing as exploitation.

It should also be noted that the United States Department of the Interior has published leasing maps of the ocean floor off Southern California in water depths ranging up to 6,000 feet. It has granted exploitation leases to oil companies off the Oregon coast in 1,800 feet of water, and exploratory leases to drill core holes off the Atlantic coast in depths of up to 5,000 feet. Do such actions mean extension of the United States' rights to the shelf out to these depths? The interactions of domestic and international law are extremely complex in this area. United States oil companies are understandably anxious to secure exclusive rights against their domestic competitors in order to carry out exploratory work on the continental slope. In leasing these rights the federal government implies it has jurisdiction over the seabed in question. But it is stretching the rationale of the Geneva Convention to assume that a decision by the United States Interior Department to issue leasing maps of the sea floor in 2,000 feet of water qualifies as a demonstration of resource exploitability at that depth.

A related problem is that of reciprocity.

---

45. In addition to problems of conserving and distributing the catch among different nations are those of conflict of gear. For an excellent discussion of such conflicts off the U.S. coast see Wedin, *Impact of Distant Water on Coastal Fisheries,* The Law of the Sea: The Future of the Sea's Resources 14 (1968).

It might be argued that if a coastal state does not extend its shelf limits seaward as quickly as possible other nations may begin commercial exploitation of the seabed just beyond the coastal state's 200-meter isobath. These fears could be allayed by the reciprocity principle.[46] If a foreign oil company, for example, found it expedient to begin operations in 2,000 feet of water off the Texas coast, the very fact it had proved exploitability was possible at this depth would automatically extend United States jurisdiction over its own adjacent shelf out to this depth.

The problem of defining "adjacency" is a real one in cases where shallow areas lie beyond the first 200-meter isobath seaward from the coast. Several jurisdictional cases have arisen concerning "interrupted" shelves, the most noted perhaps being those of the Norwegian Trough, and of Forty Mile and Cortes Banks off Southern California.[47] Two points seem clear. The first is that some shelves extend up to 200 miles and more off the coast without a significant break in depth. Such areas, at least out to the 200-meter isobath, appertain to the coastal state. The second point is that in the case of some narrow shelves, the presence of a canyon should not disqualify the "outlying" shallow areas from inclusion within the shelf regime.[48] Two additional questions are also

relevant. How great an expanse of deep water must separate an "adjacent" shelf from outlying platforms or sea mounts, in order for them no longer to qualify as coming within the coastal state's sovereignty? Associated with this is the question of shelves extending 100 to 200 miles or more from the coast. If, in these situations, it is found that narrow belts of deep water separate the areas of less than 200-meter depth from outlying shelves of shallow water, can the latter be claimed as "adjacent?"

The assumption that the framers of the Continental shelf Convention had in mind some finite limits to the shelf obviates the possibility of extensions of national jurisdiction out to the median lines of the oceans.[49] At present, such prospects are remote, but were it proved feasible to exploit the hard mineral resources of the deep ocean, some states might well lay claim to these areas in the hopes at least of receiving revenues from licenses issued to foreign exploiters.[50] Clearly it would seem to be both within our own national interest, and the interests of the international community, to establish some fixed limits to the shelf beyond which national sovereignty cannot extend, even on the basis of exploitability.

In 1969 the Geneva Convention becomes subject to a request by any contracting party for revision. Pressure is already mounting within the United States for some consideration of possible changes. It is suggested here that the exploitability criterion be deleted in any revised text, and that some fixed isobath and/or distance from shore be taken to limit the outer boundary of the legally-defined shelf.

---

46. Young, for example, states that "every coastal state would seem entitled to assert rights off its shores to the maximum depth for exploitation reached anywhere in the world, regardless of its own capability or of local conditions, other than depth, which might prevent exploitation." *The Geneva Convention on the Continental Shelf,* 52 Am J. Int'l L. 735 (1958).

47. These Banks lie off the coast of Southern California, and have been the subject of considerable debate as to whether they pertain to the California shelf. See Tubman, *The Legal Status of Minerals Located on or Beneath the Sea Floor Beyond the Continental Shelf,* Exploiting the Ocean: Transactions of the Second Annual MTS Conference and Exhibit 379 (1966); also Barry, *Administration of Laws for the Exploration of Offshore Minerals in the United States and Abroad,* and Luce, *The Development of Ocean Minerals and the Law of the Sea, supra* note 39.

48. It has been held by some writers that the criterion for inclusion of an "outlying" area as part of the shelf should be a geological one. Sea

mounts, unrelated to the shelf, could then be easily identified.

49. See Christy's map, based on the median line principle, which appears with The Law of the Sea: The Future of the Sea's Resources, *supra* note 28.

50. There is considerable debate as to the imminence of commercial deep ocean mining. This division of opinion is reflected in the following two articles: Brooks, *Deep Sea Manganese Nodules* 32 and Mero, *Mineral Deposits in the Sea,* The Law of the Sea: The Future of the Sea's Resources 94 (1968).

Any such revisions should (1) be consistent with United States interests, and (2) be acceptable to the world community of nations. Some coastal states wish exclusive rights to exploit the shelf's resources off their coasts to whatever depths prove feasible; the world community needs to maximize the freedom of the high seas overlying the shelf. The interplay of these concepts was fully recognized in the Shelf Convention. As one proceeds further from the coast, the freedom of use of the sea should become stronger.

What is proposed here is to separate the concept of the continental *shelf* from that of the continental *slope*—that portion of the sea beyond the edge of the shelf were the incline is steeper down to the deep-sea floor. Although geologically the average depth of the break in slope is 133 meters,[51] international law has picked 200 meters as the edge of the shelf. Beyond this depth is the slope. At this stage in marine geology, it is not possible to determine an average depth at which the slope meets the deep sea floor, and the light rocks of the continent abut on the darker, more solid rocks of the ocean basin. However, for purposes of clarifying regimes of the sea bed, some fixed limit must be established, and it is suggested, at least for the immediate future, that a depth of 2,500 meters be taken as the seaward limit of the continental slope.[52]

It is also suggested that the sovereign rights of the coastal state to explore the natural resources of its adjacent continental shelf be extended to the continental slope. The choice of the 2,500-meter isobath would eliminate most of the problems of intervening canyons between the contiguous shelf and outlying areas. The rights of the coastal state to grant or withhold consent for foreign research activities would pertain to the shelf only, that is, out to the 200-meter isobath. Beyond this point, foreign research concerning the shelf could be conducted independently of the coastal state's consent.

Associated with this redefinition of "shelf" and "slope" would be a compensating distance factor. For those states having a narrow shelf, similar rights for exploration, exploitation, and jurisdiction over foreign research would extend to a distance fifty miles offshore from the baseline from which the breadth of the territorial sea is measured, and for exclusive exploration and exploitation rights (but not jurisdiction over foreign research activities concerning the sea floor), to a distance 100 miles seaward of this baseline.

Decisions on the outer limits of the shelf are, of course, related to questions of jurisdiction over the resources of the deep ocean floor. Proponents of regimes for the ocean floor fall generally into five categories. A first group are those who suggest a "wait and see" attitude. These persons point out that within the foreseeable future there are no apparent uses which can be made of the deep sea floor and subsoil, and that until some problems arise we would be better off to rely on the existing convention. The proponents of change come under four headings: those favoring the "flag nation" approach; those wishing an international registry system; those who advocate internationalization of the deep seas' resources; and the "national lake" proponents, mentioned earlier, who would simply extend the jurisdiction of the coastal state out to the midpoint in the oceans.[53]

51. Emery, *Geological Aspects of Sea-Floor Sovereignty*, 149, *supra* note 35.
52. Suggestions of the depths for the outer limit of the shelf range from 1,000 meters (Emery) to 2,500 meters (Mero). Marine geologists admit that knowledge to date is insufficient to define the approximate isobath at which the lighter rocks of the continent abut on the darker rocks of the ocean basins.

53. See Bernfeld, *Developing the Resources of the Sea—Security of Investment* 2. The Int'l Lawyer 67 (1967). Several writers have pointed out that a "national lake" regime might be the logical consequence of unlimited extension of the shelf limits on the basis of exploitability. Countering this, however, is the argument that the criterion of adjacency in the Shelf Convention would prevent unlimited expansion of the outer limits of the shelf to the midpoint in the oceans.

The basis of the "flag-nation" concept is that any exploiter of deep-ocean resources would have his investment protected by the nation under whose flag he is operating. He would be guaranteed exclusive rights to the area in which he is operating against all other claimants for a given period of time. These rights would presumably be embodied in an international agreement, recognizing the rights of nations to protect their citizens in deep sea operations. Disputes could go to the World Court or to some special arbitral body.[54]

A second approach would be to establish an international registry (under the United Nations or some other body) to which exploiters of the sea floor could apply for a certain area of seabed and for a certain period of time.[55] A part of the revenue derived from the subsequent mining operations could go to the registrant body, either to compensate for administrative expenses and/or to distribute wealth to other nations of the world community which share in the ownership of the deep-sea resources.

An internationalization scheme would involve the control and management of deep-sea mineral resources by some world body. Not only would the registration provisions of the previous regime prevail here, but also such an arrangement could provide an effective overall management program and facilitate a rational allocation of exclusive rights. By regulating production rates it could prevent the wide fluctuations in mineral prices which so often accompany unrestricted development.[56]

Two problems associated with the international approach are: (1) on what basis would leases to mineral sites be distributed; and (2) how would the revenues derived from the exploitations be distributed? Christy, who perhaps more than anyone else has studied these issues, concludes that the optimal system for distributing leases would be a competitive bidding system, as is carried on for oil leases in the United States continental shelf. He prefers this to a "first-come, first-served" system (as presumably would be followed in an international registry system) on the ground that the latter might result in excessive "land grabs" and in uneconomic production by concerns seeking to make maximum use of their concessions within the time period allotted to them. So far as distribution of wealth is concerned, Christy feels that one factor which would make this international regime acceptable to a majority of the world's nations would be the provision that a portion of the revenue derived from exploitation "could be earmarked for some widely accepted goal, such as the overcoming of malnutrition."[57]

The advantages of internationalization are offset by two drawbacks: the lack of an effective organizational system at this time to handle such a regime, and the fear on the part of the United States and other maritime powers that such a regime would be contrary to their national interests, and might stultify, rather than advance, deep ocean developments. Since this nation is a leader in marine affairs, it is argued, why should it jeopardize its freedom of action by vesting even partial authority in an international agency whose membership cannot be controlled by the United States?

While in the long run, these negative features may become minimized it is suggested that for the present the United States and

54. Ely, *The Administration of Mineral Resources Underlying the High Seas, supra* note 35; *American Policy Options in the Development of Undersea Mineral Resources,* 2 The Int'l Lawyer 215 (1968).
55. Goldie, *The Geneva Conventions,* 273, *supra* note 35. See also Henkin, *Law for the Sea's Mineral Resources* (Paper available from the National Council on Marine Resources and Engineering Development, Washington 20500).
56. Christy, *Alternative Regimes for Minerals of the Sea Floor, supra* note 39; *A Social Scientist Writes to the International Lawyer on Economic Criteria for Rules Governing the Exploitations of Deep Sea Minerals,* 2 The Int'l Lawyer 224 (1968).

But see also the objections to internationalization in Burke, *A Negative View of Proposals for United Nations Ownership of Ocean Mineral Resources, supra* note 39.
57. *A Social Scientist Writes to the International Lawyer, supra* note 56.

other maritime powers seek only a regime of international registry. Under this arrangement the international machinery for dealing with deep-ocean resources could be gradually developed without being burdened at the outset with the awesome responsibilities of managing the total resource exploitation. Since it appears likely that some years will elapse before substantial development of deep-ocean minerals will begin, there is ample time to work out the operational details of such a registry system.

Scientific Inquiry

The conduct of scientific research beyond a coastal state's territorial limits might be considered in terms of two general categories: (1) the nature of the inquiry, and (2) the area of the sea in which it is carried out. So far as the nature of the inquiry is concerned there are three major types of scientific investigation. First is general research into the composition of the marine environment, the results of which are made available to all nations. Second is investigation into the exploitation of certain marine resources, the results of which may or may not become public. Third is research carried on for military purposes; the results here are almost always classified. Unfortunately, when research is being carried out by an oceanographic vessel, it is often impossible for an interested foreign state to determine to which category the operations apply, unless the state has an observer on board the vessel.

The right of scientific inquiry by one country's vessels within another country's internal or territorial waters is prohibited without the express consent of the foreign government. If the coastal state claims exclusive fisheries rights in an extra-territorial contiguous zone out to the twelve-mile limit, it may take the position that any research being carried on by foreign vessels relating to the availability and exploitation of fisheries resources in that zone is contrary to the coastal state's interests.

A more serious problem concerns research involving a foreign state's continental shelf. The Shelf Convention provides in Article 5(8) that:

The consent of the coastal State shall be obtained in respect of any research concerning the continental shelf and undertaken there. Nevertheless, the coastal State shall not normally withhold its consent if the request is submitted by a qualified institution with a view to purely scientific research into the physical or biological characteristics of the continental shelf, subject to the proviso that the coastal State shall have the right, if it so desires, to participate or to be represented in the research, and that in any event the results shall be published.

There are definitional problems here which may in time work to the disadvantage of the United States. Up to now, practically all the research carried on by scientists of one country in the waters overlying the shelf of another state has been conducted by United States scientists off foreign coasts. While the number of refusals by foreign governments to permit such research has been kept to a minimum, the possibility of future difficulties is a strong one.[58] To avoid unnecessary suspicion by foreign governments the suggestion has been made to delegate to some international scientific agency, such as the International Council of Scientific Unions, the responsibility of deciding whether requests for research by foreign vessels on a coastal state's continental shelf are indeed bonafide scientific inquiry. The agency would also undertake to ensure open publication of the results of the inquiry.[59]

58. In the summer of 1967 the Soviet Union turned down a request by the University of Washington for its oceanographic vessel, "Thomas G. Thompson," to conduct research in the West Bering and Chuckcki seas. The Pueblo incident off North Korea in January, 1968, may stimulate other countries to begin refusing research requests on the grounds that the oceanographic vessels in question might be part of a military intelligence net.
59. Schaefer, *The Changing Law of the Sea—Effects on Freedom of Scientific Investigation*, The Law of the Sea: The Future of the Sea's Resources 113 (1968).

Any attempts by nations to expand the areas of their legally-defined continental shelf could, of course, affect the freedom of research by foreigners in the overlying waters. While the United States at present would seem to have little to gain from the provisions of Article 5(8) in terms of protecting its economic interests on the shelf, it might well resist any attempts to delete this article in possible future modifications of the Convention because of our military interests.[60]

Some writers have claimed that freedom of research is, in a sense, a fifth freedom of the high seas since it qualifies under the provision of Article 2 of the High Seas Convention as one of those other freedoms "which are recognized by the general principles of international law."[61] The inter-relationships of natural phenomena throughout the world ocean make imperative the collection and dissemination of data from all parts of the marine environment. It would seem to be within the United States' interests to press for the greatest possible degree of freedom of research beyond the twelve-mile limits of all coastal states.

Military Operations

The two key goals sought by the United States military in the sea are mobility and concealment. The United States Navy has traditionally sought the maximum freedom of mobility in the sea and in the airspace

above it. To this end, its interests at times conflict with commercial fishing interests which desire an extension of national authority some distance from shore as protection against foreign competition.[62]

The Geneva Conventions did not guarantee to warships a clear right to pass through foreign territorial waters,[63] although they are accorded the freedom of innocent passage through international straits whose waters lie wholly within the territorial limits of the adjacent state or states. Navy spokesmen are keenly aware of efforts throughout the world ocean to close off or restrict the passage of United States warships or planes. Extensions of territorial limits, the delimitation of straight baselines, or efforts to intercept United States vessels more than twelve miles off a foreign coast are actions of great concern to the Department of the Navy.

There is one exception to adherence by the United States military to the principle of the free use of the sea, namely, the temporary closure of certain areas for the purpose of weapons testing. The most noted examples were the establishment of "warning areas" in the mid-Pacific during the early 1950's for hydrogen bomb testing.[64] These areas were up to 400,000 square miles in extent. In addition to these, there have been temporary closures of much smaller areas of the high seas by United States military

60. In a discussion of possible future vehicles which may operate off a country's coast, Burke notes "telechiric (remote control) systems . . . robot systems, the various inhabited submersibles capable of operating at continental shelf depths and deeper, the special structures for prolonged habitation in submerged regions, and the various types of buoy systems that are unmanned but either inert in the water or self-propelled." Burke, Ocean Sciences, Technology, and the Future International Law of the Sea 42 (1965).
61. See Burke, *Law and the New Technologies* 206, and Goldie, *The Geneva Conventions* 273, *supra* note 35. See also Burke, *A Report on International Legal Problems of Scientific Research in the Ocean* (Paper available from the National Council on Marine Resources and Engineering Development, Washington 20500).

62. An example of the conflicting military and fisheries interests was provided during the debates on U.S. policy positions during the 1958 and 1960 Geneva Law of the Sea Conferences. See, e.g., Dean, *The Second Geneva Conference on the Law of the Sea: The Fight for Freedom of the Seas*, 54 Am. J. Int'l L. 751 (1960).
63. Article 23 of the Territorial Sea Convention states: "If any warship does not comply with the regulations of the coastal State concerning passage through the territorial sea and disregards any request for compliance which is made to it, the coastal State may require the warship to leave the territorial sea." 3 *Off. Rec.* A/C 13/39 (1958). Within foreign territorial waters submarines are required to operate on the surface and to show their flag.
64. See McDougal and Schlei, *The Hydrogen Bomb Test in Perspective: Lawful Measures for Security*, 64 Yale L. J. 648 (1955).

authorities for the testing of nuclear or other weapons.

The interests of the military with respect to the seafloor are more complicated. The Navy might be expected to favor a narrow continental shelf, since this reduces the extent of area off foreign coasts within which governments might seek to preclude American scientific investigations. Yet the Navy would be very much concerned over knowledge of an attempted emplacement by some foreign country of data-gathering devices on the floor of the sea close to the United States coasts but beyond the legally-defined shelf limits. This same argument extends to sea mounts or other shallow areas beyond the physical limits of the shelf. Since no nation has jurisdiction over the floor of the deep sea, does it follow that any government can install temporary or permanent military devices on these sea mounts close to a foreign shore?

Two additional points are: (1) the suggestion of demilitarizing the deep ocean floor; and (2) the possible establishment of national defense areas on the floor beyond continental limits. Attempts have been made to find an analogy between the Antarctic Treaty of 1959, which provided that the continent be an area used exclusively for peaceful purposes, and a possible future status for the deep ocean floor. In view of the extensive undersea warfare programs of the United States, the Soviet Union, and other powers, the prospects for a demilitarization treaty for the deep sea bed do not appear strong.[65] Suggestions have been made, however, for an international treaty banning the permanent emplacement of nuclear weapons or other weapons of mass destruction on the ocean floor, perhaps beyond the shelf limits.[66] Such a treaty would be somewhat analogous to the 1963 Treaty banning the use of such weapons in space.

The designation of defense areas in the ocean floor might reduce some elements of international uncertainty and possible incidents. Goldie, for example, notes "states seeking to establish *fixed* defense installations on sea mounts and on the seabed should give notice to the effect that such are taken for defense purposes and are not to be viewed as any longer within the general regime of the seabed and its subsoil."[67] Since the number of sea mounts in the Pacific alone is estimated to be in the thousands the acceptance of this suggestion might well herald the beginnings of a far-ranging jurisdictional pattern on the ocean floor.

RELATIONSHIPS AMONG USES OF THE SEA

The principal goals of legal regimes in the sea should be to maximize net benefits received from the marine environment and to provide a means for the equitable resolution of conflicts. In terms of benefits the United States seeks five major types from its use of the sea. Four of these are largely domestic: production of wealth, maintenance of natural security, acquisition of knowledge, and promotion of public and private welfare. The fifth is international: the advancement of world community interests. In our pursuit of these objectives we often meet situations of conflicting interests—wealth vs. public welfare, security vs. world community interests, science for science's sake vs. science for military uses.

Because of the complexities of its objectives and of the role it occupies as a major world power, the United States faces difficult choices both in the ways it uses the marine environment and in the types of jurisdiction it seeks to exert over that envi-

65. For a positive approach see Michael, *Avoiding the Militarization of the Seas*, New Dimensions for the United Nations: The Problems of the Next Decade 161 (1966).
66. Thus adding still another component to the problem of defining shelf limits.

67. Goldie, *supra* note 55. See also his *Submarine Zones of Special Jurisdiction Under the High Seas—Some Military Aspects*, The Law of the Sea: The Future of the Sea's Resources 100 (1968).

ronment. In a world of political, economic, and technological change, there are various costs involved in rigid adherence to existing regimes of the sea. Yet the costs of change might be even greater. Thus, two basic problems present themselves. First, do apparent inadequacies in the present law of the sea present serious problems at this time, or are the difficulties more apparent than real? And, second, if there is need to consider revisions in existing regimes, how might this best be accomplished?

There probably are no direct answers to the first question. The economic, scientific, and military uses we make of the sea are not at the present time seriously hampered by inadequate international rules and regulations.[68] Nor, apparently, are most other countries concerned with the shortcomings of the present system.[69] But even if things are not serious now, there is reason to anticipate future problems and prepare for the day when they will be upon us. It is not so much the actual uses of the sea that will bring such problems to the surface, but the political pressures associated with these uses—expectations of food from the sea, competition among nations for data acquisition, the demands of specialized interest groups within our own country.

It is important that within the federal government we have a strong locus for debate and decision-making on jurisdictional matters in the marine environment. This organization must be able to weigh the conflicting interests, to assess possible trade-offs, and to formulate and advance new concepts in the law of the sea when necessary.

Beyond this is the international scene. Already voices are heard favoring a Third Law of the Sea Conference to handle new or unresolved problems. With the growing American emphasis on marine affairs, such a conference may in time be necessary, but among some experts there is opposition to the idea.[70] Many of our problems concerning the use of the sea are not dependent on changing existing regimes. International law is a slowly-evolving process; new concepts developed in 1958 should be thoroughly tested before still different regimes are introduced to the world community. New laws should be given a chance to develop through custom rather than more codifications.

In the face of these swirling developments we can identify two basic principles. First, our development of the marine environment must proceed as expeditiously as possible, consistent with our overall national interest. There is much to be done in the sea within existing international regimes. Second, we must remain alert to new concepts in the law of the sea. In the complex world of today the United States cannot afford to lose its leadership role in the realm of ideas and ideals.

68. With the exception that international rules have done nothing to alleviate gross economic waste in international fisheries.
69. Two-thirds of the independent countries of the world have not seen fit to ratify even one of the Geneva Conventions, nor have they officially explained the cause for their refusal to do so. Apparently they have other problems to worry about than the law of the sea.

70. "I think it may take a hundred years for the law of the sea to recover from the last two international conferences which dealt with it, and I would regard the immediate call of another conference as an unmitigated disaster," McDougal, *International Law and the Law of the Sea*, The Law of the Sea: Offshore Boundaries and Zones 3 (L. Alexander ed. 1967).

Marine Resources and the Freedom of the Seas

FRANCIS T. CHRISTY, JR.   1968

INTRODUCTION

The freedom of the seas, which has served as a fundamental principle of international law for three centuries, is a strange guide to adopt for economic enterprise. In the past, it served a useful purpose in diminishing jurisdictional impedimenta to the use of the seas and its resources. But at the same time, it removed exploitation and use from the realities of property. And today, for many purposes, the principle has been severely eroded, or has been maintained only at great cost to the world community. The continued reliance upon the principle, as it is presently understood, is both dangerous and damaging.

THE FREEDOM OF ACCESS

The freedom of the seas is generally understood to be a freedom of access, such that no single user can exclude others from participating directly and simultaneously in the same use. Thus, any stock of fish can be exploited at the same time by any number of fishermen, and any ship can move through the same narrow strait, subject only to the rules of the road. In these terms, the wealth of the seas is associated with access, and can only be obtained by exercising the right of access. Those who do not participate directly in the use or exploitation of the seas do not receive any share of the wealth.

However, it is possible to distinguish between a right of access and a right to a share in the wealth. The distinction depends in part upon one's concept of the community interest in the marine environment and in part upon political reality. If the seas are to be considered the property of the world community (*res communes*), then the distinction between wealth and access is conceivable, since it is a distinction commonly made for the enjoyment of public property intra-nationally. If, however, the seas are considered the property of no one (*res nullius*), then exclusive rights to both access and wealth can be appropriated and the distinction is invalid.

Dr. Francis T. Christy, Jr. is senior research associate with Resources of the Future, Inc., Washington. He is a member of the Executive Board of the Law of the Sea Institute. He has co-authored books on natural resources, including ocean fisheries.

From *Natural Resources Journal*, Vol. 8, No. 3, pp. 424-433, 1968. Reprinted with light editing and by permission from the author and *Natural Resources Journal* published by the University of New Mexico School of Law, Albuquerque, New Mexico.

Neither of these concepts is universally held. In certain circumstances and for certain resources, a nation may be guided by *res nullius*. Under different situations, the concept of community ownership may dominate. If there is any trend discernible, it would appear that it is toward this latter point of view. The principle of the freedom of the seas has led many nations to believe that they have an interest in the seas that goes beyond the simple right of access. This has been expressed in international fishery agreements where there is a clear obligation for states to adopt measures to maintain a resource that is the property of the world community. It has also been expressed by President Johnson when he warned that "under so circumstances, we believe, must we ever allow the prospects of rich harvest and mineral wealth to create a new form of colonial competition among the maritime nations. We must be careful to avoid a race to grab and to hold the lands under the high seas."[1]

In summary, the freedom of the seas is generally considered to be a freedom of access, and only through exercise of this freedom can one obtain a share of the seas' wealth. It is possible, however, to distinguish between the right to wealth and the right of access. And the distinction is one of the first steps necessary to the redefinition of the principle of the freedom of the seas.

CHANGES IN THE CONDITIONS OF FREEDOM

The modern development of the principle began with the publication of *Mare Liberum* by Hugo Grotius in 1608. Grotius, in defending Dutch enterprise on the high seas against the interference and claims of Spain, Portugal, and Britain, asserted that all na-

1. President Lyndon B. Johnson, *Remarks of the President at the Commissioning of the New Research Ship, the Oceanographer* (July 13, 1966).

tions should have free and equal access to the seas and its resources. He based his argument, in part, upon the assumption that the rewards of exclusive rights were not sufficient to incur the costs of obtaining and protecting exclusive rights.

Fish stocks were believed to be sufficiently abundant at that time, so that if a fisherman found congestion in one area, he could take fish from another area at no greater cost. Under these conditions, there is no particular incentive to acquire exclusive rights to a fish stock or fishery area because the rights would have no value. In addition, since most fish are highly mobile, it would be costly to attempt to protect a claim to exclusive rights.

The conditions assumed by Hugo Grotius had some validity in the 1600's. But since then, and more particularly in the past few decades, the myth of resource abundance has been totally disproven. To be sure, there are still many who talk in glowing terms about the vast untouched resources of the seas—and who, by intricate calculations, estimate that man could take fifty or more times the quantity of protein material than is now being caught—or who estimate that there are billions of tons of minerals lying in solution or on the floor of the sea. Such estimates, while they may be of interest in the long distant future, are irrelevant to the problems and opportunities of the sea over the next several decades.

The economic scarcity of fish is due to the facts that demand is limited to a few dozen species and that natural conditions limit the supply. Demand is not for fish in general, much less for undifferentiated protein material (although this may change). The demand is for cod, haddock, tuna, halibut, lobster, salmon, and a dozen or so other species. The vast quantities of plankton, blennie, and goggle-eyed scad are not of interest to man—only to other fish.

Some changes in demand may occur in response to changes in taste preferences and to changes in the ability to disguise the source of the protein. After a fish has been cut into bite-size chunks and covered with bread crumbs it is virtually impossible for the consumer to identify the kind of fish it once was. And so long as he enjoys the taste, he really shouldn't be concerned. Perhaps more important is the possible development of a market for a marine protein concentrate which may, if successful, significantly enlarge the fishery resource. Neither of these changes, however, is likely to diminish the demand for the well-recognized species of fish.

In addition to the limited demand, the supply is also restricted. In part, this occurs because of the wide range in fertility in the marine environment—ranging from virtual deserts in mid-ocean to exceptionally fertile waters, such as those off Peru, that may produce 300 or more pounds of protein per acre per year. Because of this, fishing effort is not diffused throughout the oceans, but is concentrated on those areas of high fertility producing fish for which there is a strong market.

Supply is also limited by man's inability to cultivate the seas. Any particular stock of fish is relatively fixed in size, subject only to natural fluctuations and to man's efforts to harvest it. For any newly exploited stock of fish, an increase in catching effort is accompanied by an increase in catch. However, for any stock there is a maximum annual catch that can be sustained over time. If larger amounts of fish are taken, then there will be fewer fish available in subsequent seasons.[2] When this occurs, the stock is said to be over-fished or depleted. In addition, it is significant that under these conditions, reductions in the amount of fishing effort may lead to higher total annual yields. "It has been estimated that the total effort on some of the major stocks of cod and haddock in the northeast Atlantic has increased so far that substantially with the same or possibly an even slightly greater catch could be taken

with one-half to two-thirds of the present level of fishing."[3]

The severity of depletion might be alleviated by the discovery of new stocks of the conventional species. But "at the present rate of development few substantial unexploited stocks of fish accessible to today's types of gear will remain in another 20 years."[4] The development of new kinds of gear and new techniques of fishing will also have limited value. Indeed, such innovations are likely to be applied to stocks that are already over-fished thereby depleting them even further.

The outlook is gloomy. As demand increases, there is a growing concentration of fishing effort in those areas where the high valued species occur, and with the application of increased effort, the stocks and the total seasonal catch will become smaller and smaller. It is clear that the assumption of abundance can no longer be maintained and that there has been a significant change in this condition, which is one of the fundamental ones underlying the principle of the freedom of the seas. The incentive to appropriate exclusive rights to fisheries is rising rapidly and, in some cases, has already led nations to incur the costs (economic and non-economic) of asserting and protecting exclusive claims.

The incentive to appropriate exclusive rights to the mineral resources of the marine environment could not have been anticipated by Grotius. Today, however, the incentive is particularly great in near-shore areas, as is illustrated by the recent sale of oil and gas rights in the federal waters off the coast of Santa Barbara. There, the oil companies bid more than $600,000,000 for the exclusive rights to exploit the resources under 350,000 acres. And one tract of 5,760 acres, all of which is under 1,200 or more feet of water, was obtained for a bid of $21,000,00.

2. For a more precise statement of this see J. Crutchfield & A. Zellner, *Economic Aspects of the Pacific Halibut Fishery*, 1 Fishery Industrial Research 1 (1963).

3. *Food and Agriculture Organization of the United Nations,* The State of Food and Agriculture 1967, at 124 (1967).
4. *Id.* at 120.

The incentive to obtain rights to deep sea manganese nodules is not yet clear. These nodules presumably cover vast areas of the deep sea floor, but there are differences in the depth in which they occur, in the content of manganese, copper, cobalt, and nickel, in their density, in the topography of the bottom, and probably in other characteristics that may lead to wide differences in the values of different locations.[5] When development becomes economically feasible, these differences will militate against maintaining free and equal access under the principle of the freedom of the seas.

The conditions underlying the non-extractive uses of the marine environment have also changed considerably since the 1600's. Shipping has increased to the point where some 750 vessels per day now pass through the Straits of Dover, leading to greater costs in navigation, in the provision of navigation aids, and in collision insurance. Shipping is also finding it increasingly difficult and costly to thread its way through oil rigs and fishing fleets. And several recent incidents have indicated the growth in the problem of oil pollution, both from the pumping of bilges and from accidents to oil carriers.

These and other trends indicate that ocean space and resources, far from being abundant, are actually becoming increasingly scarce. With growing scarcity comes increasing reward to the appropriation of exclusive rights. At the same time, the economic cost of appropriating and protecting such rights is being reduced by technological innovations in aircraft, surface vessels, monitoring devices and satellites. Thus, the conditions argued by Grotius to support the principle of the freedom of the seas no longer obtain for many marine resources and uses, and are becoming less relevant for others.

THE COSTS OF MAINTAINING FREEDOM

Waste

The attempts to maintain the principle of the freedom of the seas lead to highly detrimental consequences—to economic and physical waste and to conflict. An indication of the waste has already been given for fisheries in physical terms: where a stock of fish is limited and the amount of effort applied to it increases, the annual yield from the stock will reach a maximum point and then diminish. This occurs because, under free and equal access, no individual fisherman can afford to restrain his own effort in the interest of future returns, since anything he leaves in the sea for tomorrow will be taken by others today. Thus it becomes necessary to invoke controls over the fishermen—to reduce their freedom by regulating their effort.

A less apparent but more significant form of waste is that of the inability to prevent excess applications of capital and labor. The open access, under the freedom of the seas, means that as long as there is any economic rent produced in the fishery, it will attract more fishermen until all rent has been dissipated and total costs are equal to total revenues.[6] There have been a number of studies indicating the magnitude of the economic waste associated with the applications of redundant amounts of capital and labor.[7] A

6. See H. Gordon, *The Economic Theory of a Common Property Resource: The Fishery*, 62 Journal of Political Economy 124 (1954); A. Scott, *The Fishery: The Objectives of Sole Ownership*, 63 Journal of Political Economy 116 (1955); J. Crutchfield and A. Zellner, *supra* note 2; Smith, Book Review, 56 The American Economic Review 1341 (1966).

7. E. Lynch, R. Doherty, & G. Draheim, *The Groundfish Industries of New England*, Circular 121, U.S. Fish and Wildlife Service (1961); J. Crutchfield and A. Zellner, *supra* note 2; D. Fry, *Potential Profits in the California Salmon Fishery*, 48 California Fish and Game (1962); W. Royce, *et. al.*, *Salmon Gear Limitations in Northern Washington Waters*, Contribution No. 145 University of Washington, College of Fisheries (1963); and J. Crutchfield & G. Pontecorvo, The Pacific Salmon Fisheries (1969).

5. *See* J. Mero, The Mineral Resources of the Sea (1965); and D. Brooks, *Deep Sea Manganese Nodules*, The Law of the Sea: The Future of the Seas' Resources (L. Alexander ed. 1968).

recent estimate by FAO indicates that "with present levels of landings of cod from the north Atlantic having a total value equivalent to approximately U.S. $350 million, and assuming that under present conditions of overexploitation costs are equal to the value of landings, a halving of present costs would represent a saving of the magnitude of $175 million per year."[8] For the world as a whole, the waste may be on the order of several billion dollars a year—a waste caused by the maintenance of free access under the principle of the freedom of the seas.

Where the development of marine minerals has become economically feasible, the condition of open access is not maintained. The exclusive rights of coastal states have been extended to cover the minerals of the floor of the continental shelf, as these minerals began to attract capital. The provision of some form of exclusive rights to the minerals of the deep sea floor is more difficult, but nonetheless desirable. Until the freedom of the seas is restricted to the point where exclusive rights to the sea floor can be obtained, investment is not likely to be forthcoming. Or, if investment does flow in, it is likely to precipitate a race among nations and inefficient utilization of the resource.[9]

For many of the non-extractive uses of the ocean, the maintenance of the principle of the freedom of the seas may also have damaging consequences. For shipping, even though the freedom is modified by the rules of the road and the chances of collision are reduced, there are still costs associated with congestion. Since there is no market for the use of space, there is no mechanism for allocating use in an efficient manner. Thus, small coastal freighters with low value, bulk cargo may delay other vessels with higher value and more perishable cargo in congested shipping areas. Such delays may be costly to the shippers and to the economies of the

states making use of the vessels. While it is obviously difficult to evaluate the significance of such costs, in view of many factors that might be involved, this illustrates that even in shipping, free access may lead to wasteful uses of ocean space.

Some of the damages of open access, under the principle of the freedom of the seas, are more severe than others, but for all marine resources and uses of marine space, free access will become wasteful as the resources and space become scarce and demand continues to increase.

Conflict

Where the costs of open access are particularly high and obvious, the principle of the freedom of the seas will be eroded. This has already occurred in many areas and is likely to occur with increasing frequency in the near future. For the three centuries preceding World War II, most nations were content with their marine borders. In the past few decades, however, these borders have been widely extended. The Truman Proclamation in 1945 asserted United States jurisdiction over the resources of its continental shelf. Chile, Ecuador, and Peru responded by extending exclusive rights to fisheries as well as the sea floor out to 200 miles. More recently, a large number of states (including the United States) have extended their fishery limits to 12 miles.

In 1958, the Geneva Convention on the Continental Shelf permitted coastal states to extend exclusive rights to sea floor minerals out to the 200 meter isobath or "beyond that limit, to where the depth of the superjacent waters admits of the exploitation of the natural resources. . . ." This open-ended definition, limited only by exploitability and some ill-defined concept of proximity, provides no deterrent to further encroachment on the freedom of the seas. Exploitation is under way in depths greater than 300 feet. The $21 million bonus for the lease off Santa Barbara for a tract over 1200 feet deep indicates the likelihood of commercial

8. *Supra* note 3, at 144.
9. F. Christy, Jr. *Economic Criteria for Rules Governing Exploitation of Deep Sea Minerals,* 2 The International Lawyer 224-42 (RFF Reprint No. 72, March 1968).

exploitation at that depth in the near future, and beyond that, the Department of the Interior "has indicated an assertion of jurisdiction beyond the 200 meter line by publishing leasing maps for areas off the Southern California coast as far as 100 miles from the mainland, at depths as great as 6000 feet."[10]

For fisheries beyond coastal limits, there are some interesting developments that may, if successful, serve to further restrict the freedom of the seas. These are the international agreements, under which the signatories divide up a total catch quota among themselves on the presumption that other states will not enter the fishery.

The first of these and the most successful from an economic point of view is also unique in that it distinguishes between a right of access and a right to a share in the wealth. Under the North Pacific Fur Seal Treaty, signed in 1911, the harvest of fur seals is restricted to those nations on whose islands the fur seals breed (the United States and the Soviet Union), while the other signatories (the Canadians and Japanese) abstain from taking seals on the high seas in return for a share of the furs.

The other quota agreements do not make this distinction, but they too depend upon the presumption that non-signatories will not enter the fishery. These agreements include the division of the whales of the Antarctic and the salmon of the Pacific. In the western Pacific, the Soviets and the Japanese arrive at their respective shares by annual negotiations that generally work to the detriment of the Japanese.[11] In the eastern

Pacific, the United States and Canada arrived at a 50-50 split that has worked fairly well.

For the eastern to mid-Pacific (175° West) a unique form of exclusion has been achieved under the doctrine of abstention. Under this doctrine, where a resource is fully utilized (in a biological sense) and where states are investing in the regulation of the stock, other states that are a party to the treaty agree to abstain from fishing the stock. Thus, the Japanese (who signed this treaty in 1952 as a condition for our signing of the Peace Treaty) have excluded themselves from salmon fishing in the eastern Pacific. More recently, it has been proposed that the fourteen or so nations that catch the groundfish of the North Atlantic work out total catch quotas for each of the depleted stocks of fish and then divide up the quotas among themselves.

These attempts, to the extent that they are successful in excluding free access, will serve to erode the principle of the freedom of the seas. Their success, however, may not be very durable or widespread. Protection of the presumed exclusive rights of the signatories may be difficult since these stocks cover vast areas of the ocean, and since increasing demands are likely to increase the incentives to participate. The freedom of the seas gives the non-signatories every right to do so, and since under current conditions a share of the wealth can only be obtained by exercise of the right of access, there is no reason why the non-signatories should sit aside and watch a few nations divide the wealth of the high seas fisheries.

10. Frank J. Barry, *Administration of Laws for the Exploitation of Offshore Minerals in the United States and Abroad,* a paper presented at the American Bar Assn. National Institute on Marine Resources, Long Beach, Calif., June 9, 1967.

11. Shigeru Oda, *Japan and International Conventions Relating to North Pacific Fisheries,* 43 Washington Law Review 67 (1967). One of the costs of annual negotiation of quotas is that of the time required to reach agreement. Oda points out that "since 1957, the Commission has spent as much as 52, 100, 122, 107, and 105 days, discussing this matter at its annual sessions."

CONCLUSION

The principle of the freedom of the seas has been maintained, and quite satisfactorily for a long time. As long as the use of the seas and its resources by one did not significantly decrease the use by another, there was no reason to prevent free access, and no reason to separate the right of access from a right to a share in the wealth. But for many resources and uses of the marine environment, this basic condition of abundance has long since passed. To maintain free access in view of this change is to persist in the waste and inefficient use of the seas. Access must be closed, and there must be provisions under which a user can acquire a right to exclude others from participating directly in the same use.

But the association of access with a share in the wealth makes such controls extremely difficult and leads inevitably to conflict. Nations, to the extent that they feel they have an interest in the resources beyond their coastal limits (however described), are not likely to accept arrangements that exclude them not only from access, but also from a share in the seas' wealth.

Thus, on the one hand, the maintenance of the freedom of the seas can only lead to greater and greater waste, while on the other hand, the increased extension of exclusive rights will inevitably lead to conflict. A redefinition of the freedom of the seas, which disassociates the right of access from the right to the seas' wealth, may resolve this dilemma. Closing access would permit efficient operations by the entrepreneurs. Extracting ground rents, paid by the entrepreneurs for their exclusive rights, might satisfy the interests of the world community sufficiently to reduce conflict and provide for a stable regime.

9

SOME
CONSERVATION ISSUES

Can Leviathan Long Endure So Wide a Chase?

SCOTT McVAY   1971

A hundred years ago, Herman Melville asked "whether Leviathan can long endure so wide a chase, and so remorseless a havoc?" Today, in more prosaic words, the question remains: What would be lost if the whales were gone from the sea? Of what possible use are whales to men? Esthetics aside, who cares if the whale goes the way of the dinosaur?

These words sound terrible and ominous to me; yet they represent the thinking of many people, including some of the men who set the whale-kill quotas every June. Few of these men have ever seen a whale. Few of them had ever heard a whale until biologist Roger Payne, of the New York Zoological Society and Rockefeller University, played a recording at the close of the final session of the International Whaling Commission meeting on June 26, 1970, in London.

It was 2:00 P.M., and the commissioners were hungry. The chairman, Mr. I. Fujita,

noting at the outset of the final plenary session that business would have to be completed before lunch, pointed out brightly that "hunger will expedite our deliberations." But some delegates, notably the Japanese and Soviets, lingered to listen to sounds recorded at a depth of 250 fathoms— the song of the humpback whale. Spanning *six* octaves, it filled the conference room at River Walk House overlooking the Thames.

Henceforth, the commissioners' annual deliberations will take on a new dimension. These sounds have already made a profound impression on the thousands of Americans who recently heard them in New York's Philharmonic Hall, on television, and on radio. The unexpected fact revealed by this recording and many others, according to the analysis by Dr. Payne and me, is that the sounds often fall into true song forms that are predictable in broad outline.

The humpback songs have captured the

Scott McVay has led efforts by the Environmental Defense Fund to ban the importation of whale products into the United States and achieve a worldwide moratorium on whaling. He has served on the U. S. delegation to the International Whaling Commission, appeared before Congress, and written several articles on whales which have appeared in *Scientific American, Natural History, Bulletin of Atomic Scientists, Audubon,* and *The New York Times.* He has studied the communication of porpoises, co-authored the analysis of the "Songs of Humpback Whales" (*Science,* Vol. 173, 1971), and led an expedition to the Artic to study the Bowhead whale.

From *Natural History,* Vol. 80, No. 1, pp. 36-40, 68, 70-72, 1971. Reprinted with light editing and by permission of the author and *Natural History* Magazine. Copyright by *Natural History* Magazine, 1971.

imagination of composers and musicians. In a musical ballad Pete Seeger wrote:

> If we can save
> Our singers in the sea
> Perhaps there's a chance
> To save you and me. *

This thought cuts right to the heart of the matter. (See Fig. 9-1.)

The decimation of the antarctic whale fishery is a grisly story. It has been catalogued since 1920, when the Bureau of International Whaling Statistics in Sandefjord, Norway, began recording every *reported* whale kill by species, length, sex, date, and place of death. During the 1960's, the yield in barrels of whale oil dropped fivefold, from more than 2 million barrels to less than 400,000 in the 1969-70 season. The whalers might have taken more than a million barrels year after year, indefinitely. But their insatiability in the past two decades has so ravished the stocks and so decimated the large species that the sustainable yield today is but a shadow of what it could be if the stocks had a chance to rebuild.

Last year, most of the world's whale catch was taken by two nations, the Soviet Union (43 per cent) and Japan (42 per cent). The remainder was taken by Peru (5.3 per cent), South Africa (2.8 per cent), Norway (2.5 per cent), Canada (1.7 per cent), Australia (1.4 per cent), Spain (0.8 per cent), and the United States (0.5 per cent).

The grim figures for the past season (1969-70) reflect the catch of smaller and smaller whales in the warm waters of lower and lower latitudes. Twelve years ago, 65 per cent of the catch was taken in antarctic waters south of 60° south latitude. Last season, 89 per cent of the catch was taken between 40° and 60° south latitude. In the heart of the antarctic fishery, once the most bountiful whaling ground on earth and a seemingly endless resource, the harvest has

* "The Song of the World's Last Whale," by Pete Seeger. © by Stormking Music, Inc., all rights reserved, used by permission.

dropped in a dozen years from two-thirds of the total catch to one-tenth.

In the age of sail more than a century ago, when the whale hunt was directed principally at two species, the sperm and right whales, and the old-time methods were no match for the elusive and fast-swimming blue and fin whales, Melville could assert with dreamy eloquence: "The whale-bone whales can at last resort to their Polar citadels, and diving under the ultimate glassy barriers and walls there, come up among icy fields and floes; and in a charmed circle of everlasting December, bid defiance to all pursuit from man."

Melville could not have envisioned the rapacious efficiency of modern whaling, which has all but eliminated the rich Antarctic fishery. Today whales are hunted at both ends of their migratory cycle and, in the case of the sperm whale, on the way to the southern grounds.

Victor Scheffer deflates any notion of romance in the contemporary whale chase:

"In man's attempts to catch more whales more cheaply, he has tried to poison them with strychnine and cyanide and curare. He has tried to electrocute them. Spotters in airplanes and helicopters now search them out and report the position of the herds to whaling vessels below. The ships hunt them down by ASDIC, the system that can feel the whales in total darkness. A 'whale- scaring machine' frightens the beasts into flight with ultrasound and tires them so the hunter can overtake them. What will be next? Will the hunter cut a phonograph record of the mating call of the whale, or the cry of the calf for its mother, and play back the sounds beneath the bow of his ship? Will the orbiting satellite speak through space to tell the hunter where to find the last whale?"

In the past twenty-five years, 62,022 blue whales, at 85 feet and more the largest mammals on earth, and 15,025 humpbacks, perhaps the most playful of the great whales, have been taken in the Antarctic. Never very abundant, both species have been pushed to the edge of life, but are now nominally protected. The finback is the next candidate for "commercial extinction," that is, when its numbers will have been so reduced that it will no longer be profitable to send expeditions to hunt them. The finback, a smaller cousin of the blue whale, was second only to the sperm whale in abundance. During the past quarter-century, 444,262 finbacks were taken in the Antarctic, more than half of them from 1954 to 1962 when more than 27,000 finbacks were taken each year. Their population is now estimated at 67,000 to 75,000, one-fourth of its original size. If the exploiters had shown restraint—if they had

learned the lesson of the blue and humpback, had remembered the slaughter of the right and bowhead in the last century—then the Antarctic could have yielded 10,000 to 12,000 finbacks a year down the long hungry road of the future. Today the sustainable yield is estimated at less than 3,000 finbacks.

These numbers, combined with catch data, indicate the extreme pressure on the finbacks and are an indictment of the stewardship of the International Whaling Commission. The ravaged state of the whale stocks presents an essentially nonpolitical problem that could be eased enormously if the catch effort was radically reduced to

Figure 9-1. Breaching by the humpback whale is not uncommon during springtime near Bermuda, but its diminished numbers make this spectacle a rare event. (Photograph courtesy of John Dominis and Life Magazine © 1972, Time, Inc.)

allow all whale populations to rebuild. The most desirable goal of all, a ten-year moratorium—for tagging, study, and population counts—seems beyond the capacities for cooperation and restraint of the nations present at the International Whaling Convention meeting: Argentina, Australia, Canada, England, France, Iceland, Japan, Norway, Panama, South Africa, the Soviet Union, and the United States.

With the stage set, we can better appreciate what happened at the 22nd meeting of the International Whaling Commission in London at River Walk House, overlooking the Thames, last June. The actions and inactions of the commission can be gauged by four items: (1) the whale quotas set, (2) the sperm whale, (3) the International Observer Scheme, and (4) the action by the U. S. Department of the Interior listing all great whale species as endangered.

As an observer to the meetings, I would like to point out that while the United States is involved in whaling only marginally (it operates one small land station in California), the constructive influence of the United States on the commission has been considerable. Dr. J. Laurence McHugh, the United States commissioner and vice-chairman of the commission, and Dr. Douglas G. Chapman of the Center for Quantitative Science at the University of Washington, have in recent years chaired the commission's two principal committees, the Technical (McHugh) and the Scientific (Chapman).

In addition, the United States is a major importer of whale oil and whale products, making up roughly one-fifth of the world market. Hopefully this market may be closed if the whales can be kept on the endangered species list published by the Department of the Interior. Whatever the use of whale products, whether for lipsticks or lubricants, a satisfactory substitute is available in every instance.

1. On the matter of quotas, the Scientific Committee annually recommends that the blue whale unit be eliminated. Under this curious and anachronistic arrangement one blue whale unit is equal to one blue whale or two finbacks or two and a half humpbacks or six sei whales. Because it did not specify which whales may be taken, the blue whale unit contributed to the collapse of the antarctic fishery. Again this year the commission stuck by the invidious blue whale unit in the Antarctic; in fact, the commissioners did not even raise the subject. The Scientific Committee (with the exception of the Japanese scientists) generally concurred that the sustainable yield for next season was 2,600 finbacks and 5,000 sei whales. The commission set a quota of 2,700 blue whale units, which works out to be 27 per cent more than that recommended by the Scientific Committee. Even recognizing that Norway will probably not take the 200 units assigned to her, the quota does not allow any margin for the stocks to recover and probably will cause further depletion.

In the North Pacific, the Scientific Committee's studies revealed that the sustainable yield is 1,300 finbacks and 3,100 sei whales. The commission set quotas of 1,308 for the finbacks and 4,710 for the seis. Worst of all, a fudge factor of 10 per cent—reminiscent of the blue whale unit—was built into these numbers, so that whatever the whalers fail to catch of one species they can take in the other.

2. Regarding the sperm whale, the collapse of the Antarctic fishery and the strain on the baleen whales in the North Pacific has meant that the damage inflicted on sperm whale stocks—so far without any quota whatsoever from the commission—has been intensified each year. For more than twenty years the number of sperm bulls caught in the Antarctic has ranged between 2,500 and 7,000 annually, with higher numbers killed earlier and lower numbers recently. For example, the peak was fifteen years ago when 6,974 sperm whales were reported taken, a catch that produced 342,000 barrels of sperm oil. During the 1969-70 season, 3,090 sperms were taken in the Antarctic for a production of 125,000 barrels of oil. The striking fact about these figures is that they

reflect a steady decline in the yield of barrels of oil per whale over the past fifteen years. The oil yield in the Antarctic has dropped alarmingly, from 49 barrels per sperm whale to 40 barrels. In a mere fifteen years the sperm whales are 18 per cent smaller. The pattern of predation seems intractable.

The ecology for male and female sperm whales differs markedly. While the males attain lengths of 50 to 55 feet and more, the females are mature at 35 to 40 feet; indeed, females shorter than 38 feet in length are "protected" from pelagic whaling, while those less than 35 feet are protected from land station whaling. The catch data piles up at these minimum legal lengths lending credence to the general belief that the infractions are many and blatant.

An analysis and estimate of the sustainable or potential yield of the sperm whale in the North Pacific has been made by three Japanese scientists. They estimate the present sustainable yield of male sperm whales in the North Pacific at 4,290. The catch the past two years has been 12,740 and 11,329. The Japanese scientists say that "this male sperm whale stock has . . . little or no further surplus." The population has been driven to a level of about one-half of its unexploited state. Privately, the North Pacific commissioners agreed to a catch 10 per cent below last year's. This catch limit—set provisionally behind closed doors outside of the formal business of the commission—is 240 per cent of the sustainable yield estimated by the Japanese scientists. The pattern of predation is familiar—as is the capacity of the International Whaling Commission to look the other way when the chips are down.

3. The most important single item on the agenda, the International Observer Scheme (IOS), was discussed at length. It was approved in principle seven years ago and has been piously reaffirmed annually. But no effective steps have been taken to implement it.

At the meeting, Dr. McHugh stated that the commission's inability to implement an observer scheme weakens it as a conservation organization because it seems to lack the ability to enforce its regulations and quotas. The Japanese commissioner, Mr. Fujita, said that his country would support the implementation of the observer scheme for the next Antarctic season and felt that the plan should extend to land stations as well.

The Soviet commissioner, Mr. M. N. Sukhoruchenko, said that the IOS could be used at present with some small changes. He urged that two persistent problems be settled: every country has an obligation to send observers as well as receive them; the IOS will be effective only if implemented both for land stations and pelagic operations. He recommended that the commissioners meet on and settle this matter prior to the 23rd meeting in June, 1971, in Washington.

Mr. Fujita said that there was no basic disagreement on implementation, but that the commission did not have time to pursue the matter further.

All these words sound reassuring, but the IOS is still not implemented. A possibility exists that the United States and Japan may work out some modest form of exchange for their land stations that could serve as a model for other countries next year.

A beginning may yet be made. It is crucial to know when a protected species is taken and labeled something else; as, for example, when an immature blue whale, unmistakable because of its splotchy exterior, is harpooned and listed in the day's log as a finback.

Another example of the most egregious sort of violation of the regulations took place in the 1962-63 season when a factory ship and its catchers swept in on a small colony of "protected" right whales near the island of Tristan da Cunha in the South Atlantic. The few dozen rights, one of the largest grouping to be seen in any ocean in years, was completely wiped out. This well-known incident has never been aired at the International Whaling Commission meetings nor has it appeared in print, but it is a tragic example of what happens in the absence of an International Observer Scheme. And there are many other unreported tragedies. Just talk to the whalers.

4. The meeting of the International Whaling Commission barely touched on the U.S. Department of the Interior's bold action in placing all the great whale species on the endangered species list of June 2, 1970, implementing the Endangered Species Conservation Act of 1969. According to the provisions of the act, no species that is demonstrated to be threatened with extinction may be imported, alive or dead, whole or in part, into this country. By placing baleen whales, as well as the sperm whale, on the list the Department of the Interior went beyond the mere protection of species already struggling for survival. With the threat of economic boycott, perhaps the member nations of the International Whaling Commission will be spurred to take their task more seriously.

Until last November, a big question remained as to the chances of all these species remaining on the list. The sperm whale was especially vulnerable. Interior Secretary Walter J. Hickel was under great pressure from whale oil importers, from other departments within the government, and from overseas to drop the sperm whale from the list. On November 24, 1970, after six months' intensive review, Secretary Hickel affirmed that all eight threatened species of great whales will be kept on the list and banned from importation to "prevent conditions that lead to extinction."

Explaining why the department kept the fin, sei, and sperm whales on the list, Hickel said it is "clear that if the present rate of commercial exploitation continues unchecked, these three species will become as rare as the other five." He also called for a conference, jointly sponsored by the Department of the Interior and the Smithsonian Institution, to be held early in 1971 to review what can be done to restore whale populations in the oceans of the world.

The Secretary omitted mention of one aspect crucial to any effort to save whales: funding. Scientific programs to monitor the size of the whale herds and the United States share of an observer scheme both need financial support.

This break for the threatened whales was accompanied by some good news from Japan. Last August I went there on behalf of the Environmental Defense Fund and the New York Zoological Society, to discuss with Japanese scientists the initiation of a campaign to save whales. The scientists have formed a Committee for the Protection of Whales, chaired by Dr. Seiji Kaya. Along with writers Kenzaburo Oe and Sakyo Komatsu, they have taken the whale problem to the public for the first time. They are urging the Japanese government to curb the whaling industry and to strengthen the powers of the International Whaling Commission.

The big question remaining is the Soviet Union, but we have prospects of positive developments there, too. The problem of the survival and continuity of the great whales would be eased if the Soviets extended to large whales the attitude they take toward the smaller dolphins and porpoises. In March, 1966, the Soviet government banned the catching and killing of dolphins. This decision was taken, according to Alexander Ishkov, Soviet Minister of Fisheries, because research has shown that dolphins have brains "strikingly close to our own." Dr. Ishkov, therefore, regards the dolphin as the "marine brother of man," noting, "I think that it will be possible to preserve dolphins for the sake of science. Their catch should be discontinued in all seas and oceans of the world."

May the song of the humpback whale soon sound in the Bolshoi Opera House.

We know very little about whales. Until a few months ago, for instance, we did not know that some whales sing, and that these songs make a profound impression on the human listener.

What we have seen closely of whales to date—and watched with strange fascination—

are "death flurries," the tragic scene that has played to an inert, bloated conclusion 60,000 times a year for eight years (1958 to 1965) and now occurs 40,000 times a year. Today, a whale is harpooned every 12 minutes on the average. The "life flurries" remain essentially unknown because no man has stayed with a whale pod hour after hour, day and night, week after week.

Melville concluded:

"Dissect him how I may, then, I but go skin deep; I know him not, and never will. But if I know not even the tail of this whale, how understand his head? much more, how comprehend his face, when face he has none? Thou shalt see my back parts, my tail, he seems to say, but my face shall not be seen. But I cannot completely make out his back parts; and hint what he will about his face, I say again he has no face."

As a species, man is at a point in his own evolution where he cannot yet create a flea but is wholly capable of destroying the whale. The job is three-quarters completed when measured by the great whale species that are threatened with extinction.

Our survival is curiously intertwined with that of the whale. Just as all human life is interconnected (in the Monkey-Rope situation in *Moby Dick,* Ishmael declares, "I saw that this situation of mine was the precise situation of every mortal that breathes; only, in most cases, he, one way or other, has this Siamese connexion with a plurality of other mortals. . . ."), so have we finally begun to perceive the connections between all living things. The form of our survival, indeed our survival itself, is affected as the variety and abundance of life is diminished. To leave the oceans, which girdle seven-tenths of the world, barren of whales is as unthinkable as taking all music away and everything associated with music—composers and their works, musicians and their instruments—leaving man to stumble on with only the dryness of his own mutterings to mark his way.

The Great Barrier Reef Conservation Issue – A Case History

D. W. CONNELL   1971

INTRODUCTION

In 1770, Captain Cook's *Endeavour* almost met disaster on the crags and peaks of coral of the Great Barrier Reef. Those first Europeans to discover the Reef saw it as little more than a vast navigational hazard. Geographical names such as Cape Tribulation, Weary Bay, and Hope Island, are a permanent reminder of this early encounter. Settlement of Australia led to further voyages of discovery. Even Matthew Flinders, perhaps the most prominent of navigators in Australian waters, was wrecked in the *Porpoise* on the coral patches now known as Wreck Reef.

To most Australians a dawning awareness of the Reef as a biological marvel probably began late in the 19th century with the publication of a classic book by Saville-Kent (1893). Since then an avalanche of richly illustrated books has poured forth, revealing the true character and beauty of the reef. Thousands of Australians and overseas tourists now visit the Reef each year. There would be few, if any, Australians who do not know of the Reef. Whether there is legal justification or not, the Reef is known throughout the world as *Australia's* Great Barrier Reef (Figure 9-2).

For many years the Reef's vast size permitted Australians to feel complacent, satisfied that nothing could cause significant damage to it. This attitude still persists and remains perhaps the most important single obstacle to conservation efforts. However, in the last five years problems of critical importance for the future of the Reef have emerged. The Crown-of-thorns starfish (*Acanthaster planci*) outbreak first became apparent in 1960 (Anon., 1970*a*). The subsequent destruction of coral necessitated remedial action in 1965 (Pearson, 1970), and a short time later there were proposals for oil drilling and mining.

ELLISON REEF MINING

The controversy on Reef exploitation began in a small way in 1967 (Anon., 1969*a*).

Dr. D. W. Connell is an organic chemist and senior research scientist with an Australian research institution. He is a member of the Council of the Australian Conservation Foundation and vice president and former president of the Queensland Littoral Society. His research interests include the chemical aspects of pollution of the sea by petroleum products.

From *Biological Conservation*, Vol. 3, No. 4, pp. 249-254, 1971. Reprinted with light editing and by permission of the author and *Biological Conservation*, Elsevier Publishing Company, Ltd., England.

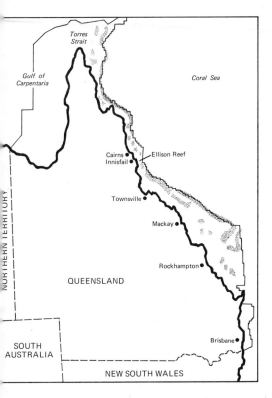

Figure 9-2. Sketch-map of the Queensland coast-line (thick line) showing the outer boundary of the off-shore oil exploration lease areas (thin line) and the general location of coral-reef areas within the Great Barrier Reef (dotted areas). (Adapted from a "titles" map released by the Queensland Department of Mines on 18 August 1970.)

Notice was served that an application to mine Ellison Reef (Figure 9-2) in the central area of the Great Barrier Reef had been received by the Queensland Department of Mines, and that hearing of objections would be held in the Innisfail Mining Warden's Court. The Wildlife Preservation Society of Queensland (Innisfail Branch) in North Queensland saw this as an event of great significance. They claimed that the granting of the lease would establish a dangerous legal precedent which could allow complete commercial exploitation of the Reef.

The North Queenslanders began their campaign by enlisting the support of the Wildlife Preservation Society of Queensland in Brisbane. As well as publicity, facts were needed to face this challenge effectively. The Queensland Littoral Society sent a team of skin-divers to examine Ellison Reef. Their evidence, which was presented in the Mining Warden's Court, indicated that the reef was not *dead* in the manner that had been claimed by the mining applicants. Although there was a high proportion of coral rubble and sand, a rich faunal population was present. Dr. D. F. McMichael, Director of the Australian Conservation Foundation at that time, presented biological evidence resulting from personal experience on Ellison Reef. Many others also presented evidence, and finally the application was rejected.

This case did not generate the high level of public interest which occurred with later issues (see Figure 9-3), but was important as the initial stage of a snowballing build-up in public interest.* The case taught conservationists in Queensland that at least two requirements are usually necessary to present a case for conservation; these are, firstly, sound logical evidence, and, secondly, widespread publicity of the issues.

A CRISIS POINT

Up until mid-1968, neither the Queensland Government nor large commercial interests had become publicly involved in the growing controversy surrounding potential mining or oil exploitation activities on the Reef. In response to accelerating public interest, and

* In a matter such as this it is important to be able to assess public interest in some meaningful way. An approximate estimate can probably be obtained by measuring the actual amount of newspaper space devoted to a particular issue, and so getting a newspaper coverage rating. This operation was carried out with three newspapers: the Australian national newspaper (*The Australian*), the Queensland daily newspaper (*The Courier Mail*), and the Queensland weekly newspaper (*The Sunday Mail*). Front-page areas were multiplied by ten, second- and third-page areas by five, and the remainder by two, and then totalled on a monthly basis for each subject of interest to give the results plotted in Figure 9-3.

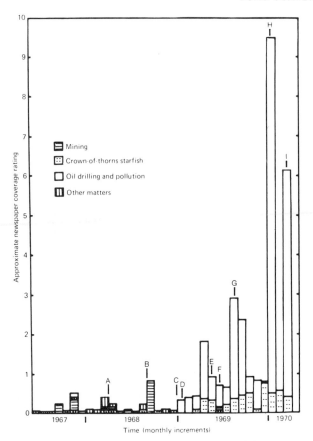

Figure 9-3. Plotting of the approximate newspaper-coverage rating of Great Barrier Reef conservation matters against time.

   A. Ellison Reef mining finally rejected.

   B. Recommendations of the Ladd report released.

   C. Tenders for oil exploration leases called.

   D. Santa Barbara oil blow-out.

   E. Planned Japex exploratory drilling announced.

   G. Japex well approved.

   H. Trade union "black" ban on supplies to the Japex well announced.

   I. Australian Academy of Science report of the Crown-of-thorns starfish problem and a spill from the tanker *Oceanic Grandeur* in Torres Strait.

to clarify the Reef exploitation position, the Queensland Government introduced Dr. Harry S. Ladd from the United States Geological Survey to survey the Reef and advise on possible exploitation. After a one-month survey, Ladd (1968) pointed out that the Reef should be protected from possible damage, and that exploitation should be controlled.

In addition he said, "Petroleum is potentially the Reef's most valuable resource but if not properly controlled it could present the greatest of all dangers. If oil is discovered I believe it would be possible to prevent damaging pollution." In commenting on mining he said, "Dead parts of the coral could be developed for agricultural lime and cement."

Thus in general terms a blueprint and go-ahead for widespread exploitation was provided.

At this time Dr. F. Grassle, an American Fulbright Scholar, was studying reef ecology at the University of Queensland. He issued a serious and widely publicized challenge to the Ladd report by pointing out the *meager scientific evidence* on which it was based. The dead coral myth was largely dispelled when Grassle (1968) commented that "During the past year, I have found that dead pieces of coral contain the greatest diversity of species yet reported in any marine environment."

The Queensland Premier, Mr. Bjelke-Petersen (1968), replied "No mining industry based on harvesting dead coral was permitted on the Reef. This would not be changed unless overwhelming evidence was produced to suggest that limited exploitation would not adversely affect the Reef in any way." Thus, by the application of specialized knowledge and publicity, public and governmental opinion had been favorably influenced.

It is noteworthy that on the same day that the Ladd Report was released, a petition organized by the Queensland Littoral Society, containing over 10,000 signatures, was presented to the Queensland Parliament. The petition called for a moratorium on drilling and mining activities and for the establishment of a National Authority to manage the Reef. This drew no official comment whatsoever and would appear to have been largely ineffectual.

AN OIL INDUSTRY ON THE REEF?

Toward the end of 1967, the Queensland Parliament passed joint legislation with the Australian Commonwealth Parliament on offshore petroleum exploration and extraction [*Petroleum (Submerged Lands) Act* of 1967]. Tenders were called under this legislation in early 1969. By 14 January 1969, forty groups of oil-search companies, some backed by giant international corporations, had applied for leases.

Conservationists and the general public were aware of the biological effects of oil and detergents used for clean-up since these matters were dramatically publicized during the *Torrey Canyon* incident in Britain. The problems of oil-well blow-outs were emphasized by a gas blow-out in Bass Strait, between mainland Australia and Tasmania, which took some weeks to cap. However, on 30 January 1969, the Santa Barbara oil blow-out occurred. This incident produced a turning point in Barrier Reef conservation, and moreover, throughout the reef oil-drilling controversy, the incidence of dramatic and repeated spillages within the oil industry itself influenced public opinion more than any of the efforts of conservationists.

At this time, the Senate Select Committee on Off-Shore Petroleum Resources was holding hearings in Queensland. These provided an excellent forum for comments on oil exploration on the Reef by conservationists and their opponents. Well-publicized submissions were made by university personnel, fishing organizations, and trade unions, as well as by conservation groups, calling for a halt to proposed oil exploration activities. Submissions by mining engineers claimed that it was problematical whether escaping oil would kill polyps: it would affect the air-oil-sea interface but not the coral beneath. In addition it was stated that oil and gas blow-outs at Santa Barbara and elsewhere had not affected marine life. In the event of any blow-out, dispersal detergents which were non-toxic, and would affect neither reef nor marine life, could be air-sprayed over the affected area within 24 hours. These statements were strongly attacked in the press by conservationists who were able to cite examples of oil damage to coral and other marine animals. Nevertheless the battle continued to rage with almost unabated publicity as statements and counter statements were issued by the various protagonists (see Figure 9-3).

POLITICAL SITUATION AND LEASES

The Queensland State Parliamentary Opoosi-
tion had expressed grave doubts about oil
drilling on the Great Barrier Reef in 1967
when the Petroleum (Submerged Lands) Act
was first debated in the Queensland Parliament. It publicly stated a policy of opposition to Reef oil exploration in March 1969,
before the State elections were held that
year. Nevertheless the incumbent government was returned to office. Within a short
time it was announced that oil drilling on
the Great Barrier Reef area was planned by
Japex, a Japanese oil exploration company,
under a farm-out agreement with an Australian company, Ampol Pty. Ltd. Final approval was given to the Japex operation in
August 1969, when it was stated that safety
precautions would be required that would be
more stringent than those applied to drilling
on land.

At this time another important political
factor entered the controversy. The Australian Prime Minster, Mr. Gorton, on a visit
to Queensland, said that despite his avowed
intention of preserving the Barrier Reef as a
national heritage, he could do nothing to
prevent the Japex well from spudding in.

The sketch-map (Figure 9-2) is based on
the map which was released by the Queensland Department of Mines in August 1970.
A similar map, which did not indicate leases
in the Gulf of Carpentaria or some Great
Barrier Reef areas, had been released in September 1969 (Anon., 1969b). Nevertheless,
this earlier map showed quite clearly that
the leases covered most of the Great Barrier
Reef region. Allen and Hogetoorn (1970)
have described the hydrocarbon potential of
off-shore basins in the Great Barrier Reef
area as ranging from poor to fair.

THE CONSERVATION CAMPAIGN

The conservation campaign for the Great
Barrier Reef has been large and complex.
Many individuals and organizations have participated and presented a variety of points of
view. From the very outset the campaign has
involved a comparatively large body of the
public and has in no way been confined to
the organized conservation groups. It is
probably because of this public involvement
that the campaign has achieved some successes.

In September 1969, a group of concerned
individuals in Brisbane formed the Save the
Barrier Reef Campaign Committee to promote Reef conservation specifically. It is
most significant that the large majority of
these people were not already members of
conservation organizations. Many of them
were disillusioned members of government
political parties, disturbed at the governmental attitude on Reef conservation. Others were simply members of the public or
were associated in some way with the tourist
industry.

In general the conservation groups have
based their case on two basic principles:
firstly, the acquisition of sound knowledge
of matters affecting the Reef; and secondly,
as wide publicity of this information as possible, the major objective being to create a
favorable social and political climate for the
introduction of measures for Reef conservation.

Fortunately, many biologists and other
scientists have been involved in gathering the
pertinent information. Much of this information has come from overseas sources, but the
Queensland Littoral Society has embarked
on its own research program. This has involved an investigation of the effects of oil
dispersants on Reef animals, and a study of
oil clean-up measures used during a recent
tanker spillage in Torres Strait. Almost all of
this information has been assembled into
reports in non-specialist language and has
been reproduced and distributed to the
public in fairly large numbers.

The Australian Conservation Foundation
held a symposium called *The Future of the
Great Barrier Reef* in Sydney in May 1969.
Papers were presented by authoritative
speakers on all aspects of Reef conservation.
A great deal of useful information was

brought forward, and the subsequent publication and distribution of the symposium proceedings has done much to bring to the fore the facts of the various issues.

During the campaign, much of the information available was issued in the form of press, radio, and television statements. As an example, the Queensland Littoral Society issued over fifty statements to the various media in 1969, and numerous other statements and appearances were made by members of the Wildlife Preservation Society of Queensland and members of the Save the Barrier Reef Campaign Committee (see above).

Apart from the more common publicity methods which have been mentioned previously, two specific activities have been undertaken which have played an important, perhaps decisive, role in the Reef issue. The first was the production and sale of Save the Barrier Reef car-bumper stickers. Over thirty thousand of these were sold, in addition to over one hundred thousand similar mini-stickers which were used on letters and other articles.

The second activity was a public opinion poll organized by the Save the Barrier Reef Campaign Committee and the Wildlife Preservation Society of Queensland. The poll was a random one conducted by non-professional interviewers, and as such would not be as accurate as a professional conducted poll. However, the Department of Mathematics of the University of Queensland advised that the expected error was considerably reduced because of the large size of the sample (1,000 people). Over 91 per cent of the people who expressed an opinion were against the establishment of an oil industry on the Reef. This result was well publicized; it was a morale-booster for conservationists and probably influenced many politicians.

THE CROWN-OF-THORNS STARFISH PROBLEM

Most of the previous discussion has centered on the mining and oil-industry controversies, but concurrently with these issues the Crown-of-thorns starfish (Figure 9-4) outbreak has also become a major public issue. Although each of these problems has been discussed as an almost entirely divorced activity, the newspaper coverage ratings illustrated in Figure 9-3 indicate that both issues have generated a rise in public interest during the same period. This would suggest that each may have helped to stimulate public interest in the other.

The tourist industry has played an active role in many aspects of the Crown-of-thorns starfish problem. The outbreak first became apparent at Green Island, a popular Barrier Reef resort off Cairns in North Queensland. The management employed a skin-diver to collect the starfish and in fifteen months from early 1965 he collected 27,000 specimens. In early 1966 the resort appealed to the Queensland State Government for assistance, and a biologist was assigned to work on the problem under the supervision of Dr. Robert Endean of the University of Queensland.

Dr. Endean submitted a report on the problem to the Queensland Government in December 1968. Months later the Government had not taken any action or released any of the report's findings. In addition to commenting adversely on this, Dr. Endean mounted a campaign to publicize the problem and initiate some decisive action.

The Premier of Queensland, Mr. Bjelke-Petersen (1969), reacted to the situation by stating, "Expert advice is that there is no great plague of the Crown-of-thorns starfish. Recent publicity has presented an unduly alarming picture of the situation. Dr. Endean obviously is sincere but many men with expert local knowledge disagree with him."

Undeterred, Dr. Endean continued to publicize the problem, bringing forward further evidence as to the spread of the starfish. Unfortunately, conservation organizations have been at a disadvantage in deciding on an effective course of action with regard to this problem, as most of the information has

Figure 9-4. A juvenile Crown-of-thorns starfish (*Acanthaster planci*) on dead coral. Photo: O. E. S. Kelly.

been in hands either of the State Government or of Dr. Endean.

Later, Pearson (1970), of the Queensland Department of Primary Industries, reported that 8.1 per cent of reefs in the total Great Barrier Reef area had suffered extensive damage. Also, the Australian Academy of Science reported that Crown-of-torns starfish destruction of the Great Barrier Reef was impossible to control except in a few selected tourist areas. To clarify the situation, the Queensland State and Commonwealth Governments have instituted an inquiry into all aspects of this problem.

DECISIVE EVENTS

The first day of 1970 opened on a grim note for conservationists. The *Brisbane Courier*

Mail announced "the 330-ft long, self-propelled oil-rig, *Navigator,* is expected to enter Queensland waters in a few weeks to prepare for oil drilling off the North Queensland coast." After many months of controversy, *The Australian* of 24 December 1969 summed up the situation as follows: "The Great Barrier Reef has been the object of a shameful exercise in buck-passing throughout this year.... The political dereliction has taken place despite the most sustained public campaign in memory on a conservation issue." At this time the conservation camp was reduced to considering a challenge to the legality of the State Government's authority to issue prospecting licenses.

The Trade Union movement now took what was to be decisive action. Key unions such as Transport Workers, Waterside, Storemen, and Packers, etc., threatened a total ban on all goods and services for the Japex oil-rig (January 6, 1970). Within a few days Ampol Pty. Ltd. announced suspension of the drilling operation and asked that an inquiry be instituted.

This offer was welcomed by the federal Prime Minister, Mr. Gorton, but the Queensland Premier, Mr. Bjelke-Petersen (1970), insisted that the drilling proceed, saying "There could be no useful purpose to a further investigation. An expert survey has been completed in a most competent way." Nevertheless the Prime Minister (Gorton, 1970) continued to press for the institution of an inquiry, explaining that "In my view the slightest danger is too much danger." Finally there was joint agreement to a Committee of Inquiry into Oil Drilling on the Great Barrier Reef which was later up-graded to Royal Commission status (Connell, 1970a).

This outcome, although a great step forward, was not altogether satisfactory for convservationists. Insufficient funds would entirely preclude representation for the conservation case at the Royal Commission which was likely to continue for many months.

At this time, once again events within the oil industry itself served to emphasize spillage dangers. The oil tanker *Oceanic Grandeur* was holed in Torres Strait and, probably only due to favorable weather, was saved from disaster with the release of relatively small quantities of oil. Also, an extensive oil-slick was released from an oil-well in the Gulf of Mexico.

Perhaps stimulated by this, a concerned group of solicitors and barristers offered legal assistance on a part-time basis. They pointed out that, while this makeshift effort would at least provide representation, it could not be regarded as a satisfactory solution to the problem.

Finally, after a number of requests to the Australian Prime Minster, the Commonwealth Government generously offered to pay all reasonable expenses for the legal representation of a number of conservation organizations. The proceedings of the Royal Commission are expected to continue for some time, and a final report cannot be expected until late in 1971.

CONCLUSIONS

The Great Barrier Reef issue has a number of important lessons for conservationists. Most importantly, every effort must be made to influence the whole community in order to introduce conservation measures. Such support can win favorable political, industrial, and commercial, decisions.

Conservationists have emphasized the community value of the controlled development of the tourist and fishing industries (Anon., 1970b). In addition, positive action leading to the formation of a Great Barrier Reef Authority has been requested, so that the Reef can be managed in a rational way that is consistent with its continued existence (Connell, 1970b).

However, with the Great Barrier Reef, the community had made a favorable value-assessment before conservation issues arose. Few would advocate the destruction of the Reef for industrial purposes. The magnitude of the various threats has been the heart of the matter.

It is quite clear that sound and detailed knowledge of the problems and surrounding circumstances was a first essential requirement. This information then needs to be disseminated to the community by means of publications, public meetings, addresses to private organizations, and all the forms of mass media that can be utilized. Even after all of these actions have been taken, results cannot be achieved overnight, and both patience and perseverance are required.

ACKNOWLEDGMENT

The substance of this paper was presented on 27 June 1970 at a symposium entitled *The Process and Problems of Seeking Conservation,* organized by, and held at, the

Centre for Continuing Education, Australian
National University, Canberra.

REFERENCES

Allen, R. J. and Hogetoorn, D. J. (1970).
Petroleum Resources of Queensland,
Queensland Department of Mines, 42 pp.,
7 maps.

Anon. (1969a). Conservation of the Great
Barrier Reef of Australia. *Biol. Conserv.*
1(3), 249-50.

Anon. (1969b). Press Report. *Brisbane
Courier Mail,* 4 September, p. 3.

Anon. (1970a). Population explosion of the
Crown-of-thorns Starfish, *Acanthaster
planci. Biol. Conserv.* 2(2), 96.

Anon. (1970b). Statement on the Great Barrier Reef. *Wildlife in Australia* 7, 1.

Bjelke-Petersen, J. (1968). Press Report.
Brisbane Courier Mail, 13 September, p.
3.

Bjelke-Petersen, J. (1969). Press Report.
Brisbane Courier Mail, 9 September, p. 3.

Bjelke-Petersen, J. (1970). Press Report.
Brisbane Courier Mail, 11 January, p. 1.

Connell, D. W. (1970a). Inquiry into advis-
ability of oil-drilling in the Great Barrier
Reef, Australia. *Biol. Conserv.* 3(1), 60-1.

Connell, D. W. (1970b). The conservation
viewpoint on Great Barrier Reef oil drilling. *Living Earth, Australia* 14, 39-40.

Gorton, J. (1970). Press Report. *The Australian,* 19 January, p. 3.

Grassle, F. (1968). Press Report. *Brisbane
Courier Mail,* 11 September, p. 3.

Ladd, H. S. (1968). *Preliminary Report on
Conservation and Controlled Exploitation
of the Great Barrier Reef.* Queensland
State Government, Brisbane, 51 pp., illustr.

Pearson, R. (1970). Studies of the Crown-
of-thorns Starfish on the Great Barrier
Reef. *Newsletter, Qd Littoral Soc.,* No.
36, pp. 1-10.

Saville-Kent, W. (1893). *The Great Barrier
Reef of Australia.* W. H. Allen, London,
xvii + 387 pp., illustr.

The Sea-Level Canal Controversy

IRA RUBINOFF   1970

INTRODUCTION

In a recent article (I. Rubinoff, 1968), I suggested that the proposed construction of a sea-level canal in Central America posed a potential threat to the ecology of the oceans in this area and at the same time provided a remarkable research opportunity. My position is that we are not certain what the consequences of faunal mixing will be; therefore I recommended appropriate studies to permit both proper evaluation of potential biological danger and exploitation of a unique scientific opportunity.

These suggestions provoked numerous comments, including critical ones (Sheffey, 1968; Weathersbee, 1968; Cusack, 1969; Hilaby, 1969; Mueller, 1969; Topp, 1969; Hubbs, MS.). Some of these criticisms were valid and/or matters of opinion about which there can be reasonable argument. Others, unfortunately, would appear to have been based upon false data or very dubious hypotheses.

The errors fall into three categories: (1) factual misinformation on what has passed, or is passing, through the present (lock and freshwater) canal; (2) conceptual misunderstandings of what may happen if and when the sea-level canal is completed and the biotas are allowed to mix; and (3) unjustified scepticism about the usefulness of a biological survey before construction of a sea-level canal.

Engineers with limited biological knowledge, who have assured us that nothing will happen, are as culpable as biologists who over-extrapolate their limited data into prophecies of doom. The fact that the sea-level canal may have far-reaching effects is no excuse for the relaxation of scientific judgment and reasoned analysis (I. Rubinoff, 1969). What I would like to consider here is the evaluation of some of these recent discussions.

Dr. Ira Rubinoff is a marine biologist and assistant director of science with the Smithsonian Tropical Research Institute, Canal Zone. His current research is on a venomous sea snake found in the Pacific but not in the Atlantic Ocean. Such studies have a direct bearing on the sea-level canal controversy. He is the author of over thirty technical articles.

From Biological Conservation, Vol. 3, No. 1, pp. 33-36, 1970. Reprinted with author's revisions and light editing by permission of the author and Biological Conservation, Elsevier Publishing Company, Ltd., England.

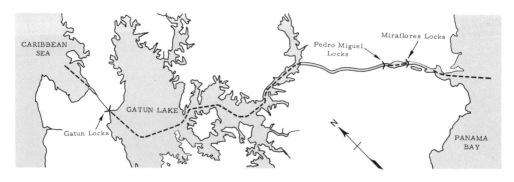

Figure 9-5. Sketch-map of Panama Canal. The isthmus is 50 miles (80 km) wide in the canal zone.

TRANSPORT THROUGH THE PRESENT CANAL

Euryhaline organisms (those which can tolerate a wide range of salinity) can migrate actively by swimming or they may be passively transported by "hitchhiking" on ships. This passive transport can occur in two forms: by attachment ("fouling") on ships' bottoms, or in ballast tanks—i.e. when the water loaded in one area of ocean is later discharged in another. This latter form of transport is also available to stenohaline organisms (those with narrow salinity tolerances). There have been some qualitative studies of these subjects (Chesher, 1968; Menzies, 1968; R. W. and I. Rubinoff, 1969), but no quantitative measurements have been made.

There are surprisingly few data on successful new colonizations of bodies of water in this region. A number of marine species (Tarpon, Snook, Jacks, etc.) are known to have invaded the fresh waters of Gatun Lake, Panama Canal (Figure 9-5). Presumably they occasionally pass through the locks in both directions. With the single exception of a small goby (R. W. and I. Rubinoff, 1968), however, there is no species which is known to have moved from one ocean to the other, and established a successful breeding colony in the new habitat; moreover, this one exception occurred in very peculiar ecological circumstances.

Many authors have failed to appreciate the fact that the occasional introduction of a few individuals into a new environment does not necessarily constitute a successful colonization. There is a minimum number (which varies with the species and situation) of individuals that is necessary to effect a colonization (i.e. breeding* and population increase). This number is called the "propagule"† (MacArthur and Wilson, 1967; Simberloff and Wilson, 1969).

The problem of achieving propagule-sized populations is also inherent in the two forms of passive transport. While the potential for transport of marine organisms as ship-fouling or in ballast tanks would seem to be enormous, there are factors which add considerably to the rigors of such a trip.

Transport as Fouling

In order to pass successfully through the present Panama Canal, fouling organisms must be able to survive an average of 5-8 hours in water of no detectable salinity (in Gatun and Miraflores Lakes, cf. Figure 9-5. For stenohaline species this trip is apt to be fatal.

Menzies (1968) towed an assortment of marine animals through the Canal in an at-

---

* Functional hermaphrodites would have an advantage as colonizers in this respect.

† This alters the sense of the term significantly from its common botanical connotation of any plant body which *can* propagate its taxon—Ed.

tempt to estimate the ability of fouling organisms to survive the sojourn in the fresh waters of Lakes Miraflores and Gatun (Figure 9-5). He concluded that there was only a "limited in-transit mortality as a result of the low salinity." His experiment is worthy of further examination. A series of intertidal organisms was selected for his experiment, these being tied in cheese-cloth and towed behind a ship at speeds of up to 18 km per hr. At the upper speeds the animals skimmed and skipped over the surface of the water, so that Menzies estimates that they were totally submerged for only 3 hours of the total 8.5 hours required to complete the transit.

Obviously, organisms attached to the bottom of a ship would be subjected to a much longer period of freshwater immersion during a passage through the canal. Furthermore, by using intertidal organisms, Menzies selected animals which are exceptionally preadapted to euryhaline situations and so the results of his experiment are of limited value in extrapolating to typical fouling organisms. Also, Menzies scored his animals as alive or dead soon after completion of the transit, whereas time should have been allowed for recovery of the organisms, so as to evaluate delayed mortality. Menzies is fully aware of the limitation of his pilot experiment: he did not endeavor to perform the definitive research but merely to demonstrate that the subject was amenable to experimental verification.

Certainly, common fouling organisms such as intertidal barnacles can survive brief periods in fresh water, and Neal Powell, of the National Museum of Canada, believes that many Bryozoa can survive moderate exposures to fresh water. In a recent survey of the Bryozoa on the buoys in both the Pacific and Atlantic entrances to the Canal, he found a few species in common (Powell, personal communication).

Transport in Ballast

Many ships, particularly unladen ones, are required to take on sea-water ballast to im-

prove their handling ability during transit. Although large volumes of water are involved (Chesher, 1968), the environment in most ballast tanks is remarkably inhospitable and frequently completely abiotic—particularly for the relatively sensitive planktonic organisms that are most likely to be taken into ballast systems (water samples from the few ballast tanks I have examined contained no living plankton). Anticorrosion paints that are used to protect these tanks are extremely toxic, and a few minutes contact with them is sufficient to kill most marine organisms. In addition, Chesher states that it takes about 6 hours to pass through the canal. A figure of 8 hours would be closer to the average time. Furthermore, ships take on ballast water before arriving in Canal Zone waters and are prohibited by law from discharging ballast in Canal Zone waters. Consequently, the time that animals must live in ballast tanks in order to effect interoceanic transport is days or weeks in most cases. The probability of organisms being discharged in a hospitable environment is diminished by the great time and distances involved.

On the other hand, modern tankers are frequently equipped with stainless steel ballast tanks which are used exclusively for sea water, and these would seem to be more or less pre-adapted to the successful carrying of marine organisms from ocean to ocean. The present Panama Canal is, however, incapable of handling most of these modern larger tankers. The actual role of ballast transport through the present Canal is a subject that *could* be properly evaluated, and a thorough study should remove this area from speculation.

Examined in the proper perspective, we see that the transfer of marine organisms through the present Canal cannot be as extensive as is claimed by Sheffey when he writes (1968) "Thus, all the small swimming and drifting marine life, that would be found in these thousands of samples of sea water taken year in and year out since 1914, have made the trip across the isthmus in salt water in both directions. . . . It follows that a large portion of the small swimming, drifting, and clinging creatures on both sides of the isthmus have long been exposed to inoculations of the same category from the opposite ocean. It seems reasonable to conclude that a sea-level canal would create little or no threat to the lower links of the ocean food-chain."

I, for one, find Sheffey's assurances unconvincing. The present canal is a highly restrictive filter.

RESULT OF BIOTIC INTERMIXING

A consequence of any sea-level canal would be a much less inhibited movement of species from one ocean to the other. Whether fauna and flora on either coast would become enriched, replaced, depauperate, extinct, or, in general, the ways in which interaction may occur between newly mingling species have been the source of some discussion. Briggs (1968, 1969), arguing strongly for a fresh-water barrier to be included in any new canal, predicts the irrevocable extinction of several thousand unique species. "For the tropical eastern Pacific, it is predicted that its fauna would be temporarily enriched but that the resulting competition would soon bring about a widespread extinction among the native species. The elimination of species would continue until the total number in the area returned to about its original level. *The fact that a large-scale extinction would take place seems inescapable.*"

Briggs's (1969) concept of numerical superiority of Atlantic fauna is based on a vertebrate: invertebrate ratio established for a small island in the Florida Keys. He then extrapolates this figure to the western Caribbean-eastern Pacific area on the basis of the "relatively" well-known numbers of fish species. One may question the validity of this 1:13 vertebrate:invertebrate ratio for the western Caribbean, but its further extrapolation into the eastern Pacific is certainly unsubstantiated.

Can we really consider the invasion po-

tential of an entire fauna or should individual phyletic groups be examined separately? Even if we accept Briggs's theory that the more diverse faunas will replace less diverse ones, this concept probably would not apply within those phyletic groups which are much more diverse in the Pacific than in the Atlantic. Present knowledge indicates that in many cases—e.g. porcellanid crabs, penaeid shrimps, sciaenid fishes, and perhaps the entire sandy-beach meiobenthos—the fauna is richer in the Pacific than in the Atlantic. Indeed, one would expect a richer intertidal fauna in the Pacific, with its much greater vertical niche differentiation (I. Rubinoff, 1968).

The question of whether species diversity is more important than physical pre-adaptations to a new environment must also be considered in evaluating a fauna's invasion potential. It is unlikely that any invading species will succeed in displacing a resident species through its entire range. Since most Panama Pacific species can be found north and south of 9°N latitude, they certainly can be expected to be adaptively superior to Atlantic invaders in at least some area along their range.

Briggs draws a parallel between linking the Atlantic and Pacific via a sea-level canal with the linkage of North and South America by the emergence of the isthmian landbridge. The latter produced a temporary enrichment of South American mammalian fauna, but this was followed by rather large-scale extinctions of indigenous fauna. The evidence for blaming these extinctions on the invaders from the north is extremely circumstantial, and indeed current information indicates that the development of human societies may be more responsible for some of these extinctions than competition from any 4-legged mammals (Jelinek, 1967; Patterson and Pascual, 1968). Whether or not one agrees with Briggs's zoogeographic conclusions, his position, that there is a potential for inestimable effects which have not been appreciated or evaluated at the present time, deserves careful consideration.

At the American Association for the Advancement of Science symposium on biological aspects of a Sea-Level Canal, held in December 1969, a number of scientists expressed the views that in many systematic groups the populations were so similar that competition was unlikely. But however similar these groups may appear, the presence of some differences indicate that evolution has taken place during the period of isolation. Therefore, by simple application of the Gaussian principle of competitive exclusion, we see that the two populations cannot conceivably come together without competing in some way. It is impossible for two different organisms (however slightly they may differ) to utilize a resource with equal efficiency. The ways that this competition will be exhibited may be subtle and not necessarily detrimental. But to claim that no competition or genetic interaction will occur, on the basis of morphological similarity or even identity, is blatantly ignoring basic ecological and evolutionary principles.

PREDICTABILITY FROM PRE-CANAL RESEARCH

Much has been written, and there has been much discussion, on possible effects of a sea-level canal. The sceptical view has been expressed that our ecological sophistication is insufficient to permit the prediction or identification of the problem organisms, and therefore that any such effort would be wasted.

I believe that many biotic interactions resulting from a sea-level canal could be more or less accurately predicted through a comprehensive pre-canal research program. For example it is possible to determine some potential genetic interactions between allopatric populations coming from different regions in controlled laboratory experiments (e.g. *Bathygobius*—R. W. Rubinoff, in prep.). It is also possible to assay the colonizing ability of potential interoceanic invaders by testing their reactions to potential predators and prey (e.g. *Pelamis platurus*—I. Rubinoff

and C. Kropach, in prep.). Examination of the vulnerability of organisms to potential parasites is also amenable to laboratory investigation.

It is not intended to design and describe a sea-level canal biological program here, but rather to point out areas of interaction for which probability limits can be established, in advance, on the basis of controlled experiments. However, I believe that if sufficient program support is available, the biological effects of a sea-level canal can be put on a firmer basis of scientific prediction. Recent developments in experimental zoogeography (Wilson and Simberloff, 1969; Simberloff and Wilson, 1969; Wilson, in press) illustrate methods of evaluating similar ecological problems by controlled experiments. If our ecological knowledge does not become sufficiently sophisticated to predict the biological effects, then the new canal certainly should include a biotic barrier to be maintained until we are confident there will be no untoward effects.

In a time of great concern for problems of environmental exploitation, we have yet to witness a tangible solution of any of these situations *on other than a local basis.* The proposed sea-level canal is a challenge—one that can be met feasibly by comprehensive study and careful non-emotional planning. Let us anticipate the problems and not have to rectify irresponsible mistakes.

Fortunately, the National Academy of Sciences of the United States has appointed a committee to evaluate the ecological problems of the proposed sea-level canal, and to recommend study strategies. Hopefully, their recommendations for further studies will be translated (in adequate time and with sufficient funds) into a comprehensive analysis of ecological consequences as well as a thorough inventory of extant Central American marine flora and fauna.

AUTHOR'S NOTE IN 1972

The Seven Volume Report of the Atlantic-Pacific Interoceanic Canal Study Commis-

sion was submitted to President Nixon on
December 1, 1970. This report recommended that nuclear excavation was neither
technically feasible nor internationally acceptable at this time. Construction of a sea-
level canal by conventional means is feasible,
and the most suitable site for such a canal is
Route 10 in the Republic of Panama, southwest of the present canal. The cost would be
approximately $2.88 billion at the 1970
price level.

The report recognized that there might be
many considerations concerning the Canal's
feasibility including political, economic, defense, engineering, navigation, medical, and
environmental which require further study.
In the area of the effect of faunal mixing the
report presented the somewhat opposing
views of the Battelle Memorial Institute and
the National Academy of Sciences' Committee on Ecological Research for the Interoceanic Canal (CERIC). The latter group
acknowledged our insufficient understanding
of this region's fauna and ecology and
recommended further research be undertaken and that a biotic barrier be included in
the proposed Canal.

REFERENCES

Briggs, J. C. (1968). Panama's sea-level canal.
    Science **162**, 511-3.
Briggs, J. C. (1969). The sea-level Panama
    Canal: potential biological catastrophe.
    BioScience **19**(1), 44-7.
Cusack, M. (1969). New Central American
    canal–threat to life? *Science World*
    17(14), 6-8.
Chesher, R. H. (1968). Transport of marine
    plankton through the Panama Canal. *Lim-
    nology and Oceanography* **13**(2), 387-8.
Hilaby, J. (1969). A risky mix. *New Scien-
    tist* **41**, 280-1.
Hubbs, C. L. (MS.). Need for thorough in-
    ventory of tropical American marine bio-
    tas before completion of an interoceanic
    sea-level canal.
Jelinek, A. J. (1967). Man's role in the ex-

tinction of Pleistocene faunas. Pp. 193-
    200 in *Pleistocene Extinctions* (Ed. P. S.
    Martin and H. E. Wright). Yale Univ.
    Press, New Haven, Conn.
MacArthur, R. H. and Wilson, E. O. (1967).
    The Theory of Island Biogeography.
    Princeton Univ. Press, Princeton, New
    Jersey, xi + 203 pp., illustr.
Menzies, R. J. (1968). Transport of marine
    life between oceans through the Panama
    Canal. *Nature (London)* **220**(5169), 802-
    3.
Mueller, M. (1969). New canal: what about
    bioenvironmental research? *Science* **163**,
    165-7.
Patterson, B. and Pascual, R. (1968). Evolu-
    tion of mammals on southern continents.
    V. The fossil mammal fauna of South
    America. *Quarterly Review of Biology*
    43(4), 409-51.
Rubinoff, I. (1968). Central American sea-
    level canal: possible biological effects.
    Science **161**, 857-61.
Rubinoff, I. (1969). Panama Canal: wide-
    spread effects. *Science* **163**, 762-3.
Rubinoff, R. W. and Rubinoff, I. (1968).
    Interoceanic colonization of a marine
    goby through the Panama Canal. *Nature
    (London)* **217**(5127), 476-8.
Rubinoff, R. W. and Rubinoff, I. (1969).
    Fisch-Austausch zweichen Atlantik und
    Pazifik durch den Panama-kanal. *Um-
    schau* **1969**(4), 121.
Sheffey, J. P. (1968). When Caribbean and
    Pacific waters mix. *Science* **162**, 1329.
Simberloff, D. S. and Wilson, E. O. (1969).
    Experimental zoogeography of islands:
    the colonization of empty islands. *Ecol-
    ogy* **50**(2), 278-96.
Topp, R. W. (1969). Interoceanic sea-level
    canal: effects on the fish faunas. *Science*
    165, 1324-7.
Weathersbee, C. (1968). Linking the oceans.
    Science News **94**, 578-81.
Wilson, E. O. and Simberloff, D. S. (1969).
    Experimental zoogeography of islands.
    Defaunation and monitoring techniques.
    Ecology **50**(2), 267-78.
Wilson, E. O. (in press). The species equi-
    librium. Brookhaven Symposium in Biol-
    ogy (1969).

6378